W0042923

Optical Signal Processing

Pankaj K. Das

Optical
Signal Processing

Fundamentals

With 319 Figures

Springer-Verlag

**Berlin Heidelberg New York
London Paris Tokyo
Hong Kong Barcelona
Budapest**

Professor Pankaj K. Das

Electrical, Computer, and Systems Engineering Department, School of Engineering,
Rensselaer Polytechnic Institute, Troy, NY 12180-3590, USA

ISBN-13:978-3-642-74964-3   e-ISBN-13:978-3-642-74962-9
DOI: 10.1007/978-3-642-74962-9

Library of Congress Cataloging-in-Publication Data. Das, Pankaj K., 1937 – Optical signal processing: fundamentals / Pankaj K. Das. p. cm. – Includes bibliographical references. ISBN-13:978-3-642-74964-3: 1. Optical data processing. 2. Signal processing. I. Title. II. Series. TA1632.D355 1990 621.382'2-dc20 89-21834

This work is subject to copyright. All rights are reserved, whether the whole or part of the material is concerned, specifically the rights of translation, reprinting, reuse of illustrations, recitation, broadcasting, reproduction on microfilms or in other ways, and storage in data banks. Duplication of this publication or parts thereof is only permitted under the provisions of the German Copyright Law of September 9, 1965, in its current version, and a copyright fee must always be paid. Violations fall under the prosecution act of the German Copyright Law.

© Springer-Verlag Berlin Heidelberg 1991
Softcover reprint of the hardcover 1st edition 1991

The use of registered names, trademarks, etc. in this publication does not imply, even in the absence of a specific statement, that such names are exempt from the relevant protective laws and regulations and therefore free for general use.

Coverdesign: W. Eisenschink, D-6805 Heddesheim
54/3140-543210 – Printed on acid-free paper

Preface

The subject "optical signal processing" can and should include all aspects of optics and signal processing. However, that is too large a scope for a textbook that, like this one, is intended as an introduction to the subject at a level suitable for first year graduate students of electrical engineering, physics, and optical engineering. Therefore, the subject matter has been restricted. The book begins with basic background material on optics, signal processing, matrix algebra, ultrasound and SAWs, and CCDs. One might argue about this choice of topics. For example, there already exist very good books on matrix algebra. However, matrix algebra is so important in signal processing, especially in connection with devices such as optical matrix processors, that it was felt that a review was essential. Also, the matrix algebra needed for systolic arrays and parallel computing has made great advances in recent years.

My original intention was to write a single-volume textbook covering most of the fundamental concepts and applications of optical signal processing. However, it soon became apparent that the large amount of material to be included would make publication in a single volume impracticable. Therefore this volume treats the "fundamentals" and a second volume will appear dealing with devices and applications.

This textbook was stimulated by a set of short courses that I have directed and lectured since 1976, as well as regular courses that I have taught at Rensselaer Polytechnic Institute since 1974. It brings together results and information available in piecemeal fashion in many other books and in professional journals and conference proceedings. I am grateful to all those authors whose material I have freely used.

I am also thankful to all my students and colleagues over the past 20 years at the Polytechnic University (previously called Brooklyn Polytech), University of Rochester, Rensselaer Polytechnic Institute, and to the short-course students, who have contributed to this book directly or indirectly. I especially thank all my Ph.D. and Master students, whose probing questions have often opened my eyes. Finally,

I should not forget those special students who, somewhat apologetically, ask what they believe to be "dumb" questions, which turn out to be not so "dumb" after all! I have learned a lot through such encounters and I gratefully acknowledge those students.

Troy, NY
January 1991

P. Das

Contents

1. Introduction

Until recently, optics had not been considered as part of the electrical engineering curriculum, even though the fundamental laws of electromagnetic waves, which include those in optics, are governed by Maxwell's equations. The main reason for this was the absence of coherent optical sources equivalent to klystrons or magnetrons for microwaves, or oscillators for lower frequencies. However, the invention of the laser has changed this situation and, as expected, there has been enormous activity in using the optical frequency region for conventional electrical engineering applications, such as optical communication, laser radar and optical signal processing. These applications are in addition to those traditionally belonging to optics, such as photography, spectroscopy, microscopy, and in telescopes, etc. which generally use incoherent light. The "optical revolution" in electrical engineering is not only fueled by the availability of the laser but also by other technical developments such as integrated optics, fiber optics, acousto-optics, electro- and magneto-optics, Fourier optics, a phenomenal need for parallel computation, hence optical computing, systolic arrays, photodetector arrays and charge-coupled devices (CCDs), charge-injected devices (CIDs), focal plane arrays, GaAs technology, very high speed integrated circuits, and the overall desire of society to perform real-time signal processing with faster speed and higher bandwidth.

The understanding of optical signal processing involves knowledge of optics, signals, processing and devices as suggested in Fig. 1.1. One should also recognize the central role played by the mathematical background, especially that

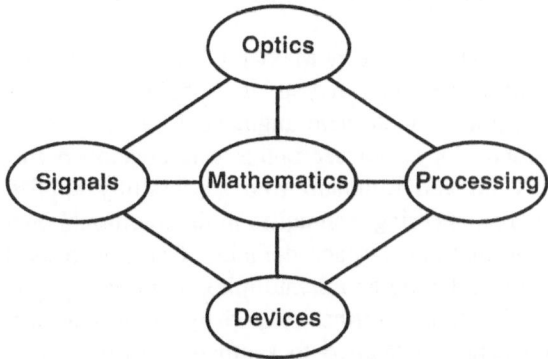

Fig. 1.1. Interrelation of disciplines required to understand optical signal processing

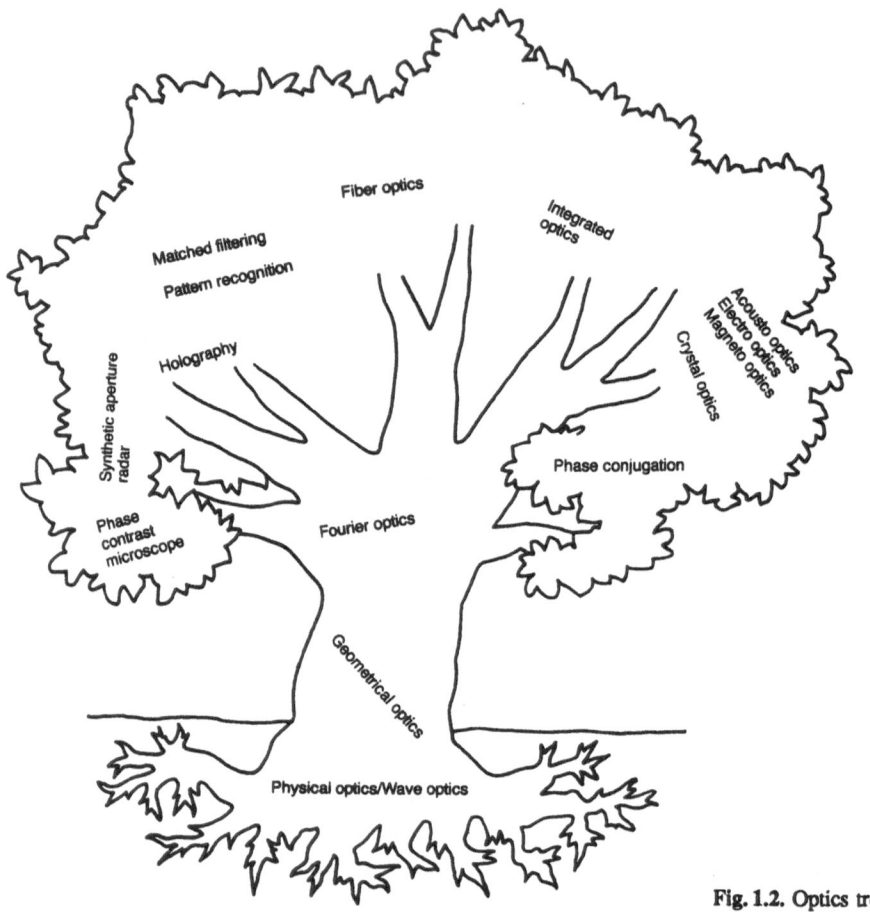

Fig. 1.2. Optics tree

offered by orthogonal functions and polynomials and matrix algebra, which serve as interconnecting links between the four disciplines. It is of interest to discuss these five topics, as shown in compact form pictorially as trees with branches in Figs. 1.2–6.

The optics tree (Fig. 1.2) root includes work by Maxwell, Fresnel and Fraunhofer as the fundamentals of physical optics or wave optics. Abbe introduced the fundamental concepts of Fourier optics, which were augmented by Zernicke in applications to phase contrast microscopy. Fourier optics was developed further by Maréchal, Tsujiuchi, O'Neill and Lohman, who successfully applied more of the conventional electrical engineering techniques in the traditional one-dimensional time domain to two-dimensional space domains. Gabor also used the concept of spatial frequency multiplexing and demultiplexing in holography which, after the invention of the laser, was further refined by Leith, Cutrona, Palermo, Porcello and Van der Lught and applied to synthetic aperture radar, matched filtering and pattern recognition. Integrated optics, optical fiber propa-

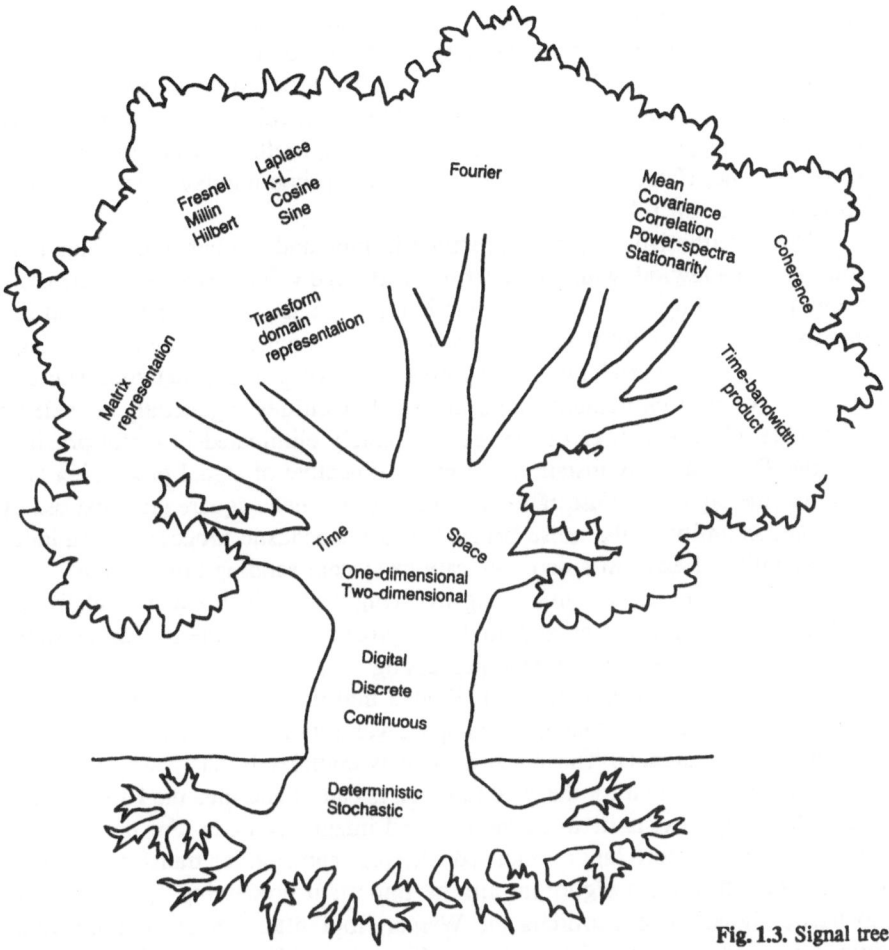

Fig. 1.3. Signal tree

gation and crystal optics are some of the other branches of the optics tree. Note that the tree is still growing very rapidly and, from the application point of view, some of the branches (e.g., optical fibers) might overshadow some of the other branches with respect to engineering applications in volume, mostly because of the eventual replacement of most telephone lines by optical fibers. Nonlinear optics is also an important branch but, for our purpose, we shall consider its use only in phase conjugation and optical computing.

The signal tree (Fig. 1.3) consists of deterministic and stochastic branches. The signal can also be subdivided as one-dimensional or two-dimensional. One-dimensional signals can be a function of time or space, whereas the two-dimensional space signals can be scanned and can be represented also as a function of time. In nature (if we neglect quantum mechanics for the moment) most of the signals one deals with are continuous, but, for the purpose of representation and manipulation, one can have continuous, discrete or digital signals when using

3

practical devices. Of course, for discrete or digital signals, one needs to worry about Nyquist sampling rate and aliasing. Also, matrix representation of discrete and digital signals might be quite convenient. Transform domain representation, especially the Fourier transform, plays an important role in many applications. Other transforms, such as Fresnel, Hilbert, Laplace, Mellin, Karhunen-Loeve (K-L), cosine, sine, Haar, Walsh, RADON, Poisson, prime number, Hadamard, etc. also play important roles.

Most deterministic signals are limited in time and frequency and thus have a finite time-bandwidth which varies from 1 to large values. Stochastic processes are characterized by their mean, correlation, covariance, power spectra, and stationarity. Note that light from a source is always stochastic, although for laser light the degree of correlation can approach that of a nearly deterministic signal. Noise (or random fluctuations) in physical quantities is generally considered undesirable. However, it can never be completely eliminated in a real practical situation. Thus, in many instances the major objective of signal processing is to maximize the signal-to-noise ratio or minimize the mean square error caused by the unpredictability of the noise because it is a stochastic process, or minimize the probability of error in a communications system sending information.

This brings us to the processing tree (Fig. 1.4), which can be subdivided into linear system processing and nonlinear processing. For the two-dimensional case, it is also referred to as image processing.

Linear systems again can be subdivided into time-variant or time-invariant cases for one-dimensional signals and space-variant or space-invariant cases for two-dimensional signals. The linear system is completely characterized by the impulse response which for the time function is causal and thus the Hilbert transform relationship exists between the real and imaginary parts. For the invariant case, convolution, correlation, matched filtering, and Fourier transformation using the chirp transform algorithm are the important processing operations. For stochastic signals, one considers the Wiener-Hopf filter. Other complex functions such as the Wigner distribution, ambiguity function, and triple correlation are also of importance and find many applications.

In nonlinear processing, homomorphic transformation and cepstrum analysis are useful signal processing tools for some cases. For the two-dimensional case, one considers nonlinear processing such as the theta modulation and optical filters with feedback.

Another factor to be included in practical signal processing applications is to consider what is called robust signal processing, i.e., how the system reacts as the noise properties or signals are varied. For the discrete or digital case, one also worries about different architectures such as serial and parallel processing, synchronous and asynchronous systolic arrays.

Adaptive processing is a very strong branch of the processing tree. The recursive algorithm for the solution of the Wiener-Hopf equation, such as Widrow-Hoff filtering or Kalman filtering, are very important tools. A somewhat different approach to adaptive processing is the use of neural nets. Another important branch is the power spectral estimation of a time series or a stochastic system

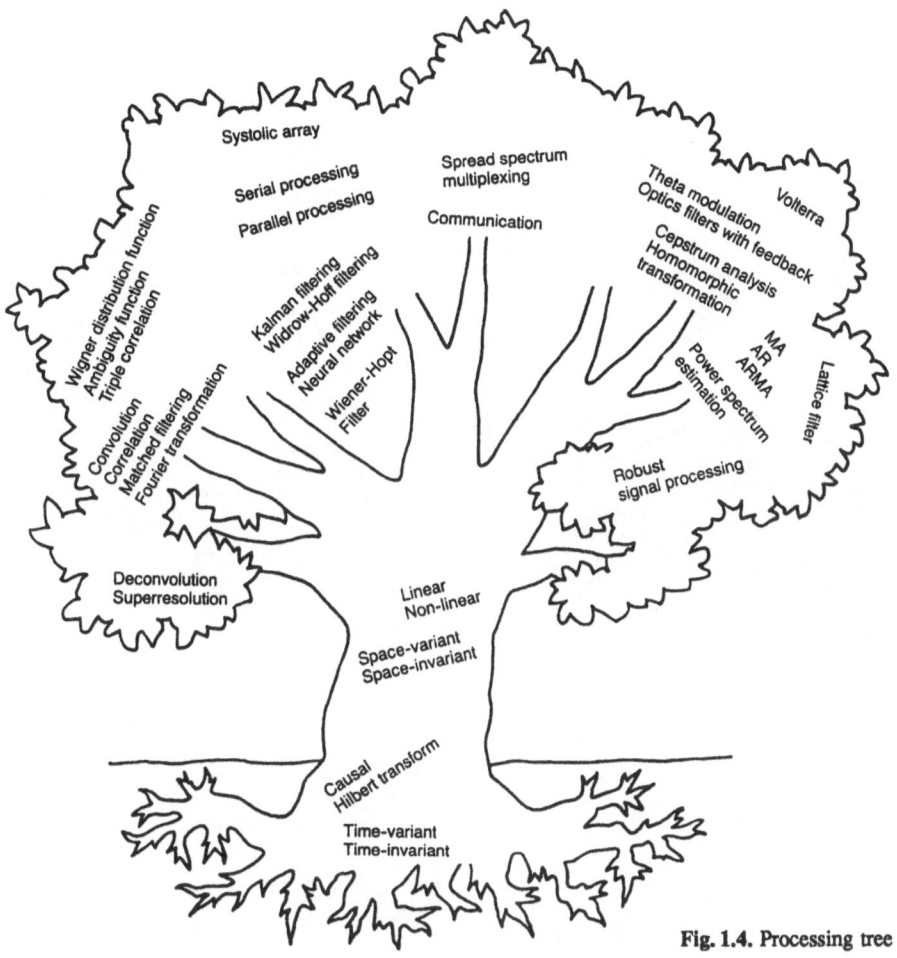

Fig. 1.4. Processing tree

using either windowing or the moving average (MA), auto-regressive (AR) or maximum entropy method, or a combination of both called the ARMA model.

The branch termed communication should be considered a tree by itself. However, for our purpose, we simply note the different multiplexing schemes such as time/space, frequency, or code and spread spectrum systems.

The problem of deconvolution or the inverse problem and its solution by different means is another important tool for image processing and other problems where data is taken using devices of finite resolution. This is also related to the technique called super-resolution where an incomplete set of data is available and we wish to obtain information beyond the obvious resolution limit. The methods used to perform this are Taylor series expansion, projection in convex space and linear programming.

Once the signal or image processing filter has been designed using the techniques presented in Fig. 1.4, one needs to implement it using different devices

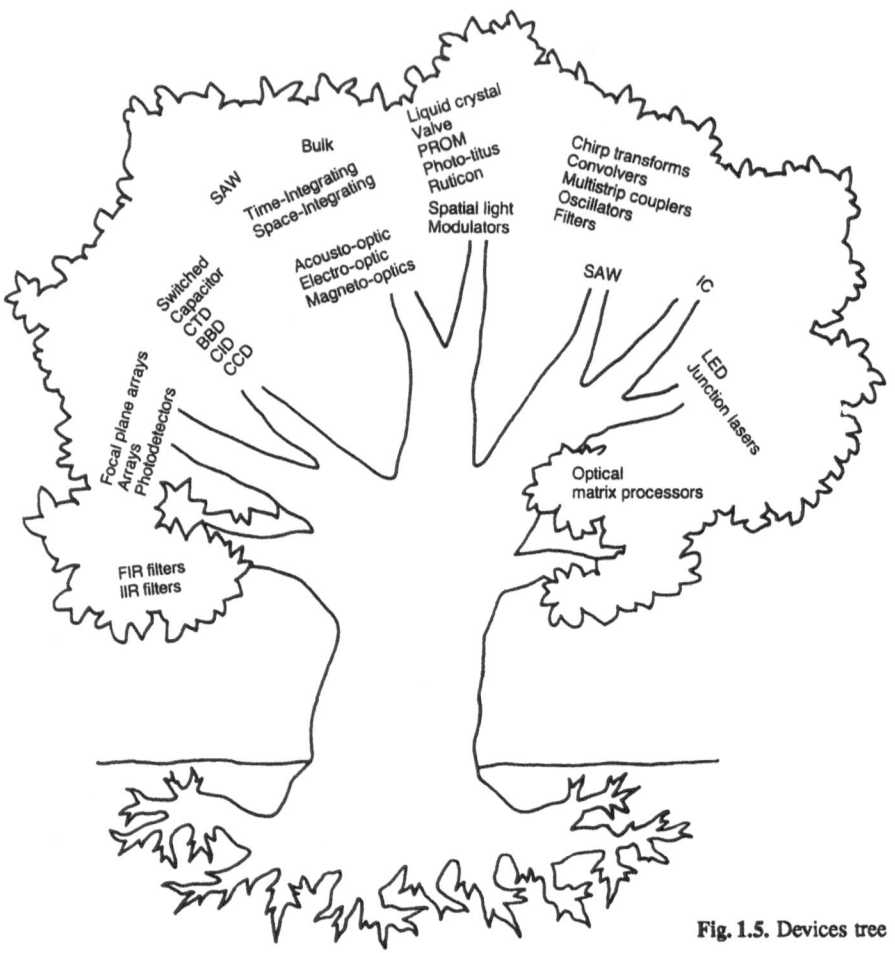

Fig. 1.5. Devices tree

(Fig. 1.5). In many cases, this can be done using an ordinary *RLC* circuit; integrated circuits; surface acoustic wave (SAW), CCD or switched capacitor filters for the analog case; or digital processors for the digital case. If real-time operation is not necessary, digital computation can also be performed using computers. However, one can also use acousto-optics, electro-optic or magneto-optic devices. These devices can be used as real time correlators, convolutors, or matched filters, etc., or as optical matrix processors. For the conversion of light signals, one has spatial light modulators, which include a host of devices, such as liquid crystal valves, e-beam potassium dihydrogen phosphate (KDP) devices, PROMs, Photo-tituses, strain-based spatial light modulation (SLM) using ceramic ferroelectrics, SLM using deformable surface tubes, Ruticons, and membrane light modulators. Of these devices, acousto-optic devices have been so far the most useful in actual applications because of their large bandwidth operation.

For the design of some of these filters, there are different implementations such as FIR (finite impulse response) filters, which are equivalent to a tapped delay line without any feedback, and the IIR (infinite impulse response) filter, which includes feedback.

For optical signal processing, one also needs a light source and light detectors. The source is usually a LED or junction laser, which have the advantage of compactness and higher efficiency, or other lasers such as He-Ne or argon. The detectors are a photodetector, photodetector arrays, focal plane arrays, and sometimes photomultipliers and ordinary photographic film if electronic output is not needed. For the electronic output case, the detectors also need amplifiers and other conventional electronics for further processing and display. For the array output one also needs CCDs or digital circuits to manipulate the large number of data coming out of the arrays.

Because of the importance of acousto-optic devices and the potential for integrating an acousto-optic device, SAW device, laser source and photo-detector array, CCD and digital circuits on a GaAs substrate, one also needs to consider all of these technologies separately and together for the system-level problems.

The last tree to consider is the mathematics tree (Fig. 1.6), where we shall only consider two branches: matrix algebra and the orthogonal polynomials and functions. In the matrix algebra branch, the important parts are eigenvalues and eigenvectors, diagonalization of a matrix, the Caley-Hamilton theorem, matrix inversion, pseudo-inverse, Gaussian elimination method, matrix factorization, Bierman's equations, successive orthogonalization by the Gram-Schmidt procedure, circulant and Fourier matrix, Toeplitz matrix, singular value decomposition, triangularization and Schur complement, Givens and modified Givens method of orthogonalization, Householder method of orthogonalization, Levinson algorithm for the inversion of a correlation matrix, CORDIC (coordinate rotation digital computer) using Schur complement, triangularization and Cholesky decomposition.

For the other branch, one has the Sturm-Liouville equation and Bessel, Hankel, Legendre, Laguerre, Gauss hypergeometric, Jacobi, and Hermite polynomials and functions.

1.1 Why Optical Signal Processing?

The signals to be processed are, in general, electrical in nature. Thus a natural question arises as to why optical signal processing is desirable, because the electrical signal has to be converted first to an optical signal before processing. To answer this question we note that very efficient high bandwidth, high time-bandwidth product modulators are available through acousto-optic or electro-optic interaction or through the direct modulation of the lasers, especially the junction laser itself. But more important are the following points:

i) For digital processing, one needs a very fast analog-to-digital (A/D) converter for analog signals. At present, A/D converters beyond sampling rates of tens of megahertz are not easily available at a reasonable price. Even for analog

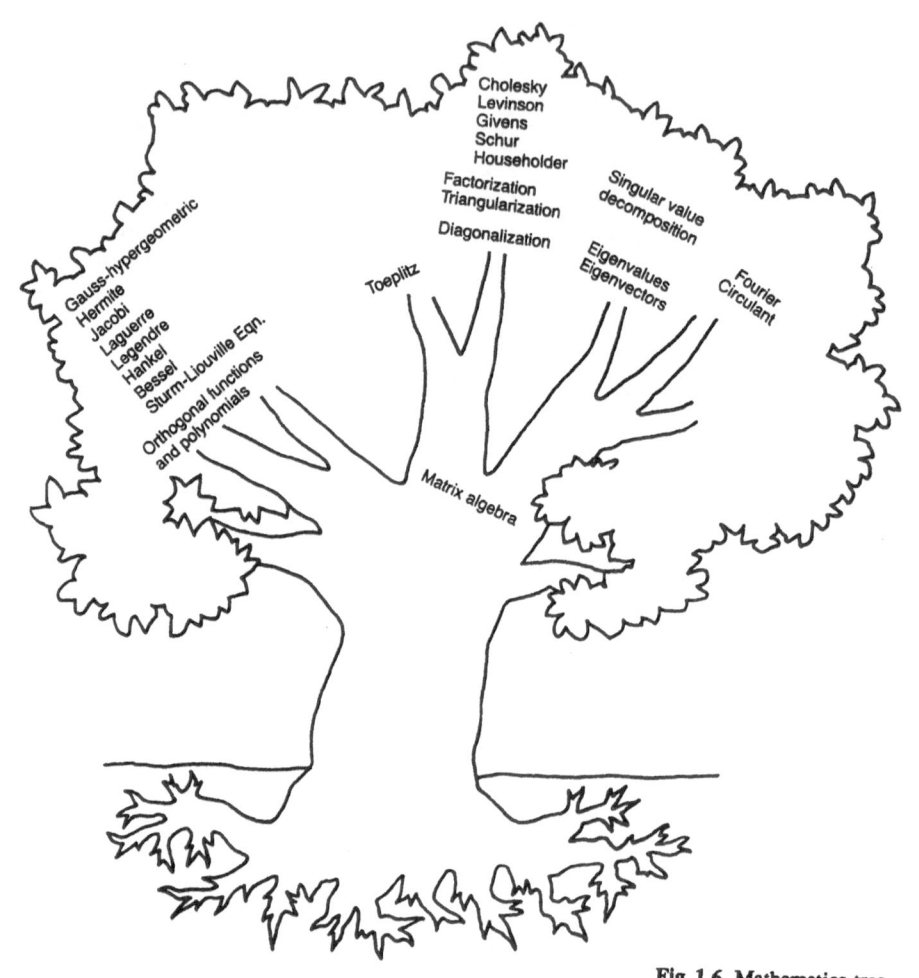

Fig. 1.6. Mathematics tree

signals, high bandwidth requirements which are very difficult to accommodate by other technologies than optical can be easily met by acousto-optic or electro-optic devices. Thus, instantaneous power spectra of an electrical signal with bandwidths exceeding 1 GHz can at present be performed easily but probably only by acousto-optic devices.

ii) The signal to be processed is not always electrical. With the increasing use of fiber-optic communication cables in the near future, there will probably be more signals already on optical carrier. For this case, one is better off processing the signals directly by optical signal processing than first converting the optical signal back to an electrical one by a detector and then processing it. A typical example might be the data from many sonars all over the submarine collected at the optical processor by transmission through fiber optic cable.

iii) The more obvious case is that the signal is itself in the optical domain, such as occurs in an optical image or transparency.

iv) Inherent parallelism in optical signal processing is probably the biggest asset of that processing method. Imagine trying to connect an electrical signal to 10^6 discrete points simultaneously. This can be easily done in optics if a junction laser placed at the focal point of a lens is modulated with the electrical signal. Note that this is the fundamental reason for the interest and importance of optical signal processors. Another example is the correlation or matched filtering of images in parallel.

One might ask why, with all these advantages of optical processing, is it not used more widely already? Except for the synthetic aperture radar, no other application has been a commercial success. The reason for this situation is that the required technology was not available before. We recall that optical communication also had to wait for high quality fiber-optic cables and efficient heterojunction lasers before it became successful. Similarly, optical signal processing needs a more mature technology to be used on a larger scale. At present, acousto-optic and electro-optic device technology has already seen enormous progress in making optical signal processing more useful, an example being the recent success of the Bragg-cell real-time power spectrum analyzer. The heterojunction laser with the intra-cavity modulator to frequency (wavelength) modulate the laser, as well as other devices, are also maturing and contributing to make optical signal processing more viable today. Furthermore, there are many proposals appearing to explore the advantages of optical devices, such as very fast optical A/D converters, optical logic (and its use to build a computer), use of optical sources or modulators in conjunction with holographic lenses for IC interconnection, and systolic processors, to name a few.

One of the main disadvantages of optical signal processing as compared to digital is the inaccuracy and dynamic range limitation associated with any analog signal processing. Residue arithmetic architecture can address these problems at the cost of introducing more complexity and, in some cases, cost. One other possibility is to use binary arithmetic in optical processors, such as in optical systolic arrays, or a digital multiplication by analog convolution (DMAC) algorithm. The heart of the DMAC algorithm is the fact that, when two binary numbers are multiplied, the result can be viewed as a convolution of two binary bit-streams and a carry propagation operation. Some of these hybrid techniques, in which the high degree of parallelism and high speed of optical processing is cleverly combined with the accuracy and flexibility of digital processing, could be success stories in the future.

Another disadvantage of using optical components for electrical signal processing is the packaging difficulty because of the clumsiness and size associated with optical devices. An example is shown in Fig. 1.7 which describes the setup used in 1974 by the author and his students to demonstrate real-time acousto-optic convolver operation using SAW delay lines. However, due to advances in integrated optics, not only the whole setup shown in Fig. 1.7 but other func-

Fig. 1.7. (a) Acousto-optic convolver. **(b)** A translational stage assembly needed for acousto-optic interaction

tions as well can now be incorporated in a small portion of a wafer, as shown in Fig. 1.8. Note that Fig. 1.8 is a schematic diagram of a futuristic system on a single semiconductor wafer using different signal processing devices such as electro-optic modulators, acousto-optic correlators, digital circuits, photodetector arrays, etc. Guided waves used in integrated optics can be used so that the optics can be planar and mechanically robust. Semiconductors such as GaAs and InP, in conjunction with their combinations (i.e., $Ga_xAl_{1-x}As$), are the most promising candidates for the planar substrate, and their properties are most suitable for implementing this newly emerging technology.

Fig. 1.7. (b) Caption see opposite page

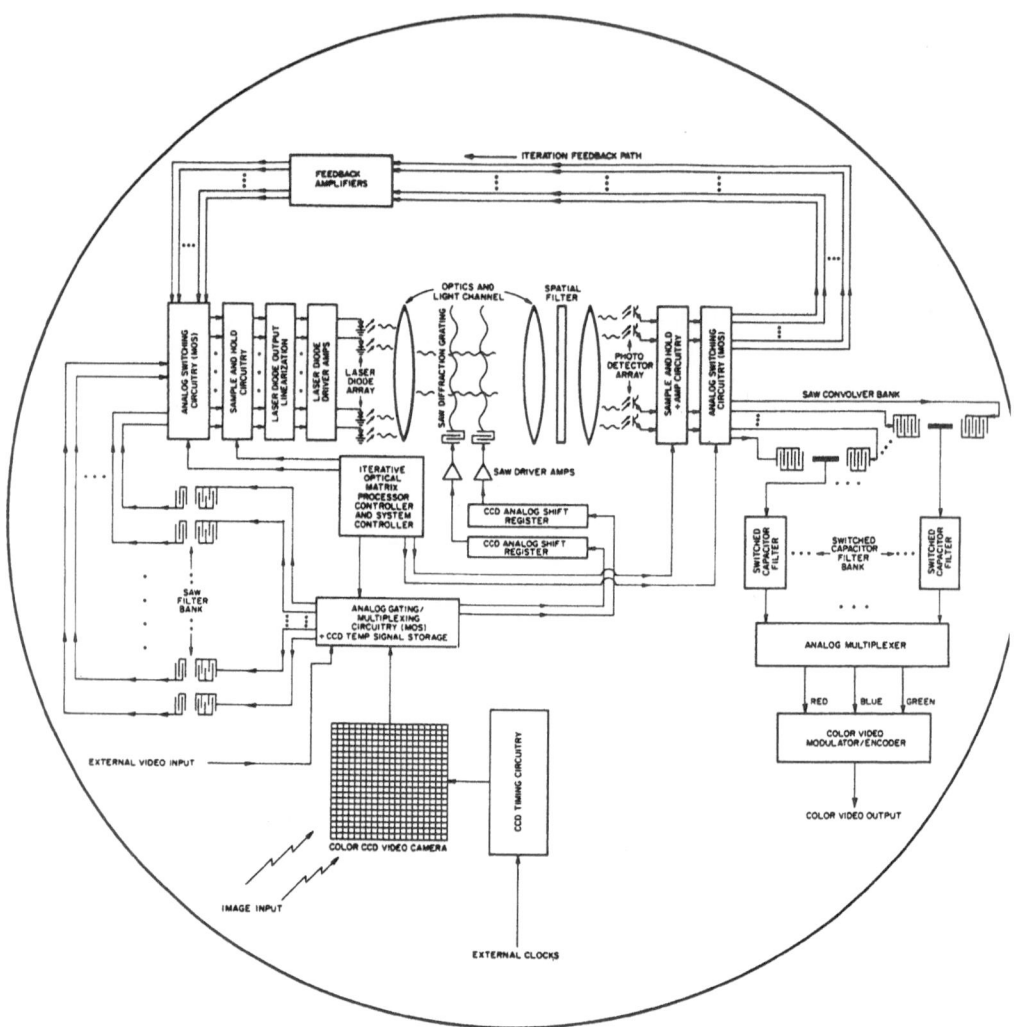

Fig. 1.8. Schematic diagram of a futuristic system on a single GaAs wafer using different signal processing devices

1.2 Signal Processing: Tools and Applications

In real life, desired electrical signals are always contaminated by random noise or by a deterministic unwanted disturbance, which might be intentional or unintentional. The objective of a signal processing device is to eliminate or reduce as much as possible these undesired components from the input so that a best estimate of the information contained in the signal can be made. Ideally, a signal processor should have unlimited bandwidth and time-bandwidth products. However, due to physical constraints, this is never possible. Thus, to perform

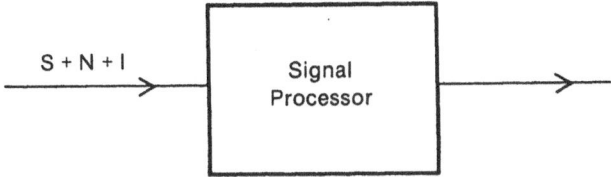

Fig. 1.9. Block diagram of a signal processor

real-time signal processing with faster speed and higher bandwidth has been an important and continuing goal for electrical engineers for many years. Recently, the revolutionary advance caused by the availability of high-speed integrated circuits, optoelectronic and SAW devices has made this hitherto elusive goal come within the horizon of technological feasibility.

A signal processor performs the improvement of the desired signal. As shown in Fig. 1.9, the signal power is denoted by S, the power of the undesired signal, disturbance, or interferer as I and the noise power as N. A typical signal processor will improve the ratio $S/(N + I)$ at the output compared to that at the input. Note that the signal and interferer can be either deterministic or random, whereas noise is always random. For two-dimensional signals, it is possible that one is looking for the presence or absence of a pattern or a character. In many of these problems, the solution is a filter which can be designed in either the time domain or in the frequency domain using varied devices such as fiber-optic cables, SAW delay lines or CCDs. Which devices to choose will depend on the frequency region in which one is operating. Figure 1.10 gives general guidelines for the choice in terms of center frequency and time-bandwidth product. In this figure, only fiber-optic, magnetic SAW, SAW including acousto-optic, CCD, acoustically coupled transport (ACT) and superconducting devices are considered. However, there are other possibilities and, depending on the application, they should also be borne in mind. Let us consider the example of transform domain processing where the signal is first Fourier transformed before being processed to improve its performance. The real-time Fourier transformation can be performed using chirp filters. These filters can be implemented using one of the following major technologies:

1. surface acoustic waves (SAWs)
2. bulk ultrasound
3. charge coupled devices (CCDs)
4. acoustically coupled CCDs
5. magneto surface wave
6. superconducting tapped delay line
7. fiber optic tapped delay line
8. acousto-optics
9. integrated circuits

These are discussed further in Sect. 4.6.

13

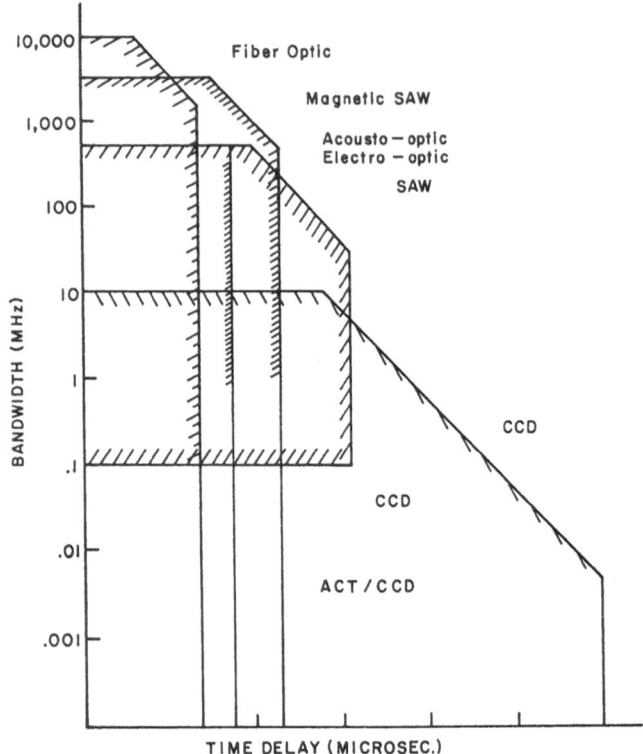

Fig. 1.10. Time-bandwidth applicability of various technologies

Adaptive filters play an important role in signal processing. A typical example for this case might be a spread-spectrum receiver, to be discussed in detail in the second volume. The general block diagram is shown in Fig. 1.11. Note that all the elements in the block diagram such as filters, correlators, convolvers, matched filters, Fourier transformers, etc., can be implemented using different technologies. The same receiver can also be implemented using a somewhat different approach, as shown in Fig. 1.12, which uses least mean square (LMS) adaptive filters based on a gradient algorithm.

Instead of fabricating subsystems such as filters and correlators separately and then packaging them to build the total system, it would be better if the major part of the system could be built in a single wafer, as shown in Fig. 1.8. Due to their properties, III-V semiconductors, such as GaAs and InP, have the greatest potential in this area. We discuss the properties of GaAs in some detail because of its importance.

GaAs is piezoelectric, so that SAW devices can be fabricated directly on it. It is a high mobility, high saturation velocity semiconductor and therefore analog and digital integrated circuits can be built using GaAs substrates. It is a direct band gap semiconductor, and thus it can serve for junction lasers. It is

14

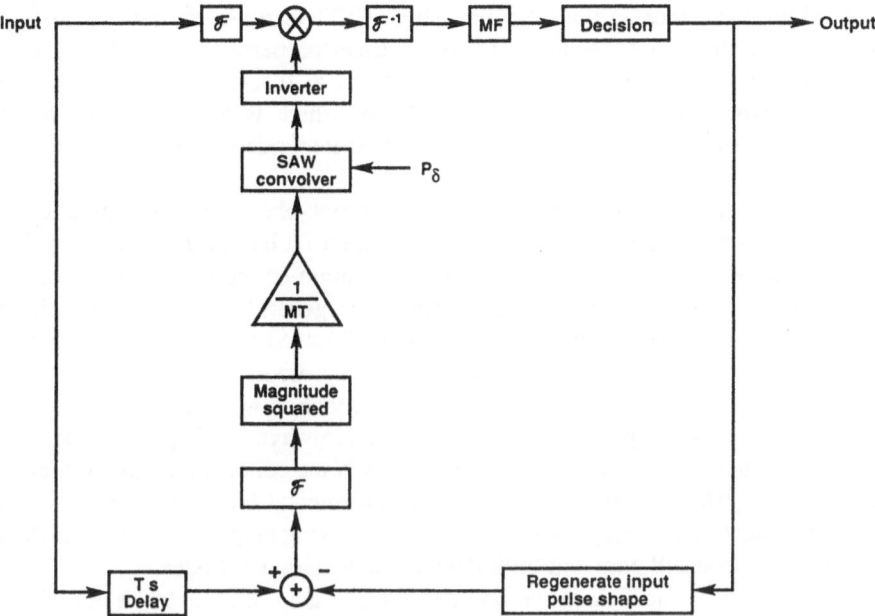

Fig. 1.11. Adaptive receiver using spectral estimation with SAW technology. (From [1.1])

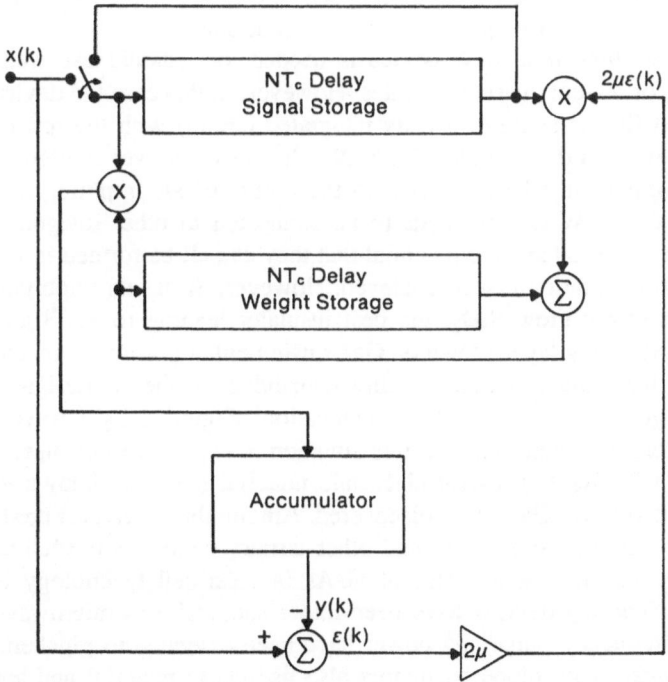

Fig. 1.12. Block diagram of burst processing adaptive filter. (From [1.2])

15

a good photodetector so that photo-sensitive CCDs can also be fabricated. In a word, it has the optical, acoustical and electrical properties needed to fabricate the complete system on a wafer shown in Fig. 1.8. It is of interest to compare the important properties of GaAs and silicon, which is the main element for all the successful achievements in the so-called integrated circuits revolution in electronics.

GaAs is a compound semiconductor with zinc-blende crystal structure and a lattice constant of 5.64321 Å at room temperature. It has a direct band gap, i.e., the minimum of the conduction band and maximum of the valence band occurs at $K = 0$, making it useful for optical devices in general, and for light emitting diodes and lasers in particular. An energy gap of 1.435 eV, 300 meV greater than Si, provides a lower leakage current for Schottky diodes. An electron mobility of $8800\, cm^2\, s^{-1}\, V^{-1}$, five times greater than silicon, low ionization energy, and high saturation velocity makes it attractive for microwave and high-speed devices. Its peculiar band structure, i.e., satellite valleys in the conduction band minima at $K = 0.6$ (in [100]) and 0.4 eV higher than fundamental band-gap energy, which have low electron mobility, makes it capable of having negative resistance, which can be used in oscillators and bulk devices such as Gunn diodes.

Although the piezoelectric properties of GaAs are not as strong as in $LiNbO_3$, for example, they can be augmented by evaporating or sputtering a thin film of ZnO. On the other hand, silicon is an elemental semiconductor, and therefore it is not piezoelectric. However, there is a possibility that someday someone may discover a process by which a SiO_2 on Si system can be made to be piezoelectric, as the crystalline form of SiO_2 is quartz, which is piezoelectric.

One might argue that, if a SAW device is needed, one should use well-established $LiNbO_3$ or ST-cut quartz substrates. However, in this case the device cannot be integrated (in the sense of a truly integrated circuit) with the rest of the necessary electronics. For example, if a SAW filter or convolver is used in a system, then the input must be connected to the output of an amplifier, and then the output of the SAW device needs to be connected to other integrated circuits. GaAs SAW devices have the potential that they can all be formed in the same wafer and true integration can be achieved. However, Si has an important advantage. One can easily grow SiO_2, the best insulator leading to MOS and other devices. The insulating layer grown on GaAs using either plasma or anodic oxidation does not have the consistent quality demanded by the IC designer. Another disadvantage of GaAs is the high attenuation of guided light waves. However, the technology is maturing and this situation is expected to change.

The potential of GaAs as a useful electronic material has been known to scientists and engineers since GaAs was discovered. Among the successful products are p-n junction lasers using GaAs and other tertiary components such as GaAlAs, and GaAs detector diodes. Use of GaAs in solar-cell technology is well known. GaAs Schottky devices have been rather successful as microwave amplifiers and the frequency range and power have been extended to gigahertz and hundreds of watts. Gunn diode oscillators also use GaAs material and are useful in the same frequency range. SAW delay lines and signal processing devices, such as correlators and convolvers, have also been built successfully as prototypes although not as a commercial product. Acousto-optic and electro-optic devices using GaAs have been proposed and some have been demonstrated.

Integrated circuits on GaAs are being fabricated and, because of the insulating layer problem, these are not MOS type. Charge coupled devices operating at around 4 GHz clock frequency have also been fabricated. This is possible because the higher mobility and higher drift velocity make the charge transfer time very small. Recently it has been shown that the SAW field can be used as a charge transfer field in place of a regular clock and potential well defined by MOS structure. These are called ACT devices and they have the potential to increase the operating frequency even further, although some flexibility of a regular CCD is lost.

In the optics area, junction lasers producing power on the order of watts cw at different optical frequencies are being manufactured routinely. These lasers can be modulated directly in the 10 GHz range. Detectors with matching bandwidth have been fabricated using GaAs and related compunds such as $Ga_xAl_{1-x}As$. The success of this new development can be traced to the quantum layered material or superlattice, which can be grown easily on GaAs substrates using molecular beam epitaxy. The ability to form very thin (~ 10 Å), controllable, high-quality layers has produced not only high-power semiconductor lasers and high bandwidth detector arrays, but also better digital circuits and analog amplifiers using quantum layered heterojunctions and optical logic devices.

The usefulness of GaAs for signal processing applications has been demonstrated in bits and pieces. A comprehensive study is needed to put together these bits and pieces into an integrated subsystem or even a system. The following examples will clarify this point. Consider a GaAs-based SAW matched filter used for a communication system and shown in Fig. 1.13. The input signal, in general at a frequency different from the operating frequency of the matched filter, needs to be mixed with another signal from an oscillator, and then it must be filtered and amplified before being applied to the matched filter. Thus, the SAW device in conjunction with the mixer, amplifier, and oscillator provides a very useful subsystem if it can be integrated in a single wafer. This is shown schematically in the figure. Individual elements have been demonstrated before and it is quite feasible to integrate this subsystem on a single GaAs wafer.

Consider also the example of a complete anti-jam spread-spectrum communication receiver discussed earlier (Fig. 1.11). The interesting point is that all these subsystems have been either experimentally demonstrated or shown to be feasible. Thus, it is reasonable to expect that complete systems will be integrated in the near future on a single wafer similarly to the different subsystems of a microprocessor that are integrated on a microprocessor chip.

The following obvious question arises in a discussion of any devices other than digital ICs for signal processing: In the face of revolutionary progress being made in digital circuits through programs like VHSIC and related to VLSI, why does one need to consider any other device? To answer this question, one needs to point out that, for a 100 MHz bandwidth signal, the sampling rate for

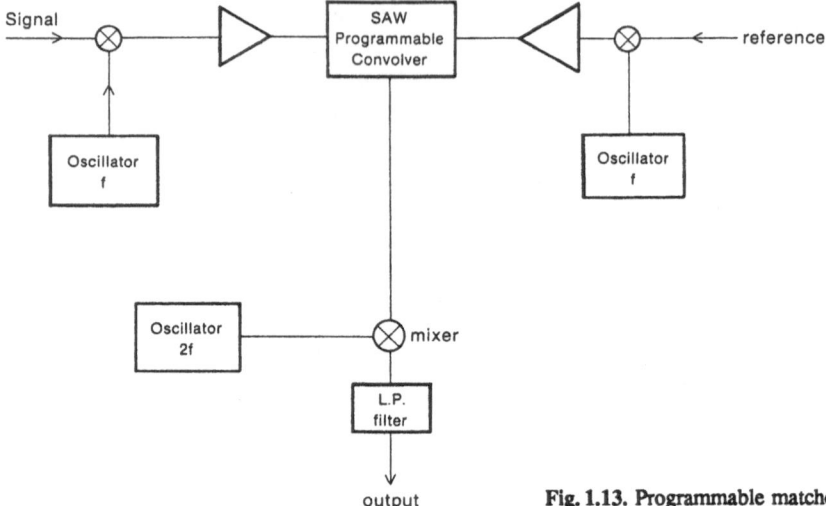

Fig. 1.13. Programmable matched filter

an A/D converter must be at least 200 MHz, and then to process the signal in real time, the clock frequency of the ICs must approach the gigahertz range. It is true that with parallel processing (at a considerable cost) one can still manage to use lower frequency clocks. However, for situations like decoding in a digital communication system, parallel processing is not always permissible. Also, with the advent of optical communication as a viable alternative and the need to process signals at higher and higher bandwidths, the devices discussed in this book will play an important role in the field of signal processing.

1.3 Arrangement of the Book

For space reasons, *Optical Signal Processing* appears as two volumes. The remainder of this volume is divided into three chapters, two dealing with fundamentals and one with devices. The second volume (*Optical Signal Processing: Applications*) has a further chapter on devices and two on applications, in addition to an introductory chapter. Each chapter of each volume is basically self-contained, but all the various parts are closely related. For example, someone interested only in understanding real-time acousto-optic convolver devices could concentrate on Sects. 2.14 and 4.1 of this volume and Sects. 2.3 and 2.4 of the second volume. For the sake of completeness, in the following we summarize the topics treated in the various chapters of both volumes, but the summary of the second volume is kept brief.

In Chap. 2, optics fundamentals are introduced, starting with Maxwell's equations and plane-wave solutions of electromagnetic wave equations. Reflection and refraction problems are considered using the transmission line approach rather than the usual solutions using boundary conditions. One section deals with

Gaussian beams, which are derived from the wave equation directly. The propagation of these Gaussian beams through lenses or through an optical system is considered using the matrix approach. Section 2.9 also reviews the matrix formulation of geometrical optics. This is followed by two sections on fiber optics and integrated optics problems of guiding waves. Both graded and step index fibers are considered. In the next four sections, the subject of light propagation in anisotropic media and the electro-optic, acousto-optic and magneto-optic effects are considered. These sections are essential for the devices discussed in Chap. 2 of the second volume. Other topics, such as double refraction, polarizers, Jones matrix representation of polarized light, Bragg and Raman-Nath diffraction, and the Faraday and Kerr effects are also introduced in these sections. In the last three sections of Chap. 2, diffraction and Fourier optics are discussed. Section 2.17 reviews Fourier optics at a very elementary level and includes Fresnel and Fraunhofer diffraction and holography.

Chapter 3 deals with the fundamentals of signal processing and includes mostly analog and discrete systems. This part is mainly for readers who are not familiar with the analytical tools involved in signal processing and with concepts of stochastic processes. Sections 3.1 and 3.2 introduce the elementary concepts of linear systems, including convolution, time invariance, frequency and Z-response. Section 3.4 discusses noise and stochastic processes, including the concept of matched filtering. Chapter 3.5 discusses the design of filters of both transversal and recursive types. The adaptive filter section, 3.6, includes least-mean-squares estimation and Wiener filters, Widrow-Hoff filters and lattice filters. Section 3.7 discusses power spectral estimation, including auto-regressive (AR), moving average (MA) and auto-regressive and moving average (ARMA) models. Section 3.8 deals with Kalman filtering and square-root filtering and with state-space formulation of the problems. In Sect. 3.9, two-dimensional spatial signal processing is introduced. Multidimensional stochastic processes are discussed in Sect. 3.10. The final section, 3.11, deals with the ambiguity function, the Wigner distribution function and other topics not discussed earlier.

Chapter 4 presents technologies that are not directly related to optics. The major objective is to discuss SAW and CCD devices. However, some discussion of magnetic SAW, ACT, superconducting and bulk ultrasound technologies is also included.

Comprehensive material on matrix algebra is included as Appendix A. This is essential for many sections, e.g. Sect. 3.8 (Kalman filtering). Topics such as the pseudo-inverse and singular value decomposition are included. Appendix B discusses orthogonal functions and polynomials starting from the Sturm-Liouville equations. Appendices C and D deal with the principle of stationary phase and elementary vector analysis, respectively. Appendix E gives the important symmetry relations of elastic, piezoelectric, photoelastic, electro-optic and magneto-optic coefficients. These appendices are helpful in providing a self-sufficient and easily readable text that does not require reference to many other books.

Chapter 2 of the second volume introduces the subject of optical devices, such as modulators, deflectors, convolvers, spatial light modulators and optical matrix processors. Chapter 3, the largest part of the second volume, presents applications of optical signal processing in subsystems and systems. It discusses the Fourier transforming and imaging properties of a lens in connection with two-dimensional signal processing, matched filtering, holographic applications, synthetic aperture radar, incoherent light spatial signal processing, fiber optic signal processors and integrated optics. A major emphasis of integrated optics is on the SAW-acousto-optic interaction. Final sections deal with communications applications and optical implementations of specific systems.

Chapter 4 covers specific topics of interest, such as super resolution, non-linear optics, phase conjugation and optical computing, and interconnections as applicable for signal processing. Discussions of optical logic devices and associative memories are also included.

1.3.1 Guide for Selective Use of the Book

Different parts of the book have been successfully used for various courses. Some of these are:

Graduate Course in Modern Optics: Most of Chap. 2 and some of Chap. 3, Vol. I, and some of Chaps. 2 and 3, Vol. II. Specifically, Sects. 4.4 and 4.9 (Vol. I), parts of Sects. 2.2, 2.4 and 2.6, and parts of Sects. 3.2 and 3.3 (Vol. II).

Graduate Course in Signal Processing Devices: SAW and CCD. Mostly Chap. 3, excluding Sects. 3.7–3.9, Chap. 4 (Vol. I) and some sections of Chap. 3 (Vol. 2).

Short Course in Optical Signal Processing: Sections 2.11.17, 3.4–3.10, Chap. 4 (Vol. I), most of Chap. 2 and parts of Chaps. 3 and 4 (Vol. II)

Of course other combinations are possible, depending on the background of the students and their interests. One can think of a course in optical computing or optical matrix processors where one starts with Appendix A, Sects. 2.4–6 (Vol. II) and some of Chaps. 3 and 4 (Vol. II).

Somebody with knowledge of signal processing but very little experience of optics should read Chap. 2 (Vol. I), to obtain some feeling for optics, then focus on Sects. 2.13–15 to become familiar with crystal optics, followed by the device and application parts.

Somebody with knowledge of optics but very little experience of signal processing should start with Chap. 3 (Vol. I), including some of Appendix A, and then go on to devices in Chap. 2 (Vol. II) and applications of interest in Chaps. 3 and 4 (Vol. II).

1.3.2 Note on References

The references cited in each chapter can be found at the end of the book. The Bibliography lists other references used in preparing this text, so that the reader may refer to them for further clarification or detailed derivations and discussions. The Bibliography also provides other references that the author feels are useful to the reader. For new topics, such as the optical matrix processor or super resolution, the number of references tends to be quite large. The author hopes that such a large number of references will be helpful to the reader interested in a particular topic.

2. Optics Fundamentals

Optics deals with light waves, which are electromagnetic waves. Electromagnetic waves include not only light waves, but also ordinary alternating current at 60 Hz, radio waves, microwaves, infrared, X-rays and γ-rays. Electromagnetic waves obey Maxwell's equations, which are introduced in Sect. 2.1, which also treats the electromagnetic wave equation followed by the plane-wave solution in homogeneous, linear isotropic space. This is followed by the derivation of Snell's law and reflection and transmission coefficients using the transmission line approach. To make the discussion self-contained, fundamentals of transmission lines are discussed in Sect. 2.5. The concepts of group and phase velocity are introduced in Sect. 2.7. The subject of Gaussian beam propagation in free space is introduced in Sect. 2.8. To consider Gaussian beam propagation through optical elements, the matrix method of geometrical optics is reviewed first and a discussion of lens aberration is included. The matrix method is then extended to introduce gradient optical fibers. The subjects of step-index fibers and integrated optics are tackled together in Sect. 2.11. Sections 2.12–16 deal with anisotropic media, including electro-optic, acousto-optic and magneto-optic effects. The subject of diffraction is considered in Sect. 2.17, which includes a review of Fresnel and Fraunhofer regions, Fourier optics, gratings, and interferometers. Fundamentals of holography and spatial filtering are also included in this review.

2.1 Maxwell's Equations

We postulate Maxwell's equations and develop the subject of optics on this assumption only. Historically, however, experimental laws such as Faraday's Law, Ampère's Law and Gauss's Law preceded Maxwell's equations. Using these experimental laws, Maxwell developed the following equations:

$$\nabla \times E = -\frac{\partial B}{\partial t} \quad ,$$

$$\nabla \times H = +\frac{\partial D}{\partial t} + J \quad ,$$

$$\nabla \cdot D = \varrho \quad ,$$

$$\nabla \cdot B = 0 \quad , \tag{2.1.1}$$

where t represents time, E is the electric field vector [units of $\mathrm{V\,m^{-1}}$], H is the magnetic field vector and has the units of magnetic induction [Wb m^{-2}], D is the electric induction (displacement) [C m^{-2}], ϱ is the free charge density [C m^{-3}], and J is the current density [A m^{-2}].

The first equation is generally known as Faraday's Law. The second one, excluding the term $(\partial D/\partial t)$, is known as Ampère's law. The last two equations are known as Poisson's equation and Gauss's law, respectively. Maxwell modified Ampère's law by including the displacement term and showed that electric and magnetic fields for the time-dependent case are intimately connected and inseparable, thus beginning the study of electromagnetic waves.

In conjunction with the above four equations, we need the so-called constitutive equations. In the most general case, these are

$$D_i = \varepsilon_{ij} E_j + \varepsilon_{ijk}^{(2)} E_j E_k + \varepsilon_{ijkl}^{(3)} E_j E_k E_l + \ldots \quad , \tag{2.1.2}$$

$$B_i = \mu_{ij} H_j + \mu_{ijk}^{(2)} H_j H_k + \mu_{ijkl}^{(3)} H_j H_k H_l \ldots \quad , \tag{2.1.3}$$

$$J_j = \sigma_{ij} E_j + \sigma_{ijk}^{(2)} E_j E_k + \ldots \quad , \tag{2.1.4}$$

where i and $j = x, y, z$; ε_{ij} is the permittivity [F m^{-1}], μ_{ij} is the permeability [H m^{-1}], σ_{ij} is the conductivity [S m^{-1}], and the nonlinear expression has been expanded in a power series. The superscript in the nonlinear coefficients $\varepsilon_{ij\ldots}^{(i)}$, $\mu_{ij\ldots}^{(i)}$, and $\sigma_{ij\ldots}^{(i)}$ represents the order of nonlinearity, and the subscripts denote the components of the E or H fields involved in the nonlinear interaction. Although in the general case, ε, μ, and σ can be functions of r and t, we shall assume that they are constant. For the linear case, (2.1.2) simplifies to

$$D_i = \varepsilon_{ij} E_j \quad \text{or} \tag{2.1.5}$$

$$\begin{pmatrix} D_x \\ D_y \\ D_z \end{pmatrix} = \begin{pmatrix} \varepsilon_{xx} & \varepsilon_{xy} & \varepsilon_{xz} \\ \varepsilon_{yx} & \varepsilon_{yy} & \varepsilon_{yz} \\ \varepsilon_{zx} & \varepsilon_{xy} & \varepsilon_{zz} \end{pmatrix} \begin{pmatrix} E_x \\ E_y \\ E_z \end{pmatrix} \quad .$$

For optical frequencies, in general, (2.1.3) and (2.1.4) can be written as

$$B = \mu H \quad , \tag{2.1.6}$$

$$J = \sigma E \quad . \tag{2.1.7}$$

The linear anisotropic case will be considered in connection with crystal optics in Sect. 2.12. For the linear, isotropic, and homogeneous case, (2.1.5) simplifies to

$$D = \varepsilon E \quad . \tag{2.1.8}$$

Equations (2.1.6) and (2.1.8) are generally rewritten as

$$D = \varepsilon_0 \varepsilon_r E \quad , \tag{2.1.9}$$

$$B = \mu_0 \mu_r H \quad , \tag{2.1.10}$$

where ε_0 and μ_0 are the permittivity and permeability of vacuum, ε_r is the dielectric constant, and μ_r is the relative permeability.

Introducing (2.1.7), (2.1.9) and (2.1.10) into (2.1.1), one obtains the wave equations for the source-free case ($J = 0$):

$$\nabla^2 E - \mu\varepsilon\frac{\partial^2 E}{\partial t^2} - \mu\sigma\frac{\partial E}{\partial t} = 0 \quad ,$$

$$\nabla^2 H - \mu\varepsilon\frac{\partial^2 H}{\partial t^2} - \mu\sigma\frac{\partial H}{\partial t} = 0 \quad . \tag{2.1.11}$$

Let us assume that the time dependence of the electromagnetic field is $\exp(j\omega t)$, where the angular frequency ω, in radians, is given by

$$\omega = 2\pi f \quad , \tag{2.1.12}$$

where f is the frequency [Hz]. This assumption is not restrictive at all, because we know that any general function of time $f(t)$ can be represented by the Fourier transform equation

$$f(t) = \int F(f) e^{-j\omega t} df \quad , \tag{2.1.13}$$

where the inverse transform is

$$F(f) = \int f(t) e^{j\omega t} dt \quad .$$

Here $F(f)$ will represent the Fourier components associated with a particular frequency ω. As Maxwell's equations are linear, superposition holds and thus a general problem can be solved.

To solve (2.1.11), we now assume

$$E(r, t) = E_0 e^{j(\omega t - k \cdot r)},$$

$$H(r, t) = H_0 e^{j(\omega t - k \cdot r)}, \tag{2.1.14}$$

where k is the wave vector associated with propagation, which is to be determined. In the Cartesian coordinate system

$$\begin{aligned}
k \cdot r &= (i_x k_x + i_y k_y + i_z k_z) \cdot (i_x x + i_y y + i_z z) \\
&= k_x x + k_y y + k_z z \quad ,
\end{aligned} \tag{2.1.15}$$

where k_x, k_y and k_z represent the x, y and z components of the wave vector, respectively. Substituting (2.1.14) into (2.1.11) we obtain

$$|k|^2 = \omega^2 \mu\varepsilon - j\omega\mu\sigma \quad . \tag{2.1.16}$$

Due to the presence of the conduction loss, $\sigma \neq 0$, and $|k|$ is complex:

$$k = k_r - jk_{im} \tag{2.1.17}$$

where k_r and k_{im} are respectively the real and imaginary parts of the wave vector magnitude, also known as the propagation constant. Thus the plane wave solution for (2.1.11) can be written as

$$E(r,t) = E_0 e^{-k_{im} i_k \cdot r} e^{j(\omega t - k_r i_k \cdot r)} \quad ,$$
$$H(r,t) = H_0 e^{-k_{im} i_k \cdot r} e^{j(\omega t - k_r i_k \cdot r)}, \tag{2.1.18}$$

where i_k represents the unit vector along the direction of propagation. The amplitude of the wave decays if k_{im} is positive, which is the most common case. However, for active cases such as a lasing medium, k_{im} is negative and the wave grows as it propagates.

To describe the wave motion, we consider how the planes of constant phase are moving. A plane of constant phase is defined by

$$\omega t - k_r i_k \cdot r = \text{const} \quad . \tag{2.1.19}$$

Thus these planes are perpendicular to i_k. If $i_k = i_x$, then they are perpendicular to the x axis for this case, i.e., they lie in the yz plane. The phase velocity of the wave v_p can be found as follows:

$$\omega \Delta t - k_r i_k \cdot \Delta r = 0 \quad , \quad \text{or}$$

$$\frac{dr}{dt} = v_p = \frac{\omega}{k} i_k \quad . \tag{2.1.20}$$

As (2.1.16) is defined in terms of $|k|^2$, k_r has two possible values of equal magnitude but opposite sign. The wave with positive sign represents the forward-going wave whereas the wave with negative sign represents the backward-going wave.

If

$$\sigma = 0 \quad , \quad \text{then} \quad k_{im} = 0 \quad , \quad \text{and}$$

$$k = k_r = \pm \omega \sqrt{\mu \varepsilon} \quad . \tag{2.1.21}$$

For this case

$$v_p = \frac{1}{\sqrt{\mu \varepsilon}} \quad . \tag{2.1.22}$$

In the general case

$$k_r = \omega \sqrt{\mu \varepsilon} \sqrt{\frac{1 + (\sigma/\omega \varepsilon)^2 + 1}{2}} \quad ,$$

$$k_{im} = \omega \sqrt{\mu \varepsilon} \sqrt{\frac{1 + (\sigma/\omega \varepsilon)^2 - 1}{2}} \quad . \tag{2.1.23}$$

For a poor conductor $\sigma/\varepsilon \ll 1$, we use the first approximation for the square root to obtain

$$k = k_r + j k_{im} \approx \omega \sqrt{\mu \varepsilon} + j \frac{1}{2} \sigma \sqrt{\frac{\mu}{\varepsilon}} \quad .$$

For a good conductor, $\sigma/\omega\varepsilon \gg 1$, (2.1.23) similarly simplifies to

$$k = k_r + jk_{im} \approx (1+j)\sqrt{\frac{\omega\mu\sigma}{2}} \quad .$$

To obtain the relationship between E_0, H_0 and k for the plane wave case, we note that

$$\nabla \cdot E = 0 \quad \text{or} \quad k \cdot E_0 = 0 \quad \text{or} \quad i_k \cdot i_E = 0 \quad , \tag{2.1.24}$$

where i_E is the unit vector along the direction of the E-field vector. Similarly one obtains

$$i_H \cdot i_k = 0 \quad . \tag{2.1.25}$$

Thus both the electric and magnetic field vectors are perpendicular to the direction of propagation. To obtain the relationship between E_0 and H_0 we note that

$$H = -\frac{1}{j\omega\mu}\nabla \times E = +\frac{|k|}{\omega\mu}|E_0|(i_k \times i_E)e^{j(\omega t - k \cdot r)}$$

$$= |H_0|i_H e^{j(\omega t - k \cdot r)} \quad .$$

Thus

$$i_H = i_E \times i_k \quad , \quad \text{and} \tag{2.1.26}$$

$$|H_0| = |E_0|\sqrt{\frac{\varepsilon}{\mu}} = \frac{|E_0|}{Z_0} \quad , \tag{2.1.27}$$

where $Z_0 = \sqrt{\mu/\varepsilon}$ has the units Ω $(= V A^{-1})$ and is the characteristic impedance of an equivalent transmission line. For free space,

$$Z = Z_0 = \sqrt{\frac{\mu_0}{\varepsilon_0}} = 120\pi = 377\,\Omega \quad . \tag{2.1.28}$$

In an isotropic linear homogeneous medium, Maxwell's equations are compatible with plane-wave solutions whose E, H and k are perpendicular to each other. These solutions are also known as the TEM (transverse electric and magnetic fields) mode of propagation.

The power flow can be obtained from the Poynting vector

$$P = \frac{1}{2}\operatorname{Re}\{E \times H\} = \frac{1}{2}\frac{|E_0|^2}{Z}i_k = \frac{1}{2}|H_0|^2 Z i_k \,[\mathrm{W\,m^{-2}}] \quad . \tag{2.1.29}$$

2.2 Boundary Conditions

To derive Snell's laws of geometrical optics and solve other problems of electromagnetic wave propagation we need to know what happens when an electromagnetic wave is incident on the interface between two different media. For this we need boundary conditions, which can be derived from the integral form of Maxwell's equations.

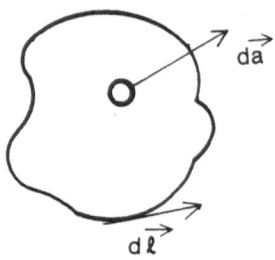

Fig. 2.1. Contour for the integral in (2.2.2)

Integrating the Maxwell's equation involving the curl of E over a closed surface, we obtain

$$\int (\nabla \times E) \cdot da = -\frac{\partial}{\partial t} \int B \cdot da \quad . \tag{2.2.1}$$

Using Stokes' theorem on the left-hand side of the equation, we obtain

$$\int_C E \cdot dl = -\frac{\partial}{\partial t} \int B \cdot da \quad , \tag{2.2.2}$$

where C denotes a contour integral as shown in Fig. 2.1. Note that the line integral is to be taken along the contour of the closed surface. Using similar arguments, one obtains for the magnetic field H

$$\int_C H \cdot dl = +\frac{\partial}{\partial t} \int D \cdot da + \int J \cdot da \quad . \tag{2.2.3}$$

Integrating over a closed volume, we also obtain

$$\int_V \nabla \cdot D \, dV = \int_V \varrho \, dV \quad . \tag{2.2.4}$$

Using Gauss's law of vector identity one has

$$\int D \cdot da = \int_V \varrho \, dV \quad , \tag{2.2.5}$$

where the surface integral is over the surface enclosing the volume.

Similarly for the magnetic induction one obtains

$$\int B \cdot da = 0 \quad . \tag{2.2.6}$$

To obtain the boundary conditions from (2.2.2), we consider the enclosed surface shown in Fig. 2.2a. The two sides are parallel to the surface of separation and the distance h between them tends to zero although the parallel lines are in different media. As the area tends to zero we have

$$E_1 \cdot dl - E_2 \cdot dl = 0 \quad , \quad \text{or}$$

$$E_{1\,\text{tan}} = E_{2\,\text{tan}} \quad . \tag{2.2.7}$$

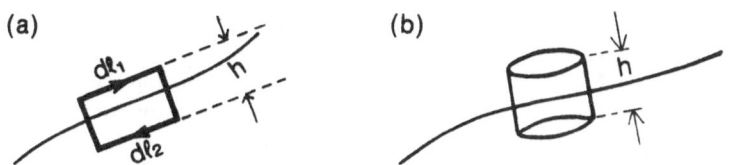

(a)

(b)

Fig. 2.2a,b. Constructions used to obtain the boundary conditions between two media

Thus the tangential component of the electric field or the electric field parallel to the surface of separation must be continuous. Using similar arguments to (2.2.3), we obtain

$$H_1 \cdot dl - H_2 \cdot dl = J_s \quad , \tag{2.2.8}$$

where J_s represents the surface current. In the absence of any surface current, one obtains

$$H_{1\,\text{tan}} = H_{2\,\text{tan}} \quad . \tag{2.2.9}$$

To obtain the boundary condition using (2.2.4), we consider the pillbox shown in Fig. 2.2b. One surface of the pillbox parallel to the interface is in medium 1 and the other in medium 2. Both these surfaces have an area da. If we let the sides of the pillbox tend to zero, we have

$$D_{2n} - D_{1n} = \sigma_s \tag{2.2.10}$$

where the subscript "n" denotes the normal component with respect to the surface and σ_s is the surface charge. In the absence of any surface charge, the normal components of D are continuous.

Similarly using (2.2.6) and Fig. 2.2b we obtain

$$B_{2n} = B_{1n} \quad .$$

In the optical frequency region, there are no surface currents and charges. Thus in the next section we shall use the following boundary conditions:

i) The tangential component of the electric field is continuous.
ii) The tangential component of the magnetic field is continuous.
iii) The normal component of the electrical displacement is continuous.
iv) The normal component of magnetic induction is continuous.

2.3 Snell's Laws

Let a plane electromagnetic wave with electric field E_i be incident from medium 1 on medium 2 as shown in Fig. 2.3. Media 1 and 2 are characterized by ε_1, μ_1 and ε_2, μ_2, respectively. Let the angle of incidence be θ_i, the angle of reflection θ_r and the angle of refraction or angle of transmitted wave be θ_t. Media 1 and 2 are separated by the plane defined by $z = 0$. Thus the unit vector normal to the

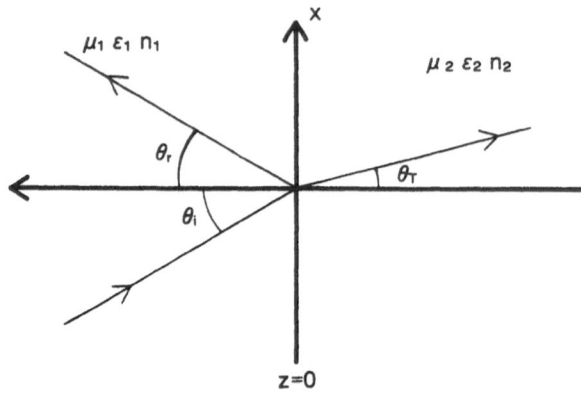

Fig. 2.3. An incident electromagnetic wave is reflected and refracted (transmitted) at the boundary between two media

boundary of separation in this case is i_z. Also, without any loss of generality, we can assume that the plane of incidence is the xz plane. Let the incident, reflected and transmitted electric fields be represented by E_i, E_r and E_t, respectively.

Thus we can write

$$E_i \propto e^{j(\omega t - k_i \cdot r)} \quad , \tag{2.3.1}$$

$$E_r \propto e^{j(\omega t - k_r \cdot r)} \quad , \tag{2.3.2}$$

$$E_t \propto e^{j(\omega t - k_t \cdot r)} \quad , \tag{2.3.3}$$

where k_i, k_r and k_t are incident, reflected, and transmitted wave vectors respectively and are given by

$$k_i = i_x k_i \sin \theta_i + i_y \cdot 0 + i_z(-k_i \cos \theta_i) \quad , \tag{2.3.4}$$

$$k_r = i_x k_{rx} + i_y k_{ry} + i_z k_{rz} \quad , \tag{2.3.5}$$

$$k_t = i_x k_{tx} + i_y k_{ty} + i_z k_{tz} \quad , \tag{2.3.6}$$

where k_{ri} and k_{ti} ($i = x, y, z$) are components of k_r and k_t that have to be determined.

As the normal components of D and B and the tangential components of E and H must be continuous across the boundary of separation at all times

$$k_i \cdot r|_{z=0} = k_r \cdot r|_{z=0} = k_t \cdot r|_{z=0} \quad . \tag{2.3.7}$$

Thus we obtain

$$[x k_i \sin \theta_i + z(-k_i \cos \theta_i)] = (k_{rx} x + k_{ry} y + k_{rz} z)|_{z=0}$$

or

$$k_{ry} = 0 \quad , \quad \text{and} \tag{2.3.8}$$

$$k_i \sin \theta_i = k_r \sin \theta_r \quad . \tag{2.3.9}$$

Similarly we also obtain

$$k_i \sin \theta_i = k_t \sin \theta_t \quad \text{and} \quad k_{ty} = 0 \quad .$$

Noting that

$$|k_i| = \omega \sqrt{\mu_1 \varepsilon_1} = |k_r| \quad \text{and} \quad |k_t| = \omega \sqrt{\mu_2 \varepsilon_2} \quad ,$$

we obtain

$$\theta_i = \theta_r \quad , \quad \text{and} \tag{2.3.10}$$

$$\sqrt{\mu_1 \varepsilon_1} \sin \theta_i = \sqrt{\mu_2 \varepsilon_2} \sin \theta_r \quad . \tag{2.3.11}$$

Denoting the refractive index of a medium by n, we obtain

$$n_1 \sin \theta_i = n_2 \sin \theta_t \quad , \quad \text{where} \tag{2.3.12}$$

$$n_1 = \sqrt{\frac{\mu_1 \varepsilon_1}{\mu_0 \varepsilon_0}} \quad \text{and} \quad n_2 = \sqrt{\frac{\mu_2 \varepsilon_2}{\mu_0 \varepsilon_0}} \quad . \tag{2.3.13}$$

Equations (2.3.10, 11) are generally known as Snell's laws.

In the optical frequency region in general $\mu_1 \approx \mu_2 \approx \mu_0$. Thus in the optical region one can write

$$\sqrt{\varepsilon_{r1}} \sin \theta_i = \sqrt{\varepsilon_{r2}} \sin \theta_t \quad . \tag{2.3.14}$$

2.4 Total Internal Reflection and Optical Tunneling

Let a wave be incident from a medium of higher refractive index on a medium of lower refractive index, i.e., $n_1 > n_2$ and $\theta_t > \theta_i$ (Fig. 2.4). As the angle of refraction is larger than the angle of incidence, if we go on increasing the angle of incidence we reach a point when $\theta_t = 90°$. This happens for $\theta_i = \theta_c$ given by

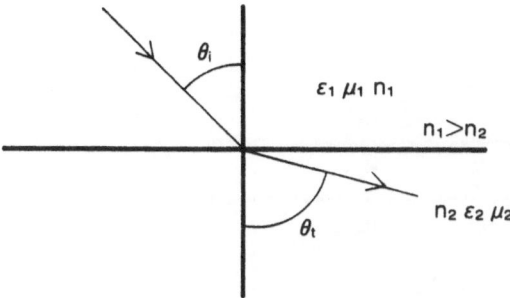

Fig. 2.4. A ray passing from a medium with a higher refractive index to one with a lower refractive index

$$\sin \theta_c = \frac{n_2}{n_1} < 1 \quad , \quad \text{or}$$

$$\theta_c = \sin^{-1}\left(\frac{n_2}{n_1}\right) \quad . \tag{2.4.1}$$

As the transmitted angle cannot be greater than 90°, all the light will be reflected for $\theta_i \geq \theta_c$. The angle θ_c is known as the critical angle of total internal reflection. This description of total internal reflection is only true when the wave front is infinite and the depth of the medium 2 is also infinite. To understand these comments we note that the transmitted wave is proportional to

$$e^{j(\omega t - \mathbf{k}_t \cdot \mathbf{r})} = e^{j(\omega t - x k_2 \sin \theta_t + z k_2 \cos \theta_t)} \quad ,$$

where

$$\cos \theta_t = \sqrt{1 - \sin^2 \theta_t}$$

$$= \sqrt{1 - \left(\frac{n_1}{n_2}\right)^2 \sin^2 \theta_i} = \sqrt{1 - \frac{\sin^2 \theta_i}{\sin^2 \theta_c}} \quad . \tag{2.4.2}$$

Thus, for $\theta_i > \theta_c$, $\cos \theta_t = j\alpha$ where α is real:

$$\alpha = \sqrt{\frac{\sin^2 \theta_i}{\sin^2 \theta_c} - 1} \quad . \tag{2.4.3}$$

The transmitted wave for this case becomes

$$e^{-\alpha k_2 z} e^{j(\omega t - x k_2 \sin \theta_t)} \quad .$$

The wave amplitude decays as a function of z. Thus if the medium 2 is not very large, as shown in Fig. 2.5, we satisfy the condition for total internal reflection at the upper boundary, but light will nevertheless be transmitted in the third medium. This phenomenon is called optical tunneling and has some practical applications in fabricating narrowband optical filters.

On the other hand, when medium 2 is infinite, if a finite wave front is incident as shown in Fig. 2.6, almost all the light energy is reflected. However,

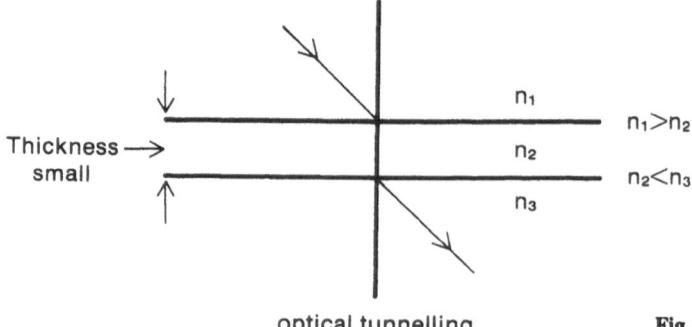

optical tunnelling

Fig. 2.5. Optical tunneling

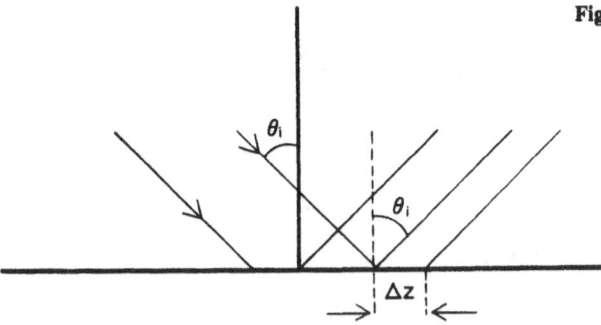

Fig. 2.6. The Goos-Hänchen effect

there is then a lateral beam displacement Δz given by

$$\Delta z \approx \frac{1}{\alpha k_2} \tag{2.4.4}$$

for the case of parallel polarization. This lateral displacement was experimentally verified by Goos and Hänchen and is generally known as the Goos-Hänchen effect. This phenomenon is discussed further in Sect. 2.7.2.

2.5 Transmission Lines

Wave propagation in transmission lines is one-dimensional and thus much simpler than general electromagnetic wave propagation. The simple transmission line equations can be used to solve complex electromagnetic-wave boundary value problems if proper care is taken in modeling those boundary value problems in terms of equivalent transmission lines. In this section, we develop the fundamental equations which can be used for solving many different problems associated with electromagnetic wave, ultrasound, surface acoustic wave and waveguide problems, including fiber optics.

The distributed nature of the transmission line is shown in Fig. 2.7a, where L is the inductance per unit length [$\mathrm{H\,m^{-1}}$] and C is the capacitance per unit length in [$\mathrm{F\,m^{-1}}$]. For a small length of the line stretching from z to $z + \Delta z$, the values of the voltage wave $V(z, t)$ and the current wave $I(z, t)$ are shown in Fig. 2.7b. From this figure we obtain

$$V(z + \Delta z, t) - V(z, t) = -L\Delta z \frac{I(z, t)}{\partial t} \quad , \quad \text{and} \tag{2.5.1}$$

$$I(z + \Delta z, t) - I(z, t) = -C\Delta z \frac{V(z, t)}{\partial t} \quad . \tag{2.5.2}$$

Expanding $V(z+\Delta z, t)$ in a Taylor series with respect to z and neglecting higher-order terms we obtain

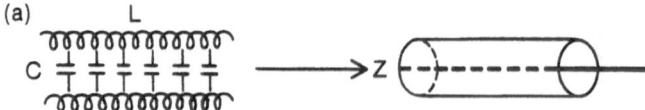

(a)

(b)

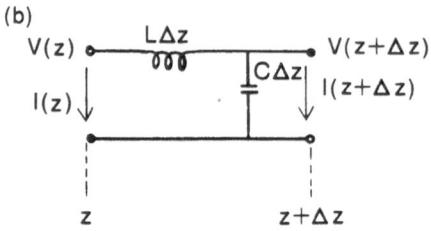

Fig. 2.7a,b. Representations of a transmission line

$$V(z,t) + \frac{\partial V(z,t)}{\partial z}\Delta z + \ldots - V(z,t) = -L\Delta z \frac{\partial I(z,t)}{\partial t} \quad , \quad \text{or}$$

$$\frac{\partial V(z,t)}{\partial z} = -L\frac{\partial I(z,t)}{\partial t} \quad . \tag{2.5.3}$$

Similarly one can obtain for the current waveform $I(z,t)$

$$\frac{\partial I(z,t)}{\partial z} = -C\frac{\partial V(z,t)}{\partial t} \quad . \tag{2.5.4}$$

Thus the voltage and current waves are intimately coupled to each other through (2.5.3, 4). We can decouple these equations by differentiating either $I(z,t)$ or $V(z,t)$ to obtain the one-dimensional wave equation for voltage and current given by

$$\frac{\partial^2 V(z,t)}{\partial z^2} = LC\frac{\partial^2 V(z,t)}{\partial t^2} \quad , \tag{2.5.5}$$

$$\frac{\partial^2 I(z,t)}{\partial z^2} = LC\frac{\partial^2 I(z,t)}{\partial t^2} \quad . \tag{2.5.6}$$

Assuming a solution of the form

$$V(z,t) \quad \text{or} \quad I(z,t) \propto e^{j(\omega t - kz)} \quad , \tag{2.5.7}$$

we find

$$k = \omega\sqrt{LC} \quad . \tag{2.5.8}$$

Thus the phase velocity of the wave is given by

$$v_{\mathrm{p}} = \frac{\omega}{k} = \pm\frac{1}{\sqrt{LC}} \quad , \tag{2.5.9}$$

the positive sign denoting the forward and the negative sign the backward propagation. Thus the general solution for the transmission line equations can be written as

$$V = V_+ e^{j(\omega t - kz)} + V_+ e^{j(\omega t + kz)} \quad , \tag{2.5.10}$$

$$I = I_+ e^{j(\omega t - kz)} + I_- e^{j(\omega t + kz)} \quad , \tag{2.5.11}$$

where the subsripts $+$ and $-$ denote the forward- and backward-propagating components, respectively. Thus I_+ is the component of the forward-propagating current wave and, although not a function of z or t, it is a complex quantity.

To obtain the relationship between V_+ and I_+, let us consider a case where only forward-propagating waves exist. For this case

$$V = V_+ e^{j(\omega t - kz)} \quad , \quad I = I_+ e^{j(\omega t - kz)} \quad . \tag{2.5.12}$$

Using (2.5.3, 4, 12) we obtain

$$\frac{V_+}{I_+} = \frac{\omega L}{k} = \sqrt{\frac{L}{C}} = Z_0 \quad , \tag{2.5.13}$$

where Z_0 is called the characteristic impedance of the transmission line.
Similarly one can show that

$$\frac{V_-}{I_-} = -Z_0 \quad . \tag{2.5.14}$$

Note the negative sign on the right-hand side of (2.5.14) for the backward-traveling wave components. Using (2.5.13 and 14), one can rewrite the expressions for V and I as

$$V = V_+ e^{j(\omega t - kz)} + V_- e^{j(\omega t + kz)} \quad , \tag{2.5.15}$$

$$I = \frac{V_+}{Z_0} e^{j(\omega t - kz)} - \frac{V_-}{Z_0} e^{j(\omega t + kz)} \quad . \tag{2.5.16}$$

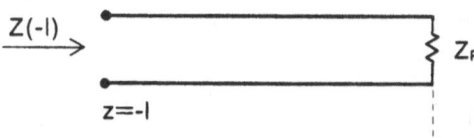

z=0   Fig. 2.8. A terminated transmission line

Consider a transmission line of length l terminated by an impedance Z_R at $Z = 0$ (Fig. 2.8). We are interested in calculating the impedance $Z(l)$ at $z = -l$. We note that

$$Z(z) = \frac{V}{I} = Z_0 \frac{V_+ e^{-jkz} + V_- e^{jkz}}{V_+ e^{-jkz} - V_- e^{jkz}} \quad . \tag{2.5.17}$$

The boundary condition at $z = 0$ is

$$Z_R = Z_0 \frac{V_+ + V_-}{V_+ - V_-} \quad , \quad \text{or} \tag{2.5.18}$$

33

$$Z_R = Z_0 \frac{1 + \Gamma}{1 - \Gamma} \quad , \tag{2.5.19}$$

where Γ is the reflection coefficient defined by

$$\Gamma = \frac{V_-}{V_+} = |\Gamma| e^{j\theta} \quad , \tag{2.5.20}$$

θ being the phase part of the complex reflection coefficient. The impedance of the line of length l terminated by Z_R is given by

$$Z(l) = Z(z = -l) = \frac{1 + \Gamma e^{-j2kl}}{1 - \Gamma e^{-j2kl}} Z_0 \quad , \quad \text{or} \tag{2.5.21}$$

$$Z(l) = Z_0 \frac{Z_R + jZ_0 \tan kl}{Z_0 + jZ_R \tan kl} \quad . \tag{2.5.22}$$

The reflection coefficient can be rewritten as

$$\Gamma = \frac{Z_R - Z_0}{Z_R + Z_0} \quad . \tag{2.5.23}$$

We note the following important cases:

i) For $Z_R = Z_0$, we have $\Gamma = 0$ and $Z(l) = Z_0$. Thus if the transmission line is terminated with the characteristic impedance (Fig. 2.9) there is no reflection and the impedance of any length of line is equal to the characteristic impedance. In other words, a line terminated with the characteristic impedance behaves like an infinite line. This is one of the great advantages of such a transmission line and is routinely used in the laboratory.

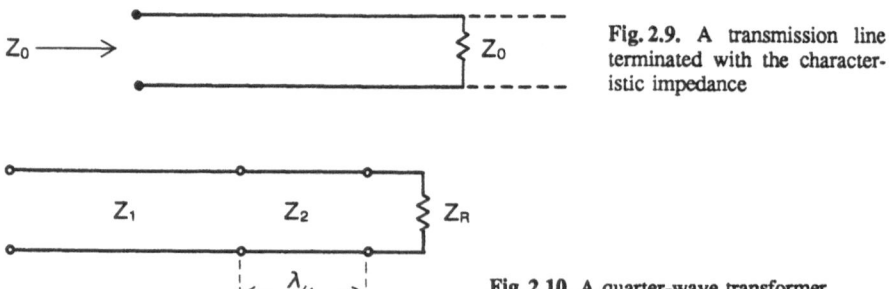

Fig. 2.9. A transmission line terminated with the characteristic impedance

Fig. 2.10. A quarter-wave transformer

ii) If $Z_R = Z_0$ and $l = \lambda/4 = \pi/2k$, then

$$Z(\lambda/4) = Z_0^2/Z_R \quad . \tag{2.5.24}$$

This quarter-wave line is also called a quarter-wave transformer and is often used to match impedances. As shown in Fig. 2.10, to match the first line with characteristic impedance Z_1 to the impedance Z_R, one must choose a characteristic impedance for the quarter-wave transformer given by

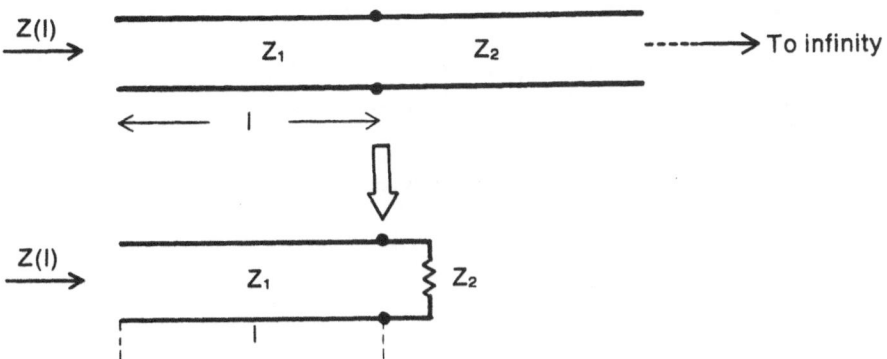

Fig. 2.11. Two transmission lines connected together

$$Z_2 = \sqrt{Z_1 Z_R} \quad . \tag{2.5.25}$$

iii) Consider two lines of characteristic impedances Z_1 and Z_2 which are connected as shown in Fig. 2.11. For this case

$$Z(l) = Z_1 \frac{Z_2 + jZ_1 \tan kl}{Z_1 - jZ_2 \tan kl} \quad . \tag{2.5.26}$$

The reflection coefficient is given by

$$\Gamma = \frac{Z_2 - Z_1}{Z_2 + Z_1} \quad . \tag{2.5.27}$$

However, the transmission coefficient defined as

$$T = \frac{V_t}{V_+} \quad , \tag{2.5.28}$$

where V_t is the transmitted voltage, is not given by the expression

$$T = 1 - \Gamma \quad .$$

The correct expression is

$$T = 1 + \Gamma \quad . \tag{2.5.29}$$

This is easily obtained from the conservation of electric power. Thus

$$\frac{V_+^2}{2Z_1} = \frac{V_-^2}{2Z_1} + \frac{V_t^2}{2Z_2} \quad , \tag{2.5.30}$$

from which (2.5.29) can be easily derived.

iv) For $l = m\lambda/2$, where m is any integer,

$$Z(l) = Z_R \quad . \tag{2.5.31}$$

v) For $l = (2m + 1)\lambda/4$, where m is any integer,

$$Z(l) = Z_0^2/Z_R \quad . \tag{2.5.32}$$

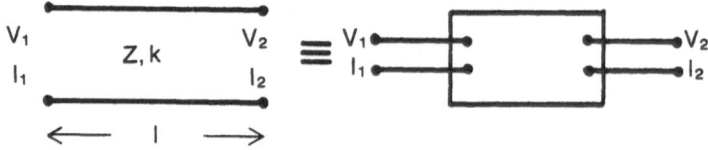

Fig. 2.12. The transmission line as a four-terminal network

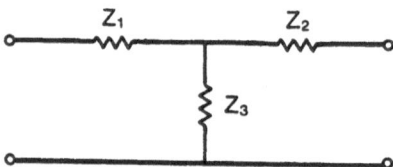

Fig. 2.13. The network equivalent to a transmission line

vi) The transmission line of length l can also be looked upon as a four-terminal network with input and output as shown in Fig. 2.12. It is easy to show that the equivalent T network shown in Fig. 2.13 represents the transmission line provided

$$Z_1 = Z_2 = -jZ_0 \cosec kl \quad ,$$

$$Z_3 = jZ_0 \tan \frac{kl}{2} \quad . \tag{2.5.33}$$

vii) Until now we have been dealing with lossless lines only. To represent a lossy line (Fig. 2.14) one introduces R, the series resistance per unit length, and G, the parallel conductance per unit length. All the formulas derived hold with $j\omega L$ and $j\omega C$ replaced by $R + j\omega L$ and $G + j\omega C$ respectively. Then k and Z become complex. As an example, k is given by

$$k = k_r + jk_{im} = [(GR - \omega^2 LC) + j\omega(GL + RC)]^{1/2} \quad , \tag{2.5.34}$$

where k_r is the real part of the wave number, related to the wavelength λ by

$$k_r = 2\pi/\lambda \quad , \tag{2.5.35}$$

and k_{im} represents the attenuation constant. Thus the forward propagation wave is represented in general as

$$e^{-k_{im}z} e^{j(\omega t - k_r z)} \quad . \tag{2.5.36}$$

From (2.5.34) one can derive

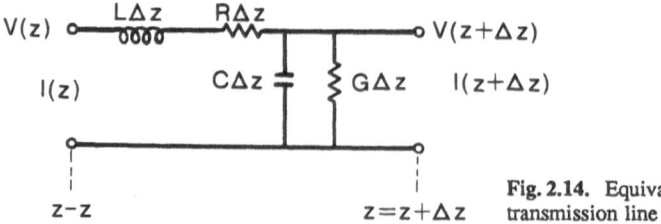

Fig. 2.14. Equivalent circuit of a lossy transmission line

$$k_r = \frac{1}{\sqrt{2}}\{[\omega^4 C^2 L^2 + (G^2 L^2 + C^2 R^2)\omega^2 + G^2 R^2]^{1/2}$$
$$+ \omega(LG + RC)\}^{1/2} \tag{2.5.37}$$

$$k_{im} = \frac{1}{\sqrt{2}}\{[\omega^4 C^2 L^2 + (G^2 L^2 + C^2 R^2)\omega^2 G^2 R^2]^{1/2}$$
$$+ (GR - \omega^2 LC)\}^{1/2} \quad . \tag{2.5.38}$$

Note that for the lossless case $G \to 0$ and $R \to 0$.

2.6 Reflection and Transmission Coefficients for Electromagnetic Waves

Let the incident, reflected and transmitted electric field vectors be denoted by E_i, E_r and E_t respectively, as in Fig. 2.3. These can be written as

$$E_i = i_{E_i} E_i e^{j(\omega t - k_i \cdot r)} \quad ,$$

$$E_r = i_{E_r} E_r e^{j(\omega t - k_i \cdot r)},$$

$$E_t = i_{E_t} E_t e^{j(\omega t - k_i \cdot r)} \quad ,$$

where i_{E_i}, i_{E_r} and i_{E_t} represent the unit vectors in the direction of the electric field vectors for the incident, reflected, and transmitted fields, respectively. We define the reflection coefficient Γ and the transmission coefficient T as follows:

$$\Gamma = \frac{E_r}{E_i} \quad , \quad \text{and} \tag{2.6.1}$$

$$T = \frac{E_t}{E_i} \quad . \tag{2.6.2}$$

To solve the general case where the incident angle is θ_i and E_i has components parallel and perpendicular to the plane of incidence, one can write the boundary conditions at the interface and solve for Γ and T. However, instead of solving these equations directly, we shall take a slightly different approach where an analogy with the transmission line is used. This analogy is obvious for the case where $\theta_i = 0$. Thus we shall first consider this case before solving the general case.

2.6.1 Normal Incidence: $\theta_i = 0$

This case is shown in Fig. 2.15a where the direction of propagation of the incident light is along the z direction. Let us choose

(a)

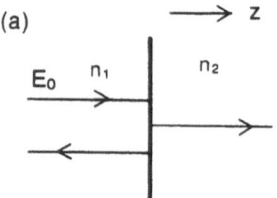

Fig. 2.15. Transmission line analogy (b) for electromagnetic waves normally incident on a boundary between two media (a)

(b)

$$E_i = E_0 i_x \quad . \tag{2.6.3}$$

Note that this choice is arbitrary. For this x-polarized incident wave, H_i will be given by

$$H_i = H_0 i_y \quad .$$

Maxwell's equations for this case become

$$\frac{\partial E_x}{\partial z} = -\mu \frac{\partial H_y}{\partial t} \quad , \tag{2.6.4}$$

$$\frac{\partial H_y}{\partial z} = -\varepsilon \frac{\partial E_x}{\partial t} \quad .$$

In the following, we rewrite the transmission line equations discussed in Sect. 2.5 as

$$\frac{\partial V}{\partial z} = -L \frac{\partial I}{\partial t} \quad , \tag{2.6.5}$$

$$\frac{\partial I}{\partial z} = -C \frac{\partial V}{\partial t} \quad .$$

If we compare the two sets of equations, we find

$$E_x \rightarrow V \quad , \quad H_y \rightarrow I$$

$$\mu \rightarrow L \quad , \quad \varepsilon \rightarrow C \quad .$$

We see immediately that the equivalent transmission line analog of the problem in Fig. 2.15a can be represented by Fig. 2.15b, where we have defined the following quantities

$$Z_1 = \sqrt{\frac{\mu_1}{\varepsilon_1}} \quad , \tag{2.6.6}$$

$$Z_2 = \sqrt{\frac{\mu_2}{\varepsilon_2}} \quad . \tag{2.6.7}$$

Thus, in this case

$$\Gamma = \frac{E_{r0}}{E_0} = \frac{Z_2 - Z_1}{Z_2 + Z_1} \quad , \quad \text{and} \tag{2.6.8}$$

$$T = \frac{E_{t0}}{E_0} = 1 + \Gamma = \frac{2Z_2}{Z_1 + Z_2} \quad . \tag{2.6.9}$$

We also note that

$$E_r = i_x E_{r0} \quad ,$$
$$E_t = i_x E_{t0} \quad . \tag{2.6.10}$$

Similar expressions for the magnetic fields can also be written down.

2.6.2 General Case

In the general case, note that E_i is given by

$$E_i = i_y E_{i\perp} + i_{\parallel} E_{i\parallel} \quad , \tag{2.6.11}$$

where the unit vector $i_{\parallel}$ is in the xz plane but perpendicular to k_i. It is convenient to subdivide the general case problem into two separate cases: perpendicular polarization and parallel polarization, with respect to the plane of incidence.

Perpendicular Polarization. For a plane wave propagating along k ($k_x = k \sin \theta$, $k_z = k \cos \theta$) as in Fig. 2.16, the E and H fields are given by

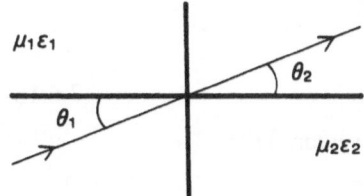

Fig. 2.16. Plane wave propagating from one medium to another

$$E = i_y E_0 e^{j(\omega t - k_i \cdot r)} \quad ,$$
$$H = (i_x H_0 \cos \theta - i_z H_0 \sin \theta) e^{j(\omega t - k_i \cdot r)} \quad , \tag{2.6.12}$$

where H_0 is the magnitude of the magnetic field vector. Using (2.1.27), we obtain

$$\frac{\partial H_z}{\partial x} = j H_0 k \sin^2 \theta = +\varepsilon \frac{\partial E_y}{\partial t} \sin^2 \theta \quad .$$

As the boundary conditions to be satisfied are the continuity of H_x and E_y, we can write Maxwell's equations as

$$\frac{\partial E_y}{\partial z} = -\mu \frac{\partial(-H_x)}{\partial t} \quad ,$$

$$\frac{\partial H_x}{\partial z} = \varepsilon \frac{\partial E_y}{\partial t} + \frac{\partial H_z}{\partial x} \quad ,$$

$$\frac{\partial(-H_x)}{\partial z} = -\varepsilon \cos^2 \theta \frac{\partial E_y}{\partial t} \quad .$$

Identifying

$$V \rightarrow E_y \quad , \quad I \rightarrow -H_x \quad ,$$

$$L \rightarrow \mu \quad , \quad C \rightarrow \varepsilon \cos^2 \theta \quad ,$$

we obtain an equivalent transmission line with characteristic impedance given by

$$Z_\perp = \sqrt{\frac{\mu}{\varepsilon \cos^2 \theta}} = Z_0 \sec \theta \quad . \tag{2.6.13}$$

Note that H_i is the total magnetic field vector for the incident wave. Then

$$\Gamma_\perp = \frac{Z_1 \sec \theta_1 - Z_2 \sec \theta_2}{Z_1 \sec \theta_1 + Z_2 \sec \theta_2} \quad , \quad \text{where} \tag{2.6.14}$$

$$n_1 \sin \theta_1 = n_2 \sin \theta_2 \quad , \quad \text{and} \tag{2.6.15}$$

$$T_\perp = 1 + \Gamma_\perp \quad .$$

The equivalent transmission line model for the perpendicular polarization case is shown in Fig. 2.17.

Parallel Polarization. For this case one can write

$$E = (i_x E_0 \cos \theta - i_z E_0 \sin \theta) e^{j(\omega t - k_i \cdot r)} \quad ,$$
$$H = i_y H_0 e^{j(\omega t - k_i \cdot r)} \quad , \tag{2.6.16}$$

where E_0 is the magnitude of the electric field vector. Then

$$\frac{\partial E_x}{\partial x} = \mu \frac{\partial H_y}{\partial t} \sin^2 \theta \quad .$$

Using (2.6.16), we obtain from Maxwell's equations

$$\frac{\partial E_x}{\partial z} = -\mu \frac{\partial H_y}{\partial t} \cos^2 \theta \quad , \quad \frac{\partial H_y}{\partial z} = -\varepsilon \frac{\partial E_x}{\partial t} \quad .$$

Identifying

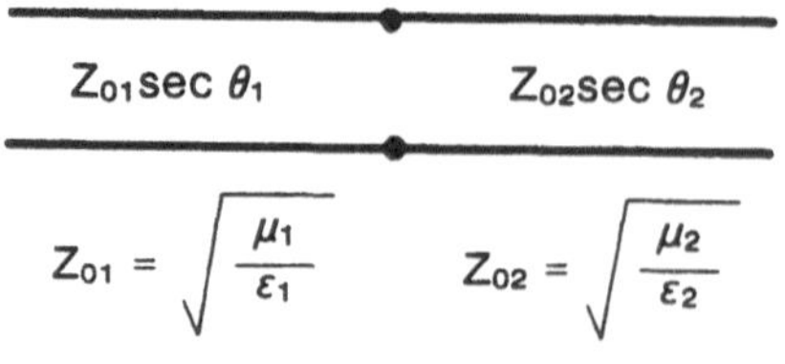

$Z_{01} \sec \theta_1$	$Z_{02} \sec \theta_2$

$$Z_{01} = \sqrt{\frac{\mu_1}{\varepsilon_1}} \qquad Z_{02} = \sqrt{\frac{\mu_2}{\varepsilon_2}}$$

Fig. 2.17. Equivalent transmission line for perpendicular polarization

$$V \longrightarrow E_x \quad , \quad I \longrightarrow H_y \quad ,$$

$$L \longrightarrow -\mu \cos^2 \theta \quad , \quad C \longrightarrow -\varepsilon \quad ,$$

we obtain an equivalent transmission line with characteristic impedance given by

$$Z_\parallel = \sqrt{\frac{\mu \cos^2 \theta_1}{\varepsilon}} = Z_0 \cos \theta_1 \quad . \tag{2.6.17}$$

The reflection coefficient is then given by

$$\Gamma_\parallel = \frac{Z_1 \cos \theta_1 - Z_2 \cos \theta_2}{Z_1 \cos \theta_1 + Z_2 \cos \theta_2} \quad . \tag{2.6.18}$$

The equivalent transmission line model for the parallel case can be represented as shown in Fig. 2.18. Note that the propagation constants to be used also scale as $k_i \cos \theta_i$.

$Z_{01} \cos \theta_1$ $Z_{02} \cos \theta_2$

Fig. 2.18. Equivalent transmission line for parallel polarization

For the general case, where both the parallel and perpendicular components are present, one calculates separately the parallel and perpendicular reflected and transmitted components and adds them up vectorially to obtain the final result. Equations (2.6.14, 15 and 18) are known as the Fresnel equations for the reflection and transmission coefficients.

Examples. Let us try to find an incident angle for the parallel case for which there is no reflection. Noting that for optical frequencies

$$Z_1 = \sqrt{\frac{\mu_1}{\varepsilon_1}} = \sqrt{\frac{\mu_0}{\varepsilon_0}} \sqrt{\frac{\mu_{r1}}{\varepsilon_{r1}}} \approx Z_0 \frac{1}{\sqrt{\varepsilon_{r1}}} = \frac{Z_0}{n_1} \quad , \quad \text{and} \tag{2.6.19}$$

$$Z_2 = \frac{Z_0}{n_2} \quad , \tag{2.6.20}$$

we obtain from (2.6.18)

$$\begin{aligned}
\Gamma_\parallel &= \frac{(\cos \theta_2)/n_2 - (\cos \theta_1)/n_1}{(\cos \theta_2)/n_2 + (\cos \theta_1)/n_1} \quad , \\
&= \frac{n_1 \cos \theta_2 - n_2 \cos \theta_1}{n_1 \cos \theta_2 + n_2 \cos \theta_1} \quad , \tag{2.6.21}
\end{aligned}$$

$$T_\parallel = 1 + \Gamma_\parallel = \frac{2 n_1 \cos \theta_2}{n_1 \cos \theta_2 + n_2 \cos \theta_1} \quad . \tag{2.6.22}$$

For $\Gamma_\parallel$ to be zero at $\theta_1 = \theta_B$, we need

41

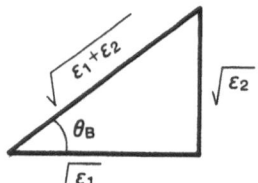

Fig. 2.19. Relationship between the Brewster angle for two media and their permittivities

$$n_1 \cos \theta_2 = n_2 \cos \theta_B \quad . \tag{2.6.23}$$

Using

$$n_2 \sin \theta_2 = n_1 \sin \theta_B, \tag{2.6.24}$$

one easily obtains

$$\sin^2 \theta_B = \frac{n_2^2}{n_2^2 + n_1^2} = \frac{\varepsilon_2}{\varepsilon_1 + \varepsilon_2} \quad . \tag{2.6.25}$$

From Fig. 2.19,

$$\tan \theta_B = \sqrt{\frac{\varepsilon_2}{\varepsilon_1}} = \frac{n_2}{n_1} \quad . \tag{2.6.26}$$

This incident angle, θ_B, for which the reflection coefficient is zero and the transmission coefficient is unity, is called the Brewster angle. It plays a very important role in lasers for which the cavity mirrors are outside the amplifying medium. To minimize losses, the Brewster angle is used at both ends of the amplifying medium. This is sketched in Fig. 2.20.

Now we might ask the obvious question whether there is an angle like the Brewster angle also for the perpendicular polarization case. For this we require

$$\Gamma_\perp = 0 \quad , \quad \text{or}$$

$$\frac{n_1 \cos \theta_1 - n_2 \cos \theta_2}{n_1 \cos \theta_1 + n_2 \cos \theta_2} = 0 \quad , \quad \text{or}$$

$$n_1 \cos \theta_1 = n_2 \cos \theta_2 \quad . \tag{2.6.27}$$

However, from Snell's law

$$n_1 \sin \theta_1 = n_2 \sin \theta_2 \quad .$$

Thus

$$n_1^2 \cos^2 \theta_1 = n_2^2 \cos^2 \theta_2$$
$$= n_2^2 (1 - \sin^2 \theta_2) \quad , \quad \text{or}$$

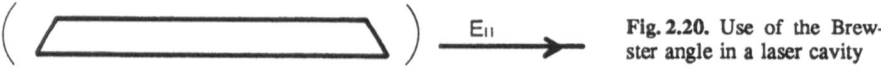

Fig. 2.20. Use of the Brewster angle in a laser cavity

Fig. 2.21. Mirrors placed in the active medium of the laser

$$n_1^2 - n_1^2 \sin^2 \theta_1 = n_2^2 - n_1^2 \sin^2 \theta_1 \quad , \quad \text{or}$$

$$n_1 = n_2 \quad . \tag{2.6.28}$$

So there is no equivalent of the Brewster angle for the perpendicular polarization case.

Note that the radiation coming out of a laser which uses the Brewster angle (shown in Fig. 2.20) is always polarized with parallel polarization because the perpendicular polarization has higher loss and is less likely to oscillate. However, if the mirrors are inside the lasing media as shown in Fig. 2.21, the output of the laser is unpolarized.

Figure 2.22 plots the reflection coefficient versus angle of incidence for two different cases. Case (a) has both the Brewster angle and critical angle for total internal reflection whereas case (b) has only the Brewster angle.

In the following we note several aspects of importance in connection with the above material.

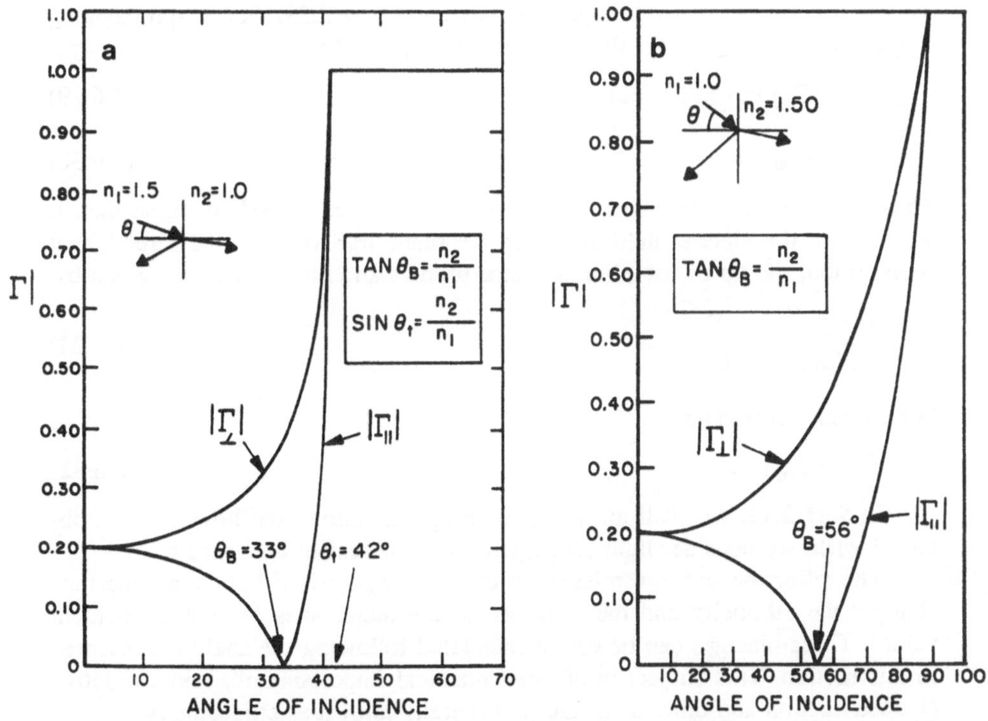

Fig. 2.22a,b. Magnitude of reflection coefficient as a function of angle of incidence. (a) $n_1 = 1.5$, $n_2 = 1.0$. (b) $n_1 = 1.0$, $n_2 = 1.50$

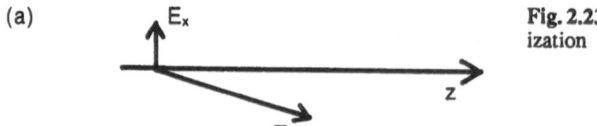

(a)

Fig. 2.23. Construction of elliptical polarization

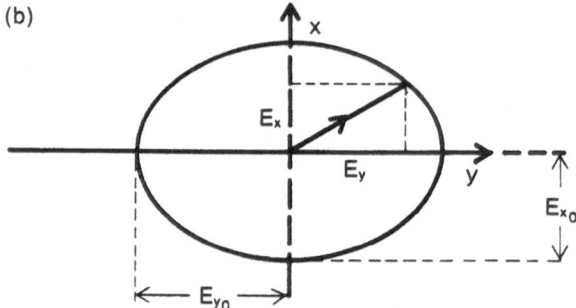

(b)

a) Polarization and Ellipsometry

We have so far considered only plane-polarized waves. As Maxwell's equations are linear and any linear combinations of elementary solutions are possible, one can easily construct general elliptical polarization (Fig. 2.23). For a z-propagating elliptically polarized wave, the electric field is given by

$$E_x = E_{x0} \cos (\omega t - kz) \quad , \tag{2.6.29}$$

$$E_y = E_{y0} \sin (\omega t - kz) \quad . \tag{2.6.30}$$

Thus the x and y components have a phase difference of 90°. We note that, if we observe the electric field vector in the plane transverse to the direction of propagation, the tip of the E-field vector follows an elliptical contour given by

$$\left(\frac{E_x}{E_{x0}} \right)^2 + \left(\frac{E_y}{E_{y0}} \right)^2 = 1 \quad . \tag{2.6.31}$$

For circular polarization

$$E_{x0} = E_{y0} \quad . \tag{2.6.32}$$

In Sect. 2.12, we shall discuss different polarizations and how they are obtained when we consider light propagation through anisotropic media.

The reflection and transmission of elliptically polarized light at an interface changes the ellipticity and the direction of the major axes from their incident values. These changes can be easily calculated following the analysis presented in this section. This subject is of great practical importance for nondestructive characterization and other areas and is generally known as ellipsometry.

b) Total Internal Reflection

The Fresnel equations are valid even if $\theta_1 > \theta_c$. However, for this case $|R| = 1$, and total reflection of light occurs. Now R is complex and can be written as

$$R = e^{j2\phi} \quad , \quad \text{where} \tag{2.6.33}$$

$$\phi_\| = \frac{n_1^2}{n_2^2} \frac{\sqrt{n_1^2 \sin^2 \theta_1 - n_2^2}}{n_1 \cos \theta_1} \quad , \tag{2.6.34}$$

$$\phi_\perp = \frac{\sqrt{n_1^2 \sin^2 \theta_1 - n_2^2}}{n_1 \cos \theta_1} \quad . \tag{2.6.35}$$

These equations are used to derive the expression for the Goos-Hänchen effect in Sect. 2.7.2.

2.7 Group and Phase Velocity

For plane electromagnetic waves propagating through an isotropic medium, the phase velocity v_p, or the velocity with which the constant phase front advances, is given by (2.1.20), which is repeated here:

$$v_p = \frac{\omega}{|k|} i_k \quad . \tag{2.7.1}$$

If the medium is dispersive, i.e., $\varepsilon = \varepsilon(\omega)$, the phase velocity will in general be a function of frequency.

Group velocity, as we shall derive shortly, is given by

$$v_g = \nabla_k \omega(k) \quad . \tag{2.7.2}$$

This group velocity describes the propagation of the envelope of a wave consisting of a group of plane waves having frequencies in the range between ω and $\omega + d\omega$. Note that group velocity is an important quantity because it represents how energy is transferred, i.e., how the information is propagated. Note that theoretically a wave having a single frequency component ω must exist for all times, i.e., for t from $-\infty$ to $+\infty$. Thus we will never know whether, in fact, it has existed or not. By the detection process, we change the wave by some small amount, which in turn makes the wave be represented by a group of waves with finite frequency spread $d\omega$.

Consider a general one-dimensional electromagnetic wave represented by the electric field $E(x,t)$ given by

$$E(x,t) = \int_{-\infty}^{+\infty} A(f_x) e^{j2\pi(ft - f_x x)} df_x \quad . \tag{2.7.3}$$

Here

$$\omega = 2\pi f \quad \text{and} \quad k_x = 2\pi f_x \quad . \tag{2.7.4}$$

For a specific case

$$A(f_x) = E_0 \delta(f_x - f_{x0}) = A(f_{x0}) \quad , \tag{2.7.5}$$

we find

$$E(x,t) = E_0 e^{j2\pi(f_0 t - f_{x0} x)} = E_0 e^{j(\omega_0 t - f_{x0} x)} \quad . \tag{2.7.6}$$

Thus, for a more general case, $A(f_x)$ represents the amplitude of different components having different k_x values. Using the inverse Fourier transform and $t = 0$, we can write

$$A(f_x) = \int\limits_{-\infty}^{+\infty} E(x,0) e^{j2\pi f_x x} df_x \quad , \quad \text{and} \tag{2.7.7}$$

$$E(x,0) = \int\limits_{-\infty}^{+\infty} A(f_x) e^{-j2\pi f_x x} df_x \quad . \tag{2.7.8}$$

Let us consider the case where $A(f_x)$ is nonzero only in a narrow band around the center spatial frequency f_{x0}, as shown in Fig. 2.24. A situation like this might occur, for example, when we consider a pulse at time $t = 0$, as shown in Fig. 2.25. The pulse has a duration of L_x and center frequency f_{x0}. For this case, $A(f_x)$ will be given by

$$A(f_x) = L_x \frac{\sin \pi (f_x - f_{x0}) L_x}{\pi (f_x - f_{x0}) L_x} = L_x \operatorname{sinc} (f_x - f_{x0}) L_x \quad . \tag{2.7.9}$$

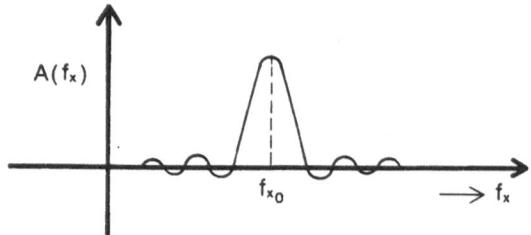

Fig. 2.24. A narrow amplitude function $A(f_x)$

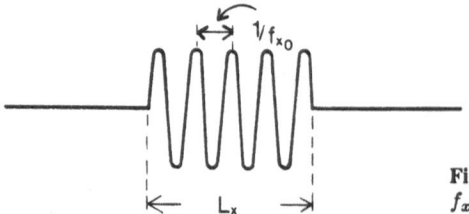

Fig. 2.25. A pulse of duration L_x and frequency f_{x0}

Thus

$$\Delta f_x \sim \frac{1}{L_x} \quad . \tag{2.7.10}$$

We can expand $\omega(k)$ and $f(k)$ as

$$2\pi f(k_x) = \omega(k_x) = \omega_0 + \left.\frac{d\omega}{dk}\right|_{k=k_{x0}} (k_x - k_{x0}) + \cdots$$

$$= 2\pi f_0 + \left.\frac{d\omega}{dk}\right|_{f_x=f_{x0}} 2\pi(f_x - f_{x0}) + \cdots \quad . \tag{2.7.11}$$

Substituting these values in (2.7.3) we obtain

$$E(x,t) \approx \int\limits_{-\infty}^{+\infty} A(f_x) \exp\left[j2\pi\left(f_0 t + \left.\frac{d\omega}{dk}\right|_{f_x=f_0} (f_x - f_{x0})t - f_x x \right) \right] df_x$$

$$= \exp\left[j2\pi\left(f_0 t - f_{x0}t\frac{d\omega}{dk}\Big|_{f_x=f_{x0}} \right) \right]$$

$$\times \int\limits_{-\infty}^{+\infty} A(f_x) \exp\left[-j2\pi f_x\left(x - \left.\frac{d\omega}{dk}\right|_{f_x=f_{x0}} \right) \right] dx$$

$$= \exp\left(j2\pi f_0 t - f_{x0}t\frac{d\omega}{dk}\Big|_{f_x=f_{x0}} \right) E\left(x - t\frac{d\omega}{dk}\Big|_{f_x=f_{x0}}, 0 \right) \quad . \tag{2.7.12}$$

In the last step above, we have used (2.7.8).

Excluding a phase factor we note that the initial envelope $E(x,0)$ propagates as $E(x - td\omega/dk|_{f_x=f_{x0}}, 0)$, so that the group velocity will be defined by

$$x - t\frac{d\omega}{dk}\Big|_{f_x=f_{x0}} = \text{const} \quad , \quad \text{or}$$

$$v_g = \frac{dx}{dt} = \left.\frac{d\omega}{dk}\right|_{f_x=f_{x0}} \quad . \tag{2.7.13}$$

If we generalize the above equation for the three-dimensional case, we obtain

$$v_g = \nabla\omega(k)\Big|_{\substack{f_x=f_{x0}\\k=k_{x0}}} \quad . \tag{2.7.14}$$

It is interesting to note that, if the relationship between ω and k is not linear, then

$$v_g \neq v_p \quad .$$

Actually it is quite possible to envision a case where the group velocity is negative whereas the phase velocity is positive (Fig. 2.26). However, for the isotropic, homogeneous case the ω versus k curve is linear (Fig. 2.27) and

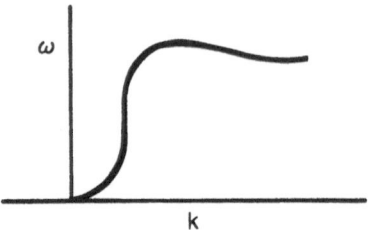

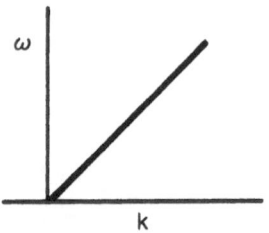

Fig. 2.26. Nonlinear relationship between ω and k giving a negative group velocity and positive phase velocity

Fig. 2.27. Dispersion curve for the isotropic, homogeneous case

$$v_g = v_p = \frac{1}{\sqrt{\mu\varepsilon}} \quad . \tag{2.7.15}$$

2.7.1 Poynting Vector, Ray Velocity, Phase Velocity, and Group Velocity

We have already introduced the Poynting vector, P, given by

$$P = E \times H \quad . \tag{2.7.16}$$

In this section, we will develop the concept of the Poynting vector and its relationship with stored energy in a formal manner. From Maxwell's equations, we have

$$H \cdot (\nabla \times E) = -H \cdot \frac{\partial B}{\partial t} \quad , \tag{2.7.17}$$

$$-E \cdot (\nabla \times H) = -E \cdot \frac{\partial D}{\partial t} - E \cdot J_c - E \cdot J_s \quad . \tag{2.7.18}$$

Here J_c and J_s are the conduction and source current densities, respectively. Adding (2.7.17 and 18), we obtain

$$\nabla \cdot P = -H \cdot \frac{\partial B}{\partial t} - E \cdot \frac{\partial D}{\partial t} - E \cdot J_c - E \cdot J_s \quad . \tag{2.7.19}$$

For a lossless medium, $J_c = 0$, and we obtain

$$\nabla \cdot P = -H \cdot \frac{\partial B}{\partial t} - E \cdot \frac{\partial D}{\partial t} - E \cdot J_s \quad . \tag{2.7.20}$$

Integrating (2.7.20) over an enclosed volume V that has a perfectly conducting surface S, one has

$$(E \times H) \cdot n \, dS = 0 = \int_V \left(H \cdot \frac{\partial B}{\partial t} + E \cdot \frac{\partial D}{\partial t} \right) dV - \int_V E \cdot J_s dV \quad . \tag{2.7.21}$$

The left-hand side of (2.7.21) is zero because the tangential component of the electric field $n \times E$ on the surface is zero and

$$(E \times H) \cdot n = (n \times E) \cdot H \quad .$$

Equation (2.7.21) states that the power supplied by the source, $\int_V \boldsymbol{E} \cdot \boldsymbol{J}_s dV$, must be equal to the rate of change of stored energy, U. Thus

$$\frac{\partial U}{\partial t} = \int_V \left(\boldsymbol{H} \cdot \frac{\partial \boldsymbol{B}}{\partial t} + \boldsymbol{E} \cdot \frac{\partial \boldsymbol{D}}{\partial t} \right) dV$$

$$= \frac{\partial}{\partial t} \int (U_H + U_E) dV \quad , \tag{2.7.22}$$

where U_H and U_E are magnetic and electric stored energy, respectively. We note that

$$\frac{\partial U_E}{\partial t} = \boldsymbol{E} \cdot \frac{\partial \boldsymbol{D}}{\partial t} = \boldsymbol{E} \cdot \left(\varepsilon \frac{\partial \boldsymbol{E}}{\partial t} \right) \quad , \quad \text{or} \tag{2.7.23}$$

$$U_E = E_i \varepsilon_{ij} E_j \quad .$$

Note also that $\partial U_E / \partial E_i = \varepsilon_{ij} E_j$. By interchanging the subscripts i and j, we obtain

$$\frac{\partial U_E}{\partial E_j} = \varepsilon_{ji} E_i \quad .$$

If the stored electric energy is dependent only on the electric field components, we must have

$$\frac{\partial^2 U_E}{\partial E_i \partial E_j} = \varepsilon_{ij} = \frac{\partial^2 U_E}{\partial E_j \partial E_i} = \varepsilon_{ji} \quad . \tag{2.7.24}$$

Thus the dielectric tensor must then be symmetrical. Similarly one can argue that the permeability tensor μ_{ij} is also symmetrical. In the general case, (2.7.21) can be written as

$$\oint_S \boldsymbol{P} \cdot \boldsymbol{n} \, dS + \frac{\partial U}{\partial t} + P_c = P_s = \int_V \boldsymbol{E} \cdot \boldsymbol{J}_c dV \quad , \tag{2.7.25}$$

where P_c is the conduction energy loss and P_s is the source energy.

If the electromagnetic field variables are represented as $\boldsymbol{E} = \boldsymbol{E} \exp(j\omega t)$, one easily obtains the complex Poynting vector

$$\boldsymbol{P} = \frac{\boldsymbol{E} \times \boldsymbol{H}^*}{2} \quad , \tag{2.7.26}$$

and (2.7.25) can be rewritten as

$$\oint_S \boldsymbol{P} \cdot \boldsymbol{n} \, dS - j\omega [(U_E)_{\text{peak}} + (U_H)_{\text{peak}}] + (P_c)_{\text{av}} = P_s \quad . \tag{2.7.27}$$

For a plane wave propagating in the $\boldsymbol{k}$ direction, we have all the field variables proportional to $\exp(-j\boldsymbol{k} \cdot \boldsymbol{r})$. For this case, (2.7.17) becomes

$$\boldsymbol{H} \cdot (\boldsymbol{k} \times \boldsymbol{E}) = \boldsymbol{k} \cdot (\boldsymbol{E} \times \boldsymbol{H}) = -j\omega(U_H)_{\text{peak}} \quad , \quad \text{or}$$

$$\boldsymbol{k} \cdot \boldsymbol{P} = -j\omega(U_H)_{\text{peak}} \quad . \tag{2.7.28}$$

Similarly, from (2.7.18) we obtain

$$k \cdot P = -j\omega (U_E)_{\text{peak}} \quad . \tag{2.7.29}$$

Thus

$$(U_H)_{\text{peak}} = (U_E)_{\text{peak}} = U_{\text{av}} \quad . \tag{2.7.30}$$

From (2.7.29) we obtain

$$k \cdot P = \omega U_{\text{av}} \quad , \quad \text{or}$$

$$i_k \cdot \frac{P}{U_{\text{av}}} = \frac{\omega}{k} = v_{\text{p}} \quad , \quad \text{or}$$

$$i_k \cdot v_e = v_{\text{p}} \quad . \tag{2.7.31}$$

We define the magnitude of the ray velocity or the energy velocity as

$$v_e = \frac{P}{U_{\text{av}}} \cos \psi \quad , \tag{2.7.32}$$

where ψ is the angle between the k vector and P. For isotropic material of course

$$\psi = 0 \quad .$$

We have also obtained the group velocity to be given by

$$v_g = \nabla_k \omega \quad .$$

For a lossless medium, v_g and v_e are identical as shown below.

If we have two sets of electromagnetic fields given by E_1, H_1 and E_2, H_2 we obtain from Maxwell's equations the following complex reciprocity relation:

$$\nabla \cdot (E_1 \times H_2^* + E_2 \times H_1) = j(E_1 \cdot D_2^* + H_1 \cdot B_2^*) \quad . \tag{2.7.33}$$

Using

$$E_1 \rightarrow E e^{j(\omega t - k \cdot r)} \quad ,$$

$$E_2 \rightarrow (E + \delta E) e^{j[(\omega + \delta \omega)t - (k + \delta k) \cdot r]} \quad ,$$

$$H_1 \rightarrow H e^{j(\omega t - k \cdot r)} \quad ,$$

$$H_2 \rightarrow (H + \delta H) e^{j[(\omega + \delta \omega)t - (k + k) \cdot r]} \quad ,$$

one obtains to first order, from (2.7.33),

$$\delta k \cdot [E \times H + E \times H^*] = \delta \omega (U_E + U_H) = 2 \delta \omega U_{\text{av}} \quad .$$

Thus

$$2 \delta k \cdot P = 2 \delta \omega U_{\text{av}} \quad , \quad \text{or} \tag{2.7.34}$$

$$v_e = \frac{P}{U_{av}} = \nabla_k \omega = v_g \quad . \tag{2.7.35}$$

2.7.2 Goos-Hänchen Effect

Consider Fig. 2.6, where the incident beam is finite and its incident angle $\theta_i > \theta_c$, where θ_c is the critical angle for total internal reflection. The incident beam can be written as

$$E_i(x, z, t) = \int\limits_{-\infty}^{+\infty} A(f_x) e^{j2\pi f_x x} e^{-jk_z z} e^{j2\pi ft} df_x \quad . \tag{2.7.36}$$

If we assume that the beam is quite wide, so that $A(f_x)$ can be described as in Fig. 2.24, then (2.7.36) becomes

$$E_i(x, z, t) \approx e^{j2\pi f_{x0} x} e^{-jk_{z0} z} \int A(f_x) e^{j2\pi (f_x - f_{x0})}$$
$$\times e^{-j(k_z - k_{z0})z} e^{j2\pi ft} df_x \quad , \tag{2.7.37}$$

where

$$2\pi f_{x0} = k_{x0} = k \cos \theta_i \quad ,$$
$$k_{z0} = k \sin \theta_i \quad .$$

The reflected beam is given by

$$E_r(x, z, t) \approx e^{-j2\pi f_{x0} x} e^{-jk_{z0} z} e^{j2\pi ft}$$
$$\times \int A(f_x) e^{j2\phi(f_x)} e^{j2\pi (f_x f_{x0})} e^{-j(k_z - k_{z0})z} df_x \quad . \tag{2.7.38}$$

In (2.7.38) we have substituted (2.6.33) for the reflection coefficient and have included the explicit dependence of ϕ and f_x, which relates to the incident angle. We can expand ϕ as

$$\phi(f_x) = \phi(f_{x0}) + \frac{\partial \phi}{\partial f_x}\bigg|_{f_{x0}} (f_x - f_{x0}) + \dots \quad . \tag{2.7.39}$$

Substituting this value for ϕ, we obtain

$$E_r(x, z, t) \approx e^{j2\phi(f_{x0})} e^{-j2\pi f_{x0} x} e^{-jk_{z0} z} e^{j2\pi ft}$$
$$\times \int A(f_x) e^{j2\pi (f_x - f_{x0})(x + \Delta x)} e^{-j(k_z - k_{z0})} df_x \quad , \tag{2.7.40}$$

where

$$\Delta x = -\frac{1}{\pi} \frac{\partial \phi}{\partial f_x}\bigg|_{f_{x0}} \quad . \tag{2.7.41}$$

We then note that

$$E_r(x, z, t) \approx e^{j2\phi(f_{x0})} E_i(-(x + \Delta x), z, t) \quad . \tag{2.7.42}$$

Thus the reflected beam is identical in shape to the incident beam but travels

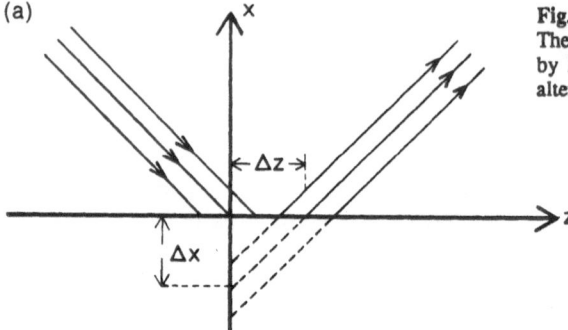

(a)

Fig. 2.28a,b. Goos-Hänchen effect. (a) The reflected beam is shifted laterally by Δz and downwards by Δx. (b) An alternative interpretation

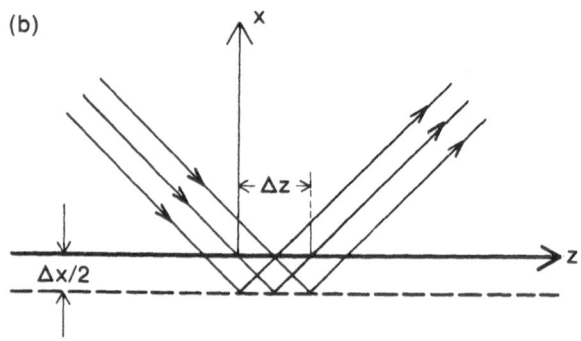

(b)

upward and away from the interface at an angle θ as shown in Fig. 2.6. The reflected beam appears to have shifted downwards with respect to x by an amount Δx as shown in Fig. 2.28. The beam also appears to be shifted laterally by an amount Δz given by

$$\Delta z = \Delta x \tan \theta_i \quad . \tag{2.7.43}$$

An alternative interpretation of the Goos-Hänchen effect is to consider the incident beam as if reflected from a fictitious boundary located at $x = -\Delta x/2$. From (2.6.34 and 35) it is easily observed that

$$\left(\frac{\Delta s}{2}\right)_{\parallel} = \frac{1}{\alpha_{\parallel}\kappa} \quad , \quad \text{parallel polarization case} \quad , \tag{2.7.44}$$

$$\left(\frac{\Delta x}{2}\right)_{\perp} = q\frac{1}{\alpha_{\perp}\kappa} \quad , \quad \text{perpendicular polarization case} \quad , \tag{2.7.45}$$

where $\alpha\kappa$ is the decay constant of the evanescent field and

$$q = \left(\frac{\kappa_{ix}^2 + \alpha_{\parallel}^2}{\kappa_{ix}^2 + (\varepsilon_1/\varepsilon_2)^2\alpha_{\perp}^2}\right)\frac{\varepsilon_1}{\varepsilon_2} \quad . \tag{2.7.46}$$

2.8 Gaussian Beam Propagation

Gaussian beams play a very important role in modern optics because the profile of laser beams is very nearly Gaussian. Such beams are also found in fiber optics and integrated optics. Because of their importance, we shall discuss Gaussian fields here and in Sect. 2.9.

Let us consider waves whose direction of propagation is z; however, the waves are not plane waves. As shown in Fig. 2.29, the amplitude in the transverse direction of these waves is somewhat confined. For this case, we can write in general

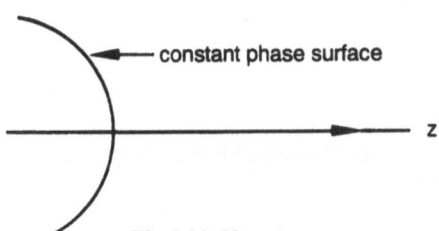

constant phase surface

z

Fig. 2.29. Non-plane wave propagating in the z direction

$$E(x, y, z) = \psi(x, y, z)e^{-jkz} \quad .$$
(2.8.1)

In the above equation, ψ is a slowly varying (weak) function of z. For this problem, the wave equation can be written as

$$\nabla^2 E + \frac{\omega^2}{v^2} E = 0 \quad , \quad \text{or}$$

$$\nabla_t^2 E + \frac{\partial^2 E}{\partial z^2} + \frac{\omega^2}{v^2} E = 0 \quad , \quad \text{where}$$
(2.8.2)

$$\nabla_t = i_x \frac{\partial}{\partial x} + i_y \frac{\partial}{\partial y} \quad .$$
(2.8.3)

We note

$$\nabla_t^2 E(x, y, z) = \nabla_t^2 \psi e^{-jkz} \quad ,$$
(2.8.4)

$$\frac{\partial E}{\partial z} = \left(\frac{\partial \psi}{\partial z} - jkz \right) e^{-jkz} \quad ,$$
(2.8.5)

$$\frac{\partial^2 E}{\partial z^2} = \left(\frac{\partial^2 \psi}{\partial z^2} - 2jk\frac{\partial \psi}{\partial z} - k^2 \psi \right) e^{-jkz} \quad .$$
(2.8.6)

Thus substituting (2.8.4 and 6) into the wave equation for this case yields

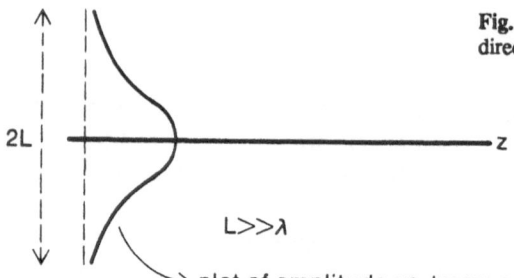

2L

z

$L \gg \lambda$

plot of amplitude vs transverse direction

$$\nabla_t^2 \psi + \left(\frac{\partial^2 \psi}{\partial z^2} - 2jk\frac{\partial \psi}{\partial z} - k^2 \psi \right) + k^2 \psi = 0 \quad , \quad \text{or}$$

$$\nabla_t^2 \psi + \frac{\partial^2 \psi}{\partial z^2} - 2jk\frac{\partial \psi}{\partial z} = 0 \quad . \tag{2.8.7}$$

As ψ is a weak function of z and as $k = 2\pi/\lambda$ is in general a very large quantity for optical frequencies, we approximate (2.8.7) as

$$\nabla_t^2 \psi - 2jk\frac{\partial \psi}{\partial z} = 0 \quad . \tag{2.8.8}$$

There is another consequence of considering ψ to be a weak function of z or the wave being confined in the transverse direction to dimensions on the order of $L \gg \lambda$ (Fig. 2.30):

$$\nabla \cdot E = 0 \quad , \quad \text{or}$$

$$\nabla_t \cdot E_t + \frac{\partial E_z}{\partial z} = 0 \quad , \quad \text{or}$$

$$\nabla_t \cdot E_t - jkE_z = 0 \quad . \tag{2.8.9}$$

As

$$\nabla_t \cdot E_t \sim \frac{|E_t|}{L} \quad ,$$

we obtain

$$\frac{|E_t|}{L} + \frac{2\pi}{\lambda}|E_z| = 0 \quad , \quad \text{or}$$

$$|E_z| \sim \frac{\lambda}{2\pi L}|E_t| \quad , \quad \text{or} \tag{2.8.10}$$

$$E_z \ll E_t \quad . \tag{2.8.11}$$

Going to the cylindrical coordinate system and noting that

$$\nabla_t^2 \psi(r, \phi, z) = \frac{1}{r}\frac{\partial}{\partial r}\left(r\frac{\partial \psi}{\partial r} \right) + \frac{1}{r^2}\frac{\partial^2 \psi}{\partial \phi^2} \tag{2.8.12}$$

we can rewrite (2.8.8) as

$$\frac{1}{r}\frac{\partial}{\partial r}\left(r\frac{\partial \psi}{\partial r}\right) + \frac{1}{r^2}\frac{\partial^2 \psi}{\partial \phi^2} - 2jk\frac{\partial \psi}{\partial z} = 0 \quad . \tag{2.8.13}$$

Before we solve the above equation, let us consider the simplest case when ψ (denoted by ψ_0) is a function of r and z, but not of ϕ. For this case, (2.8.13) becomes

$$\frac{1}{r}\frac{\partial}{\partial r}\left(r\frac{\partial \psi_0}{\partial r}\right) - j2k\frac{\partial \psi_0}{\partial z} = 0 \quad . \tag{2.8.14}$$

As a trial solution, we write

$$\psi_0(r, z) = \psi_0 = A\exp\left[-j\left(p(z) + \frac{kr^2}{2q(z)}\right)\right] \quad , \tag{2.8.15}$$

where $p(z)$ and $q(z)$ are unknown functions to be determined. We note that

$$\frac{\partial \psi_0}{\partial z} = -j\left(p'(z) - \frac{kr^2}{2q^2(z)}q'(z)\right)\psi_0 \quad , \tag{2.8.16}$$

$$\frac{\partial \psi_0}{\partial r} = -j\frac{kr}{q(z)}\psi_0 \quad , \tag{2.8.17}$$

$$\frac{\partial}{\partial r}\left(r\frac{\partial \psi_0}{\partial r}\right) = -j\frac{zkr}{q(z)}\psi_0 - \frac{k^2r^3}{q^2(z)}\psi_0 \quad ; \tag{2.8.18}$$

$p'(z)$ and $q'(z)$ denote differentiation with respect to z. Substituting (2.8.16–18) into (2.8.15), we obtain

$$\frac{k^2}{q^2(z)}[q'(z) - 1]r^2 - 2k\left(j\frac{1}{q(z)} + p'(z)\right) = 0 \quad . \tag{2.8.19}$$

As the above equation has to be true for all values of r, we obtain

$$q'(z) = 1 \quad , \quad \text{and} \tag{2.8.20}$$

$$p'(z) = -\frac{j}{q(z)} \quad . \tag{2.8.21}$$

The solution of (2.8.20) gives

$$q(z) = q(0) + z \quad . \tag{2.8.22}$$

Substituting this value of $q(z)$ into ψ_0 we obtain

$$\psi_0(r, z) = A\exp\left[-jp(z)\right]\exp\left(-j\frac{kr^2}{2q(z)}\right) \quad , \tag{2.8.23}$$

so that

$$\psi_0(r, 0) = A\exp\left[-jp(0)\right]\exp\left(-j\frac{kr^2}{2q(0)}\right) \quad . \tag{2.8.24}$$

55

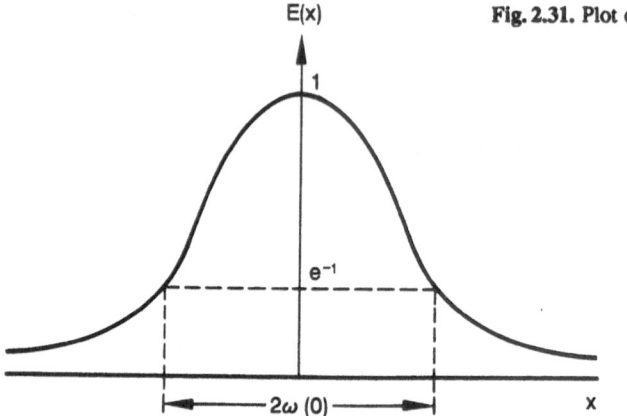

E(x)

Fig. 2.31. Plot of $\psi(r, 0)$

1

e^{-1}

$2\omega\,(0)$

x

We define $q(0) = jz_R$, where z_R is real and called the Rayleigh distance. It is given by

$$z_R = \pi \frac{\omega^2(0)}{\lambda} \quad , \qquad (2.8.25)$$

where $\omega(0)$ is the initial beam width of the Gaussian beam, to be discussed shortly. Thus

$$\psi_0(r, 0) \propto \exp\left(-\frac{kr^2}{z_R}\right) = \exp\left(-\frac{x^2 + y^2}{\omega^2(0)}\right) \quad . \qquad (2.8.26)$$

A plot of $\psi(r, 0)$ is shown in Fig. 2.31, where we note that the beam profile is Gaussian with the radius of the spot width given by $\omega(0)$. If we normalize the amplitude A with the condition

$$\int_{-\infty}^{+\infty} |E|^2 dx = 1 \quad , \qquad (2.8.27)$$

we obtain

$$A = \sqrt{\frac{2}{\pi}} \frac{1}{\omega(0)} \quad . \qquad (2.8.28)$$

For the normalized case, one thus obtains

$$\psi_0(r, z) = \exp\left[-jp(z)\right] \sqrt{\frac{2}{\pi}} \frac{1}{\omega(0)} \exp\left(-j\frac{kr^2}{2(z + jz_R)}\right) \quad ,$$

$$= \exp\left[-jp(z)\right] \sqrt{\frac{2}{\pi}} \frac{1}{\omega(0)} \exp\left(-j\frac{kr^2}{2q(z)}\right) \quad . \qquad (2.8.29)$$

To obtain an expression for $p(z)$, we solve (2.8.21) to obtain

$$jp(z) = \ln(z + jz_R)\Big|_0^z$$

$$= \ln\left(1 - j\frac{z}{z_R}\right) \quad , \quad \text{or} \tag{2.8.30}$$

$$e^{-jp(z)} = \frac{1}{1 - jz/z_R} = \left[1 + \left(\frac{z}{z_R}\right)^2\right]^{-1/2} e^{j\tan^{-1}(z/z_R)}$$

$$= \left[1 + \left(\frac{z}{z_R}\right)^2\right]^{-1/2} e^{j\psi(z)} \quad , \quad \text{where} \tag{2.8.31}$$

$$\psi(z) = \tan^{-1}(z/z_R) \quad . \tag{2.8.32}$$

Thus we obtain the final solution as

$$\begin{aligned} E_0(r, z, t) &= \psi_0(r, z)e^{j(\omega t - kz)}\zeta \\ &= \sqrt{\frac{2}{\pi}}\frac{1}{\omega(z)}\exp\left(-\frac{x^2 + y^2}{q(z)}\right)e^{j[\omega t - kz + \psi(z)]} \quad , \end{aligned} \tag{2.8.33}$$

where we have defined

$$\omega(z) = \omega(0)\left[1 + \left(\frac{z}{z_R}\right)^2\right]^{1/2} \quad . \tag{2.8.34}$$

We note that

$$\frac{1}{q(z)} = \frac{1}{z + jz_R} = \frac{1}{R(z)} - j\frac{\lambda}{\pi\omega^2(z)} \quad , \quad \text{where} \tag{2.8.35}$$

$$R(z) = z[1 + (z_R/z)^2] \quad . \tag{2.8.36}$$

Thus, if a Gaussian beam intially had a beam radius $\omega(0)$ and no radius of curvature in the wave front at $z = 0$, as it propagates its beam size increases to $\omega(z)$ and it develops a radius of curvature of the wave front, $R(z)$, as shown in Fig. 2.32.

These Gaussian modes are of importance for the laser cavity mode solution. All the laser output is in the form of Gaussian beam (or, to be exact, Hermite-Gaussian) and the same solution obtained in (2.8.33) can also be obtained from the diffraction integral. This will be discussed later.

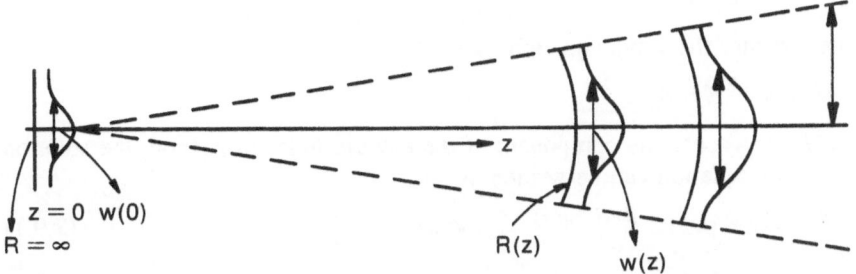

Fig. 2.32. Propagation of a Gaussian wave front. At $z = 0$ the radius of curvature $R = \infty$. Note the spot width $w(z)$ increases with z whereas $R(z)$ decreases for $z < z_R$ and increases for $z > z_R$

It is noted that the phase velocity in (2.8.33) is not a constant, but a function of spot size. The phase ϕ is given by

$$\phi(z, t) = \omega t - kz + \tan^{-1}\left(\frac{z}{z_R}\right) = \text{const} \quad . \tag{2.8.37}$$

Thus the phase velocity is given by

$$v_p = \frac{dz}{dt} = \frac{\omega/k}{1 - \frac{\lambda}{2\pi}\frac{d}{dz}[\tan^{-1}(\frac{z}{z_R})]} = \frac{v_{p0}}{1 - \frac{\lambda}{2\pi}\frac{d}{dz}[\tan^{-1}(\frac{z}{z_R})]} \quad , \tag{2.8.38}$$

where v_{p0} denotes the plane wave phase velocity. We recall that ψ_0 is the solution of (2.8.13) when we neglect the ϕ dependence. If we solve the more general case, we can show that

$$E(x, y, z, t) = \sqrt{\frac{2}{2^{m+n}m!n!\pi}} \frac{1}{\omega(z)} H_m\left(\frac{\sqrt{2}x}{\omega(x)}\right)$$
$$\times H_n\left(\frac{\sqrt{2}y}{\omega(x)}\right)\exp\left(-j\frac{k(x^2 + y^2)}{2q(z)}\right)$$
$$\times \exp\left[j\omega t - jkz + j(m + n + 1)\psi(z)\right] \quad , \tag{2.8.39}$$

where m and n are integers (including zero) and H_m denotes Hermite polynomials of order m. Hermite polynomials and their important properties are discussed in Appendix B.

2.9 Geometrical Optics

So far, we have considered the propagation of a Gaussian wave only in free space. We would like to consider the situation where it passes through lenses. Much simplification is possible if we analyze such a situation using matrix methods that are suitable for use in geometrical optics. In this section, we review these matrix methods, but we first discuss the foundations of geometrical optics, as derived from the wave equation.

2.9.1 Eikonal Equation

Let us consider the wave equation

$$\nabla^2 E + k^2 E = 0 \quad ,$$

where E represents any component of the electric field. In general, the solution of the wave equation can be written as

$$E = E_0(x, y, z)e^{-jk_0 S(x,y,z)} \quad , \quad \text{where} \tag{2.9.1}$$

$$k = nk_0 = n\left(\frac{2\pi}{\lambda_0}\right) \quad .$$

Note that E_0 and S are real functions of x, y, and z. The phase function $S(x, y, z)$ is generally referred to as the eikonal and includes the effect of the medium.

Substituting the expression for E in the wave equation, we obtain

$$\nabla^2(E_0 e^{-jk_0 S}) + k^2 E_0 e^{-jk_0 S} = 0 \quad . \tag{2.9.2}$$

We note

$$\nabla^2(e^{-jk_0 S}) = [-k_0^2(\nabla S)^2 - jk_0 \nabla^2 S]e^{-jk_0 S} \quad , \quad \text{and} \tag{2.9.3}$$

$$(\nabla S)^2 = \nabla S \cdot \nabla S = \left(\frac{\partial S}{\partial x}\right)^2 + \left(\frac{\partial S}{\partial y}\right)^2 + \left(\frac{\partial S}{\partial z}\right)^2 \quad , \tag{2.9.4}$$

$$\nabla^2(E_0 e^{-jk_0 S}) = E_0[-k_0^2(\nabla S)^2 - jk_0 \nabla^2 S]e^{-jk_0 S}$$
$$+ e^{-jk_0 S}\nabla^2 E_0 - j2k_0 e^{-jk_0 S}\nabla S \cdot \nabla E_0 \quad . \tag{2.9.5}$$

Substituting the above expressions into (2.9.2), we obtain

$$E_0[-k_0^2(\nabla S)^2 - jk_0 \nabla^2 S] + \nabla^2 E_0 - 2jk_0 \nabla S \cdot \nabla E_0 + k^2 E_0 = 0 \quad .\tag{2.9.6}$$

Equating real and imaginary parts in the above equation, we obtain

$$-E_0 k_0^2(\nabla S)^2 + \nabla^2 E_0 + k^2 E_0 = 0 \quad , \quad \text{and} \tag{2.9.7}$$

$$E_0 \nabla^2 S + 2\nabla S \cdot \nabla E_0 = 0 \quad . \tag{2.9.8}$$

From (2.9.7) we obtain

$$(\nabla S)^2 - \frac{\nabla^2 E_0}{k_0^2 E_0} = \frac{k^2}{k_0^2} = n^2 \quad . \tag{2.9.9}$$

However, we note that

$$\frac{\nabla^2 E_0}{k_0^2 E_0} = \frac{\lambda_0^2}{(2\pi)^2}\frac{\nabla^2 E_0}{E_0} \to 0 \quad \text{as} \quad \lambda_0 \to 0 \quad . \tag{2.9.10}$$

Thus, in the geometrical optics approximation ($\lambda_0 \to 0$), one obtains the following eikonal equation:

$$(\nabla S)^2 = n^2 \quad . \tag{2.9.11}$$

Solving the eikonal equation, one can obtain the phase front as the wave propagates through a medium with $n(x, y, z)$.

A ray is defined to be a line normal to the phase front. In Fig. 2.33 we show a ray propagating through an arbitrary medium. The unit vector a_s defines the instantaneous direction of wave propagation. Hence a_s is given by

$$a_s = \frac{dr}{d\alpha} \quad , \tag{2.9.12}$$

when the ray is represented by the function $r(\alpha)$ where α denotes the distance along the path of the ray. In this case

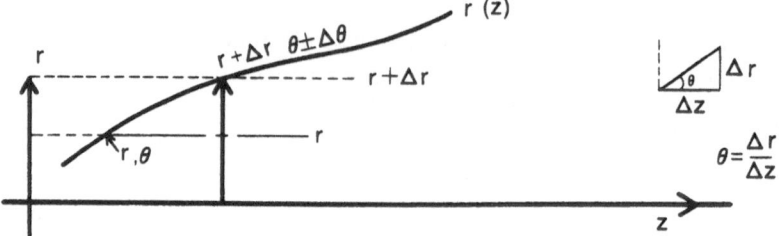

Fig. 2.33. Ray propagating through an arbitrary medium

$$\nabla S = n\boldsymbol{a}_s = n\frac{d\boldsymbol{r}}{d\alpha} \quad . \tag{2.9.13}$$

Now

$$\frac{d}{d\alpha} = \frac{d\boldsymbol{r}}{d\alpha}\cdot\nabla = \boldsymbol{a}_s\cdot\nabla$$

$$= (\boldsymbol{i}_x a_x + \boldsymbol{i}_y a_y + \boldsymbol{i}_z a_z)\cdot\left(\frac{\partial}{\partial x}\boldsymbol{i}_x + \boldsymbol{i}_y\frac{\partial}{\partial y} + \boldsymbol{i}_z\frac{\partial}{\partial z}\right) \quad . \tag{2.9.14}$$

Thus

$$\frac{d}{d\alpha}(\nabla S) = \boldsymbol{a}_s\cdot\nabla(\nabla S) = \boldsymbol{a}_s\cdot(\nabla\nabla S) \tag{2.9.15}$$

where $\nabla\nabla$ is the dyadic tensor operator

$$\begin{pmatrix} \frac{\partial^2}{\partial x^2} & \frac{\partial^2}{\partial x\partial y} & \frac{\partial^2}{\partial x\partial z} \\ \frac{\partial^2}{\partial y\partial x} & \frac{\partial^2}{\partial y^2} & \frac{\partial^2}{\partial y\partial z} \\ \frac{\partial^2}{\partial z\partial x} & \frac{\partial^2}{\partial z\partial y} & \frac{\partial^2}{\partial z^2} \end{pmatrix} \begin{pmatrix} \boldsymbol{i}_x \\ \boldsymbol{i}_y \\ \boldsymbol{i}_z \end{pmatrix} \quad .$$

From the eikonal equation (2.9.11) we obtain

$$\nabla S \cdot \nabla\nabla S = n\nabla n \quad , \quad \text{or} \tag{2.9.16}$$

$$n\boldsymbol{a}_s \cdot \nabla\nabla S = n\nabla n \quad , \quad \text{or}$$

$$\boldsymbol{a}_s \cdot \nabla\nabla S = \nabla n \quad .$$

Thus

$$\frac{d}{d\alpha}(\nabla S) = \nabla n \quad , \quad \text{or} \tag{2.9.17}$$

$$\frac{d}{d\alpha}\left(n\frac{d\boldsymbol{r}}{d\alpha}\right) = \nabla n \quad .$$

Written in Cartesian component form, (2.9.17) becomes

$$\frac{d}{d\alpha}\left(n\frac{dn}{d\alpha}\right) = \frac{\partial n}{\partial x} \quad ,$$

$$\frac{d}{d\alpha}\left(n\frac{dy}{dx}\right) = \frac{\partial n}{\partial y} \quad , \tag{2.9.18}$$

$$\frac{d}{d\alpha}\left(n\frac{dz}{d\alpha}\right) = \frac{\partial n}{\partial z} \quad .$$

In cylindrical coordinates it becomes

$$\frac{d}{d\alpha}\left(n\frac{dr}{d\alpha}\right) - nr\left(\frac{d\phi}{d\alpha}\right)^2 = \frac{\partial n}{\partial r} \quad , \tag{2.9.19}$$

$$\frac{d}{d\alpha}\left(nr^2\frac{d\phi}{d\alpha}\right) = \frac{\partial n}{\partial \phi} \quad ,$$

$$\frac{d}{d\alpha}\left(n\frac{dz}{d\alpha}\right) = \frac{\partial n}{\partial z} \quad .$$

In the paraxial approximation case,

$$d\alpha = dz \tag{2.9.20}$$

as the angle of the ray with respect to the optical axis (z) is small, and the eikonal equation simplifies to

$$n\frac{d^2 r}{dz^2} = \nabla n \quad . \tag{2.9.21}$$

If $n = n(r)$, i.e. the refractive index is a function of radius only, then

$$n\frac{d^2 r}{dz^2} = \frac{dn}{dr} \quad . \tag{2.9.22}$$

If $n(r) = n_0$, we have

$$\frac{dr}{dz} = \text{const.} \tag{2.9.23}$$

If z and θ are defined as in Fig. 2.34, we have

$$r = \theta z \quad . \tag{2.9.24}$$

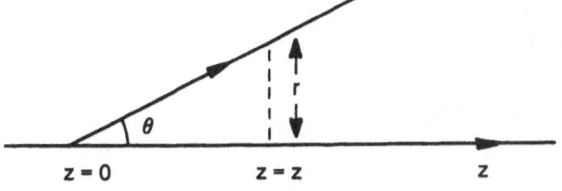

$z = 0$ $\qquad\qquad$ $z = z$ $\qquad\qquad$ z

Fig. 2.34. Propagation of a ray from $z = 0$ to $z = z$

2.9.2 Matrix Formulation of Geometrical Optics

This section provides a brief review of the matrix method, starting from the paraxial approximation and Snell's law. Readers are referred to [2.1, 2] for full details.

In the plane of incidence a ray is defined by X as shown in Fig. 2.35a:

$$X \rightarrow \begin{pmatrix} x \\ \theta \end{pmatrix} \quad .$$

Note that X is a function of z along the optical axis. Propagation of rays is denoted by optical matrices of optical systems with input x_1 and output x_2 as shown in Fig. 2.35b. One can define the following matrices:

Propagation through

a distance D along the $\rightarrow \begin{vmatrix} 1 & D \\ 0 & 1 \end{vmatrix} = T(D)$

z direction

Refraction through

a radius of curvature $\rightarrow \begin{vmatrix} 1 & 0 \\ -\dfrac{n_2 - n_1}{n_1 R} & \dfrac{n_1}{n_2} \end{vmatrix} = R(R)$

R with refractive

indices n_1 and n_2

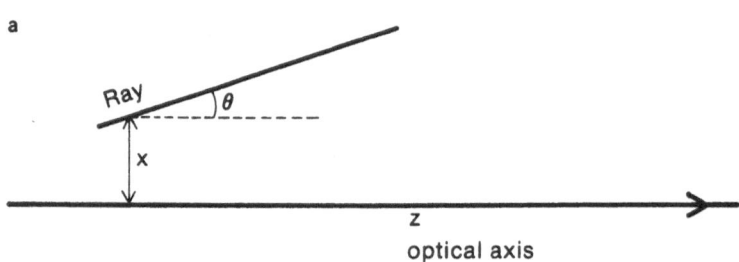

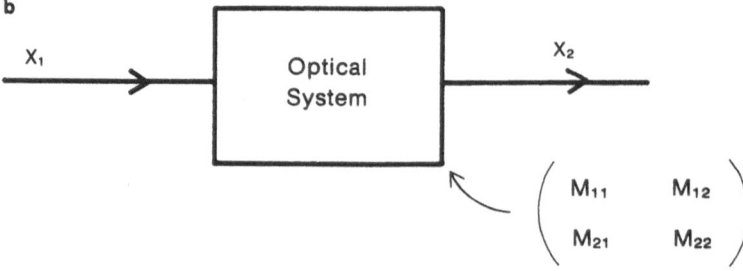

Fig. 2.35. (a) A ray defined by X in the plane of incidence. (b) An optical system with input X_1 and output X_2 can be represented by a matrix

Lens with focal

length f $\qquad \rightarrow \begin{vmatrix} 1 & 0 \\ -\frac{1}{f} & 1 \end{vmatrix} = M(f)$

For any optical matrix with input and output space having refractive indices n_1 and n_2, respectively, one has

$$\det(M) = \frac{n_1}{n_2} \quad . \tag{2.9.25}$$

For a combination of systems in tandem, the individual matrices are multiplied to obtain the effective matrix.

2.9.3 Gaussian Optics Including Lenses

A Gaussian wave is completely characterized by the quantity q, which is a complex quantity. The real part is related to the radius of curvature of the wave front and the imaginary part to the spot size. For an optical system whose matrix is given by M as shown in Fig. 2.35b, we note that

$$\begin{aligned} x_2 &= M_{11}x_1 + M_{12}\theta_1 \quad , \\ \theta_{12} &= M_{21}x_1 + M_{22}\theta_1 \quad . \end{aligned} \tag{2.9.26}$$

Thus, the radius of curvature R_2 is defined as

$$R_2 = \frac{x_2}{\theta_2} = l\frac{M_{11}x_1 + M_{12}\theta_1}{M_{21}x_1 + M_{22}\theta_1} \quad , \quad \text{or}$$

$$R_2 = \frac{M_{11}R_1 + M_{12}}{M_{21}R_1 + M_{22}} \quad . \tag{2.9.27}$$

Thus (2.9.27) gives the relationship between an input wave of radius of curvature R_1 and the output wave with radius of curvature R_2. As an example, for a simple lens with focal length f we have

$$M = M(f) = \begin{pmatrix} 1 & 0 \\ -\frac{1}{f} & 1 \end{pmatrix} \quad . \tag{2.9.28}$$

Thus, for this case

$$R_2 = \frac{R_1}{R_1/f + 1} = \frac{fR_1}{f - R_1} \quad . \tag{2.9.29}$$

As shown in Fig. 2.36a, if the input wave is plane, i.e. $R_1 = \infty$, then

$$R_2 = -f \quad .$$

That is, the output wave front will have a curvature such that it focuses at the focal point. Or, as shown in Fig. 2.36b, if $R_1 = f$, then

$$R_2 = \infty \quad .$$

That is, if the source is at the focal point, the wave front incident on the lens

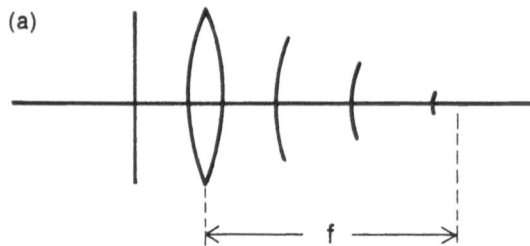

(a)

Fig. 2.36. (a) A plane wave is focused by a convex lens. (b) A spherical wave emitted by a source at the focal point becomes a plane wave on passing through the lens

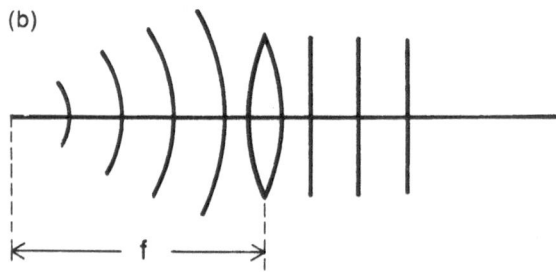

(b)

has $R_1 = f$ and the output wave front is plane, i.e., has an infinite radius of curvature.

This gives us a clue as to the desired relationship between q_2 and q_1:

$$q_2 = \frac{M_{11}q_1 + M_{12}}{M_{21}q_1 + M_{22}} \quad . \qquad (2.9.30)$$

Although no formal proof of this equation exists as yet, it has so far been found to be true.

If we use the simple lens matrix again and consider the propagation of a Gaussian wavefront, then we have

$$q_2 = \frac{f q_1}{f - q_1} \quad . \qquad (2.9.31)$$

If $q_1 = q(0) = jz_R$, then it can be shown that the spot size is not minimum at $z = f$ but at z_m given by

$$z_m = \frac{f}{1 + (f/z_R)^2} \quad . \qquad (2.9.32)$$

2.9.4 Optical Fiber

Next we consider the case of an optical fiber having a refractive index n that is a function of r only, as shown in Fig. 2.37, namely,

$$n(r) = n_0 \left(1 - \frac{r^2}{2l^2} \right) \quad . \qquad (2.9.33)$$

Note that we are using a cylindrical coordinate system. The eikonal equation

64

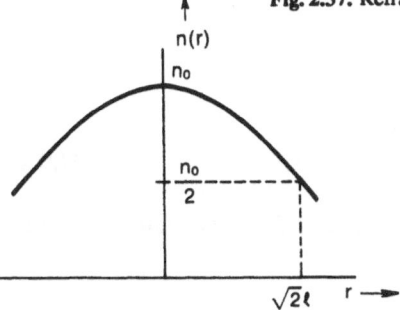

Fig. 2.37. Refractive index profile of a parabolic gradient index fiber

applicable for this case is (2.9.21), which is repeated here:

$$n(r)\frac{d^2 r(z)}{dz^2} = \frac{\partial n}{\partial r} \quad , \tag{2.9.34}$$

where $r(z)$ represents the propagation of the ray. For this case, we note that

$$\frac{\partial n}{\partial r} = -n_0 \frac{r}{l^2} \quad , \tag{2.9.35}$$

and (2.9.34) becomes

$$\frac{\partial^2 r(z)}{\partial z^2} = \frac{1}{n(r)}\left(-n_0 \frac{r}{l^2}\right) \quad . \tag{2.9.36}$$

If we assume that l is large, i.e., the variation in refractive index is small, then one can write

$$\frac{d^2 r(z)}{dz^2} \approx -\frac{r(z)}{l^2} \quad . \tag{2.9.37}$$

The solution of the above equation is

$$r(z) = A \cos\left(\frac{z}{l}\right) + B \sin\left(\frac{z}{l}\right) \quad , \tag{2.9.38}$$

where A and B are constants to be determined. Let us consider

$$r(z) = r_0 \quad \text{at} \quad z = 0 \quad , \quad \text{and}$$
$$\theta = \theta_0 \quad \text{at} \quad z = 0 \quad . \tag{2.9.39}$$

Thus we obtain

$$A = \theta_0 \quad .$$

We also note that $\theta(z)$ is given by

$$\theta(z) = \frac{dr(z)}{dz} = -\frac{1}{l} r_0 \sin\left(\frac{z}{l}\right) + \frac{B}{l} \cos\left(\frac{z}{l}\right) \quad . \tag{2.9.40}$$

From this we find

$$B = \theta_0 l \quad .$$

65

Thus the ray propagation is represented by

$$r(z) = r_0 \cos\left(\frac{z}{l}\right) + \theta_0 l \sin\left(\frac{z}{l}\right) \quad , \tag{2.9.41}$$

$$\theta(z) = -\frac{r_0}{l} \sin\left(\frac{z}{l}\right) + \theta_0 \cos\left(\frac{z}{l}\right) \quad . \tag{2.9.42}$$

We thus immediately identify the equivalent matrix for a medium of length z (Fig. 2.38) as

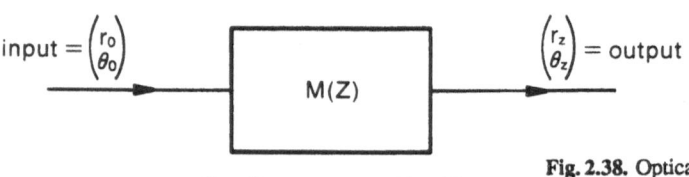

$$\text{input} = \begin{pmatrix} r_0 \\ \theta_0 \end{pmatrix} \qquad \begin{pmatrix} r_z \\ \theta_z \end{pmatrix} = \text{output}$$

$$M(Z)$$

$$Z = 0 \qquad\qquad Z = Z$$

Fig. 2.38. Optical matrix for ray propagation from $z = 0$ to $z = z$

$$M(z) = \begin{pmatrix} \cos z/l & l \sin z/l \\ -\frac{1}{l} \sin z/l & \cos z/l \end{pmatrix} \quad . \tag{2.9.43}$$

We note that if the refractive index variation is given by

$$n = n_0\left(1 + \frac{r^2}{2l^2}\right) \quad , \tag{2.9.44}$$

we obtain for the equivalent optical matrix

$$M(z) = \begin{pmatrix} \cosh z/l & \sinh z/l \\ \frac{1}{l} \sinh z/l & \cosh z/l \end{pmatrix} \quad . \tag{2.9.45}$$

We want to find out when the beam will propagate without diffraction in an optical fiber with refractive index variation given by (2.9.33). Alternatively, is it possible that under certain conditions we will obtain q_0 if we start with q_0? An optical fiber with this property is called self-focusing. Thus we must have

$$q(z) = \frac{M_{11}q_0 + M_{12}}{M_{21}q_0 + M_{22}} = q_0 \quad , \quad \text{or} \tag{2.9.46}$$

$$M_{21}q_0^2 + (M_{22} - M_{11})q_0 - M_{12} = 0 \quad , \quad \text{or}$$

$$\frac{1}{q_0^2}M_{12} + (M_{11} - M_{22})\frac{1}{q_0} - M_{21} = 0 \quad , \tag{2.9.47}$$

which leads to

$$\frac{1}{q_0} = -\frac{M_{11} - M_{22}}{2M_{12}} + \sqrt{\frac{(M_{11} - M_{22})^2 - 4M_{12}M_{21}}{4M_{12}^2}}$$

$$= -\frac{M_{11} - M_{22}}{2M_{12}} - j\left[1 - \left(\frac{M_{11} + M_{22}}{2}\right)^2\right]^{1/2} . \tag{2.9.48}$$

Note that the above equation is true for any optical matrix, not just that of an optical fiber. Substituting the following values for the self-focusing optical fiber,

$$M_{11} = M_{22} = \cos z/l \quad , \quad M_2 = -\frac{1}{l}\sin z/l \quad , \quad \text{and}$$

$$M_{12} = l \sin z/l \quad ,$$

we obtain

$$\frac{1}{q_0} = -\frac{j}{l} = -j\frac{\lambda}{\pi\omega^2(0)} \quad , \quad \text{or} \tag{2.9.49}$$

$$\omega(0) = \sqrt{\frac{\lambda}{\pi l}} . \tag{2.9.50}$$

As the relationship between $q(z)$ and $p(z)$ is given by (2.8.21) we have

$$p'(z) = -j\frac{1}{q(z)} = -j\left(-\frac{j}{l}\right) = -\frac{1}{l} \quad , \quad \text{or}$$

$$p(z) = A - \frac{z}{l} \quad , \tag{2.9.51}$$

where A is a constant. If we choose the origin of z such that $A = 0$, we obtain

$$p(z) = -z/l \quad . \tag{2.9.52}$$

Thus the phase factor of the Gaussian wave for this case will be given by

$$\exp\left[-j\left(k - \frac{1}{l}\right)z\right] . \tag{2.9.53}$$

Hence the electric field inside the fiber can be represented as

$$E(x, y, z) = E_0\frac{1}{\omega^2(0)}\exp\left(-\frac{x^2 + y^2}{\omega^2(0)}\right)\exp(-j\beta z) \quad , \quad \text{where} \tag{2.9.54}$$

$$\beta = k\left(1 - \frac{2}{kl}\right) . \tag{2.9.55}$$

It is interesting to note that the spot size does not increase due to diffraction. The diffraction effect is suppressed by the gradient refractive index.

2.10 Gradient Optical Fiber

In the previous section, the discussion of the gradient optical fiber used geometrical optics and Gaussian beam propagation. However, we could solve the wave equation directly to obtain a complete solution. We should also use our knowledge about self-focusing, i.e., we know that diffraction is eliminated. Thus, for the solution of the wave equation given by

$$E \propto \psi(x, y, z)e^{-jkz} \quad ,$$

we need to consider the solution as

$$E \propto \psi(x, y)e^{-jkz} = E(x, y)e^{-jkz} \quad . \tag{2.10.1}$$

The wave equation for our case is given by

$$\nabla_t^2 E = \frac{\partial^2 E}{\partial z^2} + \frac{\omega^2}{c^2}n^2(r)E = 0 \quad . \tag{2.10.2}$$

As

$$n(r) = n_0 \left(1 - \frac{r^2}{2l^2}\right) \quad , \tag{2.10.3}$$

we can write

$$n^2(r) = n_0^2 \left(r - \frac{r^2}{2l^2}\right)^2$$

$$\approx n_0^2 \left(r - \frac{r^2}{2l^2}\right) \tag{2.10.4}$$

where we have neglected the term r^4/l^4 as small. Thus the wave equation to be solved becomes

$$\frac{\partial^2 E(x, y)}{\partial n^2} + \frac{\partial^2 E(x, y)}{\partial y^2} + \left(k_0^2 - k^2 - \frac{k_0^2}{l^2}(x^2 + y^2)\right)E(x, y) = 0 \quad ,$$

$$\tag{2.10.5}$$

where

$$k_0 = \frac{\omega}{c}n_0 \quad \text{and} \quad k = \frac{\omega}{c}n \quad . \tag{2.10.6}$$

Using the method of separation of variables, we try the solution

$$E(x, y) = X(x)Y(y) \quad . \tag{2.10.7}$$

Equation (2.10.5) becomes after substitution

$$X''Y + Y''X + (k_0^2 - k^2)XY - \frac{k_0^2}{l^2}(x^2 + y^2)XY = 0 \quad ,$$

$$\left(\frac{X''}{X} - \frac{k_0^2 x^2}{l^2}\right) + \left(\frac{Y''}{Y} - \frac{k_0^2 y^2}{l^2}\right) + (k_0^2 - k^2) = 0 \quad . \tag{2.10.8}$$

Thus

$$\frac{X''}{X} - \frac{k^2 x^2}{l^2} = -a \quad , \tag{2.10.9}$$

$$\frac{Y''}{Y} - \frac{k^2 y^2}{l^2} = -b \quad , \quad \text{and} \tag{2.10.10}$$

$$-a - b + k_0^2 - k^2 = 0 \quad , \quad \text{or} \tag{2.10.11}$$

$$k_0^2 - k^2 = a + b \quad .$$

From (2.10.9), we obtain

$$\frac{d^2 X(x)}{dx^2} + \left(a - \frac{k_0^2 x^2}{l^2} \right) = 0 \quad . \tag{2.10.12}$$

Substituting

$$u = \left(\frac{k_0}{l} \right)^{1/2} x \quad , \quad \text{or} \tag{2.10.13}$$

$$x = \left(\frac{l}{k_0} \right)^{1/2} u$$

we obtain

$$\frac{d^2 X(u)}{du^2} + \left(\frac{al}{k_0} - u^2 \right) X(u) = 0 \quad . \tag{2.10.14}$$

This equation is recognized as the equation for the Hermite polynomial discussed in Appendix B. Thus the solution of (2.10.14) can be written as

$$X(u) = \exp \left(\frac{u^2}{2} \right) H_m(u) = \exp \left(-\frac{k_0}{l} x^2 \right) H_m \left(\left(\frac{k_0}{l} \right)^{1/2} x \right) \quad , \tag{2.10.15}$$

where $al/k_0 = 2m + 1$ and m is an integer. Similarly, for the y dependence we obtain

$$Y(y) = \exp \left(-\frac{k_0}{l} y^2 \right) H_n \left(\left(\frac{k_0}{l} \right)^{1/2} y \right) \quad . \tag{2.10.16}$$

Equation (2.10.11) can be rewritten as

$$k_0^2 - k_{m,n}^2 = \frac{2 k_0}{l} (m + n + 1) \quad , \quad \text{or}$$

$$k_{m,n} = k_0 \left(1 - \frac{2}{lk} (m + n + 1) \right)^{1/2} \quad . \tag{2.10.17}$$

Thus the final solution for the electric field for the gradient-index fiber is given by

$$E_{m,n}(x, y, z) = E_{0m,n}$$
$$\times \exp \left(-\frac{(x^2 + y^2) k_0}{2l} \right) H_m \left(\left(\frac{k_0}{l} \right)^{1/2} x \right) H_n \left(\left(\frac{k_0}{l} \right)^{1/2} y \right)$$
$$\times \exp (-j k_{m,n} z) \quad . \tag{2.10.18}$$

Note that $E_{m,n}$ denotes the (m, n)th order mode propagation possible in the fiber. These modes will be confined and will not be attenuated by diffraction.

Unfortunately the group velocity of each mode is different. To obtain the group velocity, we note that the phase factor in the wave is

$$e^{j(\omega t - k_{m,n} z)} \quad . \tag{2.10.19}$$

Thus

$$v_{pm,n} = \frac{\omega}{k_{m,n}} \quad , \tag{2.10.20}$$

$$v_{gm,n} = \left(\frac{dk_{m,n}}{d\omega}\right)^{-1} = \left(\frac{dk_{m,n}}{dk_0} \frac{dk_0}{d\omega}\right)^{-1} \quad , \tag{2.10.21}$$

$$\frac{dk_{m,n}}{dk_0} = \frac{1 - \frac{2}{k_0 l}(m + n + 1) + \frac{1}{kl}(m + n + 1)}{[1 - \frac{2}{k}(m + n + 1)]^{1/2}} \quad , \tag{2.10.22}$$

$$k_0 = \frac{\omega}{c} n(\omega) \quad , \quad \text{or} \tag{2.10.23}$$

$$\begin{aligned} \frac{dk_0}{d\omega} &= \frac{n(\omega)}{c} + \frac{\omega}{c} \frac{dn(\omega)}{d\omega} \\ &= \frac{n(\lambda)}{c} \left(1 - \frac{\lambda}{n(\lambda)} \frac{dn(\lambda)}{d\lambda}\right) \quad . \end{aligned} \tag{2.10.24}$$

Here $n(\lambda)$ represents the refractive index of the optical fiber material as a function of λ. Hence

$$v_{gm,n} \approx \frac{c}{n(\lambda)} \left(1 - \frac{\lambda}{n(\lambda)} \frac{dn(\lambda)}{d\lambda}\right)^{-1} \left(1 - \frac{(1 + m + n + 1)^2}{2(k_0 l)^2}\right)^{-1} \quad . \tag{2.10.25}$$

In an actual gradient-index optical fiber, the outer coating will have a certain value, n_1, as shown in Fig. 2.39. Thus the radius of the gradient fiber part is limited to a value r_0. It is then of interest to calculate the maximum value of m or n that is allowed for mode confinement. To obtain an expression for m_{max}, we note that the following equation must have certain restrictions so that it represents the equation for Hermite polynomials:

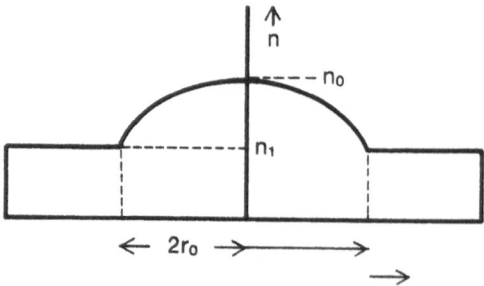

Fig. 2.39. Refractive index profile of a graded-index optical fiber

$$\frac{d^2 X}{dx^2} + \left(a - \frac{k_0^2 x^2}{l^2}\right) X = 0 \quad . \tag{2.10.26}$$

For large a, the solutions are sinusoidally varying for small x. However, for very large x, the solutions decay exponentially along x as $a - k_0^2 x^2/l^2$ becomes negative.

For a guided wave, the field must decay for $x > r_0$ and thus

$$a - \frac{k_0^2 r_0^2}{l^2} \leq 0 \quad . \tag{2.10.27}$$

As

$$a = (2m + 1)\frac{k_0}{l} \quad ,$$

we obtain

$$(2m_{max} + 1)\frac{k_0}{l} = \frac{k_0^2 r_0^2}{l^2} \quad , \quad \text{or} \tag{2.10.28}$$

$$2m_{max} + 1 \sim \frac{k_0 r_0^2}{l} \quad , \quad \text{or}$$

$$2m_{max} + 1 \approx \frac{r_0^2}{\omega^2(0)} \quad . \tag{2.10.29}$$

In general, the E field inside the optical fiber will be a linear combination of all the modes given by

$$E(x, y, z) = \sum_{m=0}^{m_{max}} \sum_{n=0}^{n_{max}} E_{m,n}(x, y, z)$$

$$= \sum_{m=0}^{m_{max}} \sum_{n=0}^{n_{max}} E_{0m,n} \exp\left(-\frac{(x^2 + y^2)k_0}{2l}\right) \exp\left(-j k_{m,n} z\right)$$

$$\times H_m\left(\left(\frac{k_0}{l}\right)^{1/2} x\right) H_n\left(\left(\frac{k_0}{l}\right)^{1/2} y\right) \quad . \tag{2.10.30}$$

To obtain the values of $E_{0m,n}$ one can use the boundary condition at $z = 0$ given by

$$E(x, y)\bigg|_{z=0} = \sum E_{0m,n} H_m\left(\left(\frac{k_0}{l}\right)^{1/2} x\right) H_n\left(\left(\frac{k_0}{l}\right)^{1/2} y\right)$$

$$\times \exp\left(-\frac{(x^2 + y^2)k_0}{2l}\right) \quad . \tag{2.10.31}$$

If $E(x, y)|_{z=0}$ is known, one can use the following orthogonality equation for the Hermite polynomials to obtain $E_{0m,n}$:

$$\int_{-\infty}^{+\infty} \exp(-x^2) H_m(x) H_n(x) = 0 \quad , \qquad m \neq n \quad ,$$

$$= \sqrt{\pi} 2^n n! \quad , \quad m = n \quad . \tag{2.10.32}$$

Thus, $E_{0m,n}$ is given by

$$E_{0m,n} = \frac{1}{\pi n! m! 2^{m+n}} \iint \exp[-(x^2 + y^2)]E(x', y')\Big|_{z=0}$$
$$\times H_m(x)H_n(y)dx\,dy \quad ,$$
(2.10.33)

where

$$x' = \left(\frac{l}{k_0}\right)^{1/2} x \quad \text{and} \quad y' = \left(\frac{l}{k_0}\right)^{1/2} y \quad .$$

It is to be noted that analytical solutions exist for the two following refractive index profiles.

a) *Cosh Profile:*

$$n^2(x) = n_0^2 + 2n_0\Delta n/\cosh^2(2x/h),$$

$$E_y \propto U_m(2x/h)\cosh^2(2x/h) \quad ,$$

where U_m is the mth order hypergeometric function.

b) *Exponential Profile:*

$$n^2(x) = n_0^2 + 2n_0\Delta n \exp(-2|x|/h) \quad ,$$

$$E_y \propto J_m[V\exp(-x/h)] \quad \text{for} \quad x > 0 \quad ,$$

$$J_m[V\exp(x/h)] \quad \text{for} \quad x < 0 \quad ,$$

where J_m is the mth-order Bessel function and $V = k_0 h\sqrt{2n_0\Delta n}$.

2.11 Integrated Optics and Step-Index Optical Fibers

A refractive index profile of the step-index optical fiber is shown in Fig. 2.40a. It has a core of refractive index n_1 surrounded by a layer of refractive index n_2. Before we consider this step-index fiber problem, let us consider a simpler planar case encountered in integrated optics. This is shown in Fig. 2.40b. Here the center layer is adjacent to two media of different refractive indices, n_2 and n_3. As an example, the n_1 medium could be out-diffused LiNbO$_3$, whereas the n_2 and n_3 materials correspond to LiNbO$_3$ and air, respectively. As the solutions of this problem and the step-index fiber optic problem are somewhat more complex than that of a metal microwave waveguide, we shall consider first the microwave waveguide problem. However, before that we must develop the general waveguide equations.

(a)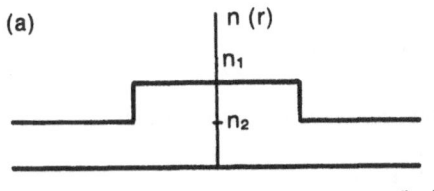

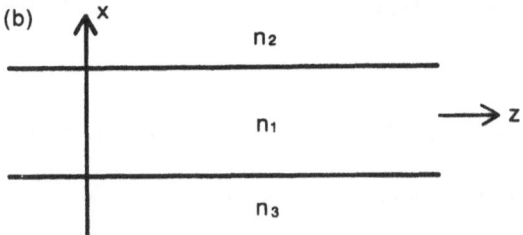

Fig. 2.40. (a) Refractive index profile of a step-index optical fiber. (b) Simpler case of a planar center layer between two materials of different refractive index

2.11.1 Electromagnetic Waveguide Solutions

In many cases of electromagnetic wave propagation we are faced with a problem where the wave propagates in the z direction but, due to the guiding refractive index profile, the wave is confined near the center in the transverse plane. For these cases, one can write the E and H fields as

$$E = E_t + E_z i_z \quad ,$$
$$H = H_t + H_z i_z \quad , \tag{2.11.1}$$

where E_t and H_t are the transverse components and E_z and H_z are the longitudinal components. We can denote the ∇ operator as

$$\nabla = \nabla_t + i_z \frac{\partial}{\partial z} = \nabla_t - i_k jk \quad , \tag{2.11.2}$$

where we have assumed that the z dependence of the wave is given by $\exp(-jkz)$ where k is the propagation constant which determines the wave velocities. Separating the longitudinal and transverse components, Maxwell's equations can be written as

$$\nabla_t \times E_t = -j\omega\mu H_z \quad , \tag{2.11.3}$$

$$i_z \times \nabla_t E_z + jki_z \times E_t = j\omega\mu H_t \quad , \tag{2.11.4}$$

$$\nabla_t \times H_t = j\omega\varepsilon E_z \quad , \tag{2.11.5}$$

$$i_z \times \nabla_t H_z + jki_z \times H_t = j\omega\varepsilon E_t \quad , \tag{2.11.6}$$

$$\nabla_t \cdot H_t = jk H_z \quad , \tag{2.11.7}$$

$$\nabla_t \cdot E_t = jk E_z \quad . \tag{2.11.8}$$

Taking the curl of (2.11.3), one obtains

$$\nabla_t \times (\nabla_t \times E_t) = \nabla_t(\nabla_t \cdot E_t) - \nabla_t^2 E_t = -j\omega\mu \nabla_t \times H_z \quad . \qquad (2.11.9)$$

Using $\nabla_t^2 E_t = -(k_0^2 - k^2)E_t$ obtained from the wave equation, we find

$$(k_0^2 - k^2)E_t + \nabla_t(jk E_z) - j\omega\mu \nabla_t \times H_z \quad , \quad \text{or}$$

$$E_t = \frac{1}{k_0^2 - k^2}(-jk\nabla_t E_z - j\mu\omega \nabla_t \times H_z) \quad . \qquad (2.11.10)$$

Similarly, one can obtain

$$H_t = \frac{1}{(k_0^2 - k^2)}(-jk\nabla_t H_z + j\omega t \nabla_t \times E_z) \quad . \qquad (2.11.11)$$

Thus we see that for waveguide modes one needs to solve for E_z and H_z only. Once E_z and H_z are known, E_t and H_t can always be determined from (2.11.10 and 11).

In some cases, one can separate all the possible solutions as follows:

i)   TEM (transverse electric and magnetic field) mode:

$$E_z = H_z = 0 \quad .$$

ii) TE (transverse electric) mode:

$$E_z = 0 \quad ; \quad H_z \neq 0 \quad .$$

iii) TM (transverse magnetic) mode:

$$H_z = 0 \quad ; \quad E_z \neq 0 \quad .$$

We shall consider all these cases separately.

i) TEM Solution

$$\nabla_t \times E_t = 0 \quad \text{or} \quad E_t = -\nabla_t\phi \quad . \qquad (2.11.12)$$

As

$$\nabla_t \cdot E_t = 0$$

we have

$$\nabla_t^2\phi = 0 \quad . \qquad (2.11.13)$$

Here ϕ is an electric potential and is a function of x and y only. Let us consider the example of a coaxial transmission line shown in Fig. 2.41.

Writing (2.11.13) in cylindrical coordinates, we obtain

$$\frac{1}{r}\frac{\partial}{\partial r}\left(r\frac{\partial\phi}{\partial r}\right) + \frac{1}{r^2}\frac{\partial^2\phi}{\partial\phi^2} = 0 \quad . \qquad (2.11.14)$$

If we neglect the ϕ variation, we have

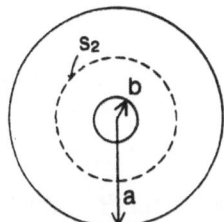

Fig. 2.41. Coaxial transmission line

$$\frac{1}{r}\frac{\partial}{\partial r}\left(r\frac{\partial\phi}{\partial r}\right) = 0 \quad , \quad \text{or} \qquad (2.11.15)$$

$$\frac{\partial\phi}{\partial r} = \frac{A}{r} \quad , \quad \text{or}$$

$$\phi = A \ln r + B \quad , \qquad (2.11.16)$$

where A and B are arbitrary constants. Applying the boundary condition

$$\phi(a) = 0 = A \ln a + B \quad ,$$

we find $B = -A \ln a$, or

$$\phi = A \ln\frac{r}{a} \quad . \qquad (2.11.17)$$

As

$$\phi(b) = V_0 \quad ,$$

where V_0 is the applied voltage, we have

$$A = \frac{V_0}{\ln(b/a)} \quad . \qquad (2.11.18)$$

Thus

$$\phi = \frac{V_0}{\ln(b/a)}\ln\frac{r}{a} \quad , \qquad (2.11.19)$$

$$E_r = -\frac{\partial\phi}{\partial r} = -\frac{V_0}{\ln(b/a)}\frac{1}{r} \quad , \qquad (2.11.20)$$

$$H_\phi = \frac{E_r}{Z} \quad , \qquad (2.11.21)$$

where $Z = \sqrt{\mu/\varepsilon}$. The quantities μ and ε are the values for the medium of the transmission line.

The current flowing, I_0, is given by

$$I_0 = \int_{s_2} \boldsymbol{H}\cdot d\boldsymbol{l} = \int_0^{2\pi} H_\varphi r\, d\phi = \frac{2\pi}{z}\frac{V_0}{\ln(b/a)} \quad . \qquad (2.11.22)$$

Thus the transmission line characteristic impedance will be given by

$$Z_{\text{tr. line}} = \frac{V_0}{I_0} = Z \frac{\ln(b/a)}{2\pi} \quad . \tag{2.11.23}$$

ii) TE Waves

For the TE case, $E_z = 0$ and all the quantities are determined from H_z, which is obtained from the wave equation. Once H_z is found, the other quantities can be obtained from

$$E_t = \frac{1}{k_0^2 - k^2}(-j\omega\mu\nabla_t \times i_z H_z) \quad , \tag{2.11.24}$$

$$H_t = \frac{1}{k_0^2 - k^2}(-jk\nabla_t H_z) = -\frac{jk}{k_0^2 - k^2}\nabla_t H_z \quad . \tag{2.11.25}$$

For this case also, we have

$$jk i_z \times E_z = j\omega\mu H_t \quad , \quad \text{or} \tag{2.11.26}$$

$$jk i_z \times (i_z \times E_t) = j\mu\omega i_z \times H_t \quad .$$

Using the vector identity

$$A \times (B \times C) = B(A \cdot C) - C(A \cdot B) \quad , \tag{2.11.27}$$

we obtain

$$jk[(i_z \cdot E_t)i_z - (i_z \cdot i_z)E_t] = j\omega\mu i_z \times H_t \quad , \quad \text{or}$$

$$E_t = \frac{\omega\mu}{k} i_z \times H_t = Z_e i_z \times H_t \quad . \tag{2.11.28}$$

Here Z_e represents the equivalent transmission line impedance for the TE mode.

iii) TM Waves

For the TM wave we have $H_z = 0$, $E_z \neq 0$. The transverse components of electric and magnetic fields for this case are

$$E_t = \frac{1}{k_0^2 - k^2}(-jk\nabla_t E_z) \quad , \tag{2.11.29}$$

$$H_t = \frac{1}{k_0^2 - k^2}(j\omega\varepsilon\nabla_t \times E_z) \quad , \tag{2.11.30}$$

$$H_t = Z_m^{-1} i_z \times E_t = \frac{k_0}{k} \frac{1}{Z} i_z \times E_t \quad , \tag{2.11.31}$$

where Z_m represents the equivalent transmission line characteristic impedance for the TM mode. Note that

$$Z_e Z_m = Z^2 = \frac{\mu}{\varepsilon} \quad . \tag{2.11.32}$$

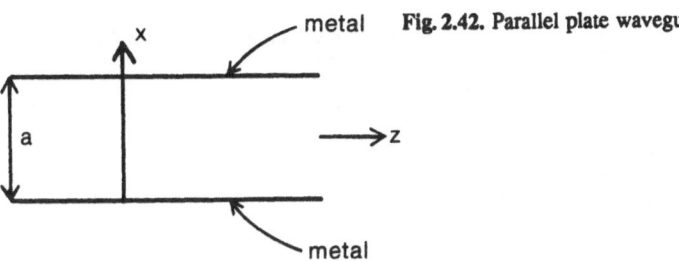

Fig. 2.42. Parallel plate waveguide

metal

x

a

z

metal

2.11.2 Parallel Plate Waveguide: TE Solution

Let us consider two metal plates separated by a distance a as shown in Fig. 2.42. Consider the y extension to be infinite so that

$$\frac{\partial}{\partial y} = 0 \quad .$$

For this two-dimensional waveguide problem, the wave equation is

$$\frac{d^2 H_z}{dx^2} + (k_0^2 - k^2)H_z = 0 \quad . \tag{2.11.33}$$

Defining $k_x^2 = (k_0^2 - k^2)$, one obtains

$$H_z = A \cos k_x x + B \sin k_x x \quad , \tag{2.11.34}$$

where A and B are constants to be determined. We note that for unattenuated propagation the component of the propagating constant k_x has to be real. Otherwise we obtain evanescent or attenuating waves. Note that

$$H_x = -\frac{jk}{k_0^2 - k^2}\frac{\partial H_z}{\partial x} \quad , \quad \text{and} \tag{2.11.35}$$

$$E_y = -Z_e H_x \quad . \tag{2.11.36}$$

The boundary conditions are

$$H_x = 0 \quad \text{for} \quad x = 0 \quad \text{and} \quad x = a \quad .$$

As

$$\frac{\partial H_z}{\partial x} = -k_x A \sin k_x x + k_x B \cos k_x x \quad , \tag{2.11.37}$$

we obtain

$$B = 0 \quad \text{and} \quad \sin k_x a = 0 \quad , \quad \text{or}$$

$$k_x = k_0^2 - k^2 = n\frac{\pi}{a} \quad . \tag{2.11.38}$$

Thus

$$k_n^2 = \omega^2 \mu\varepsilon - \frac{n^2\pi^2}{a^2} \quad , \quad \text{or}$$

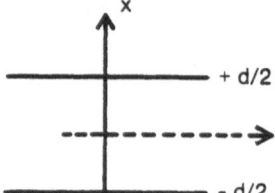

Fig. 2.43. Plots of electric field for various *TE* modes

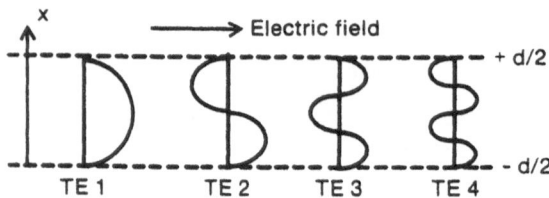

$$k_n = \left(\omega^2 \mu \varepsilon - \frac{n^2 \pi^2}{a^2} \right)^{1/2} \quad . \tag{2.11.39}$$

If we call the cutoff frequency for the nth order mode ω_{cn}, then

$$\omega_{cn}^2 \mu \varepsilon = \frac{n^2 \pi^2}{a^2} \quad , \quad \text{or}$$

$$\omega_{cn} = \frac{1}{\sqrt{\mu \varepsilon}} \frac{n \pi}{a} \quad . \tag{2.11.40}$$

For $\omega < \omega_{cn}$, that mode cannot propagate. Some plots of E field for different modes as a function of x are shown in Fig. 2.43.

If we redefine the geometry of the problem as shown in Fig. 2.44 we obtain

$$E_y = A_1 \sin k_x x + A_2 \cos k_x x \quad . \tag{2.11.41}$$

For this case we can define odd and even modes as follows:

$$E_y = A_1 \sin k_x x \quad , \quad \text{odd mode} \quad , \tag{2.11.42}$$

$$E_y = A_2 \cos k_x x \quad , \quad \text{even mode} \quad . \tag{2.11.43}$$

For the odd mode we have

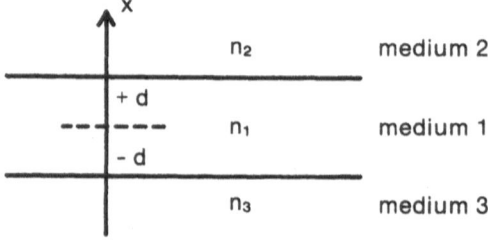

Fig. 2.44. Redefinition of the geometry of the problem

$$\sin k_x d = 0 \quad \text{or} \quad k_x = \frac{n\pi}{d} \quad . \tag{2.11.44}$$

For the even mode

$$\cos k_x d = 0 \quad \text{or} \quad k_x = \frac{(2n+1)}{d}\pi \quad . \tag{2.11.45}$$

Substituting $a = 2d$, one obtains the characteristic equation

$$2 \sin k_x d \cos k_x d = \sin k_x a = 0 \quad . \tag{2.11.46}$$

The even mode is also called the symmetrical mode because for this case the E field is symmetrical with respect to the center of the guide. The odd mode is called the antisymmetrical mode for a similar reason.

Note that the phase and group velocities of the wave depend on the mode number. The phase velocity v_{pn} is given by

$$v_{pn} = \frac{\omega}{k_n} = \frac{1}{\sqrt{\mu\varepsilon}} \frac{1}{\sqrt{1 - (\omega_{cn}/\omega)^2}} \quad . \tag{2.11.47}$$

Similarly, the group velocity v_{gn} is given by

$$v_{gn} = \frac{d\omega}{dk_n} = \frac{1}{\sqrt{\mu\varepsilon}} \sqrt{1 - \left(\frac{\omega_{cn}}{\omega}\right)^2} \quad . \tag{2.11.48}$$

Thus

$$v_{gn} v_{pn} = \frac{1}{\mu\varepsilon} = v^2$$

where v is the velocity of the wave in the infinite medium.

2.11.3 Integrated Optics Problem

a) TE Modes

The problem to be solved is shown in Fig. 2.45. For confinement of the modes we must have $n_1 > n_2$ and $n_1 > n_3$. Let us define

$$\alpha_1^2 = k_1^2 - k^2 = \frac{\omega^2}{c^2} n_1^2 - k^2 \quad , \tag{2.11.49a}$$

$$-\alpha_2^2 = k^2 - \frac{\omega^2}{c^2} n_2^2 \quad , \tag{2.11.49b}$$

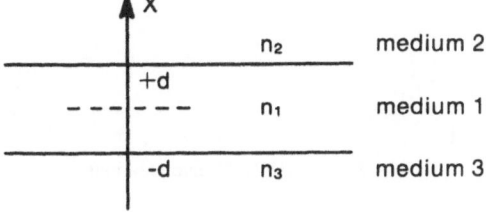

Fig. 2.45. Geometry of the symmetrical waveguide problem

$$-\alpha_3^2 = k^2 - \frac{\omega^2}{c^2}n_3^2 \quad . \tag{2.11.49c}$$

The positive values of α_1 will denote propagating waves in medium 1, which will be represented by

$$E_y = [A_1 \cos \alpha_1 x + B_1 \sin \alpha_1 x]e^{-jkz} \quad , \tag{2.11.50}$$

Similarly the positive values of α_2 and α_3 will denote attenuating waves in media 2 and 3, respectively. These can be written as

$$E_y = A_2 e^{-\alpha_2(x-d)}e^{-jkz} \quad \text{for} \quad x \geq d \quad , \tag{2.11.51}$$

$$E_y = A_3 e^{\alpha_2(x+d)}e^{-jkz} \quad \text{for} \quad x \leq -d \quad . \tag{2.11.52}$$

In (2.11.51 and 52) we have already used the boundary conditions

$$E_y(d \rightarrow \infty) = 0 \quad , \quad \text{and} \quad E_y(d \rightarrow -\infty) = 0 \quad .$$

The above-mentioned solution of the E_y field is the desired one. However, depending on the value of ω, other solutions are possible, even if one chooses $n_1 > n_2$ and $n_1 > n_3$. These are shown in Fig. 2.46.

If we assume a symmetrical profile for the time being, i.e., $n_2 = n_3$, we have $B_1 = 0$ for the even mode. To determine the other constants A_2 and A_3, we use the following boundary conditions:

$$E_{y1}\Big|_{d^+} = E_{y2}\Big|_{d^-} \quad \text{at} \quad x = d \quad , \quad \text{thus}$$

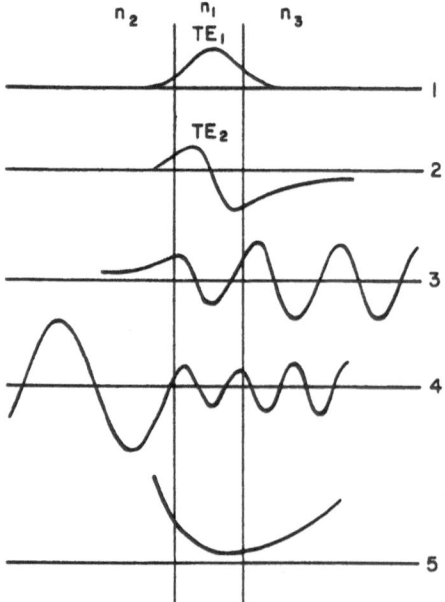

Fig. 2.46. Possible E-field solutions

$$A_1 \cos \alpha_1 d = A_2 \quad , \quad \text{and}$$

$$H_{z1}\bigg|_{d^+} = H_{z2}\bigg|_{d^-} \quad \text{at} \quad x = d \quad . \tag{2.11.53}$$

As

$$H_{z1} = -\frac{j\alpha_1}{\omega\mu_0} A_1 \sin \alpha_1 x \quad , \quad \text{and}$$

$$H_{z2} = -\frac{j\alpha_2}{\omega\mu_0} A_2 e^{-\alpha_2(x-d)} \tag{2.11.54}$$

we have

$$A_2/A_1 = \frac{\alpha_1}{\alpha_2} \sin \alpha_1 d \quad . \tag{2.11.55}$$

Using (2.11.35 and 54) we obtain the characteristic equation given by

$$\tan \alpha_1 d = \frac{\alpha_2}{\alpha_1} \quad . \tag{2.11.56}$$

Similarly for the odd modes, considering $A_1 = 0$, one can easily show that the relevant characteristic equation is given by

$$\tan \alpha_1 d = -\frac{\alpha_1}{\alpha_2} \quad . \tag{2.11.57}$$

Using (2.11.47–49) we have

$$\alpha_1^2 + \alpha_2^2 = k_1^2 - k_2^2 = \frac{\omega^2}{c^2}(n_1^2 - n_2^2) \quad , \quad \text{or}$$

$$(\alpha_1 d)^2 + (\alpha_2 d)^2 = \frac{\omega^2 d^2}{c^2}(n_1^2 - n_n^2) \quad . \tag{2.11.58}$$

If we define

$$\alpha_1 d = X \quad , \tag{2.11.59}$$

$$\alpha_2 d = Y \quad , \quad \text{and} \tag{2.11.60}$$

$$R^2 = \frac{\omega^2 d^2}{c^2}(n_1^2 - n_2^2) \quad , \tag{2.11.61}$$

the characteristic equations for the odd and even modes become

$$\tan X = \frac{Y}{X} \quad \text{or} \quad Y = X \tan X \quad , \tag{2.11.62}$$

$$\tan X = \frac{X}{Y} \quad \text{or} \quad Y = -X \cot X \quad , \quad \text{and} \tag{2.11.63}$$

$$X^2 + Y^2 = R^2 \quad . \tag{2.11.64}$$

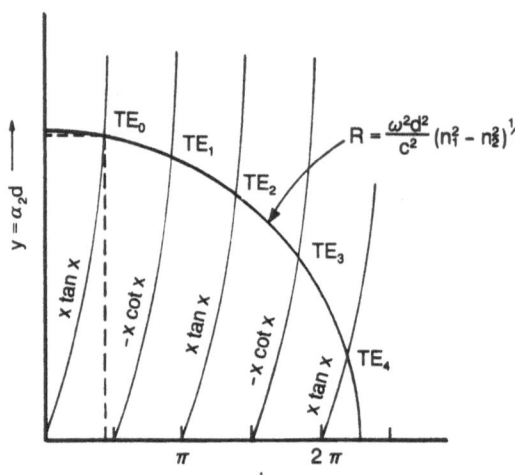

Fig. 2.47. Characteristic equation diagram for TE modes of a dielectric slab waveguide

Note that for the parallel plate waveguide case $\alpha_2 \to \infty$ or $y \to 0$ for $|x| \geq d$. Thus analytical solutions given by (2.11.44 and 45) are easily found from (2.11.56, 57). However, for finite α_2, we must solve the transcendental equation given by (2.11.64). These equations are plotted in Fig. 2.47. For a particular value of R, we obtain the values of Y and X where the circle with radius R intersects the set of curves shown in Fig. 2.47. We note that the number of propagating modes increases as R is increased. Thus, only the TE1 mode propagates for $R < \frac{\pi}{2}$.

For $R \gg 2\pi$ many modes can propagate. Thus to design a multimode slab one must have

$$d \gg \frac{\lambda_0}{\sqrt{n_1^2 - n_2^2}} = \frac{\lambda_0}{\text{NA}} \qquad (2.11.65)$$

Where NA (numerical aperture) $= \sqrt{n_1^2 - n_2^2}$. Figure 2.48 shows the E-field variation along the x axis for TE_1, TE_2, TE_3, and TE_4 modes.

For an asymmetrical waveguide, $n_2 \neq n_3$, one can show that the characteristic equation is

$$\tan 2\alpha_1 d = \frac{\alpha_2 + \alpha_3}{\alpha(1 - \alpha_2\alpha_3/\alpha_1 2)} \qquad . \qquad (2.11.66)$$

For $n_2 = n_3$, the above equation can be rewritten as

$$\left(\tan \alpha_1 d + \frac{\alpha_1}{\alpha_2}\right)\left(\tan \alpha_1 d - \frac{\alpha_2}{\alpha_1}\right) = 0 \quad , \qquad (2.11.67)$$

which leads to the same characteristic equations (2.11.55 and 56) in a combined form.

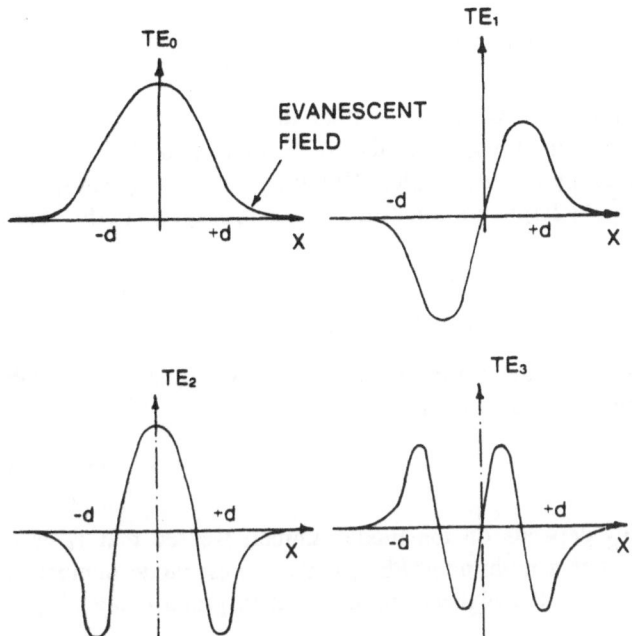

Fig. 2.48. TE modes in a dielectric slab waveguide as a function of x

b) TM Modes

The results for TM wave propagation are quite similar to those for TE mode propagation. Thus we will not discuss TM waves in detail. First let us consider the parallel plane metal waveguide of size a shown in Fig. 2.42. For this case we solve for the z component of the electric field. It is easy to show that for the nth mode we have

$$E_{zn} = A_n \sin \frac{n\pi}{a} \qquad (2.11.68)$$

and the cutoff frequency ω_{cn} is given by

$$\omega_{cn} = \frac{1}{\sqrt{\mu\varepsilon}} \frac{n\pi}{a} \quad . \qquad (2.11.69)$$

Note that (2.11.69) is identical to (2.11.40), the cutoff frequency for the TE waves. The expressions for the group and phase velocities, given by (2.11.47 and 48) are also valid for the TM case.

Coming to the integrated optics problem, let us consider TM wave propagation in the symmetrical waveguide shown in Fig. 2.44, where $n_2 = n_3$. Following the derivations for the TE case, one obtains the following characteristic equations for the TM modes:

$$\tan \alpha_1 d = \frac{n_1^2}{n_2^2} \frac{\alpha_2}{\alpha_1} \quad \text{for even modes} \quad , \qquad (2.11.70)$$

$$\tan \alpha_1 d = -\frac{n_2^2}{n_1^2} \frac{\alpha_1}{\alpha_2} \quad \text{for odd modes} \quad . \tag{2.11.71}$$

The general features of TM modes are similar to those of the TE modes discussed earlier. Because of the presence of the (n_1/n_2) factor in (2.11.70,71), the decay constants defined by α_2 are smaller. This means that, in general, TM modes are less confined and a larger portion of the power propagates in the outer media.

2.11.4 Multimode Group Delay in a Dielectric Waveguide

It is of interest to calculate the delay time for a length L of the dielectric slab waveguide. The delay time τ_g is defined as

$$\tau_g = \frac{L}{v_g} \quad . \tag{2.11.72}$$

As the group velocity v_g depends on the mode number, we see that τ_g is a function of that number. For a multimode fiber, if the lowest mode number is denoted by a subscript L and the highest one by H, then group delay Δt_g is defined as

$$\Delta t_g = \tau_{gL} - \tau_{gH} \quad . \tag{2.11.73}$$

Thus, a pulse introduced at one end of a line will be spread by the group delay Δt_g when it arrives at the other end.

We know that, for propagating waves, (2.11.61) implies that the following condition must be satisfied:

$$0 < \alpha_1 < \sqrt{n_1^2 - n_2^2 k_0} \quad . \tag{2.11.74}$$

Also, the propagation constant k is given by

$$k = \sqrt{n_1^2 k_0 - \alpha_1^2} \quad . \tag{2.11.75}$$

Thus, for the lowest order mode propagation we have

$$k_L = n_1 k_0 \tag{2.11.76}$$

whereas for the highest order mode propagation

$$k_H = n_2 k_0 \quad . \tag{2.11.77}$$

As

$$v_g = \left(\frac{dk}{d\omega} \right)^{-1} \tag{2.11.78}$$

we have for τ_{gL} and τ_{gH}

$$\tau_{gL} = \frac{L}{c} \left(n_1 + \omega \frac{dn_1}{d\omega} \right) \quad , \tag{2.11.79}$$

$$\tau_{gH} = \frac{L}{c}\left(n_2 + \omega \frac{dn_2}{d\omega}\right) \quad . \tag{2.11.80}$$

Thus

$$\Delta\tau_g = \frac{L}{c}\left[(n_1 - n_2) + \left(\frac{dn_1}{d\omega} - \frac{dn_2}{d\omega}\right)\right] \quad . \tag{2.11.81}$$

In general

$$\frac{dn_1}{d\omega} - \frac{dn_2}{d\omega} \approx 0 \quad .$$

For this case (2.11.81) simplifies to

$$\Delta\tau_g \approx \frac{L}{c}(n_1 - n_2) \quad . \tag{2.11.82}$$

2.11.5 Cylindrical Waveguide

As for the slab waveguide, we consider first the cylindrical metal waveguide and then the stepped-index optical fiber, which is really a dielectric cylindrical waveguide.

The metal waveguide has a radius a (Fig. 2.49). Let us consider the TM modes first, for which we need to solve for E_z from the wave equation in the cylindrical coordinates system:

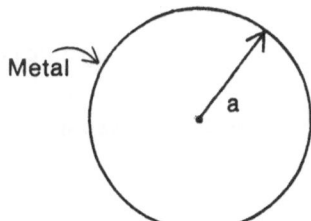

Metal

a

Fig. 2.49. Cylindrical metal waveguide

$$\nabla_t^2 E_z = (k_0^2 - k^2)E_z = 0 \quad , \quad \text{or}$$

$$\frac{\partial^2 E_z}{\partial r^2} + \frac{1}{r}\frac{\partial E_z}{\partial r} + \frac{1}{r}\frac{\partial^2 E_z}{\partial \phi^2} = -(k_0^2 - k^2)E_z \quad . \tag{2.11.83}$$

We try the separation of variables

$$E_z(r, \phi) = f(r)g(\phi) \quad . \tag{2.11.84}$$

Substituting this expression for E_z in (2.11.83) we obtain

$$\frac{f''}{f} + \frac{1}{r}\frac{f'}{f} + \frac{1}{r^2}\frac{g''}{g} = -(k_0^2 - k^2) \quad , \quad \text{or} \tag{2.11.85}$$

$$\frac{r^2 f''}{f} + \frac{r f'}{f} + (k_0^2 - k^2)r^2 = -\frac{g''}{g} = n^2 = \text{const} \quad . \tag{2.11.86}$$

Thus we obtain two separate differential equations:

$$\frac{d^2 g}{d\phi^2} + n^2 g = 0 \quad , \quad \text{and} \tag{2.11.87}$$

$$\frac{d^2 f}{dr^2} + \frac{1}{r}\frac{df}{dr} + \left(k_0^2 - k^2 - \frac{n_2}{r^2}\right)f = 0 \quad . \tag{2.11.88}$$

Solving (2.11.87) for $g(\phi)$, we have

$$g(\phi) = A \cos n\phi + B \sin n\phi \quad . \tag{2.11.89}$$

The solution for (2.11.88) can be written in terms of Bessel functions discussed in Appendix B:

$$f(r) = C J_n\left(\sqrt{k_0^2 - k^2}\,r\right) + D Y_n\left(\sqrt{k_0^2 - k^2}\,r\right) \quad , \tag{2.11.90}$$

where J_n is the nth-order Bessel function of the first kind and Y_n is the nth-order Bessel function of the second kind. We also note that

$$J_n(r) \to 0 \quad \text{as} \quad r \to 0 \quad , \quad \text{and}$$

$$Y_n(r) \to \infty \quad \text{as} \quad r \to 0 \quad .$$

As the electric field must be finite at $r = 0$, we have

$$E_{zn} = J_n\left(r\sqrt{k_0^2 - k^2}\right)(A \cos n\phi + B \sin n\phi) \quad , \tag{2.11.91}$$

where A and B are arbitrary constants. The boundary condition is given by

$$E_z = 0 \quad \text{at} \quad r = a \quad .$$

Thus

$$J_n\left(r\sqrt{k_0^2 - k^2}\right) = 0 = J_n(p_{nm}) \quad , \tag{2.11.92}$$

where p_{nm} is the mth root of the nth-order Bessel function. Table 2.1 lists some of the values of p_{nm}. We note that the propagation constant k_{nm} and the cutoff frequency ω_{cnm} are given by

Table 2.1. Values of p_{nm}

n	p_{n1}	p_{n2}	p_{n3}
0	2.405	5.520	8.654
1	3.832	7.016	10.174
2	5.135	8.417	11.620

$$k_{nm} = \left[k_0^2 - \left(\frac{p_{nm}}{a} \right)^2 \right]^{1/2} \quad , \quad \text{and} \tag{2.11.93}$$

$$\omega_{cnm} = c\left(\frac{p_{nm}}{a} \right) \quad . \tag{2.11.94}$$

For the TE mode, using similar arguments one can show that

$$H_z = J_n\left(r\sqrt{k_0^2 - k^2} \right)(A' \cos n\phi + B' \sin n\phi) \quad . \tag{2.11.95}$$

For this case the boundary condition is given by

$$E_r = 0 \quad \text{at} \quad r = a \quad .$$

As $E_r \propto \partial H_z / \partial r$, we obtain

$$J_n'\left(a\sqrt{k_0^2 - k^2} \right) = 0 = J_n'(p_{nm}') \quad , \tag{2.11.96}$$

where $J_n'(x)$ denotes differentiation of $J_n(x)$ with respect to x and p_{nm}' is the mth root of the derivative of the nth-order Bessel function. For TM modes we have

$$k_{nm} = \left[k_0^2 - \left(\frac{p_{nm}'}{a} \right)^2 \right]^{1/2} \quad , \tag{2.11.97}$$

$$\omega_{cnm} = c\left(\frac{p_{nm}'}{a} \right) \quad . \tag{2.11.98}$$

Some of the values of p_{nm}' are listed in Table 2.2.

Table 2.2. Values of p_{nm}'

n	p_{n1}	p_{n2}	p_{n3}
0	0.000	2.832	7.0156
1	1.841	5.331	8.536
2	3.054	6.706	9.969

2.11.6 Stepped-Index Optical Fiber

The difference between metallic and dielectric waveguides is in the reflection mechanism responsible for confining the energy. The metallic guide does it by reflection from a good conductor at the boundary. In the fiber optic cable, this is accomplished by total internal reflection. As discussed in the previous section, the metallic guide has two sets of solutions, the transverse electric and transverse magnetic modes. However, in the dielectric guide, all but the cylindrically symmetric modes (TE_{0m}, TM_{0m}) are hybrid. The term hybrid implies that the modes have both electric and magnetic components along the z direction. As we shall see shortly, there are two sets of hybrid modes. Because the boundary conditions give a characteristic equation which is quadratic in the Bessel function in terms of the ray analogy for the stepped-index fiber, the hybrid modes correspond to propagating skew rays, and TE and TM modes to meridional rays. However, for the zeroth-order mode, only meridional rays propagate in the guide.

The stepped-index optical fiber problem to be solved is shown in Fig. 2.50. We consider the outermost radius to be infinite for mathematical simplification.

The z component of the electric and magnetic field vectors in the core region can be written as [1]

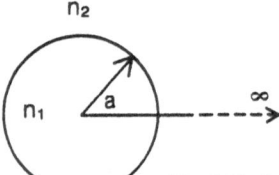

Fig. 2.50. Stepped-index optical fiber

$$r < a \begin{cases} E_z = A J_\nu(\alpha_1 r) e^{j\nu\phi} \quad , & (2.11.99) \\ \\ H_z = B J_\nu(\alpha_1 r) e^{j\nu\phi} \quad . & (2.11.100) \end{cases}$$

Here $\alpha_1^2 = n_1^2 k_0^2 - k^2$ and we have discarded the solution containing Y_ν because the electric field has to be finite at $r = 0$.

In region 2, $b \le r \le \infty$, we must have attenuating waves. Thus the solutions will consist of modified Bessel functions. However, modified Bessel functions of first and second kind have the following properties:

$$K_\nu(r) \to 0 \quad \text{as} \quad r \to \infty \quad ,$$

$$I_\nu(r) \to \infty \quad \text{as} \quad r \to \infty \quad .$$

Thus the solution in region 2 can be written as

$$r > a \begin{cases} E_z = C K_\nu(\alpha_2 r) e^{j\nu\phi} \quad , & (2.11.101) \\ \\ H_z = D K_\nu(\alpha_2 r) e^{j\nu\phi} \quad . & (2.11.102) \end{cases}$$

Here

$$\alpha_2^2 = k^2 - n_2^2 k_0^2 \quad . \tag{2.11.103}$$

To obtain the characteristic equation for k, we must eliminate the constants A, B, C, and D using the following boundary conditions: E_z, E_ϕ, H_z, and H_ϕ are continuous at $r = a$. We then obtain expressions for E_ϕ and H_ϕ in both regions:

---

[1] As n is generally used for the refractive index, we use the subscript ν ($\nu = 0, 1, 2 \ldots$) as mode number in this section.

$$
r < a
\begin{cases}
E_\phi = -\dfrac{j}{\alpha_1^2}\left(jk\dfrac{\nu}{r}AJ_\nu(\alpha_1 r) - \alpha_1\omega\mu B J_\nu'(\alpha_1 r)\right)e^{j\nu\phi} & (2.11.104) \\[4mm]
H_\phi = -\dfrac{j}{\alpha_1^2}\left(\alpha_1\omega\varepsilon_1 A J_\nu'(\alpha_1 r) + jk\dfrac{\nu}{r}B J_\nu(\alpha_1 r)\right)e^{j\nu\phi} & (2.11.105)
\end{cases}
$$

$$
r > a
\begin{cases}
E_\phi = \dfrac{1}{\alpha_2^2}(k\dfrac{\nu}{r}C K_\nu(\alpha_2 r) - \alpha_2\omega\mu_0 D K_\nu'(\alpha_2 r))e^{j\nu\phi} & (2.11.106) \\[4mm]
H_\phi = \dfrac{1}{\alpha_2^2}\left(\alpha_2\omega\varepsilon_2 C K_\nu'(\alpha_2 r) + k\dfrac{\nu}{r}D K_\nu(\alpha_2 r)\right)e^{j\nu\phi} & . \quad (2.11.107)
\end{cases}
$$

Substituting the expression for the boundary conditions and rearranging, we obtain the characteristic equation

$$
\left(\frac{\varepsilon_1}{\varepsilon_2}\frac{a\alpha_2^2}{\alpha_1}\frac{J_\nu'(\alpha_1 a)}{J_\nu(\alpha_1 a)} + \alpha_2 a\frac{K_\nu'(\alpha_2 a)}{K_\nu(\alpha_2 a)}\right)
$$
$$
\times\left(\frac{a\alpha_2^2 J_\nu'(\alpha_1 a)}{\alpha_1 J_\nu(\alpha_1 a)} + \alpha_2 a\frac{K_\nu'(\alpha_2 a)}{K_\nu(\alpha_2 a)}\right) = \left[\nu\left(\frac{n_1^2}{n_2^2} - 1\right)\frac{kk_2}{\alpha_1^2}\right] \quad , \quad (2.11.108)
$$

where

$$
k_2^2 = k_0^2 n_2^2 \quad . \tag{2.11.109}
$$

If $\nu = 0$, the equation simplifies as follows for the TE mode, i.e., $E_z = 0$:

$$
\frac{a\alpha_2^2}{\alpha_1}\frac{J_0'(\alpha_1 a)}{J_0(\alpha_1 a)} + \alpha_2 a\frac{K_0'(\alpha_2 a)}{K_0(\alpha_2 a)} = 0 \quad . \tag{2.11.110}
$$

Similarly for the TM mode ($H_z = 0$) and $\nu = 0$ we obtain

$$
\frac{\varepsilon_1}{\varepsilon_2}\frac{a\alpha_2^2}{\alpha_1}\frac{J_0'(\alpha_1 a)}{J_0(\alpha_1 a)} + \frac{\alpha_2 a K_0'(\alpha_2 a)}{K_0(\alpha_2 a)} = 0 \quad . \tag{2.11.111}
$$

It is of interest to obtain asymptotic values, by allowing $\alpha_2 \to 0$. For this case we find the following equations for cutoff:

$$
\text{TE} \quad \nu = 0 \quad J_0\left(a\sqrt{k_0^2 - k^2}\right) = 0 \quad , \tag{2.11.112}
$$

$$
\text{TM} \quad \dot{\nu} = 0 \quad J_0'\left(a\sqrt{k_0^2 - k^2}\right) = 0 \quad . \tag{2.11.113}
$$

It is noted that for $\nu \neq 0$, only hybrid modes (i.e., both H_z and $E_z \neq 0$) can appear.

Let us define

$$
\eta_1 = \frac{J_\nu'(\alpha_1 a)}{\alpha_1 a J_\nu(\alpha_1 a)} \quad , \tag{2.11.114}
$$

$$\eta_2 = \frac{K_\nu'(\alpha_2 a)}{\alpha_2 a K_\nu(\alpha_2 a)} \quad . \tag{2.11.115}$$

Equation (2.11.108) can be written as

$$a^4 \left(\frac{n_1^2}{n_2^2} \alpha_2^2 \eta_1 + \alpha_2^2 \eta_2 \right) (\alpha_2^2 \eta_1 + \alpha_2^2 \eta_2)$$

$$= \left[\nu \left(\frac{n_1^2}{n_2^2} - 1 \right) \frac{k k_2}{\alpha_1^2} \right]^2 \quad . \tag{2.11.116}$$

Equation (2.11.116) determines the value of k, the propagation constant for the hybrid mode. The quantity α_1 enters (2.11.116) both explicitly and as the argument of η_1. As η_1 is a rapidly varying oscillatory function of $\alpha_1 a$, (2.11.116) can be considered roughly as a quadratic equation in η_1. The two sets of solutions are the two sets of hybrid modes denoted by HE_{mn} and EH_{mn}. From the set of four boundary condition equations, R, the relative amount of E_z and H_z in a hybrid mode, can be found as

$$R = \left(\frac{E_z}{H_z} \right) \propto \frac{\nu(n_1^2/n_2^2 - 1)1/\alpha_1^2}{\eta_1 + \eta_2} \quad . \tag{2.11.117}$$

The solution of (2.11.116) to obtain η_1 is quite complex. For convenience, let us rewrite (2.11.116) in terms of two new variables ξ_1 and ξ_2 as follows:

$$\xi_1 = \frac{J_{\nu-1}}{\alpha_1 a J_\nu} \quad , \tag{2.11.118}$$

$$\xi_2 = \frac{K_{\nu-1}}{\alpha_2 a K_\nu} \quad , \tag{2.11.119}$$

$$\xi_1^2 - \xi_1 \left[\frac{n_1^2 + n_2^2}{n_1^2} \xi_2 + \nu \left(\frac{2}{\alpha_1^2 a^2} - \frac{n_1^2 + n_2^2}{n_1^2} \frac{1}{\alpha_2^2 a^2} \right) \right]$$

$$+ \left[\frac{n_2^2}{n_1^2} \xi_2^2 + \xi_2 \nu \left(\frac{n_1^2 + n_2^2}{n_1^2} \frac{1}{\alpha_1^2 a^2} - 2 \frac{n_2^2}{n_1^2} \frac{1}{\alpha_2^2 a^2} \right) \right] = 0 \quad . \tag{2.11.120}$$

For $\nu = 1$ and in the limit $(\alpha_2 a)^2 \rightarrow 0$, the two roots are

$$\xi_1 = \left(\frac{n_1^2 + n_2^2}{n_1^2} \right) \frac{1}{\alpha_2^2 a^2} \longrightarrow \infty \quad , \tag{2.11.121}$$

$$\xi_1 = \frac{2 n_2^2}{n_1^2 + n_2^2} \ln \left(\frac{2}{\alpha_2 a} \right) \rightarrow \infty \quad . \tag{2.11.122}$$

Both the above equations have a cutoff given by

$$J_1(\alpha_1 a) = 0 \quad . \tag{2.11.123}$$

The root of (2.11.123) at $\alpha_1 a = 0$ corresponds to the well-known HE_{11} mode

Table 2.3. Summary of cutoff conditions. The Bessel function of order n and argument u is given by $J_n(u)$, and n_1 and n_2 are the refractive indices for core and cladding. R gives the relative amount of H_z to E_z in a mode. (From [2.3])

First set of solutions			Second set of solutions		
Cutoff condition	R at cutoff	Mode designation	Cutoff condition	R at cutoff	Mode designation
$n=0$ $\quad J_0(u)=0$	0	$TM_{0m}\ m=1,2\ldots$	$J_0(u)=0$	∞	$TE_{0m}\ m=1,2\ldots$
$n=1$ $\quad J_1(u)=0$	-1	$HE_{1m}\ m=1,2\ldots$	$J_1(u)=0$	n_1^2/n_2^2	$EH_{1m}\ m=1,2\ldots$
$n\geq2$ $\quad \dfrac{uJ_{n-2}(u)}{J_{n-1}(u)}=-(n-1)\dfrac{n_1^2-n_2^2}{n_2^2}$	-1	$HE_{nm}\ m=1,2\ldots$	$J_n(u)=0$	n_1^2/n_2^2	$EH_{nm}\ m=1,2\ldots$

which does not have a cutoff. Note that (2.11.123) specifies two sets of modes whose cutoff conditions are identical. HE_{11} is the first mode of only one of the sets.

For $\nu \geq 2$, from (2.11.121) for $(\alpha_2 a)^2 \to 0$ one obtains

$$\xi_1 = \left(\frac{1}{\nu} - 1\right)\left(\frac{n_2^2}{n_1^2 + n_2^2}\right) \quad , \tag{2.11.124}$$

$$\xi_1 = \nu\left(\frac{n_1^2 + n_2^2}{n_1^2}\right)\frac{1}{\alpha_2^2 a^2} \to \infty \quad . \tag{2.11.125}$$

For $(n_2^2 - n_1^2) \to 0$, (2.11.125) becomes equivalent to $J_{\nu-2}(\alpha_1 a) = 0$. For $\nu = 2$, this gives another set of modes whose cutoffs are close to those of TE_{0m} and TM_{0m} modes. Table 2.3 lists some properties of these modes.

Let $(\alpha_1 a)_{nm}$ be the value $\alpha_1 a$ assumed at cutoff for the mth root of the cutoff condition involving the nth-order Bessel function. At cutoff $\alpha_2 a = 0$ and $k = n_2 k_0$ and we have

$$(\alpha_1 a)_{nm} = \frac{2\pi}{a}\lambda_0(n_1^2 - n_2^2)^{1/2} = V \quad . \tag{2.11.126}$$

The modes which can propagate are those for which $(\alpha_1 a)_{nm}$ is less than V. Hence $(\alpha_1 a)_{nm}$ forms an increasing sequence for fixed n and increasing m or for fixed m and increasing n. Thus the number of allowed modes increases as the square of the radius a. Figure 2.51 shows the number of propagating modes versus V. We note that for $V < 2.405$ we get a single-mode fiber, i.e., only the HE_{11} mode propagates.

Another simplification is obtained by considering weakly guiding fibers or the situation for the case $n_1 \to n_2$. For this case, (2.11.116) simplifies and the propagation constants k_{mn} can be easily obtained. One important point about this simplification is that HE modes of order $\nu = \nu' + 1$ are almost degenerate with EH modes of $\nu = \nu' - 1$. Linear combination of these HE and EH modes can result in linearly polarized (LP) modes. These LP modes have only four field components (E_x, H_y, E_z and H_z, or E_y, H_x, E_z and H_z) rather than the six components discussed earlier.

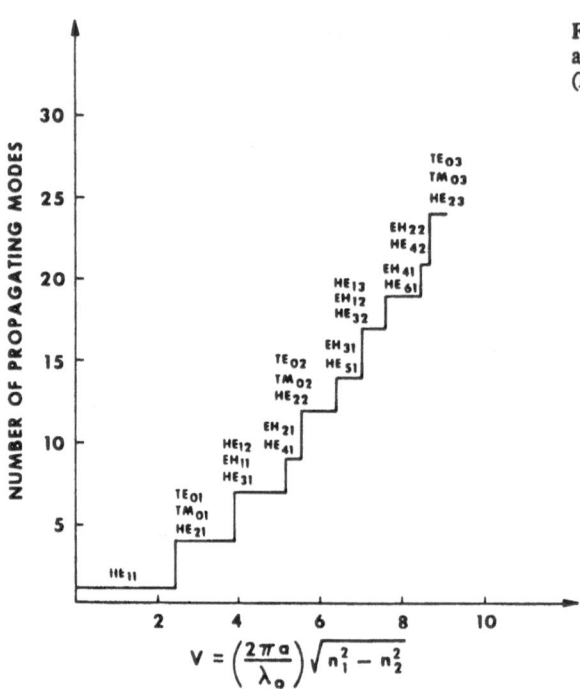

Fig. 2.51. Plot of number of propagating modes vs fiber V number. (From [2.14])

To understand LP modes, we note that for $n_2 \to n_1$ the waves in the guide propagate at small angles with respect to z and we can construct modes whose transverse fields are essentially polarized. Thus for $r < a$ we assume (for the case of E_y, H_x, E_z, and H_z)

$$\left.\begin{aligned}
E_y &= \frac{A J_\nu(\alpha_1 r)}{J_\nu(\alpha_1 a)} \cos \nu\phi \\
H_x &= \frac{E_y}{Z_1} = \frac{E_y}{\sqrt{\mu_0/\varepsilon_1}}
\end{aligned}\right\} \quad r < a \quad, \tag{2.11.127}$$

$$\left.\begin{aligned}
E_y &= \frac{A K_\nu(\alpha_2 r)}{K_\nu(\alpha_2 a)} \cos \nu\phi \\
H_x &= \frac{E_y}{Z_2} = \frac{E_y}{\sqrt{\mu_0/\varepsilon_2}}
\end{aligned}\right\} \quad r > a \quad, \tag{2.11.128}$$

The components E_z and H_z are obtained from Maxwell's equations and can be written

$$E_z = \frac{j}{\omega\varepsilon_1} \frac{\partial H_x}{\partial y} \quad r < a \quad, \tag{2.11.129}$$

$$E_z = \frac{j}{\omega\varepsilon_2} \frac{\partial H_x}{\partial y} \quad r > a \quad, \quad \text{and} \tag{2.11.130}$$

$$H_z = \frac{j}{\omega\mu_0} \frac{\partial E_y}{\partial x} \quad . \tag{2.11.131}$$

Applying the boundary conditions, one obtains the following characteristic equation, which is much simpler than the exact one given by (2.11.116):

$$\alpha_1 a \left(\frac{J_{\nu-1}(\alpha_1 a)}{J_\nu(\alpha_1 a)} \right) = \alpha_2 a \left(\frac{K_{\nu-1}(\alpha_2 a)}{K_\nu(\alpha_2 a)} \right) \quad . \tag{2.11.132}$$

Numerically, it is found that for $\Delta = (n_1 - n_2)/n_1 < 0.1$ the result is accurate within 1 %. The cutoff condition is obtained from (2.11.132) by demanding $\alpha_2 a = 0$. For this case we have

$$J_{\nu-1}(\alpha_1 a) = 0 \quad . \tag{2.11.133}$$

For the lowest order mode, $\nu = 0$, we have

$$J_{-1}(\alpha_1 a) = -J_1(\alpha_1 a) = 0 \quad . \tag{2.11.134}$$

As the first root of the above equation is zero, the lowest-order mode LP_{01} has no cutoff and represents approximately the mode HE_{11} we discussed before. In general, the modes are denoted by $LP_{\nu m}$ and the cutoff frequency corresponds to the mth root of the corresponding νth-order Bessel function.

In the following, we note some important results without derivation.

i) The approximate total number of modes that can exist in a stepped-index fiber is given by $N = 4V^2/\pi^2$.

ii) The power density distribution of $LP_{\nu m}$ is given by

$$P \propto \begin{cases} (\cos^2 \nu\phi) J_\nu^2(\alpha_{1m}r) & r \le a \quad , \\ (\sin^2 \nu\phi) J_\nu^2(\alpha_{1m}r) & r \le a \quad , \\ (\cos^2 \nu\phi) \dfrac{J_\nu^2(\alpha_{1m}a)}{K_\nu^2(\alpha_{2m}a)} K_\nu^2(\alpha_{2m}r) & r > a \quad , \\ (\sin^2 \nu\phi) \dfrac{J_\nu^2(\alpha_{1m}a)}{K_\nu^2(\alpha_{2m}a)} K_\nu^2(\alpha_{2m}r) & r > a \quad . \end{cases} \tag{2.11.135}$$

Normalized intensity plots are shown in Fig. 2.52 for LP_{01} (HE_{11}), LP_{11}, and LP_{21} modes for two cases: frequency far away from cutoff and frequency very near cutoff.

iii) The total power flowing in the core (P_{core}) and that in the cladding (P_{clad}) are given by

$$P_{core} = D \left[1 + \left(\frac{\alpha_2}{\alpha_1} \right)^2 \frac{1}{E} \right] \quad , \tag{2.11.136}$$

$$P_{clad} = D \left(\frac{1}{E} - 1 \right) \quad , \tag{2.11.137}$$

$$P_{core} + P_{clad} = \frac{D}{E} \left(\frac{V^2}{\alpha_1^2 a^2} \right) \quad ; \tag{2.11.138}$$

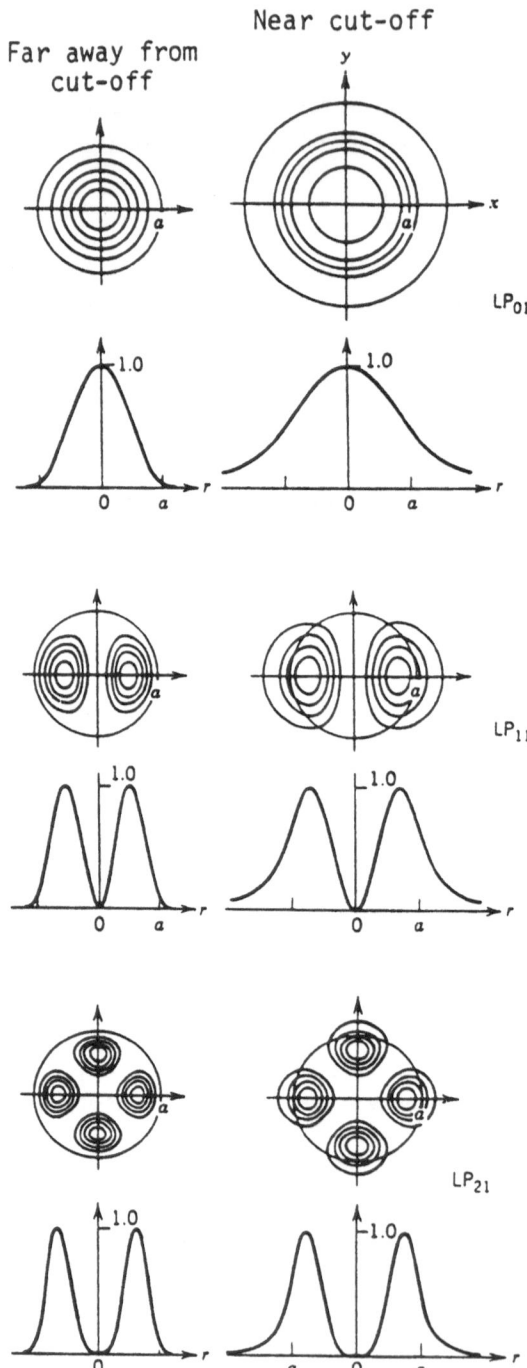

Fig. 2.52. Normalized intensity plot for several LP modes for frequencies far from and near cutoff. (From [2.15])

$$D = \frac{[K_\nu(\alpha_2 a)]^2}{K_{\nu+1}(\alpha_2 a)K_{\nu-1}(\alpha_2 a)} \quad ,$$

$$E = \frac{\pi n_2 A^2}{4Z_0} \quad ,$$

where A^2 is the proportionality constant for (2.11.135),

$$\frac{P_{clad}}{P_{tot}} = \frac{\alpha_1^2}{V^2}(1 - E) \quad . \tag{2.11.139}$$

Very near cutoff, (2.11.139) becomes

$$\frac{P_{clad}}{P_{tot}} \approx \frac{1}{\sqrt{\nu^2 + 1}} \quad . \tag{2.11.140}$$

Figure 2.53 plots (2.11.139) for different modes with V as abscissa.

iv) Note that even for single mode (HE_{11}) propagation, two orthogonal polarization modes are possible. These polarization modes have identical characteristics except for their electric and magnetic field directions.

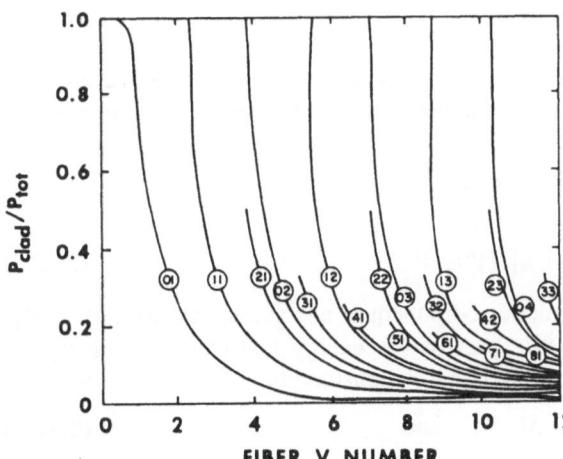

Fig. 2.53. Plot of P_{clad}/P_{tot} vs fiber V number. (From [2.14])

2.12 Propagation in Anisotropic Media

Most optical devices use light propagation in anisotropic media. The fundamentals of this propagation leading to the index ellipsoid are considered first. This is followed by discussion of group velocity in anisotropic media, double refraction, and internal conical refraction.

To consider light propagation in crystals, we must consider the dielectric tensor given by (2.1.5) and repeated here:

$$D_k = \varepsilon_{kl} E_l \quad , \quad \text{or}$$

$$
\begin{pmatrix} D_x \\ D_y \\ D_z \end{pmatrix} = \begin{pmatrix} \varepsilon_{xx} & \varepsilon_{xy} & \varepsilon_{xz} \\ \varepsilon_{yx} & \varepsilon_{yy} & \varepsilon_{yz} \\ \varepsilon_{zx} & \varepsilon_{zy} & \varepsilon_{zz} \end{pmatrix} \begin{pmatrix} E_x \\ E_y \\ E_z \end{pmatrix} \quad .
\tag{2.12.1}
$$

The most general case has 9 independent elements ε_{ij}. However, using arguments about stored energy and Poynting vector, we shall show that the dielectric tensor is real and symmetric. (See also Sect. 2.7.1.) That means

$$\varepsilon_{kl} = \varepsilon_{kl}^* \quad , \quad \text{and} \tag{2.12.2}$$

$$\varepsilon_{kl} = \varepsilon_{lk} \quad . \tag{2.12.3}$$

Thus, in the most general case, we have only six independent elements.

The stored electric energy W_e in a medium is given by

$$W_e = \tfrac{1}{2} \boldsymbol{E} \cdot \boldsymbol{D} = \tfrac{1}{2} E_k \varepsilon_{kl} E_l \quad . \tag{2.12.4}$$

Note that here we are using the convention of summation over repeated indices. If we take the derivative of W_e with respect to time we obtain

$$
\begin{aligned}
\dot{W}_e &= \tfrac{1}{2} \dot{E}_k \varepsilon_{kl} E_l + \tfrac{1}{2} E_k \varepsilon_{kl} \dot{E}_l \\
&= \tfrac{1}{2} \varepsilon_{kl} (\dot{E}_k E_l + E_k \dot{E}_l) \quad .
\end{aligned}
\tag{2.12.5}
$$

The Poynting vector $\boldsymbol{P}$ denotes the total power flow through a surface. If we consider a small volume ΔV, then the net power lost from the volume is given by

$$
\begin{aligned}
-\nabla \cdot \boldsymbol{P} \Delta V &= [-\nabla \cdot (\boldsymbol{E} \times \boldsymbol{H})] \Delta V \\
&= (\boldsymbol{E} \cdot \dot{\boldsymbol{D}} + \boldsymbol{H} \cdot \dot{\boldsymbol{B}}) \Delta V \\
&= (E_k \varepsilon_{kl} \dot{E}_l + \mu |\dot{H}|^2) \Delta V \quad .
\end{aligned}
\tag{2.12.6}
$$

We have assumed that the magnetic component of energy W_m is given by $\mu |H|^2$ when μ is isotropic. The electric part of the energy lost calculated using the Poynting vector and using the stored energy considerations must be equal. Thus

$$-\nabla \cdot \boldsymbol{P} \Delta V = (\dot{W}_e + \dot{W}_m) \Delta V \quad , \quad \text{and} \tag{2.12.7}$$

$$E_k \varepsilon_{kl} \dot{E}_l = \tfrac{1}{2} \varepsilon_{kl} (\dot{E}_k E_l + E_k \dot{E}_l) \quad . \tag{2.12.8}$$

The above equation cannot be satisfied unless

$$\varepsilon_{kl} = \varepsilon_{lk} \quad \text{and} \quad \varepsilon_{kl} = \varepsilon_{kl}^* \quad .$$

Thus, in the most general case, (2.12.1) becomes

$$
\begin{pmatrix} D_x \\ D_y \\ D_z \end{pmatrix} = \begin{pmatrix} \varepsilon_{xx} & \varepsilon_{xy} & \varepsilon_{xz} \\ \varepsilon_{yx} & \varepsilon_{yy} & \varepsilon_{yz} \\ \varepsilon_{zx} & \varepsilon_{zy} & \varepsilon_{zz} \end{pmatrix} \begin{pmatrix} E_x \\ E_y \\ E_z \end{pmatrix} \quad .
\tag{2.12.9}
$$

The axes we have chosen for x, y, and z are not unique. We can choose a new set of axes, represented by x', y', z'. In this new coordinate system, the symmetrical real dielectric matrix can be always made to be diagonal. Thus in this new system, denoted here by x, y, and z again for simplicity from now on, (2.12.9) simplifies to

$$
\begin{pmatrix} D_x \\ D_y \\ D_z \end{pmatrix} = \begin{pmatrix} \varepsilon_x & 0 & 0 \\ 0 & \varepsilon_y & 0 \\ 0 & 0 & \varepsilon_z \end{pmatrix} \begin{pmatrix} E_x \\ E_y \\ E_z \end{pmatrix} \quad . \tag{2.12.10}
$$

The axes in the crystal along which the dielectric matrix becomes diagonal are called the principal axes. In this coordinate system, W_e becomes

$$
\begin{aligned}
W_e &= \frac{1}{2} \boldsymbol{E} \cdot \boldsymbol{D} = \frac{1}{2}(\varepsilon_x E_x^2 + \varepsilon_y E_y^2 + \varepsilon_z E_z^2) \\
&= \frac{1}{2}\left(\frac{D_x^2}{\varepsilon_x} + \frac{D_y^2}{\varepsilon_y} + \frac{D_z^2}{\varepsilon_z}\right) \quad .
\end{aligned} \tag{2.12.11}
$$

Thus we have along the principal axes

$$
\frac{D_x^2}{\varepsilon_x} + \frac{D_y^2}{\varepsilon_y} + \frac{D_z^2}{\varepsilon_z} = 2W_e \quad . \tag{2.12.12}
$$

The above equation is generally known as the index ellipsoid because the contours of constant W_e plotted in D_x, D_y, and D_z space yield an ellipsoid. Some values of ε_x, ε_y, and ε_z appropriate for different crystals are given in Table 2.4.

Table 2.4. Dielectric constants for various crystals (at $\lambda \sim 0.5\,\mu m$)

	$\varepsilon_x/\varepsilon_0$	$\varepsilon_y/\varepsilon_0$	$\varepsilon_z/\varepsilon_0$
KDP	2.28	2.28	2.16
LiNbO$_3$	5.24	5.24	4.84
LiTaO$_3$	4.73	4.73	4.75
BaTiO$_3$	5.94	5.94	6.94

We are interested in electromagnetic wave propagation in an anisotropic dielectric medium. Let us define the unit vector along the direction of propagation i_k. Thus

$$
\exp[j(\omega t - \boldsymbol{k} \cdot \boldsymbol{r})] = \exp\left[j\omega\left(t - \frac{n}{c}\boldsymbol{r} \cdot i_k\right)\right] \quad . \tag{2.12.13}
$$

Note that phase velocity v_p is given by

$$
v_p = \frac{c}{n} i_k \quad . \tag{2.12.14}
$$

For isotropic media, $n = \sqrt{\varepsilon}$. However, for the anisotropic problem, the effective value of the refractive index n has to be determined.

From Maxwell's equations, using (2.12.13), we find

$$\nabla \times H = \frac{\partial D}{\partial t} \quad \text{yields} \quad j\omega D = -j\omega \frac{n}{d} i_k \times H \quad . \tag{2.12.15}$$

As $\nabla \times E = -\partial B/\partial t$, we have

$$-j\omega \mu H = -j\omega \frac{n}{d} i_k \times E \quad . \tag{2.12.16}$$

Thus

$$H = \frac{n}{c\mu} i_k \times E \quad , \quad \text{and} \tag{2.12.17}$$

$$D = -\frac{n}{c} i_k \times H \quad , \quad \text{or} \tag{2.12.18}$$

$$D = -\frac{n^2}{c^2\mu} i_k \times (i_k \times E) = n^2 \varepsilon_0 i_k \times (i_k \times E) \quad . \tag{2.12.19}$$

Using the vector identity

$$A \times (B \times C) = B(A \cdot C) - C(A \cdot B) \tag{2.12.20}$$

in (2.12.19) we obtain

$$D = n^2 \varepsilon_0 [E - i_k(i_k \cdot E)] \quad . \tag{2.12.21}$$

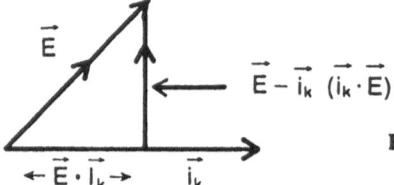

Fig. 2.54. Construction for (2.12.22)

Using Fig. 2.54, we note that

$$D = n^2 \varepsilon_0 E_\perp \quad , \tag{2.12.22}$$

where $E_\perp$ is the component of the electric field perpendicular to i_k. We also note that D, i_k and E are coplanar, i.e., they all lie the same plane. From the figure, we also observe that D is perpendicular to i_k.

Writing (2.12.21) in expanded form we have for the x component

$$\varepsilon_x E_x = n^2 \varepsilon_0 [E_x - i_{kx}(i_k \cdot E)] \quad , \tag{2.12.23}$$

$$E_x(n^2 \varepsilon_0 - \varepsilon_x) = n^2 \varepsilon_0 i_{kx}(i_k \cdot E) \quad ,$$

$$E_x = \frac{n^2 i_{kx}(i_k \cdot E)}{n^2 - \varepsilon_x'} \quad , \tag{2.12.24}$$

where we have defined $\varepsilon_x' = \varepsilon_x/\varepsilon_0$. Multiplying both sides of (2.12.24) by i_{kx} we have

$$i_{kx}E_x = \frac{n^2 i_{kx}^2}{n^2 - \varepsilon_x'}(i_{kx}E_x + i_{ky}E_y + i_{kz}E_z) \quad . \tag{2.12.25}$$

Similarly we have

$$i_{ky}E_y = \frac{n^2 i_{ky}^2}{n^2 - \varepsilon_y'}(i_{kx}E_x + i_{ky}E_y + i_{kz}E_z) \quad , \quad \text{and} \tag{2.12.26}$$

$$i_{kz}E_z = \frac{n^2 i_{kz}^2}{n^2 - \varepsilon_z'}(i_{kx}E_x + i_{ky}E_y + i_{kz}E_z) \quad . \tag{2.12.27}$$

If we add up the last three equations we obtain

$$\boldsymbol{i}_k \cdot \boldsymbol{E} = \left(\frac{n^2 i_{kx}^2}{n^2 - \varepsilon_x'} + \frac{n^2 i_{ky}^2}{n^2 - \varepsilon_y'} + \frac{n^2 i_{kz}^2}{n^2 - \varepsilon_z'} \right) \boldsymbol{i}_k \cdot \boldsymbol{E} \quad \text{or}$$

$$\frac{i_{kx}^2}{n^2 - \varepsilon_x'} + \frac{i_{ky}^2}{n^2 - \varepsilon_y'} + \frac{i_{kz}^2}{n^2 - \varepsilon_z'} = \frac{1}{n^2} \quad . \tag{2.12.28}$$

Equation (2.12.28) is known as Fresnel's equation. Note that it is fourth order in n and not sixth order, because if we multiply out we find

$$n^2 i_{kx}^2 (n^2 - \varepsilon_y')(n^2 - \varepsilon_z') + n^2 i_{ky}^2 (n^2 - \varepsilon_x')(n^2 - \varepsilon_z')$$

$$+ n^2 i_{kz}^2 (n^2 - \varepsilon_x')(n^2 - \varepsilon_y') = (n^2 - \varepsilon_x')(n^2 - \varepsilon_y')(n^2 - \varepsilon_z') \quad . \tag{2.12.29}$$

If we collect all the terms with n^6 we find on the left-hand side

$$n^6(i_{kx}^2 + i_{ky}^2 + i_{kz}^2) = n^6 \quad . \tag{2.12.30}$$

The right-hand side also has n^6 and thus those terms cancel each other. Solving Fresnel's equation, we obtain

$$n = \pm n_1 \quad \text{or} \quad \pm n_2 \quad . \tag{2.12.31}$$

Thus, in general, the effective refractive index in any direction has two values n_1 and n_2, which can be found by solving Fresnel's equation for n.

The problem of wave propagation in crystals, although it can be solved using Fresnel's equation, is in general easier to address using the index ellipsoid. We shall describe below how this is done before we prove the results later.

The equation of the index ellipsoid given by (2.12.12) can be rewritten as

$$\frac{x^2}{\varepsilon_x'} + \frac{y^2}{\varepsilon_y'} + \frac{z^2}{\varepsilon_z'} = 1 \quad , \tag{2.12.32}$$

where we have substituted

$$\frac{\boldsymbol{D}}{\sqrt{2W_e\varepsilon_0}} = \boldsymbol{r} = i_x x + i_y y + i_z z \quad . \tag{2.12.33}$$

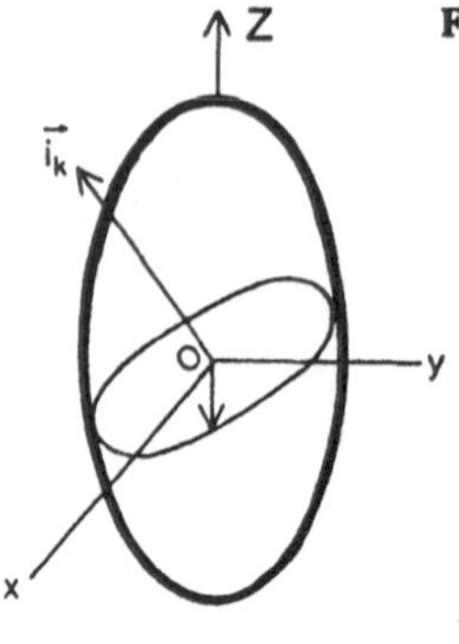

Fig. 2.55. Plot of (2.12.32)

Equation (2.12.32) is plotted in Fig. 2.55. To obtain the value of n_1 and n_2 for a particular direction of propagation i_k, find the plane passing through the origin of the ellipsoid and perpendicular to i_k. The intersection of this plane and the index ellipsoid gives us an ellipse. The two major axes of this ellipse correspond to $2n_1$ and $2n_2$ respectively. The corresponding D_1 and D_2 are parallel to these major axes of the ellipse.

To prove the result discussed above we note that the ellipse is formed by the intersection of (2.12.32) and

$$\boldsymbol{r} \cdot \boldsymbol{i}_k = x i_{kx} + y i_{kx} + z i_{kx} = 0 \quad . \tag{2.12.34}$$

The principal semi-axes of this ellipse are given by the extrema of

$$r^2 = x^2 + y^2 + z^2 \tag{2.12.35}$$

subject to satisfying (2.12.32 and 34). This is a problem which can be solved using the Lagrange multiplier method. Form a function F defined as

$$F = x^2 + y^2 + z^2 + \lambda_1 (x i_{kx} + y i_{ky} + z i_{kz})$$
$$+ \lambda_2 \left(\frac{x^2}{\varepsilon'_x} + \frac{y^2}{\varepsilon'_y} + \frac{z^2}{\varepsilon'_z} - 1 \right) \quad , \tag{2.12.36}$$

where λ_1 and λ_2 are arbitrary constants. For extrema, we note that

$$\frac{\partial F}{\partial x} = \frac{\partial F}{\partial y} = \frac{\partial F}{\partial z} = 0 \quad . \tag{2.12.37}$$

The method of Lagrange multipliers demands that

$$\frac{\partial F}{\partial \lambda_1} = \frac{\partial F}{\partial \lambda_2} = 0 \quad . \tag{2.12.38}$$

Using these conditions, we obtain

$$2x + \lambda_1 i_{kx} + \frac{2\lambda_2}{\varepsilon'_x} x = 0 \tag{2.12.39}$$

or, multiplying this by $i_{kx}/2$,

$$i_{kx}x + \frac{\lambda_1}{2}i_{kx}^2 + \frac{\lambda_2}{\varepsilon_x'}xi_{kx} = 0 \quad .$$

(2.12.40)

Similarly,

$$i_{ky}y + \frac{\lambda_1}{2}i_{ky}^2 + \frac{\lambda_2}{\varepsilon_y'}yi_{ky} = 0 \quad ,$$

(2.12.41)

$$i_{kz}z + \frac{\lambda_1}{2}i_{kz}^2 + \frac{\lambda_2}{\varepsilon_z'}zi_{kz} = 0 \quad .$$

(2.12.42)

If we add the above three equations and use (2.12.34),

$$\frac{\lambda_1}{2} + \lambda_2\left(\frac{i_{kx}x}{\varepsilon_x'} + \frac{i_{ky}y}{\varepsilon_y'} + \frac{i_{kz}z}{\varepsilon_z'}\right) = 0 \quad .$$

(2.12.43)

From (2.12.40–43) we find

$$x^2 + \frac{\lambda_1}{2}i_{kx}x + \frac{\lambda_2}{\varepsilon_x'}x^2 = 0 \quad ,$$

(2.12.44)

$$y^2 + \frac{\lambda_1}{2}i_{ky}y + \frac{\lambda_2}{\varepsilon_y'}y^2 = 0 \quad ,$$

(2.12.45)

$$z^2 + \frac{\lambda_1}{2}i_{kz}z + \frac{\lambda_2}{\varepsilon_z'}z^2 = 0 \quad .$$

(2.12.46)

Adding the above three equations gives

$$r^2 + \lambda_2\left(\frac{x^2}{\varepsilon_x'} + \frac{y^2}{\varepsilon_y'} + \frac{z^2}{\varepsilon_z'}\right) = 0$$

(2.12.47)

$$r^2 + \lambda_2 = 0 \quad , \quad \text{or}$$

$$\lambda_2 = -r^2 \quad .$$

(2.12.48)

Substituting this value of λ_2 in (2.12.43) we have

$$\frac{\lambda_2}{2} = r^2\left(\frac{i_{kx}x}{\varepsilon_x'} + \frac{i_{ky}y}{\varepsilon_y'} + \frac{i_{kz}z}{\varepsilon_z'}\right) \quad .$$

(2.12.49)

Substituting for the values λ_1 and λ_2 in (2.12.39) and rearranging, we have

$$x\left(1 - \frac{r^2}{\varepsilon_x'}\right) + r^2i_{kx}\left(\frac{i_{kx}x}{\varepsilon_x'} + \frac{i_{ky}y}{\varepsilon_y'} + \frac{i_{kz}z}{\varepsilon_z'}\right) = 0 \quad .$$

(2.12.50)

As

$$r = \frac{D}{\sqrt{2\varepsilon_0 W_e}} \quad ,$$

(2.12.51)

we note

$$r^2 = \frac{|D|^2}{2\varepsilon_0 W_e} = \frac{|D|^2}{2\varepsilon_0 E \cdot D} = n^2 \quad .$$

(2.12.52)

Also

$$\frac{x}{\varepsilon'_x} = \frac{E_x}{\sqrt{\varepsilon_0 \mathbf{E} \cdot \mathbf{D}}} \quad . \tag{2.12.53}$$

Thus (2.12.50) becomes

$$\varepsilon'_x E_x \left(1 - \frac{n^2}{\varepsilon'_x}\right) + i_{kx} n^2 (\mathbf{i}_k \cdot \mathbf{E}) = 0 \quad . \tag{2.12.54}$$

As this equation is identical to (2.12.24) used to obtain Fresnel's equation, it completes the proof.

We can thus use the index ellipsoid to obtain the phase velocities given by n_1 and n_2. Note that once n_1 and n_2 are known, we can calculate $\mathbf{D}_1$, $\mathbf{E}_1$, $\mathbf{D}_2$ and $\mathbf{E}_2$ using (2.12.21, 24). It is of interest to obtain some relationships between these quantities.

We note that

$$\mathbf{D}_1 \cdot \mathbf{D}_2 = 0 \quad . \tag{2.12.55}$$

Thus $\mathbf{D}_1$ is perpendicular to $\mathbf{D}_2$. As mentioned before, $\mathbf{D}_1$ and $\mathbf{D}_2$ are parallel to the major axes of the ellipse. We now give a formal proof of this result.

$$\begin{aligned}
\mathbf{E}_2 \cdot \mathbf{D}_1 &= \sum_k E_{2k} D_{1k} \\
&= \sum_k \sum_l E_{2k} \varepsilon_{kl} E_{1kl} \\
&= \sum_l D_{2l} E_{1l} \\
&= \mathbf{E}_1 \cdot \mathbf{D}_2 \quad .
\end{aligned} \tag{2.12.56}$$

However, from (2.12.22),

$$\mathbf{E}_2 \cdot \mathbf{D}_1 = \mathbf{E}_2 \cdot (n_1^2 \varepsilon_0 \mathbf{E}_{1\perp}) \quad ,$$

$$\mathbf{E}_1 \cdot \mathbf{D}_2 = \mathbf{E}_1 \cdot (n_2^2 \varepsilon_0 \mathbf{E}_{2\perp}) \quad . \tag{2.12.57}$$

Subtracting the above two equations and using (2.12.56), we have

$$(n_1^2 - n_2^2)(\mathbf{E}_{2\perp} \cdot \mathbf{E}_{1\perp}) = 0 \quad .$$

Thus $\mathbf{E}_{2\perp}$ is perpendicular to $\mathbf{E}_{1\perp}$, as $n_1 \neq n_2$. From (2.12.22)

$$\mathbf{D}_2 \perp \mathbf{D}_1 \quad .$$

In an anisotropic crystal, the phase velocity is not equal to the group velocity. It is therefore quite possible that the phase and group velocity are not colinear, which is actually the usual case for such crystals. The magnetic field vector $\mathbf{H}$ is given by

$$\mathbf{H} = \frac{n}{\mu c} \mathbf{i}_k \times \mathbf{E} \quad . \tag{2.12.58}$$

Thus H is perpendicular to the E field and to the direction of propagation. The Poynting vector P is given by

$$P = E \times H \quad .$$

Thus P is perpendicular to E and H, and is also in the plane defined by i_k, E, and D. However, it is not parallel to i_k:

$$P = E \times H$$
$$= \frac{n}{\mu c} [E \times (i_k \times E)] \quad .$$

Using the vector identity, we obtain

$$P = \frac{n}{\mu c} [i_k (E \cdot E) - E(i_k \cdot E)] \quad .$$

Denoting $E = |E|e$, where e denotes the unit vector along the electric field direction, we obtain

$$P = \frac{n}{\mu c} |E|^2 [i_k - e(i_k \cdot e)] \quad . \tag{2.12.59}$$

Using Fig. 2.54, we can see that

$$P = \frac{n}{\mu c} |E|^2 i_{k\perp} \quad , \tag{2.12.60}$$

where $i_{k\perp}$ is the component of i_k perpendicular to the direction of the E field. Thus we see that the direction of power flow, P, is not the same as the phase velocity direction. As power flow is related to the group velocity we have, see (2.7.35),

$$v_p = v_g \cos \psi \quad , \tag{2.12.61}$$

where ψ is the angle between i_k and P.

Another important result to note is that E is normal to the tangential plane at the intersection of the D vector and the index ellipsoid. This is easily seen by observing that the direction parallel to the normal to the tangential plane is given by

$$n \propto \nabla f = i_x \frac{\partial f}{\partial x} + i_y \frac{\partial f}{\partial y} + i_z \frac{\partial f}{\partial z} \quad , \quad \text{where} \tag{2.12.62}$$

$$f(x, y, z) = \frac{x^2}{\varepsilon'_x} + \frac{y^2}{\varepsilon'_y} + \frac{z^2}{\varepsilon'_z} - 1 = 0 \quad . \tag{2.12.63}$$

Thus

$$n \propto i_x \frac{x}{\varepsilon'_x} + i_y \frac{y}{\varepsilon'_y} + i_z \frac{z}{\varepsilon'_z} \quad . \tag{2.12.64}$$

Noting from (2.12.33) that

$$\frac{x}{\varepsilon'_x} = \frac{D_x}{\sqrt{2W_e \varepsilon_0 \varepsilon'_x}} \propto E_x \quad ,$$

103

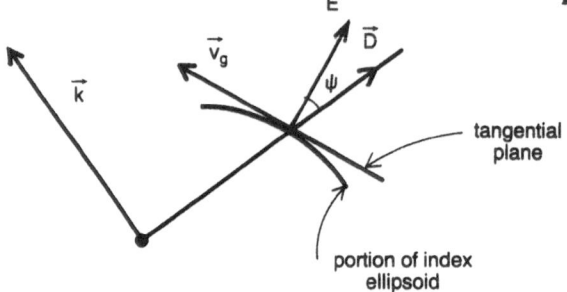

Fig. 2.56. Illustration of (2.12.65)

tangential
plane

portion of index
ellipsoid

we obtain

$$n \propto i_x E_x + i_y E_y + i_z E_z = E \quad . \tag{2.12.65}$$

This is schematically shown in Fig. 2.56. As the group velocity is in a direction normal to the E field, we note that all necessary values for D, E, and v_g can be obtained from the index ellipsoid.

2.12.1 Wave Vector Surface, Phase Velocity Surface, and Ray Velocity Surface

The wave vector surface and phase velocity surface of an anisotropic crystal can be obtained by Fresnel's equation (2.12.28) and noting that

$$|k| = \frac{\omega n}{c} \quad , \quad \text{and}$$

$$v_p = \frac{c}{n} \quad . \tag{2.12.66}$$

Thus the wave-vector surface is the four-dimensional plot of the following equation with k_x, k_y and k_z as axes:

$$\frac{k_x^2}{k^2 - \frac{\omega^2}{c^2}\varepsilon_x'} + \frac{k_y^2}{k^2 - \frac{\omega^2}{c^2}\varepsilon_y'} + \frac{k_z^2}{k^2 - \frac{\omega^2}{c^2}\varepsilon_z'} = 1 \quad . \tag{2.12.67}$$

However, more insight is obtained by writing (2.12.25, 26) as

$$E_x \left(\frac{\omega^2}{c^2}\varepsilon_x' - k_y^2 - k_z^2 \right) + E_y(k_x k_y) + E_z(k_x k_z) = 0 \quad , \tag{2.12.68}$$

$$E_x(k_y k_x) + E_y \left(\frac{\omega^2}{c^2}\varepsilon_y' - k_x^2 - k_z^2 \right) + E_z(k_y k_z) = 0 \quad , \tag{2.12.69}$$

$$sE_x(k_x k_z) + E_y(k_y k_z) + E_z \left(\frac{\omega^2}{c^2}\varepsilon_z' - k_x^2 - k_y^2 \right) = 0 \quad . \tag{2.12.70}$$

The following determinant has to be zero for the above equations to have a valid solution:

$$\begin{vmatrix} \frac{\omega^2}{c^2}\epsilon_x' - k_y^2 - k_z^2 & k_x k_y & k_x k_z \\ k_y k_x & \frac{\omega^2}{c^2}\epsilon_y' - k_x^2 - k_z^2 & k_y k_z \\ k_z k_x & k_z k_y & \frac{\omega^2}{c^2}\epsilon_z' - k_x^2 - k_y^2 \end{vmatrix} = 0 \quad . \quad (2.12.71)$$

Note that (2.12.71) and (2.12.67) are equivalent. If we consider the yz plane, then $k_x = 0$ and (2.12.71) simplifies to

$$\begin{vmatrix} \frac{\omega^2}{c^2}\epsilon_x' - k_y^2 - k_z^2 & 0 & 0 \\ 0 & \frac{\omega^2}{c^2}\epsilon_y' - k_z^2 & k_y k_z \\ 0 & k_y k_z & \frac{\omega^2}{c^2}\epsilon_z' - k_y^2 \end{vmatrix} = 0 \quad ,$$

or

$$\left(\frac{\omega^2}{c^2}\epsilon_x' - k_y^2 - k_z^2 \right) \left[\left(\frac{\omega^2}{c^2}\epsilon_y' - k_z^2 \right) \left(\frac{\omega^2}{c^2}\epsilon_z' - k_y^2 \right) - k_y^2 k_z^2 \right] = 0 \quad ,$$

or

$$k_y^2 + k_z^2 = \frac{\omega^2}{c^2}\epsilon_x' \quad , \quad \text{and} \tag{2.12.72}$$

$$\frac{k_y^2}{(\omega^2/c^2)\epsilon_z'} + \frac{k_z^2}{(\omega^2/c^2)\epsilon_y'} = 1 \quad . \tag{2.12.73}$$

Thus the wave-vector surface in the $k_y k_z$ plane consists of a circle given by (2.12.72) and an ellipse given by (2.12.73). The wave-vector surface is a double surface and it is expected from our previous result that n has two possible values, n_1 and n_2. A general plot of (2.12.71) is shown in Fig. 2.57. The two wave surfaces touch at one point, P, which defines the optic axis. Any anisotropic

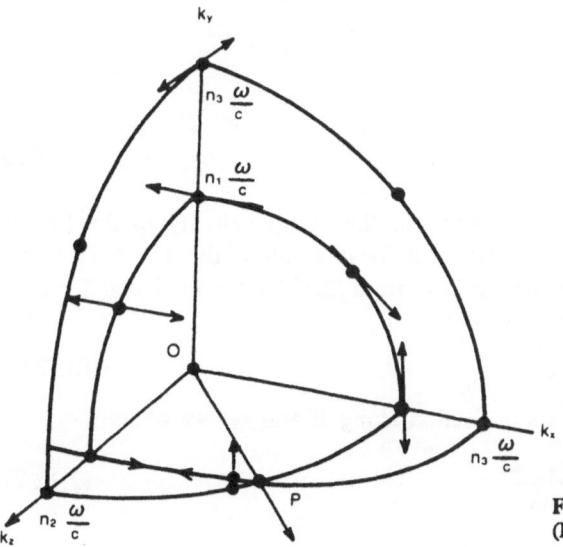

Fig. 2.57. The wave-vector surface. (From [2.16])

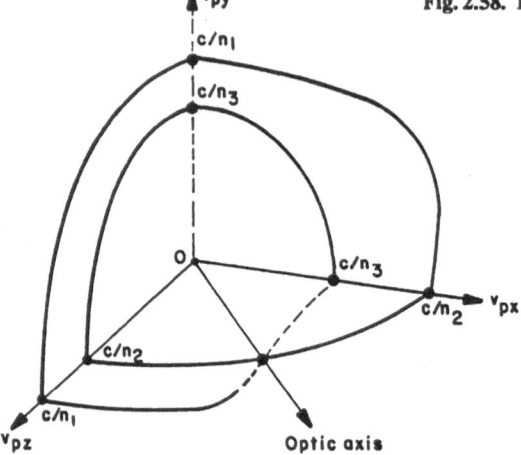

Fig. 2.58. The phase velocity surface. (From [2.16])

crystal will have only one value of n along the optic axis. If $\varepsilon'_x \neq \varepsilon'_y \neq \varepsilon'_z$, then the crystal will have two optic axes and it is then known as a biaxial crystal. If any of the two dielectric constants are the same, then the crystal is uniaxial and has only one optic axis. The phase velocity surface is given by

$$\begin{vmatrix} \varepsilon'_x v_p^4/c^2 - v_{py}^2 - v_{pz}^2 & v_{px} v_{py} & v_{px} v_{pz} \\ v_{py} v_{px} & \varepsilon'_y v_p^4/c^2 - v_{px}^2 - v_{pz}^2 & v_{py} v_{pz} \\ v_{pz} v_{px} & v_{pz} v_{py} & \varepsilon'_z v_p^4/c^2 - v_{px}^2 - v_{py}^2 \end{vmatrix} = 0 \quad .$$

(2.12.74)

The phase velocity surface is very similar to the wave-vector surface and is shown in Fig. 2.58. Again there are two surfaces and in the yz plane the equations are given by

$$v_{py}^2 + v_{pz}^2 = c^2/\varepsilon'_x \quad ,$$

$$\frac{v_{py}^2}{\varepsilon'_z} + \frac{v_{pz}^2}{\varepsilon'_y} = \frac{v_p^4}{c^2} \quad .$$

(2.12.75)

To obtain the ray velocity (v_e) surface or the group velocity in the lossless medium, we note that v_e is perpendicular to the direction of the E field (2.12.60). We also know that k is perpendicular to D from (2.12.22). If we denote the angle between E and D by ψ, then we have

$$v_p = v_e \cos \psi = v_g \cos \psi \quad .$$

(2.12.76)

If we express D in terms of its projection along E and v_e, we obtain

$$D = E \frac{|D|}{|E|} \cos \psi + v_e \frac{v_e \cdot D}{v_e^2} \quad .$$

(2.12.77)

From the wave equation, we have

$$k(k \cdot E) - k^2 E = -\frac{\omega^2}{c^2} D \quad . \tag{2.12.78}$$

As $k \cdot D = 0$, by taking the dot product with D we obtain

$$E \cdot D = ED \cos \psi$$

$$= \frac{v_p^2}{c^2} D^2$$

$$= \frac{v_e^2 \cos^2 \psi}{c^2} D^2 \quad . \tag{2.12.79}$$

Thus

$$\frac{D}{E} \cos \psi = \frac{c^2}{v_e^2} \quad . \tag{2.12.80}$$

Substituting (2.12.80) into (2.12.77)

$$D = E \frac{c^2}{v_e^2} + v_e \frac{v_e \cdot D}{v_e^2} \quad . \tag{2.12.81}$$

If the principal axes are the coordinate axes, D is related to E by (2.12.10) and we finally find

$$D_x(c^2/\varepsilon_x' - v_{ey}^2 - v_{ez}^2) + D_y v_{ex} v_{ey} + D_z v_{ex} v_{ez} = 0 \quad , \tag{2.12.82}$$

$$D_x v_{ey} v_{ey} + D_y(c^2/\varepsilon_y' - v_{ex}^2 v_{ez}^2) + D_2 v_{ey} v_{ez} = 0 \quad , \tag{2.12.83}$$

$$D_x v_{ez} v_{ex} + D_y v_{ey} v_{ez} + D_z(c^2/\varepsilon_z' - v_{ex}^2 v_{ey}^2) = 0 \quad . \tag{2.12.84}$$

For a valid solution, the determinant of the above equation system must be zero. Thus

$$\begin{vmatrix} c^2/\varepsilon_x' - v_{ey}^2 - v_{ez}^2 & v_{ex} v_{ey} & v_{ex} v_{ez} \\ v_{ey} v_{ex} & c^2/\varepsilon_y' - v_{ex}^2 - v_{ez}^2 & v_{ey} v_{ez} \\ v_{ez} v_{ex} & v_{ez} v_{ey} & c^2/\varepsilon_z' - v_{ex}^2 - v_{ey}^2 \end{vmatrix} = 0 \quad . \tag{2.12.85}$$

Equation (2.12.85) defines the ray velocity surface and a typical case is shown in Fig. 2.59. Along the direction of the ray axis the two velocities are equal. For a biaxial crystal there are two ray axes, which are different from the optic axes. However, for uniaxial crystals the ray axis and optic axis are identical. Moreover, the ray and phase velocities are the same along the principal axes.

One need not draw all three surfaces for each crystal in solving problems. All the useful quantities can be found from one surface only. For example, if we consider the wave-vector surface, then in any direction of propagation we find the two relevant refractive indices by drawing the propagation direction from the origin, as shown in Fig. 2.60. At the two points of intersection, we draw a tangent surface, the normal to which is the direction of ray velocity. The plane containing

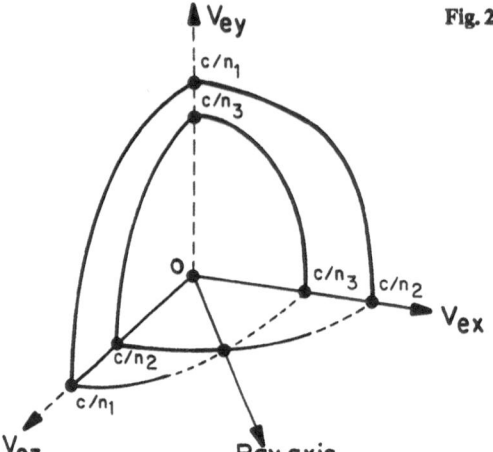

Fig. 2.59. The ray velocity surface. (From [2.16])

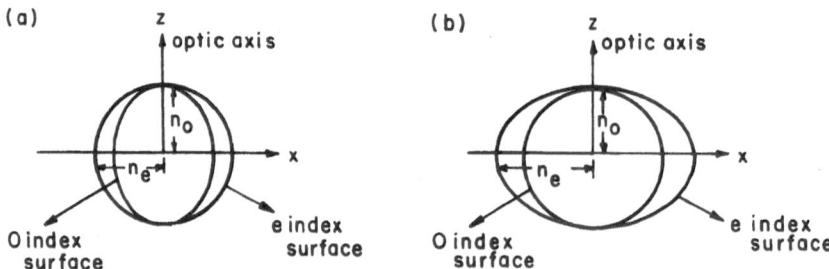

Fig. 2.60a,b. Index surfaces for a uniaxial crystal. (a) Negative uniaxial crystal ($n_e < n_o$). (b) Positive uniaxial crystal ($n_e > n_o$). (From [2.1])

the ray velocity and the phase velocity (direction of propagation) also contains E and D. The displacement D is perpendicular to the direction of propagation and E is perpendicular to v_e. The wave vector surface is very useful in the discussion of double refraction.

a) Group Velocity

It is of interest to discuss some explicit formulas for the group velocity v_g given by

$$v_g = \nabla_k \omega(k) \quad . \tag{2.12.86}$$

It can be shown that

$$v_g = \frac{c/n}{1 + \frac{\omega}{n} \frac{\partial n}{\partial \omega}} \left(i_k - \frac{1}{n} \nabla_s n \right) \quad , \quad \text{where} \tag{2.12.87}$$

$$\nabla_s n = i_\theta \frac{\partial n}{\partial \theta} + i_\phi \frac{1}{\sin \theta} \frac{\partial n}{\partial \phi} \quad . \tag{2.12.88}$$

In the expression for group velocity, the effect of frequency dispersion due to $\partial n/\partial \omega \neq 0$ is included. If one writes

$$v_g = v_g i_{v_g} \quad , \quad \text{where} \tag{2.12.89}$$

$$i_{v_g} \cdot i_k = \cos \psi \quad . \tag{2.12.90}$$

Then

$$\cos \psi = \frac{i_k \cdot v_g}{|v_g|} = \frac{v_p}{v_g(1 + \frac{\omega}{n} \frac{\partial n}{\partial \omega})} = \left(1 + \frac{1}{n^2} \nabla_s n \cdot \nabla_s n\right) \tag{2.12.91}$$

and

$$i_{v_g} = \cos \psi \left(i_k - \frac{1}{n} \nabla_s n\right) \quad . \tag{2.12.92}$$

It can also be shown that v_g can be written as

$$v_g = \frac{c^2 |k| [i_k - i_E(i_k \cdot i_E)]}{\omega i_E(\overset{\leftrightarrow}{\varepsilon} + \frac{\omega}{c} \frac{\partial}{\partial \omega} \overset{\leftrightarrow}{\varepsilon}) i_k} \quad . \tag{2.12.93}$$

For a uniaxial crystal,

$$\tan \psi = \frac{(\varepsilon_{11} - \varepsilon_{33}) \sin 2\theta}{2(\varepsilon_{11} \cos^2 \theta + \varepsilon_{33} \sin^2 \theta)} \quad , \tag{2.12.94}$$

where θ is the angle from the optic axis, and ψ is the angle between v_g and v_p. For θ given by

$$\tan \theta = \sqrt{\frac{n_e}{n_o}} = \left(\frac{\varepsilon_{33}}{\varepsilon_{11}}\right)^{1/4} \quad , \tag{2.12.95}$$

ψ_m is maximum and given by

$$\tan \psi_m = \frac{\varepsilon_{11} - \varepsilon_{33}}{2\sqrt{\varepsilon_{11} \varepsilon_{33}}} = \frac{n_e^2 - n_o^2}{2 n_o n_e} \quad . \tag{2.12.96}$$

2.12.2 Double Refraction

Let us consider the problem of incident light on an anisotropic crystal as shown in Fig. 2.61. At the boundary of separation, (2.3.7) must be satisfied, i.e.,

$$k_i \cdot r \bigg|_{z=0} = k_t \cdot r \bigg|_{z=0} \quad . \tag{2.12.97}$$

As the transmitted wave-vector k_t can have two values given by the wave-vector surface, we have in general two transmitted beams given by

$$k_i \sin \theta = k_{t1} \sin \theta_{t1} \quad \text{or} \quad \sin \theta = n_1(\theta_{t1}) \sin \theta_{t1} \quad , \quad \text{and} \tag{2.12.98}$$

$$k_i \sin \theta = k_{t2} \sin \theta_{t2} \quad \text{or} \quad \sin \theta = n_2(\theta_{t2}) \sin \theta_{t2} \quad . \tag{2.12.99}$$

Note that n_1 and n_2 are functions of the direction of propagation itself. Thus,

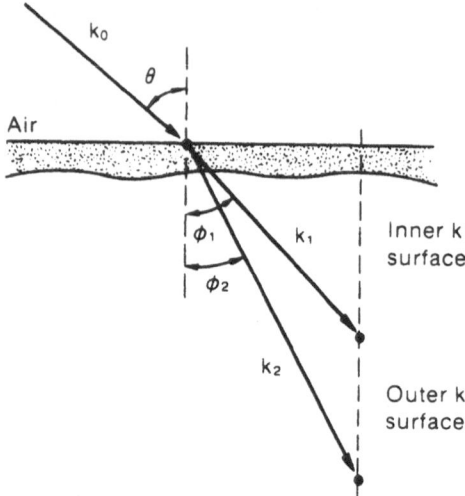

Fig. 2.61. Wave vectors for double refraction at the boundary of a crystal

in general, Snell's law must be modified. For a biaxial crystal, one has to solve (2.12.98 and 99) numerically to obtain θ_{t1} and θ_{t2}. This is done using the wave-vector surface in the plane of incidence as shown in Fig. 2.57.

For a uniaxial crystal the problem becomes somewhat simpler. In this case

$$\varepsilon'_x = \varepsilon'_y = n_0^2 \quad , \quad \text{and} \quad \varepsilon'_z = n_{ge}^2 \tag{2.12.100}$$

where n_0 and n_e are defined as the ordinary and extraordinary refractive indices, respectively. If $n_0 < n_e$ the crystal is called positive uniaxial. Otherwise, if $n_0 > n_e$ it is called a negative uniaxial crystal. The wave-vector surface corresponding to n_0 is a sphere whereas the one corresponding to n_e is a spheroid. Some wave-vector surface cross sections are shown in Fig. 2.62. Because the

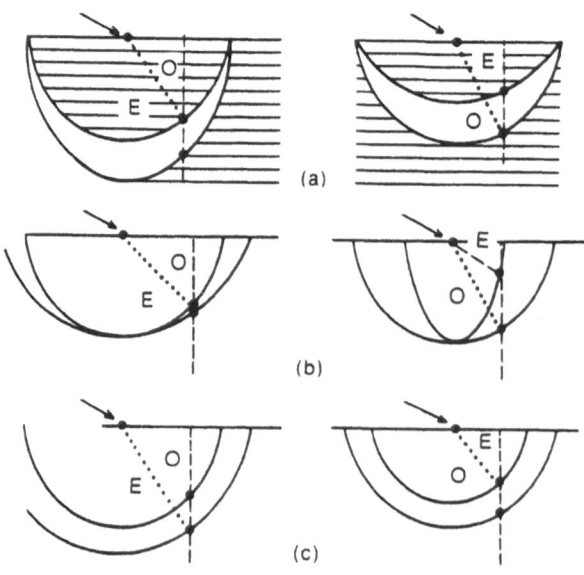

Fig. 2.62a–c. Wave vectors for double refraction in uniaxial crystals. (a) Optic axis parallel to the boundary and parallel to the plane of incidence. (b) Optic axis perpendicular to the boundary and parallel to the plane of incidence. (c) Optic axis parallel to the boundary and perpendicular to the plane of incidence. (From [2.16])

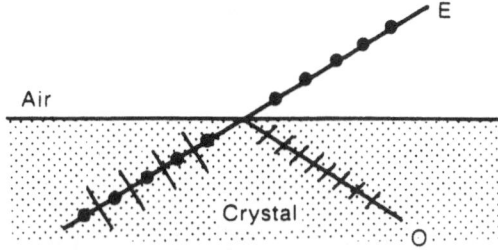

Fig. 2.63. Separation of the ordinary and extraordinary rays at the boundary of a crystal in the case of internal refraction

ordinary wave-vector surface is a sphere, it has a constant value of refractive index n_o; the transmitted ray for this case is called the ordinary ray and it obeys the normal Snell's law. However, the extraordinary ray must be obtained numerically. Some cases of double refraction for uniaxial crystals are shown in Fig. 2.62. The two refracted beams are polarized orthogonal to each other. This property and the fact that $n_1(\theta_1) \neq n_2(\theta_2)$ are used to make practical polarizers.

Let us consider the case of a negative uniaxial crystal through which an unpolarized beam propagates and is incident on an air interface, as shown in Fig. 2.63. We note that the total internal reflection for the ordinary ray takes place beyond the critical angle, θ_{co} given by

$$n_o \sin \theta_{co} = 1 \quad . \tag{2.12.101}$$

However, the critical angle for the extraordinary ray, θ_{ce}, is larger than θ_{co} as

$$n_e \sin \theta_{ce} = 1 \quad \text{and} \quad n_o > n_e \quad . \tag{2.12.102}$$

Thus, if the incident light has its angle of incidence θ between θ_{co} and θ_{ce}, the ordinary ray will be totally reflected, whereas part of the extraordinary ray will be transmitted. The transmitted wave is thus plane polarized even though the incident light is unpolarized. Using this property, one can fabricate polarizers such as the Glan prism and the Nicol prism.

The Glan prism consists of two identical calcite prisms mounted as shown in Fig. 2.64. The space between the two prisms contains either air or a transparent material such that the ordinary ray suffers total internal reflection. The output consists of only the extraordinary ray, which is linearly polarized. Note that the prisms have their optic axes parallel to the corner edges. The Nicol prism shown in Fig. 2.65 is in the form of a rhomb and the principal of operation is very similar to that of the Glan prism.

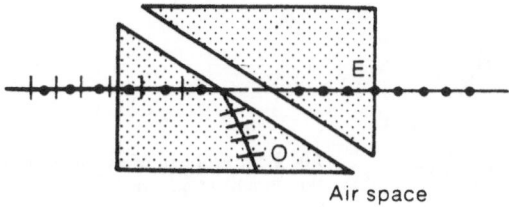

Air space

Fig. 2.64. Construction of the Glan polarizing prism

Canada balsam cement **Fig. 2.65.** The Nicol prism

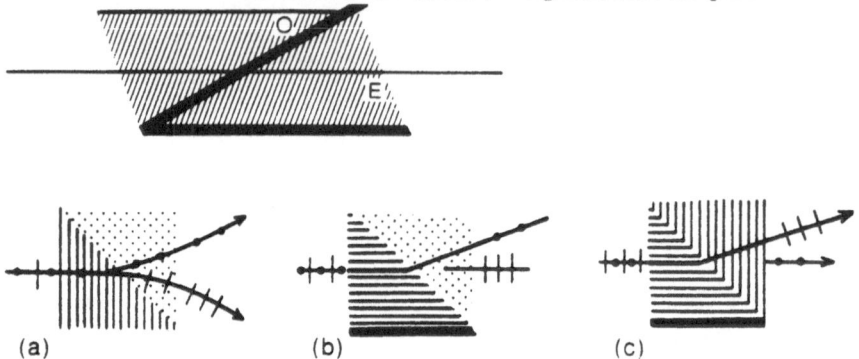

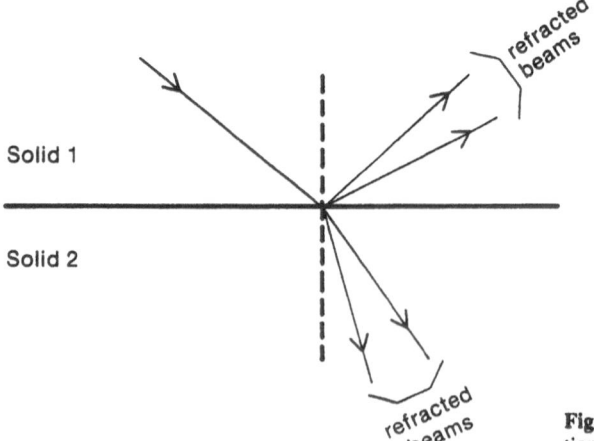

Fig. 2.66. The Wollaston (a), Rochon (b) and Sevarmont (c) prisms. All prisms shown are made with uniaxial positive material (quartz)

The Glan and Nicol prisms use total internal reflection of the ordinary ray as the main function of the polarizer, whereas the Wollaston, Rochon and Sevarmont prisms use the separation between the ordinary and extraordinary rays. Because the angles of refraction are different for the two rays, as shown in Figs. 2.66a–c, the ordinary and extraordinary rays will emerge from the prisms separated, and thus the prisms can be used as polarizers.

Finally, in Fig. 2.67, the most general problem of double refraction is shown, where light is incident from medium 1 to medium 2 when both media are anisotropic. Here, there will be two refracted and two reflected beams, the directions of which can be determined using the wave-vector surfaces for the two media. However, to calculate the amount of light reflected or refracted in the individual beams, one needs to apply the boundary conditions discussed in Sects. 2.2 and 2.3.

Solid 1

Solid 2

refracted beams

refracted beams

Fig. 2.67. Problem of double refraction in anisotropic media

a) Explicit Expressions for n_1 and n_2

We note that

$$\frac{1}{n^2} = \frac{1}{n^2}\sum_i i_{ki} \quad , \quad i = x, y, z \quad . \tag{2.12.103}$$

Substituting this equation into (2.12.28) we obtain

$$\sum i_{ki}^2 \left(\frac{1}{n^2 - \varepsilon_i'} - \frac{1}{n^2} \right) = 0 \quad \text{or}$$

$$\sum \frac{i_{ki}^2 \varepsilon_i'}{n^2 - \varepsilon_i'} = 0 \quad . \tag{2.12.104}$$

Defining

$$v_i = c/\sqrt{\varepsilon_i} \tag{2.12.105}$$

we have

$$\sum \frac{i_{ki}^2}{v_p^2 - v_i^2} = 0 \quad , \quad \text{or}$$

$$v_p^4 - c_1 v_p^2 + c_2 = 0 \quad , \quad \text{where} \tag{2.12.106}$$

$$c_1 = i_{kx}^2(v_y^2 + v_z^2) + i_{ky}^2(v_z^2 + v_x^2) + i_{kz}^2(v_x^2 + v_y^2) \quad , \quad \text{and} \tag{2.12.107}$$

$$c_2 = i_{kx}^2 v_y^2 v_z^2 + i_{ky}^2 v_z^2 v_x^2 + i_{kz}^2 v_x^2 v_y^2 \quad . \tag{2.12.108}$$

Therefore

$$v_p = \{\tfrac{1}{2}[c_1 \pm (c_1^2 - 4c_2)^{1/2}]\}^{1/2} = \frac{c}{n} \quad \text{or} \tag{2.12.109}$$

$$n_1 = c/\{\tfrac{1}{2}[c_1 - \sqrt{c_1^2 + 4c_2}]\}^{1/2} \quad , \quad \text{and} \tag{2.12.110}$$

$$n_2 = c/\{\tfrac{1}{2}[c_1 - \sqrt{c_1^2 - 4c_2}]\}^{1/2} \quad .$$

For isotropic solids

$$v_x = v_y = v_z \quad .$$

Thus

$$c_1^2 = 4c_2 \quad \text{and} \quad n_1 = n_2 \quad .$$

For uniaxial crystals

$$\varepsilon'_x = \varepsilon'_y \quad , \quad v_x = v_y = v_0 \quad , \quad v_z = v_e \quad .$$

Also denoting i_k at an angle θ with respect to the z axis, we note

$$i_{kx} = \sin \theta \cos \phi \quad , \quad i_{ky} = \sin \theta \sin \phi \quad , \quad i_{kz} = \cos \theta \quad ,$$

where ϕ is the azimuthal angle in the xy plane, so

$$v = \{\tfrac{1}{2}[2v_0^2 - (v_0^2 - v_e^2)\sin^2 \theta \pm (v_0^2 - v_e^2)\sin^2 \theta]\}^{1/2} \quad , \quad \text{or}$$

$$v_1 = v_0 \quad , \tag{2.12.111}$$

$$v_2 = \{v_0^2 \cos^2 \theta + v_e^2 \sin^2 \theta\}^{1/2} \quad , \tag{2.12.112}$$

$$n_1 = \sqrt{\varepsilon'_x} \quad , \quad \text{and}$$

$$n_2 = \frac{\varepsilon'_x \varepsilon'_z}{\varepsilon'_x \sin^2 \theta + \varepsilon'_z \cos^2 \theta} \quad . \tag{2.12.113}$$

Note $n_1 = n_2$ for $\theta = 0$ and $v_1 = v_2$. Thus the z axis is the optic axis.

Biaxial Crystals

For biaxial crystals, let us denote the principal axes such that

$$\varepsilon'_x < \varepsilon'_y < \varepsilon'_z \quad .$$

For this case

$$c_1^2 - 4c_2 = \{i_{ky}^2(v_x^2 - v_z^2) + [i_{kz}(v_x^2 - v_y^2) + i_{kx}(v_y^2 - v_z^2)^{1/2}]\}$$
$$\times \{i_{ky}^2(v_x^2 - v_z^2) + [i_{kz}(v_x^2 - v_y^2)^{1/2} - i_{kx}(v_y^2 - v_z^2)^{1/2}]^2\} \quad . \tag{2.12.114}$$

Thus $c_1^2 - 4c_2$ is always positive and from (2.12.109) we observe that the refractive index is either real or imaginary, but not complex. The optic axis is defined by

$$c_1^2 - 4c_2 = 0 \quad . \tag{2.12.115}$$

This happens for

$$i_{ky} = 0 \quad , \quad \text{and} \tag{2.12.116}$$

$$\frac{i_{kx}}{i_{kz}} = \pm \frac{v_x^2 - v_y^2}{v_y^2 - v_z^2} \quad . \tag{2.12.117}$$

Thus the optic axis is in the xz plane, as shown in Fig. 2.68 where

$$\tan \beta = \left(\frac{v_x^2 - v_y^2}{v_y^2 - v_z^2}\right)^{1/2} \quad . \tag{2.12.118}$$

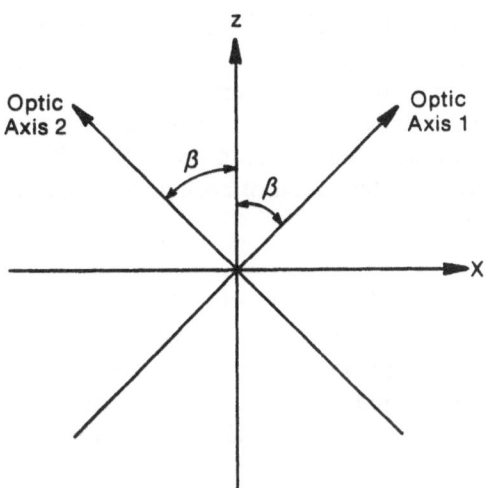

Fig. 2.68. Optic axes of a biaxial crystal in the xz plane

If the angles between i_k and the optic axes are denoted by θ_1 and θ_2 respectively, it can be shown that

$$v_{1,2} = \{\tfrac{1}{2}[v_x^2 + v_z^2 + (v_x^2 - v_z^2)\cos{(\theta_1 \pm \theta_2)}]\}^{1/2} \quad , \quad \text{and} \qquad (2.12.119)$$

$$n_{1,2} = \left[\frac{\varepsilon_x' \varepsilon_z'}{\varepsilon_x' \sin^2 \tfrac{1}{2}(\theta_1 \pm \theta_2) + \varepsilon_z' \cos^2 \tfrac{1}{2}(\theta_1 \pm \theta_2)}\right]^{1/2} . \qquad (2.12.120)$$

Note that for $\theta_1 = \theta_2$, (2.12.119 and 120) become equal to (2.12.112, 113), representing the uniaxial crystals.

b) Internal Conical Refraction

There are two fixed polarizations of light which can propagate along any direction in a crystal except along the optic axis. These polarizations are determined by the axes of the ellipse formed by the normal to the direction of propagation and the index ellipsoid. However, if the propagation direction is along the optical axis, then the ellipse is really a circle as shown in Fig. 2.69 and all polarization

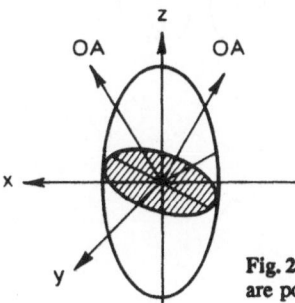

Fig. 2.69. For propagation along the optical axis all polarization directions are possible

directions are possible. Thus, if a narrow, unpolarized light beam is incident normally on a plane parallel, crystalline plate with normal along the optic axis, the light ray inside the plate forms a hollow cone and emerges as a hollow cylinder. The light at each point on the cylinder is linearly polarized. Note that the cone and the cylinder are formed because the direction of group velocity or ray is different from the phase velocity. This formation of cone and cylinder is known as internal conical refraction.

c) General Solution of Wave Propagation in an Anisotropic Medium

To solve the electromagnetic wave propagation problem in an anisotropic medium, we first diagonalize the dielectric tensor using the reference axes as the principal axes. In many of the problems to be considered (e.g. optical activity, magneto-optics), it is more convenient to solve the wave equation along any set of axes in preference to the principal axes. For that purpose,

$$\nabla \times (\nabla \times E) = -\omega^2 \mu \bar{\bar{\varepsilon}} E \quad , \tag{2.12.121}$$

where $\bar{\bar{\varepsilon}}$ is given by (2.12.1). Using a vector identity we obtain

$$\nabla(\nabla \cdot E) - \nabla^2 E = -\omega^2 \mu \bar{\bar{\varepsilon}} E \quad . \tag{2.12.122}$$

Note that $\nabla \cdot E$ is not equal to zero in the anisotropic case. Thus we have

$$k(k \cdot E) - |k|^2 E = -\omega^2 \mu \bar{\bar{\varepsilon}} E \quad . \tag{2.12.123}$$

Equation (2.12.123) can be written in matrix form by taking

$$\bar{\bar{k}} = \begin{pmatrix} k_x \\ k_y \\ k_z \end{pmatrix} \quad , \quad \text{and} \tag{2.12.124}$$

$$k(k \cdot E) = (\bar{\bar{k}} \bar{\bar{k}}^T) E \quad . \tag{2.12.125}$$

We finally obtain

$$(|k|^2 I - \omega^2 \mu \bar{\bar{\varepsilon}} - \bar{\bar{k}} \bar{\bar{k}}^T) E = 0 \quad . \tag{2.12.126}$$

Written in expanded form

$$\left\{ |k|^2 \begin{vmatrix} 1 & 0 & 0 \\ 0 & 1 & 0 \\ 0 & 0 & 1 \end{vmatrix} \begin{pmatrix} E_x \\ E_y \\ E_z \end{pmatrix} - \omega^2 \mu \begin{pmatrix} \varepsilon_{xx} & \varepsilon_{xy} & \varepsilon_{xz} \\ \varepsilon_{yx} & \varepsilon_{yy} & \varepsilon_{yz} \\ \varepsilon_{zx} & \varepsilon_{zy} & \varepsilon_{zz} \end{pmatrix} \begin{pmatrix} E_x \\ E_y \\ E_z \end{pmatrix} \right.$$
$$\left. - \begin{pmatrix} k_x^2 & k_x k_y & k_x k_z \\ k_y k_x & k_y^2 & k_y k_z \\ k_z k_x & k_z k_y & k_z^2 \end{pmatrix} \begin{pmatrix} E_x \\ E_y \\ E_z \end{pmatrix} \right\} = 0 \quad .$$

For the case

$$k_x = k_y = 0 \quad ,$$
$$|k|^2 = |k_z|^2$$

and we have

$$(|k|^2 E - \omega^2 \mu \overline{\overline{\varepsilon}}) E = 0 \quad . \tag{2.12.127}$$

As the refractive index n is given by

$$n = \frac{|k|}{\omega \sqrt{\mu \varepsilon_0}} = \frac{|k \cdot k|^{1/2}}{\omega \sqrt{\mu \varepsilon_0}} \tag{2.12.128}$$

(2.12.126) can be written as

$$\left(n^2 \overline{\overline{I}} - \frac{1}{\varepsilon_0} \overline{\overline{\varepsilon}} - \frac{1}{\omega^2 \mu \varepsilon_0} \overline{\overline{k}} \, \overline{\overline{k}}^T \right) E = 0 \quad . \tag{2.12.129}$$

The value of n for a particular case is obtained when the determinant of the matrix vanishes:

$$\det \left(n^2 \overline{\overline{I}} - \frac{1}{\varepsilon_0} \overline{\overline{\varepsilon}} - \frac{1}{\omega^2 \mu \varepsilon_0} \overline{\overline{k}} \, \overline{\overline{k}}^T \right) = 0 \quad . \tag{2.12.130}$$

The solution of the equations yields two values of refractive index. Substituting these values (2.12.129), one obtains the E field corresponding to each refractive index. The waves with these particular polarizations and refractive indices are the normal modes that exist in that particular medium. Some examples are discussed in Sect. 2.15 in connection with optical activity and magneto-optics.

So far we have considered propagation in a medium where no attenuation of the wave takes place. However, the equations are still valid if the wave does attenuate. In this case k is complex and has a real part k_r and an imaginary part k_{im}:

$$k = k_r + j k_{im} \quad ,$$

where k_r gives the normal to the wavefront and k_{im} gives the normal to the planes of constant intensity. For the case where k_r is not parallel to k_{im} the wave is said to be inhomogeneous. In our absorbing medium, if the wave is incident at an oblique angle, inhomogeneous waves will result. For the case of complex k in (2.12.128), the square root has to be taken such that k_r is positive.

2.13 Electro-optic Effect

In this section we discuss the effect of the applied electric field on the propagation of light through a crystal. This is followed by the effect of ultrasound propagation (Sect. 2.14) and of the magnetic field (Sect. 2.15). However, it is of interest to discuss in general the effects of a small perturbation in the index ellipsoid.

The index ellipsoid in matrix notation is given by

$$x \overline{\overline{\varepsilon}}^{-1} x^T = I \quad , \quad \text{or}$$
$$x A x^T = I \quad , \quad \text{or}$$

$$x \frac{1}{n^2} x^{\mathrm{T}} = I \quad . \tag{2.13.1}$$

Due to any perturbing field, $\bar{\bar{\varepsilon}}^{-1}$, A, and $1/n^2$ change. This change is written as

$$(\Delta \varepsilon)^{-1} = \Delta \left(\frac{1}{n^2} \right) = \Delta A \quad . \tag{2.13.2}$$

To first order, we have

$$(\Delta \varepsilon)^{-1} = -\varepsilon^{-1} \Delta \varepsilon \varepsilon^{-1} \quad . \tag{2.13.3}$$

Using perturbation analysis, the perturbation in the refractive indices, Δn, is given by either

$$\Delta n = \frac{n_0}{2} D_0 (\Delta \varepsilon)^{-1} D_0^{\mathrm{T}} \quad , \quad \text{or} \tag{2.13.4}$$

$$\Delta n = -\frac{(n_0)^3}{3} d_0 (\Delta \varepsilon)^{-1} d_0^{\mathrm{T}} \quad , \tag{2.13.5}$$

where D_0 is the normalized displacement vector in the absence of perturbation and $d_0 = D_0/n_0$. Consider the following examples.

i) *Isotropic Solid*[2]

$$D_0^1 = n_0 i_E \quad , \tag{2.13.6}$$

$$D_0^2 = n_0 i_k \times i_E \quad ,$$

$$i_k \cdot i_E = 0 \quad ,$$

$$i_E^1 \cdot (\Delta \varepsilon)^{-1} \cdot (i_k \times i_E^1) = 0 \quad , \tag{2.13.7}$$

$$(\Delta n^{(1)}) = -\frac{n_0^3}{2} i_E^1 \cdot (\Delta \varepsilon)^{-1} i_E^1 \quad , \tag{2.13.8}$$

$$(\Delta n^{(2)}) = -\frac{n_0^3}{2} (i_k \times i_E^1)(\Delta \varepsilon)^{-1} (i_k \times i_E^1) \quad . \tag{2.13.9}$$

Note that $i_E^1 \cdot i_k = 0$.

ii) *Uniaxial Crystals*[3]

$$E_0^o = \frac{i_k \times i_z}{n_0^o \sqrt{1 - (i_k \cdot i_z)^2}} \quad , \tag{2.13.10}$$

$$D_0^o = (n_0^o)^2 E_0^o \quad , \tag{2.13.11}$$

---

[2] The superscripts 1 and 2 refer to the two orthogonal propagating modes.

[3] Superscripts o and e correspond to the ordinary and extraordinary waves, respectively.

$$i_3 = i_z \quad ,$$

$$\Delta n^o = -\frac{(n_0^o)^3}{2}\left[\frac{i_y^2(\Delta\varepsilon)_{xx}^{-1} + i_z^2(\Delta\varepsilon)_{yy}^{-1} - 2i_x i_y(\Delta\varepsilon)_{xy}^{-1}}{i_x^2 + i_y^2}\right]$$

$$= \frac{(n_0^o)^3}{2}[(\Delta\varepsilon)_{xx}^{-1}\sin^2\phi + (\Delta\varepsilon)_{yy}^{-1}\cos^2\phi - (\Delta\varepsilon)_{xy}^{-1}\sin 2\phi] \quad ,$$

$$(2.13.12)$$

where

$$i_x = \sin\theta\cos\phi \quad , \quad i_y = \sin\theta\sin\phi \quad , \quad i_z = \cos\theta \quad ,$$

$$n_0^e(\theta) = \frac{\varepsilon_x'\varepsilon_z'}{\varepsilon_x'\sin^2\theta + \varepsilon_z'\cos^2\theta} \quad , \tag{2.13.13}$$

$$D_0^e = \frac{i_k \times (i_k \times i_z)}{\sqrt{1 - (i_k \cdot i_z)^2}}n_0^e(\theta) \quad , \tag{2.13.14}$$

$$\Delta n^e = -[n_0^e(\theta)]^3[(\Delta\varepsilon)_{xx}^{-1}\cos^2\theta\cos^2\phi + (\Delta\varepsilon)_{yy}^{-1}\cos^2\theta\sin^2\phi$$
$$+ (\Delta\varepsilon)_{zz}^{-1}\sin^2\theta + (\Delta\varepsilon)_{xy}^{-1}\cos^2\theta\sin^2 2\phi$$
$$- (\Delta\varepsilon)_{yz}^{-1}\sin 2\theta\sin\phi - (\Delta\varepsilon)_{xz}^{-1}\sin 2\theta\cos\phi] \quad . \tag{2.13.15}$$

The above equations are not valid along the optical axis, as degenerate perturbation theory should be applied for this case. The value of Δn^e

$$(\Delta n)^{e/o} = \tfrac{1}{4}(n_0^o)^3\left((\Delta\varepsilon)_{xx}^{-1} + +(\Delta\varepsilon)_{yy}^{-1}\right.$$
$$\left.\pm\{[(\Delta\varepsilon)_{xx}^{-1} - (\Delta\varepsilon)_{yy}^{-1}]^2 + 4[(\Delta\varepsilon)_{xy}^{-1}]^2\}^{1/2}\right) \quad . \tag{2.13.16}$$

The plus sign is applicable for Δn^e and the minus sign for Δn^o.

iii) Biaxial Crystals

The refractive indices are given by (2.12.119) and α corresponds to two propagating modes designated as 1 and 2:

$$D_{0i}^\alpha = \frac{\varepsilon_i' i_{ki}}{[(n_0^\alpha)^2 - \varepsilon_i']N^\alpha} \quad , \quad i = x, y, z \quad , \quad \text{where} \tag{2.13.17}$$

$$N^\alpha = \sum_{i=1}^{3}\frac{i_{ki}^2\varepsilon_i'}{[(n_0^\alpha)^2 - \varepsilon_i']^2} \tag{2.13.18}$$

$$[\Delta n]^\pm = -\frac{n_0^\pm}{2(N^\pm)^2}\sum_{i,j=1}^{3}\frac{i_i\varepsilon_i'(\Delta\varepsilon)_{ij}^{-1}i_{kj}\varepsilon_j'}{[(n_0^\pm)^2 - \varepsilon_i'][(n_0^\pm)^2 - \varepsilon_j']} \quad , \tag{2.13.19}$$

where $(\Delta n)^+$ and $(\Delta n)^-$ correspond to $(\Delta n)^1$ and $(\Delta n)^2$ respectively.

2.13.1 General Discussion

Application of an electric field changes the dielectric tensor of a material, however small. The electro-optic effect is in general defined by the change in the refractive index rather than the change in the dielectric constant because of the usefulness of the index ellipsoid method in solving problems. Thus the change in the index ellipsoid due to an applied electric field is written as

$$\Delta\left(\frac{1}{n^2}\right)_{ij} = r_{ijq}E_q + R_{ijpq}E_pE_q \quad , \quad i, j, p, q \to x, y, z \quad , \tag{2.13.20}$$

where r_{ijq} is the linear (Pockels) and R_{ijpq} the quadratic (Kerr) electro-optic coefficient. The summation convention for repeated indices is used in (2.13.20).

For a centrosymmetric crystal, the Pockels coefficient goes to zero. Because of the inversion symmetry in a centrosymmetric crystal, we must have a change in refractive index when the sign of the applied field is reversed. Thus

$$r_{ijq}E_q = r_{ijq}(-E_q) \quad . \tag{2.13.21}$$

The above equation can only be satisfied for $r_{ijq} = 0$. Thus, for an isotropic material, only the Kerr effect can exist.

Piezoelectric materials also lack inversion symmetry (Appendix E). Thus all piezoelectric materials have nonzero Pockels coefficients. Actually the form of the piezoelectric coefficients e_{ijk} can be obtained from r_{ijk} by conjugation, i.e., $e_{ijk} \to r_{kij}$. The index ellipsoid given by (2.11.32) can be written as

$$\frac{x^2}{n_x^2} + \frac{y^2}{n_y^2} + \frac{z^2}{n_z^2} = 1 \quad . \tag{2.13.22}$$

In the presence of the electro-optic effect, the above equation is modified to

$$x^2\left[\frac{1}{n_x^2} + \Delta\left(\frac{1}{n^2}\right)_{xx}\right] + y^2\left[\frac{1}{n_y^2} + \Delta\left(\frac{1}{n^2}\right)_{yy}\right] + z^2\left[\frac{1}{n_z^2} + \Delta\left(\frac{1}{n^2}\right)_{zz}\right]$$

$$+xy\left[\Delta\left(\frac{1}{n^2}\right)_{xy} + \Delta\left(\frac{1}{n^2}\right)_{xy}\right] + xz\left[\Delta\left(\frac{1}{n^2}\right)_{xz}\right.$$

$$\left.+\Delta\left(\frac{1}{n^2}\right)_{zx}\right] + yz\left[\Delta\left(\frac{1}{n^2}\right)_{yz} + \Delta\left(\frac{1}{n^2}\right)_{zy}\right] = 1 \quad . \tag{2.13.23}$$

It is customary to rewrite (2.13.23) as

$$x_1^2\left[\frac{1}{n_1^2} + \Delta\left(\frac{1}{n^2}\right)_1\right] + x_2^2\left[\frac{1}{n_2^2} + \Delta\left(\frac{1}{n^2}\right)_2\right]$$

$$+ x_3^2\left[\frac{1}{n_3^2} + \Delta\left(\frac{1}{n^2}\right)_3\right] + 2x^2x_3\Delta\left(\frac{1}{n^2}\right)_4$$

$$+2x_1x_3\Delta\left(\frac{1}{n^2}\right)_5 + 2x_1x_2\Delta\left(\frac{1}{n^2}\right)_6 = 1 \quad , \tag{2.13.24}$$

where $x \rightarrow x_1$, $y \rightarrow x_2$, $z \rightarrow x_3$, $xx \rightarrow 1$, $yy \rightarrow 2$, $zz \rightarrow 3$, $yz \rightarrow 4$, $zx \rightarrow 5$, and $xy \rightarrow 6$. In this notation (2.13.20) becomes

$$\Delta\left(\frac{1}{n^2}\right)_i = r_{ij} E_j + R_{ipq} E_p E_q \quad . \tag{2.13.25}$$

The Pockels coefficients are uniquely determined by the point group symmetry of the crystal. This is illustrated in Appendix E for all crystal symmetry classes. It is of interest to consider some typical examples. For LiNbO$_3$, considering only the linear term, (2.13.25) becomes

$$
\begin{pmatrix}
\Delta(\frac{1}{n^2})_1 \\
\Delta(\frac{1}{n^2})_2 \\
\Delta(\frac{1}{n^2})_3 \\
\Delta(\frac{1}{n^2})_4 \\
\Delta(\frac{1}{n^2})_5 \\
\Delta(\frac{1}{n^2})_6
\end{pmatrix}
=
\begin{pmatrix}
r_{11} & 0 & r_{33} \\
-r_{11} & 0 & r_{13} \\
0 & 0 & r_{33} \\
0 & r_{51} & 0 \\
r_{51} & 0 & 0 \\
0 & -r_{11} & 0
\end{pmatrix}
\begin{pmatrix}
E_1 \\
E_2 \\
E_3
\end{pmatrix}
\quad . \tag{2.13.26}
$$

For potassium dihydrogen phosphate (KH$_2$PO$_4$) one has

$$
\begin{pmatrix}
\Delta(\frac{1}{n^2})_1 \\
\Delta(\frac{1}{n^2})_2 \\
\Delta(\frac{1}{n^2})_3 \\
\Delta(\frac{1}{n^2})_4 \\
\Delta(\frac{1}{n^2})_5 \\
\Delta(\frac{1}{n^2})_6
\end{pmatrix}
=
\begin{pmatrix}
0 & 0 & 0 \\
0 & 0 & 0 \\
0 & 0 & 0 \\
r_{41} & 0 & 0 \\
0 & r_{41} & 0 \\
0 & 0 & r_{63}
\end{pmatrix}
\begin{pmatrix}
E_1 \\
E_2 \\
E_3
\end{pmatrix}
\quad . \tag{2.13.27}
$$

We have observed that the index ellipsoid equation given by (2.13.22) defined with the principal axes as the coordinate axes becomes (2.13.24) under the influence of the electro-optic effect. To obtain the new principal axes, one uses the matrix technique discussed in Appendix A8. For this case, one needs to diagonalize the matrix

$$
A =
\begin{pmatrix}
\frac{1}{n_1^2} + \Delta(\frac{1}{n^2})_1 & \Delta(\frac{1}{n^2})_6 & \Delta(\frac{1}{n^2})_5 \\
\Delta(\frac{1}{n^2})_6 & \frac{1}{n_2^2} + \Delta(\frac{1}{n^2})_6 & \Delta(\frac{1}{n^2})_4 \\
\Delta(\frac{1}{n^2})_5 & \Delta(\frac{1}{n^2})_4 & \frac{1}{n_3^2} + \Delta(\frac{1}{n^2})_3
\end{pmatrix}
\quad . \tag{2.13.28}
$$

The new principal axes X' are given by

$$X' = U X \quad , \tag{2.13.29}$$

where the matrix U corresponds to the eigenvectors of A.

Let us consider an example where an electric field is applied in the z direction ($E_3 = E_z$) for KH$_2$PO$_4$. For this case, the index ellipsoid equation becomes

$$\frac{x^2}{n_o^2} + \frac{y^2}{n_o^2} + \frac{z^2}{n_e^2} + 2r_{63} E_z xy = 1 \quad . \tag{2.13.30}$$

Note that KH_2PO_4 is positive uniaxial. Thus, for this case we have

$$A = \begin{pmatrix} \frac{1}{n_0^2} & r_{63}E_z & 0 \\ r_{63}E_z & \frac{1}{n_0^2} & 0 \\ 0 & 0 & \frac{1}{n_e^2} \end{pmatrix} . \qquad (2.13.31)$$

Equation (2.13.31) can be written as

$$A = U\Lambda U \quad , \quad \text{where} \qquad (2.13.32)$$

$$U = \begin{pmatrix} \frac{1}{\sqrt{2}} & \frac{1}{\sqrt{2}} & 0 \\ \frac{1}{\sqrt{2}} & -\frac{1}{\sqrt{2}} & 0 \\ 0 & 0 & 1 \end{pmatrix} \quad , \qquad (2.13.33)$$

$$\Lambda = \begin{pmatrix} \frac{1}{n_0^2} + r_{63}E_z & 0 & 0 \\ 0 & \frac{1}{n_0^2} - r_{63}E_z & 0 \\ 0 & 0\frac{1}{n_e^2} \end{pmatrix} . \qquad (2.13.34)$$

The new directions for the major axes are given by

$$X' = UX \quad , \quad \text{or} \qquad (2.13.35)$$

$$x' = \frac{1}{\sqrt{2}}(x+y) = x \cos 45° + y \sin 45° \quad ,$$

$$y' = \frac{1}{\sqrt{2}}(x-y) = x \sin 45° - y \cos 45° \quad ,$$

$$z' = z \quad . \qquad (2.13.36)$$

The new principal axes, x' and y', are at an angle of 45° to the crystal axes. In the new system of axes (X'), the index ellipsoid equation is given by

$$X'\Lambda X'^{\mathrm{T}} = I \quad , \quad \text{or} \qquad (2.13.37)$$

$$x'^2\left(\frac{1}{n_0^2} + r_{63}E_z\right) + y'^2\left(\frac{1}{n_0^2} - r_{63}E_z\right) + \frac{z'^2}{n_e^2} = 1 \quad , \quad \text{or} \qquad (2.13.38)$$

$$\frac{x'^2}{n_{x'}^2} + \frac{y'^2}{n_{y'}^2} + \frac{z'^2}{n_e^2} = 1 \quad . \qquad (2.13.39)$$

Let us consider the problem where light polarized in the x direction and propagating in the z direction is incident on a plate of KH_2PO_4 of thickness l as shown in Fig. 2.70. We are interested in the output light. The input light can be decomposed into two components given by

$$E = E_{x'}i_{z'} + E_{y'}i_{y'} \quad , \quad \text{where} \qquad (2.13.40)$$

$$E_{x'} = Ae^{j(\omega t - \frac{\omega}{c}n_{x'}z)} \quad , \quad \text{and} \qquad (2.13.41)$$

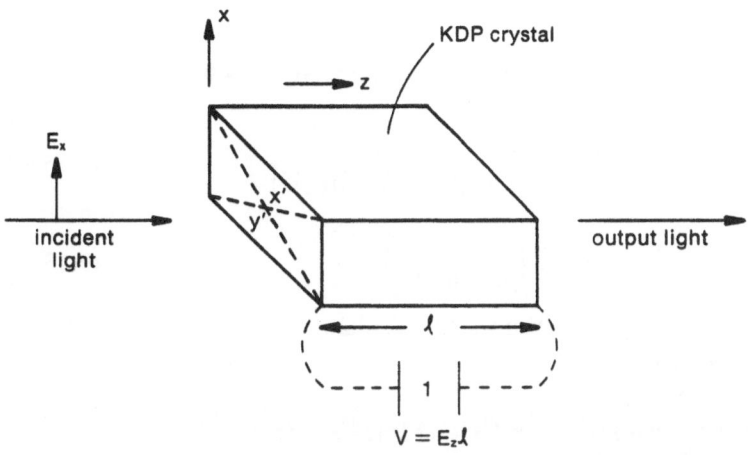

Fig. 2.70. Geometry for the electro-optic interaction in a KDP crystal plate of thickness l. The symbols x' and y' denote the rotated principal axes of the index ellipsoid

$$E_{y'} = A e^{j(\omega t - \frac{\omega}{c} n_{y'} z)} \quad ,$$

(2.13.42)

A being the input amplitude. Note that

$$\frac{1}{n_{x'}^2} = \frac{1}{n_0^2} + r_{63} E_z \quad , \quad \text{or}$$

(2.13.43)

$$n_{x'} \approx n_0 - \frac{n_0^3}{2} r_{63} E_z \quad ,$$

(2.13.44)

provided

$$r_{63} E_z \ll n_0^{-2} \quad .$$

Similarly

$$n_{y'} \approx n_0 + \frac{n_0^3}{2} r_{63} E_z \quad .$$

(2.13.45)

The phase difference between E'_x $(z = l)$ and E'_y $(z = l)$ is given by

$$
\begin{aligned}
\Gamma &= \frac{\omega}{c}(n_{y'} - n_{x'}) E_z l \\
&= \frac{\omega}{c} n_0^3 r_{63} V
\end{aligned}
$$

(2.13.46)

where $V = E_z l$ is the applied voltage across the crystal. If we define the voltage required to change the phase by π as V_π (often referred to as the half-wave voltage) then

$$V_\pi = \frac{\pi c}{\omega} \frac{1}{n_0^3 r_{63}} \quad .$$

(2.13.47)

Thus (2.13.46) can be rewritten as

$$\Gamma = \pi V/V_\pi \quad . \tag{2.13.48}$$

In terms of the Jones matrix discussed in Sect. 2.15, we can write the electro-optic effect for this case as

$$J^M_{x',y'} = \begin{pmatrix} \exp\left(j\frac{\omega}{2c}n_o^3 r_{63} V\right) & 0 \\ 0 & \exp\left(-j\frac{\omega}{2c}n_o^3 r_{63} V\right) \end{pmatrix} \tag{2.13.49}$$

$$= \exp\left(j\frac{\omega}{2c}n_o^3 r_{63} V\right) \begin{pmatrix} 1 & 0 \\ 0 & e^{-j\Gamma} \end{pmatrix} \quad . \tag{2.13.50}$$

Consider another example where LiNbO$_3$ (2.13.26) is used. The index ellipsoid for this case is given by

$$\frac{x^2}{n_o^2} + \frac{y^2}{n_o^2} + \frac{z^2}{n_e^2} + (r_{22}E_y + r_{12}E_z)x^2 + (r_{12}E_z - r_{22}E_y)y^2$$
$$+ r_{33}E_z z^2 + 2r_{22}E_x xy + 2r_{42}E_x xz$$
$$+ 2r_{42}E_y yz = 1 \quad . \tag{2.13.51}$$

The A matrix for this case becomes

$$A = \begin{pmatrix} \frac{1}{n_o^2} + r_{22}E_y + r_{12}E_z & r_{22}E_x & r_{42}E_x \\ r_{22}E_x & \frac{1}{n_o^2} + r_{12}E_z - r_{22}E_y & r_{42}E_y \\ r_{42}E_x & r_{42}E_y & \frac{1}{n_e^2} + r_{33}E_z \end{pmatrix} \quad . \tag{2.13.52}$$

If $E_y = E_z = 0$ and $E_x \neq 0$, then the A matrix is diagonal and is given by

$$A = \begin{pmatrix} \frac{1}{n_x^2} & 0 & 0 \\ 0 & \frac{1}{n_y^2} & 0 \\ 0 & 0 & \frac{1}{n_z^2} \end{pmatrix} \quad , \quad \text{where} \tag{2.13.53}$$

$$n_x \approx n_o - \tfrac{1}{2}n_o^3 r_{12}E_z \quad , \tag{2.13.54}$$

$$n_y \approx n_o - \tfrac{1}{2}n_o^3 r_{12}E_z \quad , \tag{2.13.55}$$

$$n_z \approx n_e - \tfrac{1}{2}n_e^3 r_{33}E_z \quad . \tag{2.13.56}$$

Thus, for this case, there is no rotation of the principal axes by the applied electric field. However, if $E_x = E_z = 0$ and $E_y \neq 0$, then the A matrix becomes

$$A = \begin{pmatrix} \frac{1}{n_o^2} + r_{22}E_y & 0 & 0 \\ 0 & \frac{1}{n_o^2} - r_{22}E_y & r_{42}E_y \\ 0 & r_{42}E_y & \frac{1}{n_e^2} \end{pmatrix} \quad . \tag{2.13.57}$$

Writing again $A = U\Lambda U$, one obtains

$$U = \begin{pmatrix} 1 & 0 & 0 \\ 0 & \cos\theta & \sin\theta \\ 0 & \sin\theta & -\cos\theta \end{pmatrix} \quad , \quad \text{where} \tag{2.13.58}$$

$$\tan 2\theta = \frac{2r_{42}E_y}{\frac{1}{n_0^2} - r_{22}E_y - \frac{1}{n_e^2}} \quad , \tag{2.13.59}$$

$$\Lambda = \begin{pmatrix} \frac{1}{n_x^2} & 0 & 0 \\ 0 & \frac{1}{n_y^2} & 0 \\ 0 & 0 & \frac{1}{n_z^2} \end{pmatrix} \quad , \tag{2.13.60}$$

$$n_x \approx n_0 - \tfrac{1}{2}n_0^3 r_{22}E_y \quad , \tag{2.13.61}$$

$$\frac{1}{n_y^2} = \left(\frac{1}{n_0^2} - r_{22}E_y\right)\cos^2\theta + r_{42}E_z\sin 2\theta$$

$$+ \frac{\sin^2\theta}{n_e^2} \quad , \quad \text{and} \tag{2.13.62}$$

$$\frac{1}{n_z^2} = \frac{\cos^2\theta}{n_e^2} - r_{42}E_z\sin 2\theta + \frac{\sin^2\theta}{n_0^2} \quad . \tag{2.13.63}$$

The new principal axes y' and z' in this case are rotated by an angle θ with respect to the crystalline axes y and z.

2.13.2 Kerr Effect

The quadratic component of the electro-optic effect, the Kerr effect from (2.13.20), is given by

$$\Delta\left(\frac{1}{n^2}\right)_{ij} = R_{ijpq}E_pE_q \quad , \tag{2.13.64}$$

where R_{ijpq} is a fourth-order tensor that has symmetry properties that are similar to the elasto-optic coefficients P_{ijkl} (given in Appendix E). Experimentally, it is observed that the R_{ijpq} coefficients are highly dependent on temperature. However, if (2.13.64) is rewritten in terms of induced polarization, it becomes

$$\Delta\left(\frac{1}{n^2}\right)_{ij} = h_{ijpq}P_pP_q \quad , \quad \text{where} \tag{2.13.65}$$

$$P_k = \chi_{lk}E_k \quad , \quad \text{and} \tag{2.13.66}$$

$$R_{ijpq} = h_{ijkl}\chi_{kp}\chi_{lq} \quad . \tag{2.13.67}$$

Using contracted notation we have

$$\Delta\left(\frac{1}{n^2}\right)_i = h_{ij}(PP)_j \quad , \quad i,j = 1,\dots,6 \quad . \tag{2.13.68}$$

For example $(PP)_5 = P_x P_z$. The R and h coefficients have the same symmetry properties. The h coefficients for cubic crystals in the point group symmetry $m3m$ are given by

$$h = \begin{pmatrix} h_{11} & h_{12} & h_{12} & 0 & 0 & 0 \\ h_{12} & h_{11} & h_{12} & 0 & 0 & 0 \\ h_{12} & h_{12} & h_{11} & 0 & 0 & 0 \\ 0 & 0 & 0 & h_{44} & 0 & 0 \\ 0 & 0 & 0 & 0 & h_{44} & 0 \\ 0 & 0 & 0 & 0 & 0 & h_{44} \end{pmatrix} . \tag{2.13.69}$$

The h coefficients for an isotropic material are given by (2.13.69) except that

$$h_{44} = \tfrac{1}{2}(h_{11} - h_{12}) \quad . \tag{2.13.70}$$

For an applied electric field in the z direction, the index ellipsoid can be written as

$$\frac{x^2}{n_0^2} + \frac{y^2}{n_0^2} + \frac{z^2}{n_0^2} + h_{12}P_z^2 x^2 + h_{12}P_z^2 y^2$$
$$+ h_{11}P_z^2 z^2 = 1 \quad . \tag{2.13.71}$$

The above equation can be rewritten as

$$\frac{x^2 + y^2}{n_x^2} + \frac{z^2}{n_z^2} = 1 \quad , \quad \text{where} \tag{2.13.72}$$

$$n_x = n_y \approx n_0 - \tfrac{1}{2}n_0^3 h_{12}P_z^2 \quad , \quad \text{and} \tag{2.13.73}$$

$$n_z = n_0 - \tfrac{1}{2}n_0^3 h_{11}P_z^2 \quad . \tag{2.13.74}$$

Thus the isotropic material behaves like an uniaxial crystal. Note that the principal axes of the index ellipsoid remain the same. From

$$P_z = \chi E_z = |\varepsilon - 1|E_z \tag{2.13.75}$$

we have

$$\Delta n_z \approx \tfrac{1}{2}n_0^3 h_{11}|\varepsilon - 1|^2 E_z^2 \quad , \quad \text{and} \tag{2.13.76}$$

$$\Delta n_x = \Delta n_y \approx \tfrac{1}{2}n_0^3 h_{12}|\varepsilon - 1|^2 E_z^2 \quad . \tag{2.13.77}$$

2.13.3 Indirect Electro-optic Effect

If the frequency of the applied electric field is low enough that a strain field is produced in the electro-optic material, which in turn perturbs the index ellipsoid, then one should use the stress-free electro-optic tensor given by

126

LiNbO$_3$	$r_{33}^T = 32$	$r_{42}^T = 32$
	$r_{33}^S = 31$	$r_{42}^S = 28$
LiTaO$_3$	$r_{33}^T = 22$	
	$r_{33}^S = 30$	
BaTiO$_3$	$r_{33}^T = 108$	$r_{42}^T = 1640$
	$r_{33}^S = 23$	$r_{42}^S = 820$

$$r_{ijk}^T = r_{ijk}^s + p_{ijmn}e_{kmn} \quad , \tag{2.13.78}$$

where the superscript T denotes stress free; s denotes the strain-free, high frequency, or clamped condition; p_{ijmn} is the elasto-optic tensor as defined in the next section and e_{kmn} are the piezoelectric coefficients. Here r_{ijk}^T and r_{ijk}^s have the same crystal symmetry and, as shown in Table 2.5, the values are quite different for strongly piezoelectric materials like LiNbO$_3$ and BaTiO$_3$.

2.14 The Acousto-optic or Elasto-optic Effect

Devices using the acousto-optic effect play an important role in optical signal processing. This effect is discussed in detail in this section; acousto-optic coefficients are considered in Sect. 2.14.1. The thin grating approximation is the subject of Sect. 2.14.2. This is followed by the thick grating approximation, which includes the Bragg and Raman-Nath regimes and the effect on light polarization when a shear wave is used.

2.14.1 Acousto-optic Coefficients

Similar to the electro-optic effect, any elastic deformation causes change in the dielectric tensor. Initially, Pockel hypothesized that only the strain, or the symmetric part of the deformation, is involved in this change in dielectric tensor[4]:

$$\Delta\left(\frac{1}{n^2}\right)_{ij} = (\Delta\varepsilon)_{ij}^{-1} = p_{ijkl}S_{kl} \quad , \tag{2.14.1}$$

where p_{ijkl} are called the Pockels elasto-optic tensor elements and S_{kl} is defined as

---

[4] Note that the usual summation convention is used that any subscript which is repeated must be summed over all possible values of that subscript. For example,

$$\Delta\left(\frac{1}{n^2}\right)_{xy} = p_{xykl}S_{kl} = \sum_k \sum_l p_{xykl}S_{kl} \quad , \quad \text{where} \quad k, l = x, y, z \quad .$$

$$S_{kl} = \frac{1}{2}\left(\frac{\partial u_k}{\partial x_l} + \frac{\partial u_l}{\partial x_k}\right) \quad ;$$

u is the displacement vector of the material under strain and is measured from its equilibrium position. Because of the symmetry of $(\Delta\varepsilon)^{-1}$ and S tensors we have

$$p_{ijkl} = p_{ijlk} = p_{jilk} = p_{jikl} \quad . \tag{2.14.2}$$

Using (2.13.4) we can also write

$$(\Delta\varepsilon)_{ij} \approx -\frac{1}{2}\varepsilon_{im}\varepsilon_{jn}p_{mnkl}S_{kl} \quad . \tag{2.14.3}$$

As for the case of electro-optic effect, one uses contracted notation for p_{ijkl} given by p_{ij} ($i, j = 1, \ldots, 6$). The quantities p_{ij} for different crystals with their symmetry properties and values are given in Appendix E.

In (2.14.1), the antisymmetric part of the displacement gradient is not used. However, in a biaxial or uniaxial crystal, one should also consider the contribution of the antisymmetric part which is known as the roto-optic effect. This was first pointed out by *Toupin* [2.5] and *Tiersten* and *Tsai* [2.6] and later elaborated and experimentally verified by *Lax* and *Nelson* [2.7]. The antisymmetric part of the displacement gradient is related to rotation of the crystal. For an anisotropic crystal, any rotation with respect to the direction of light propagation will change the effective index ellipsoid. For a homogeneous rotation, this effect is the same as that due to a rigid body rotation and thus can be compensated for experimentally. However, for an inhomogeneous rotation, as for a general elastic traveling wave, this roto-optic effect and elasto-optic effect will not be separable. As the shear wave has these rotational components, roto-optic effects may be important in the acousto-optic interaction in an anisotropic crystal through which a shear wave is propagating.

As in (2.14.1), the roto-optic effect can be defined as

$$\Delta\left(\frac{1}{n^2}\right)_{ij} = (\Delta\varepsilon)_{ij}^{-1} = p^r_{ij[kl]}u_{[k,l]} \tag{2.14.4}$$

where $u_{[k,l]} = 1/2(\partial u_k/\partial x_l - \partial u_l/\partial x_k)$ and u is the displacement from the equilibrium position. It can be shown that

$$p^r_{ij[kl]} = \delta_{i[k}\varepsilon^{-1}_{\overline{l}]j} + \delta_{j[k}\varepsilon^{-1}_{\overline{l}]j} \quad , \qquad \text{where} \tag{2.14.5}$$

$$\delta_{i[k}\varepsilon^{-1}_{\overline{l}]j} = \delta_{ki}\varepsilon^{-1}_{lj} - \varepsilon^{-1}_{kj}\delta_{il} \quad . \tag{2.14.6}$$

We note that, for an isotropic or cubic crystal, as

$$\varepsilon^{-1}_{ij} = \varepsilon^{-1}\delta_{ij} \quad , \tag{2.14.7}$$

we have $p^r_{ijkl} = 0$. For a uniaxial crystal, the only nonvanishing components are

$$p^r_{yz(yz)} = p^r_{xz(xz)} = \frac{1}{2}\left(\frac{1}{n_e^2} - \frac{1}{n_o^2}\right) \quad . \tag{2.14.8}$$

Table 2.6. Biaxial Crystals

---

a) Orthorhombic crystal

The only nonvanishing components are

$$p_{yz(yz)} = \frac{1}{2}\left(\frac{1}{\epsilon'_z} - \frac{1}{\epsilon'_y}\right)$$

$$p_{xz(xz)} = \frac{1}{2}\left(\frac{1}{\epsilon'_z} - \frac{1}{\epsilon'_x}\right)$$

$$p_{xy(x,y)} = \frac{1}{2}\left(\frac{1}{\epsilon'_y} - \frac{1}{\epsilon'_x}\right)$$

b) Triclinic crystal
All components can be nonzero.

Note that

$$p_{yz(yz)} = p_{zy(yz)} = -p_{zy(zy)} = -p_{yz(zy)} \quad . \tag{2.14.9}$$

For biaxial crystals, the roto-optic coefficients are given in Table 2.6.

If the crystal is piezoelectric, then the acoustic wave produces a piezoelectric field which, in turn, changes the index ellipsoid through electro-optic coefficients. When this effect is included, the effective Pockels-like tensor p^{eff}_{ijkl} can be written as

$$p^{\text{eff}}_{ijkl} = p_{ijkl} + p_{ij(kl)} - \frac{r^s_{ijm} i_{km} i_{kn} e_{nkl}}{\varepsilon_0 i_k \overset{\leftrightarrow}{\varepsilon} i_k} \tag{2.14.10}$$

where the terms on the right hand side of (2.14.10) are the acousto-optic Pockel's coefficient, the roto-optic and the indirect elasto-optic coefficients, respectively.

Let us consider a few examples of $\Delta(1/n^2)$ due to acousto-optic interaction. We note that, once $\Delta(1/n^2)$ is obtained, it must be substituted in the index ellipsoid equation as discussed in Sect. 2.13 for the electro-optic case. From this modified index ellipsoid equation one obtains the change (Δn) in refractive index. Equation (2.14.1) written in expanded form is given by

$$
\begin{pmatrix}
\Delta\left(\frac{1}{n^2}\right)_1 \\
\Delta\left(\frac{1}{n^2}\right)_2 \\
\Delta\left(\frac{1}{n^2}\right)_3 \\
\Delta\left(\frac{1}{n^2}\right)_4 \\
\Delta\left(\frac{1}{n^2}\right)_5 \\
\Delta\left(\frac{1}{n^2}\right)_6
\end{pmatrix}
=
\begin{pmatrix}
p_{11} & p_{12} & p_{13} & p_{14} & p_{15} & p_{16} \\
p_{21} & p_{22} & p_{23} & p_{24} & p_{25} & p_{26} \\
p_{31} & p_{32} & p_{33} & p_{34} & p_{35} & p_{36} \\
p_{41} & p_{42} & p_{43} & p_{44} & p_{45} & p_{46} \\
p_{51} & p_{52} & p_{53} & p_{54} & p_{55} & p_{56} \\
p_{61} & p_{62} & p_{63} & p_{64} & p_{65} & p_{66}
\end{pmatrix}
\begin{pmatrix}
S_1 \\
S_2 \\
S_3 \\
S_4 \\
S_5 \\
S_6
\end{pmatrix} \quad .
$$

and

$$\Delta\left(\frac{1}{n^2}\right) = \Delta\left(\frac{1}{n^2}\right)_1 x^2 + \Delta\left(\frac{1}{n^2}\right)_2 y^2 + \Delta\left(\frac{1}{n^2}\right)_3 z^2$$

$$+\Delta\left(\frac{1}{n^2}\right)_4 yz + \Delta\left(\frac{1}{n^2}\right)_5 xz + \Delta\left(\frac{1}{n^2}\right)_6 xy \quad .$$

i) Isotropic Solid: Longitudinal Ultrasound

$$u_x(x,t) = u_0 e^{j(\omega t - kz)} \qquad (2.14.11)$$

$$S_1 \neq 0 \quad ; \quad S_2 = S_3 = S_4 = S_5 = S_6 = 0$$

$$\Delta\left(\frac{1}{n^2}\right) = p_{11} S_1 x^2 + p_{12} S_1 y^2 + p_{12} S_1 z^2 \qquad (2.14.12)$$

ii) Isotropic Solid: Shear Ultrasound

$$u_y(z,t) = u_0 e^{j(jvt - kx)} \qquad (2.14.13)$$

$$S_6 \neq 0 \quad ; \quad S_1 = S_2 = S_3 = S_4 = S_5 = 0$$

$$\Delta\left(\frac{1}{n^2}\right) = (p_{11} - p_{12}) S_6 xy \quad . \qquad (2.14.14)$$

iii) Lithium Niobate: (LiNbO$_3$)

$$u_z(z,t) = u_0 e^{j(\omega t - kz)} \qquad (2.14.15)$$

$$\frac{\partial}{\partial x} = \frac{\partial}{\partial y} = 0 \quad .$$

Thus in the contracted notation,

$$S_1 = S_2 = S_6 = S_5 = S_4 = 0$$

$$S_3 = S_{zz} \neq 0$$

$$\Delta\left(\frac{1}{n^2}\right) = p_{13} S_3 x^2 + p_{13} S_3 y^2 + p_{33} z^2 S_3 \qquad (2.14.16)$$

iv) Shear Wave, LiNbO3, y cut z propagation

$$u_y(z,t) = u_0 e^{j(\omega t - kz)} \qquad (2.14.17)$$

$$S_4 \neq 0 \quad ; \quad S_1 = S_2 = S_3 = S_5 = S_6 = 0$$

$$S_4 = -jku_y \quad ; \quad u_{(4)} = -jku_y \quad . \qquad (2.14.18)$$

$$\Delta\left(\frac{1}{n^2}\right) = p_{14} S_4 x^2 - p_{14} S_4 y^2 + p_{44} S_4 yz$$

$$+ p_{yz(yz)}^{r} u_{(4)} yz \quad . \qquad (2.14.19)$$

v) Fluid

$$u_x(x,t) = u_0 e^{j(\omega t - kx)}$$

$$S_1 \neq 0 \quad ; \quad S_2 = S_3 = S_4 = S_5 = S_6 = 0$$

$$\Delta\left(\frac{1}{n^2}\right) = p S_1 x^2$$

2.14.2 Acousto-optic Interaction: Thin Grating

Before we consider the more complex case, let us first consider the simple case shown in Fig. 2.71, where the ultrasound is propagating in the z direction and the material is fluid. For this case we can write the ultrasonically induced change in refractive index, $\Delta n(x,t)$ as

$$\Delta n(x,t) = \Delta n_0 \sin(\omega_s t - k_s z + \delta) \quad , \tag{2.14.20}$$

where δ is a constant phase difference and Δn_0, the change in refractive index, is given by

$$\Delta n_0 \approx \frac{n_0^3}{2} p s \tag{2.14.21}$$

where p is the relevant photo-elastic constant and s is the compressional strain.

If we assume that the width L of the ultrasonic beam is quite small, then we can consider that the effect of the ultrasonic beam is to form a very thin equivalent phase grating whose phase dependence is given by

$$\Delta\phi(x,t) = \frac{2\pi}{\lambda_0}(\Delta n_0 L) \sin(\omega_s t - k_s x + \delta) \quad . \tag{2.14.22}$$

If the incident light is represented by an electric field, E_i, given by

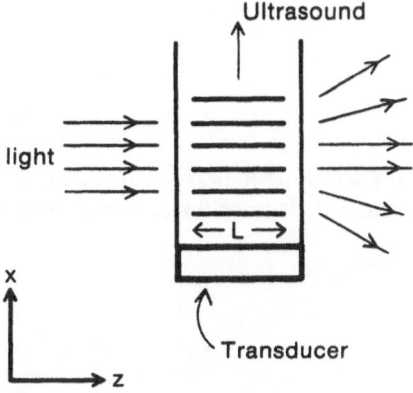

Fig. 2.71. Acousto-optic interaction. The ultrasound propagates in the z direction and the material is fluid. The thin grating case is considered

131

$$E_i = E_0 e^{j(\omega_1 t - k_1 z)} \quad , \tag{2.14.23}$$

then the output light is given by

$$\begin{aligned} E_{out} &= E_i e^{j\Delta\phi(x,t)} \\ &= E_0 e^{j[\omega_1 t - k_1 z + \Delta\phi(x,t)]} \quad . \end{aligned} \tag{2.14.24}$$

Equation (2.14.24) can be expanded in Bessel functions given by

$$e^{j\Delta\phi(x,t)} = \sum_{q=-\infty}^{+\infty} J_q\left(\frac{2\pi}{\lambda_0}\Delta n_0 L\right) e^{j[q(\omega_s t - k_s x + q)]} \quad . \tag{2.14.25}$$

Thus E_{out} can be written as

$$\begin{aligned} E_{out} &= E_0 \sum_{q=-\infty}^{+\infty} J_q\left(\frac{2\pi}{\lambda_0}\Delta n_0 L\right) e^{j(\omega_1 + q\omega_s t)} e^{jq\delta} e^{j(k_1 z - qk_s x)} \\ &= E_1 + E_2 + E_3 + \dots \\ &= \sum_{-\infty}^{+\infty} E_q \quad , \quad \text{where} \end{aligned} \tag{2.14.26}$$

$$E_q = E_0 J_q\left(\frac{2\pi}{\lambda}\Delta n_0 L\right) e^{j(\omega_1 + q\omega_s)t} e^{jq\delta} e^{-j(k_0 z + qk_s x)} \quad . \tag{2.14.27}$$

For E_q we note that its amplitude is proportional to the incident electric field amplitude as also to the qth order Bessel function. The frequency of the qth order diffracted light is given by

$$\omega_q = \omega_1 + q\omega_s \tag{2.14.28}$$

and the wave-vector is given by

$$k_q = i_z k_1 + i_z qk_s = i_k k_1 \quad . \tag{2.14.29}$$

From the wave vector diagram shown in Fig. 2.72 we obtain

$$k_1 \sin\theta_q = qk_s \quad , \quad \text{or}$$

$$\sin\theta_q = q\left(\frac{\lambda_1}{\lambda_s}\right) \quad . \tag{2.14.30}$$

Note that in (2.14.30) we have assumed $\omega_s \ll \omega_1$, so that $|k_q| \sim |k_1|$. Also, the polarization of the light is maintained at diffraction. It is of interest to consider the first order diffraction given by

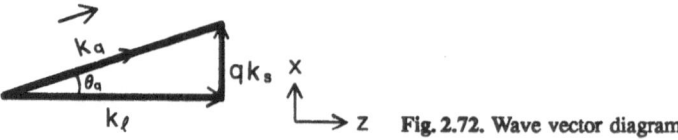

Fig. 2.72. Wave vector diagram

$$E_{\pm 1} = E_0 J_1\left(\frac{2\pi}{\lambda}\Delta n_0 L\right) e^{j(\omega_1 \pm \omega_s)t} e^{jk_1(i_z z \cos\theta_{\pm 1} z + i_x x \sin\theta_{\pm 1})} \quad . \tag{2.14.31}$$

The ratio of intensities between the zeroth order and first order are given by

$$\frac{I_1}{I_0} = \frac{J_1((2\pi/\lambda_0)\Delta n_0 L)}{J_0((2\pi/\lambda_0)\Delta n_0 L)} \quad . \tag{2.14.32}$$

Thus for

$$\frac{2\pi}{\lambda_0}\Delta n_0 L \ll 1 \quad ,$$

$$\begin{aligned}
\frac{I_1}{I_0} &\approx \left(\frac{2\pi}{\lambda_0}\Delta n_0 L\right)^2 \\
&= \left(\frac{2\pi}{\lambda_0}\frac{n_0^3}{2}psL\right)^2 \\
&= \pi^2\left(\frac{L}{\lambda_0}\right)^2 (n_0^6 p^2 s^2) \quad .
\end{aligned} \tag{2.14.33}$$

The acoustic power density, P_A, is given by

$$P_A = \tfrac{1}{2}\varrho v_s^3 s^2 \quad , \tag{2.14.34}$$

where ϱ is the density, thus we obtain

$$\begin{aligned}
\frac{I_1}{I_0} &= \pi^2\left(\frac{L}{\lambda_0}\right)^2 n_0^6 p^2 \frac{\lambda^2 P_A}{\varrho v_s^3} \\
&= 2\pi^2\left(\frac{L}{\lambda_0}\right)^2 \frac{n_0^6 p^2}{\varrho v_s^3} P_A \quad .
\end{aligned} \tag{2.14.35}$$

The total acoustic power, P_{tot}, is given by

$$P_{\text{tot}} = P_A(LH) \tag{2.14.36}$$

where H is the other dimension of the transducer.

Thus, (2.14.36) can be rewritten as

$$\begin{aligned}
\frac{I_1}{I_0} &= 2\pi^2 \frac{L}{H} \frac{n_0^6 p^2}{\varrho v_s^3} \frac{P_{\text{tot}}}{\lambda^2} \\
&= 2\pi^2 \frac{L}{H} \frac{P_{\text{tot}}}{\lambda^2} M_2 \quad ,
\end{aligned} \tag{2.14.37}$$

where M_2 is defined as a figure of merit and is given by

$$M_2 = \frac{n_0^6 p^2}{\varrho v_s^3} \quad . \tag{2.14.38}$$

2.14.3 Acousto-optic Interaction: Thick Grating

In the last section we assumed a thin grating. However, in reality it can be thick and is often made intentionally thick for stronger interaction. To analyze the thick grating problem, we should solve the following wave equation:

$$\nabla^2 E = \frac{[n(x,t)]^2}{c^2} \frac{\partial^2 E}{\partial t^2} \quad , \quad \text{where} \tag{2.14.39}$$

$$n(x,t) = n_0 - \Delta n_0 \sin(\omega_s t - k_s x + \delta) \quad . \tag{2.14.40}$$

The electric field can be given by the following Fourier series:

$$E = E_0 \sum_{q=-\infty}^{+\infty} a_q(z) e^{j(\omega_q t - \mathbf{k}_q \cdot \mathbf{r})} \quad , \quad \text{where} \tag{2.14.41}$$

$$\mathbf{k}_q \cdot \mathbf{r} = k_1(z \cos\theta + x \sin\theta) + q k_s x \quad . \tag{2.14.42}$$

We have assumed that the incident light is in the xz plane at an angle θ with respect to the z axis as shown in Fig. 2.73. Substituting the expression for E in the wave equation we obtain

$$\frac{da_q}{dz} - \frac{\pi \Delta n_0}{\lambda_0}(a_{q-1} - a_{q+1}) = j\frac{qQ}{2L}(q - 2\alpha)a_q \quad , \quad \text{where} \tag{2.14.43}$$

$$Q = \frac{2\pi L \lambda_0}{\lambda_s^2 n_0} = 2\pi\frac{L\lambda}{\lambda_s^2} \quad , \tag{2.14.44}$$

$$\alpha = -\frac{n_0 \lambda_s}{\lambda_0} \sin\theta \tag{2.14.45}$$

and we have assumed that the second derivative of a_q with respect to z is negligible.

Equation (2.14.43) defines a set of coupled-difference differential equations which relate the amplitudes of the plane wave expansion. The plane wave amplitudes, a_q for $q \neq 0$, represent the diffracted modes and the coupling coefficient $(\pi \Delta n_0/\lambda_0)$, represents the strength of the coupling coefficient between the adjacent modes. The amount of energy transfer between modes depends not only on $(\pi \Delta n_0/\lambda)$ but also on the degree of phase matching between the modes. The degree of this synchronization is given by the right-hand side of (2.14.43). As

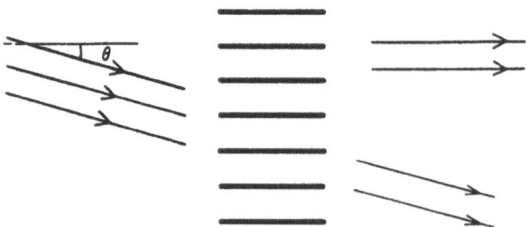

Fig. 2.73. Bragg diffraction geometry

the zeroth order is only coupled to the $q = \pm 1$ order, an appreciable amount of light can be transferred out of the zeroth order only if the right-hand side is small for $q = \pm 1$. This can be accomplished in two ways:

i)   by working at $\theta = 0$ and $Q \ll 1$
ii) by working at angles of incidence given by $\alpha = \pm \frac{1}{2}$ with $Q \gg 1$. The first case is known as the Raman-Nath regime and the second is known as Bragg diffraction.

a) Raman-Nath Regime ($Q \ll 1$)

We note that, as λ_s/λ_0 is very large, $\alpha \gg q$ for $q \leq 10$. For this case, the right-hand side of (2.14.43) simplifies to

$$j \frac{qQ}{2L}(q - 2\alpha)a_q \approx -j \frac{qQ}{L}\alpha a_q \quad . \tag{2.14.46}$$

Substituting (2.14.46) into (2.14.43) we obtain

$$\frac{\alpha a_q}{dz} - \frac{\pi \Delta n_0}{\lambda}(a_{q-1} - a_{q+1}) = -j \frac{qQ}{L}\alpha a_q \quad . \tag{2.14.47}$$

The solution of the above equation is

$$a_q = e^{jqQ\alpha z/L} J_q \left[\frac{4\pi \Delta n_0 L}{\lambda Q \alpha} \sin\left(\frac{Q\alpha z}{2L} \right) \right] \quad . \tag{2.14.48}$$

At $z = L$, the intensity of the qth order diffracted light is given by

$$I_q = J_q^2 \left[\frac{2\pi \Delta n_0 L}{\lambda} \frac{\sin (Q\alpha/2)}{Q\alpha/2} \right] \quad . \tag{2.14.49}$$

At normal incidence, $\theta = 0$, and the above equation reduces to

$$I_q = J_q^2(2\pi \Delta n_0 L/\lambda) \quad , \tag{2.14.50}$$

which is equivalent to (2.14.27). As discussed previously, $Q \ll 1$ means that there is no amplitude modulation so that the acousto-optic interaction acts like a phase-grating. Note that

$$I_{tot} = \left| \sum_{q=-\infty}^{+\infty} J_q^2 \left(\frac{2\pi \Delta n_0 L}{\lambda} \right) e^{jq(\omega_s t - k_s x)} \right|^2 = 1 \quad . \tag{2.14.51}$$

We note the following properties of Raman-Nath diffraction:

a)   The diffraction pattern is symmetric with respect to q, i.e., $I_q = I_{-q}$.
b)   Many diffraction orders can be observed simultaneously.
c)   The dependence on the incidence angle is proportional to the sinc function centered at normal incidence.
d)   In the equations above, we did not consider the effect of the finite sound beam. If this effect is important, i.e., if $L \approx \lambda_s$, it should be included in addition to the effects discussed above. For this case, the total intensity can be written as

$$I_{tot} = \sum_q J_q^2(v) W_q^2(x' - q) \quad , \quad \text{where} \tag{2.14.52}$$

$$W_q(x') = \frac{\sin(\pi L_{x'}/\lambda_s)}{\pi L_{x'}/\lambda_s} \tag{2.14.53}$$

and x' is measured from the position of maxima.

b) Bragg Region ($Q \gg 1$)

In (2.14.43), we note that for $Q \gg 1$, only the two first orders can couple to the zeroth order if $\alpha \approx \pm\frac{1}{2}$. Under this condition, light from the zeroth order transfers to the first order. However, light cannot be transferred from the first order to the second order as it is not synchronous. Thus there is an interchange of light intensity between the zero and the first order only. Physically, this means that the phase grating is thick and there is multiple diffraction at every plane, and only the components which are phase matched form the final diffracted beam at the proper angle. This is also understood from Fig. 2.74 where the phase grating is shown. Note that this grating is moving with a much slower velocity compared to light. If we consider the incident wavefront AB and the diffracted waveform CD, then the optical path difference $AC - BD$ is given by

$$AC - BD = x(\cos \theta_i - \cos \theta_r) \quad . \tag{2.14.54}$$

For constructive interference, the path difference must be an integer multiple of λ or

$$x(\cos \theta_i - \cos \theta_r) = m\lambda = \frac{m\lambda_0}{n} \tag{2.14.55}$$

where m is an integer. The above equation can be satisfied for all values of x if the following condition obtains:

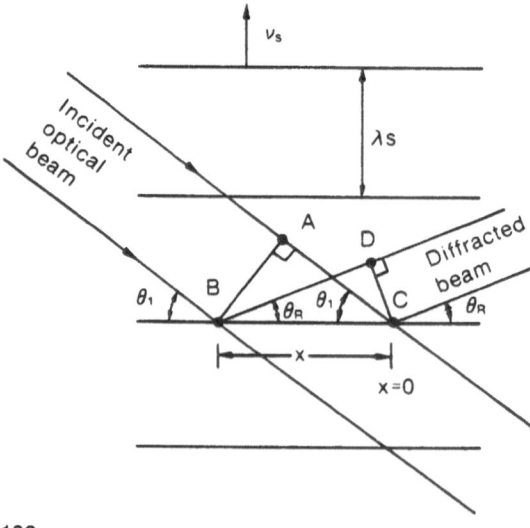

Fig. 2.74. A thick phase grating. (From [2.17])

$$\theta_i = \theta_r \quad . \tag{2.14.56}$$

The diffraction from planes parallel to the acoustic phase front and separated by λ_s must add up. The path difference $(A0+0B)$ in Fig. 2.75 must be equal to λ/n. Thus

$$2\lambda_s \sin \theta_B = \frac{\lambda_0}{n} = \lambda_1 \tag{2.14.57}$$

where θ_B is the Bragg angle. Note that for electron or X-ray diffraction in crystals, the diffraction from discrete elements rather than planes is considered. Thus the right-hand side of (2.14.57) is not $m\lambda/n$, but λ/n.

Returning to (2.14.43) and considering the case $\alpha \approx \frac{1}{2}$, we have

$$\frac{da_0}{dz} + \frac{\pi \Delta n_0}{\lambda} a_1 = 0 \quad , \tag{2.14.58}$$

$$\frac{da_1}{dz} - \frac{\pi \Delta n_0}{\lambda} a_0 = j\frac{Q}{2L}(1 - 2\alpha)a_1 \quad . \tag{2.14.59}$$

Solving the above two equations, we obtain

$$I_0 = |a_0|^2 = 1 - I \quad , \quad \text{where} \tag{2.14.60}$$

$$I_1 = |a_1|^2 = \left(\frac{\pi \Delta n_0 L}{\lambda T} \sin \sigma\right)^2 \quad \text{and} \tag{2.14.61}$$

$$\sigma = \frac{1}{2}\left\{[Q(1 - 2\alpha)]^2 + \left(\frac{2\pi \Delta n_0 L}{\lambda}\right)^2\right\}^{1/2} \quad . \tag{2.14.62}$$

If $\alpha = \pm\frac{1}{2}$, then we have

$$I_0(L) = I_0 \cos^2\left(\frac{\pi \Delta n_0 L}{\lambda}\right) \quad , \tag{2.14.63}$$

$$I_1(L) = I_0 \sin^2\left(\frac{\pi \Delta n_0 L}{\lambda}\right) \tag{2.14.64}$$

where I_0 is the incident power.

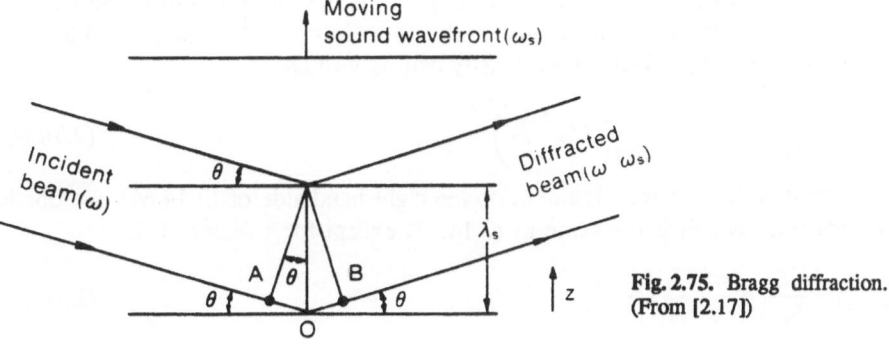

Fig. 2.75. Bragg diffraction. (From [2.17])

137

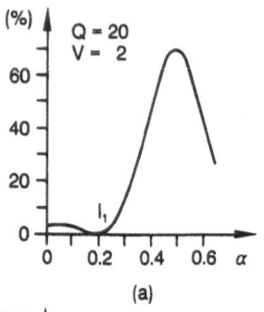

(a)

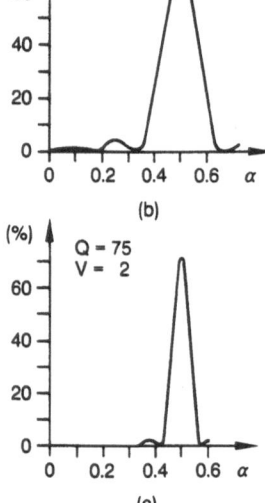

(b)

(c)

Fig. 2.76. First order light intensities vs incidence angle with Q as a parameter. (From [2.18])

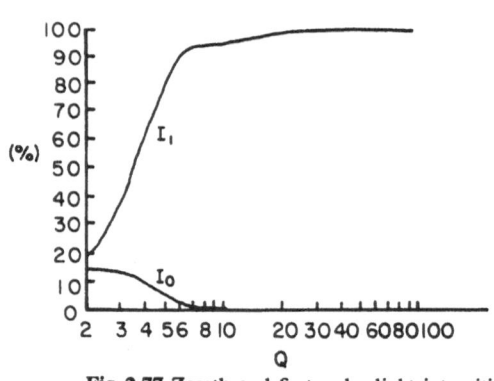

Fig. 2.77 Zeroth and first order light intensities vs Q at Bragg incidence. (From [2.18])

The dependence of I_1 on α is shown in Fig. 2.76 for different values of Q and $(\pi \Delta n_0 L/\lambda)$. It is observed that, as Q increases, the angular spread decreases. The dependence of I_0 and I_1 on Q is also shown in Fig. 2.77, where it is seen that $Q > 10$ for I_1 to be 100 %.

c) Transition Region ($Q \approx 1$)

In this region no analytical solution is possible, so that one must resort to numerical computation. Typical curves are shown in Fig. 2.78. Using (2.14.61), we note that the Bragg diffracted intensity can be written as

$$\frac{I_1(L)}{I_0} = \sin^2\left(\frac{\pi^2}{2}\frac{L}{H}M_2\frac{P_{tot}}{\lambda^2}\right)^{1/2} . \tag{2.14.65}$$

For small values of the argument on the right-hand side of (2.14.65), we obtain an equation which is the same as (2.14.37) except for a factor of 4:

$$\frac{I_1(L)}{I_0} = \frac{\pi^2}{2}\frac{L}{H}M_2\frac{P_{tot}}{\lambda^2} . \tag{2.14.66}$$

If we now include the effect of the finite sound beam width in the Bragg diffraction we obtain

$$\frac{I(\theta)}{I_0} \propto \left\{ \frac{\sin \left[\pi L/\lambda_s(\theta - \theta_B)\right]}{\frac{1}{2}\pi L/\lambda_s(\theta - \theta_B)} \right\}^2 . \tag{2.14.67}$$

The 3dB point of (2.14.67) occurs for

$$\frac{\pi L}{\lambda_s}(\Delta\theta)_s \approx 0.45\pi , \tag{2.14.68}$$

where $\Delta\theta_s$ is the value of $\theta - \theta_B$ for a half power point. By differentiating the Bragg condition (2.14.57) given by

$$\sin \theta_B = \frac{\lambda_1}{2\lambda_s} = \frac{\lambda_1}{2v_s}f_s , \tag{2.14.69}$$

we obtain the spread in the Bragg angle, $\Delta\theta_B$,

$$\Delta\theta_B = \frac{\lambda_1}{2v_s \cos \theta_B}\Delta f_s . \tag{2.14.70}$$

The bandwidth of the acousto-optic deflector will be given by the condition

$$\Delta\theta_B = \Delta\theta_s \quad \text{or} \tag{2.14.71}$$

$$\Delta f_s = \frac{1.8 n v_s^2 \cos \theta_B}{L f_{s0} \lambda_0} , \tag{2.14.72}$$

where f_{s0} is the center frequency of the acoustic beam. Thus the product of the bandwidth and the efficiency is

$$2 f_{s0}\Delta f_s\frac{I_1}{I_0} = \frac{1.8\pi^2}{\lambda_0^3 H \cos \theta_B}\frac{n^7 p^2}{\varrho v_s}P_{\text{tot}} \tag{2.14.73}$$

$$= \frac{1.8\pi^2}{\lambda_0^3 H \cos \theta_B}M_1 P_{\text{tot}} , \tag{2.14.74}$$

where $M_1 = n^7 p^2/\varrho v_s$ is the figure of merit related to the bandwidth and efficiency. The acoustic power P_{tot} needed for a certain diffraction efficiency can be minimized by reducing H to the limiting value, $h_{\min}$, the size of the optic beam. For this case, the bandwidth Δf_s will be given by

$$\Delta f_s = \frac{1}{\tau_r} \tag{2.14.75}$$

where τ_r is the transit time of the acoustic beam through the optic beam. Thus

$$\Delta f_s = \frac{h_{\min}}{v_s} \quad \text{or} \quad h_{\min} = \frac{v_s}{\Delta f_s} . \tag{2.14.76}$$

For this case, the expression for bandwidth efficiency is given by

$$2 f_{s0}\Delta f_s\frac{I_1}{I_0} = \frac{1.8\pi^2}{\lambda_0^3 \cos \theta_B}(\Delta f_s)\frac{n^7 p^2}{\varrho v^2}P , \quad \text{or} \tag{2.14.77}$$

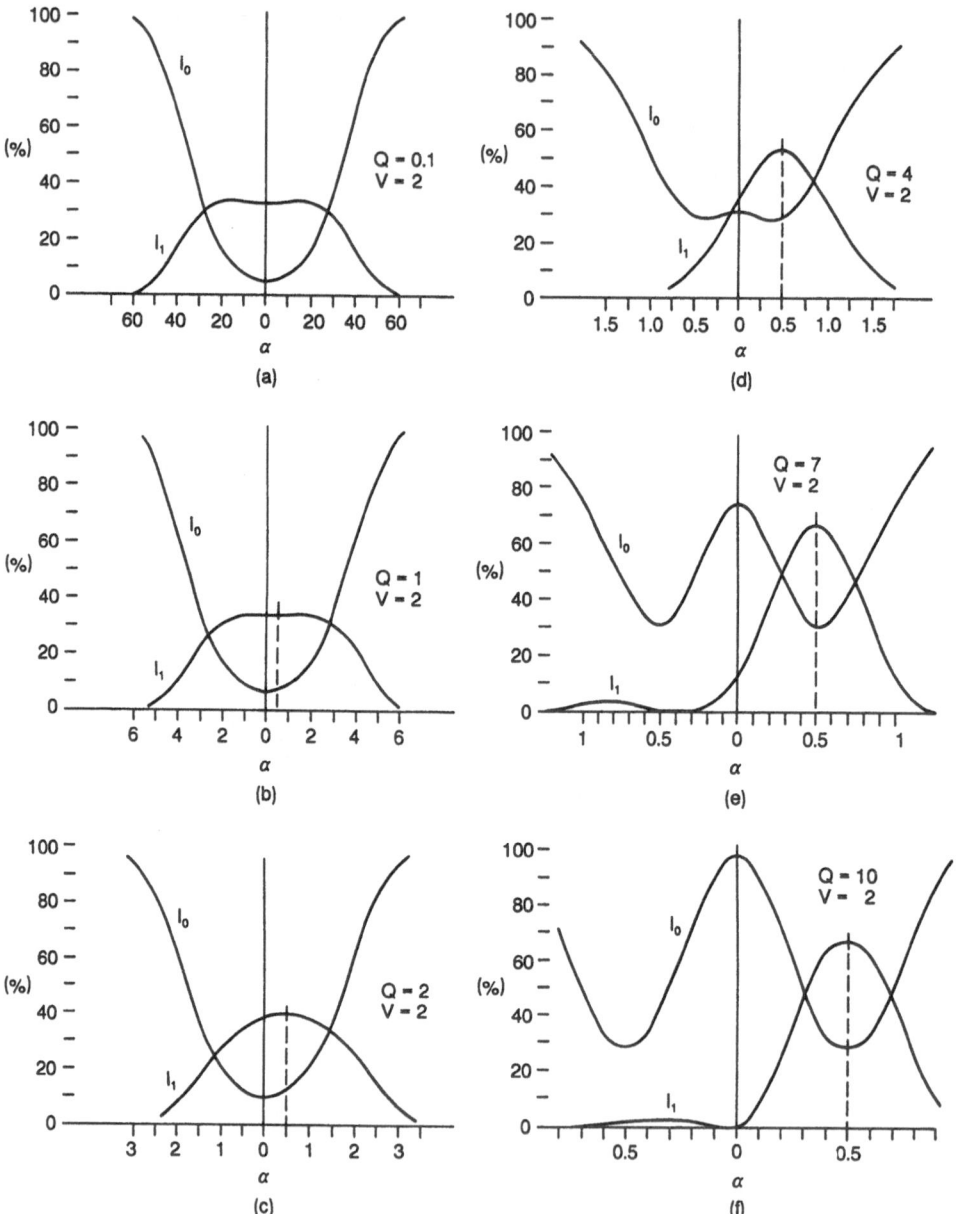

Fig. 2.78. Zeroth and first order light intensities vs incidence angle with Q as a parameter. (From [2.18])

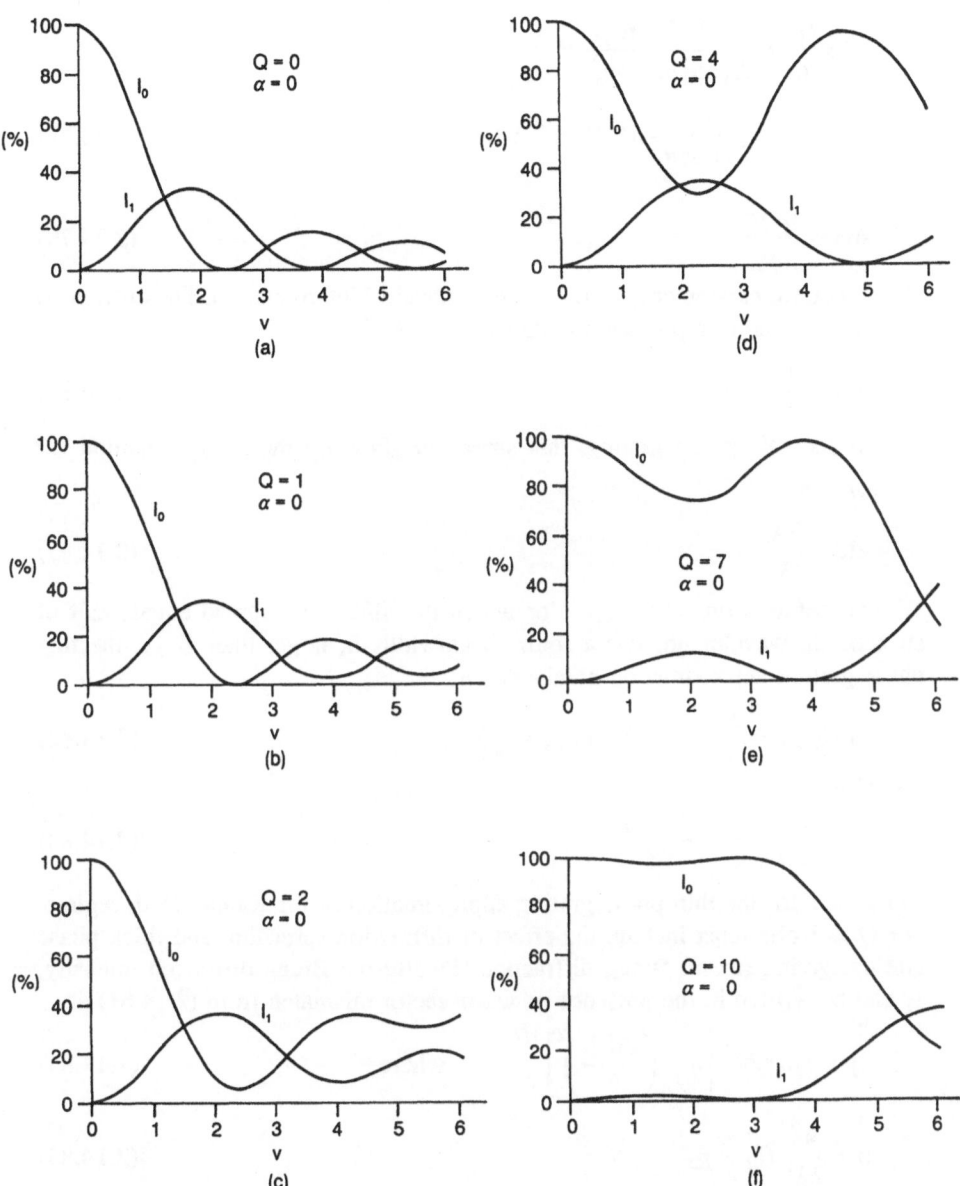

Fig. 2.78. (cont.)

$$2f_{s0}\frac{I_1}{I_0} = \frac{1.8\pi^2}{\lambda_0^3 \cos\theta_B}\frac{n^7 p^2}{\varrho v_s^2}P$$

$$= \frac{1.8\pi^2}{\lambda_0^3 \cos\theta_B}M_3 P \quad , \quad \text{where} \tag{2.14.78}$$

$$M_3 = \frac{n^7 p^2}{\varrho v_s^2} \quad . \tag{2.14.79}$$

The important parameter Q also has a nice physical interpretation. The diffraction spread, $\Delta\theta$, due to a propagation distance, z, is given by

$$\Delta\theta = \frac{\lambda_l}{z} \quad . \tag{2.14.80}$$

For an acoustic phase grating, this spread is given by the above equation for $z = \lambda_s/2$, or

$$\Delta\theta = \frac{2\lambda_l}{\lambda_s} \quad . \tag{2.14.81}$$

We can define a quantity L_{max}, for which the diffraction spread equals half of an acoustic wavelength. For a sound beamwidth L, larger than L_{max}, the thin phase grating approximation breaks down:

$$\Delta\theta L_{max} = \frac{\lambda_s}{2} \quad , \quad \text{or} \quad L_{max} = \frac{\lambda_s^2}{4\lambda_l} \quad . \tag{2.14.82}$$

We note

$$Q = \frac{\pi}{2}\frac{L}{L_{max}} = \frac{2\pi L\lambda_l}{\lambda_s^2} \tag{2.14.83}$$

and $Q \ll 1$ for the thin phase grating approximation or the Raman-Nath regime. For $Q \gg 1$ one must include the effect of diffraction spreading and thick phase grating, giving rise to Bragg diffraction. Finally the Bragg diffracted intensity, I_1 can be written in the presence of wave vector mismatch from (2.14.61) as

$$I_1 = I_0\eta \, \text{sinc}^2\left[\eta + \left(\frac{\Delta k L}{2}\right)^2\right]^{1/2} \quad , \quad \text{where} \tag{2.14.84}$$

$$\eta = \frac{\pi^2}{2\lambda^2}H_2\frac{L}{\lambda_s}p_s \tag{2.14.85}$$

and Δk is the momentum mismatch between the incident light and the acoustic propagation vector.

2.14.4 Acousto-optic Interaction Including Light Polarization: Isotropic Solids

The wave equation in the presence of a polarization, ΔP, can be written as

$$\nabla^2 E = \frac{n_0^2}{c^2}\frac{\partial^2 E}{\partial t^2} + \mu\frac{\partial^2}{\partial t^2}(\Delta P) \quad . \tag{2.14.86}$$

For the acousto-optic interaction, ΔP is given by

$$\Delta P = (\Delta \varepsilon)_{ij} E_j \quad . \tag{2.14.87}$$

For the case of a fluid,

$$\Delta P = (\Delta \varepsilon) E \quad . \tag{2.14.88}$$

As the acoustic frequency $f_s \ll f_l$, we can rewrite (2.14.84) for the isotropic case as

$$\nabla^2 E = \frac{[n(x,t)]^2}{c^2} \frac{\partial^2 E}{\partial t^2} \quad , \quad \text{where} \tag{2.14.89}$$

$$\begin{aligned} n(x,t) &= n_0 + \Delta n_0 \sin (\omega_s t - k_s x + \delta) \\ &= n_0 + \Delta n \quad . \end{aligned} \tag{2.14.90}$$

Here Δn is the change in refractive index and must be evaluated by the procedure discussed in Sect. 2.14.1.

So far, we have not discussed the electric field polarization of the incident wave. Even when we consider only an isotropic solid, this complicates the situation significantly for the diffracted amplitude, even though the angle of diffraction remains the same. Firstly, consider the light propagating in the z direction and the longitudinal wave propagating in the x direction. Thus we have

$$E = i_x E_x + i_y E_y \quad . \tag{2.14.91}$$

The only nonzero component of strain tensor is S_1 or S_{xx}. Thus the index ellipsoid for this case becomes

$$\frac{x^2 + y^2 + z^2}{n_0^2} + x^2 p_{11} S_1 + y^2 p_{12} S_1 + z^2 p_{12} S_1 = 0 \quad . \tag{2.14.92}$$

Writing the above expression as

$$\frac{x^2}{n_x^2} + \frac{y^2}{n_y^2} + \frac{z^2}{n_z^2} = 1 \quad , \tag{2.14.93}$$

we have n_x and n_y given by

$$n_x = n_0 - \frac{n_0^3}{2} p_{11} S_1 \quad , \tag{2.14.94}$$

$$n_y = n_0 - \frac{n_0^3}{2} p_{12} S_1 = n_z \quad . \tag{2.14.95}$$

If we are in the Raman-Nath regime and using the thin grating approximation, then an incident field given by

$$E(0) = i_x E_x(0) + i_y E_y(0) \tag{2.14.96}$$

will at the output be given by

$$E(L) = i_x E_x(L) + i_y E_y(L) \quad , \quad \text{where} \tag{2.14.97}$$

143

$$E_x(L) = E_x(0)e^{j\Delta\phi_x(x,t)} \quad , \tag{2.14.98}$$

$$E_y(L) = E_y(0)e^{j\Delta\phi_y(x,t)} \quad , \tag{2.14.99}$$

$$\Delta\phi_x = \frac{2\pi}{\lambda_0}(\Delta n)_{x0}L \sin(\omega_s t - k_s x + \delta) \quad , \tag{2.14.100}$$

$$\Delta\phi_y = \frac{2\pi}{\lambda_0}(\Delta n)_{y0}L \sin(\omega_s t - k_s x + \delta) \quad , \tag{2.14.101}$$

$$(\Delta n)_{x0} = -\frac{n_0^3}{2}p_{11}(S_1)_0 \quad , \tag{2.14.102}$$

$$(\Delta n)_{y0} = -\frac{n_0^3}{2}p_{12}(S_1)_0 \quad \text{and} \tag{2.14.103}$$

$$(S_1)_0 = \text{the amplitude of } S \quad , \tag{2.14.104}$$

The x and y polarized light is diffracted differently as $(\Delta n)_x \neq (\Delta n)_y$. Following the method used previously, we obtain

$$E_x(L) = E_x(0)\sum_{q=-\alpha}^{+\alpha} I_q\left[\frac{2\pi}{\lambda_0}(\Delta n)_{x0}L\right]e^{j(\omega_q t - k_q \cdot r)}e^{jq\delta} \quad , \tag{2.14.105}$$

and

$$E_y(L) = E_y(0)\sum_{q=-\alpha}^{+\alpha} I_q\left[\frac{2\pi}{\lambda_0}(\Delta n)_{y0}L\right]e^{j(\omega_q t - k_q \cdot r)}e^{jq\delta} \quad . \tag{2.14.106}$$

It is interesting to note that if the incident light is polarized parallel to the direction of acoustic propagation, it is diffracted more efficiently than when it is polarized perpendicular to the direction of propagation. This happens because $p_{11} > p_{12}$. The calculations can be extended to include the equivalent cases presented by (2.14.12).

For a shear wave propagation polarized in the y direction, we only have the nonzero S_5 (S_{xy} or S_{yx}) component. For this case the index ellipsoid can be written as

$$\frac{x^2 + y^2 + z^2}{n_0^2} = 2xyp_5S_5 = 0 \quad . \tag{2.14.107}$$

Using the coordinate transformation

$$x' = x\cos 45° + y\sin 45° \quad , \quad \text{and} \tag{2.14.108}$$

$$y' = x\sin 45° + y\cos 45° \quad , \tag{2.14.109}$$

we obtain

$$x'^2\left(\frac{1}{n_0^2} + p_5S_5\right) + y'^2\left(\frac{1}{n_0^2} - p_5S_5\right) + \frac{z^2}{n_0^2} = 1 \quad , \quad \text{or}$$

$$\frac{x'^2}{n_{x'}^2} + \frac{y'^2}{n_{y'}^2} + \frac{z^2}{n_0^2} = 1 \quad , \quad \text{where} \tag{2.14.110}$$

$$n_{x'} = n_0 - \frac{n_0^3}{2} p_5 S_5 \quad , \quad \text{and} \tag{2.14.111}$$

$$n_{y'} = n_0 + \frac{n_0^3}{2} p_5 S_5 \quad . \tag{2.14.112}$$

For this case we write the incident field as

$$E(0) = i_{x'} E_{x'}(0) + i_{y'} E_{y'}(0) \tag{2.14.113}$$

and the output electric field as

$$E(L) = i_{x'} E_{x'}(L) + i_{y'} E_{y'}(L) \quad , \quad \text{where} \tag{2.14.114}$$

$$E_{x'}(L) = E_{x'}(0) e^{j\Delta\phi_{x'}(x,t)} \quad , \tag{2.14.115}$$

$$E_{y'}(L) = E_{y'}(0) e^{j\Delta\phi_{y'}(x,t)} \quad , \tag{2.14.116}$$

$$\Delta\phi_{x'} = \frac{2\pi}{\lambda_0} (\Delta n)_{x'0} L \sin(\omega_s t - k_s x + \delta) \quad , \tag{2.14.117}$$

$$\Delta\phi_{y'} = \frac{2\pi}{\lambda_0} (\Delta n)_{y'0} L \sin(\omega_s t - k_s x + \delta) \quad , \tag{2.14.118}$$

$$(\Delta n)_{x'0} = -\frac{n_0^3}{2} p_5 S_{50} \quad , \tag{2.14.119}$$

$$(\Delta n)_{y'0} = +\frac{n_0^3}{2} p_5 S_{50} \quad , \quad \text{and} \tag{2.14.120}$$

$$S_{50} = |S_5| \quad . \tag{2.14.121}$$

We also obtain

$$E_{x'}(L) = E_{x'}(0) \sum_{2=-\infty}^{+\infty} J_q \left[\frac{2\pi}{\lambda_0} (\Delta n)_{x'0} \right] e^{j(\omega_q t - k_q \cdot r)} e^{jq\delta} \quad . \tag{2.14.122}$$

and

$$E_{y'}(L) = E_{y'}(0) \sum_{2=-\infty}^{+\infty} J_q \left[-\frac{2\pi}{\lambda_0} (\Delta n)_{x'0} \right] e^{j(\omega_q t - k_q \cdot r)} e^{jq\delta} \quad . \tag{2.14.123}$$

Note that $J_q(-m) = J_q(m)$ for $q = $ even integer and $J_q(-m) = -J_q(m)$ for $q = $ odd integer. Thus if the incident beam is polarized in the x direction only, then

$$E(0) = \frac{1}{\sqrt{2}}(i_{x'}E_0 + i_{y'}E_0) \quad . \tag{2.14.124}$$

For this case

$$
\begin{aligned}
E(L) = {} & i_x E_0 \sum J_{2q}\left[\frac{2\pi}{\lambda_0}(\Delta n)_{x'0}L\right]e^{j(\omega_{2q}-k_{2q}\cdot r)}e^{j2q\delta} \\
& + i_y E_0 \sum J_{2q+1}\left[\frac{2\pi}{\lambda_0}(\Delta n)_{y'0}L\right]e^{j(\omega_{2q+1}-k_{2q+1}\cdot r)}e^{j(2q+1)\delta} \quad .
\end{aligned}
\tag{2.14.125}
$$

Thus all the even orders are x polarized and the odd orders are y polarized.

2.14.5 Bragg Acousto-optic Interaction: Light Polarization Included

We already know that there will be only one diffracted order in this case. The wave equation is given by

$$\nabla^2 E_i = \mu \varepsilon_i \frac{\partial^2 \varepsilon_i}{\partial t^2} + \mu \frac{\partial^2}{\partial t^2}(\Delta P)_i \tag{2.14.126}$$

where $(\Delta P)_i$ is the change in polarization due to an acoustic field

$$= (\Delta \varepsilon)_{ij} E_j = -\tfrac{1}{2}\varepsilon_{im}\varepsilon_{in}p_{mnkl}S_{kl}E_j \quad . \tag{2.14.127}$$

Note that E_i is the incident electric field polarized in the i direction. Without any acoustic field, the wave equation is given by

$$\nabla^2 E_i = \mu \varepsilon_i \frac{\partial^2 E_i}{\partial t^2} \quad . \tag{2.14.128}$$

The term $(\Delta P)_i$ acts as a source term and thus provides necessary coupling for the diffracted electric field denoted by E_i^d.

$$\nabla^2 E_i^d = \mu \varepsilon_i^d \frac{\partial^2 E_i^d}{\partial t^2} + \mu \frac{\partial^2}{\partial t^2}(\Delta P)_i \quad . \tag{2.14.129}$$

For Bragg diffraction to occur, we must have phase matching for the forcing term $(\Delta P)_i$. As

$$E_i = E_i(r_i)e^{j(\omega_i t - k_i \cdot r)} \quad , \quad \text{and} \tag{2.14.130}$$

$$E_i^d = E_i^d(r_d)e^{j(\omega_d t - k_d \cdot r)} \tag{2.14.131}$$

we obtain

$$
\begin{aligned}
(\Delta P)_i \propto {} & e^{j(\omega_s t - k_s \cdot r)}[E_i(r_i)e^{j(\omega_i t - k_i \cdot r)} \\
& + E_i^d(r_d)e^{j(\omega_d t - k_d \cdot r)}] \quad .
\end{aligned}
\tag{2.14.132}
$$

For the incident field equation we have[5]

---

[5] In general E_i and E_i^d, can be polarized in different directions. However, for simplification we will assume that they are polarized in the same direction and denote $E_i^d(r)$ simply by $E_d(r)$.

$$(\Delta P)_i^i \propto E_d e^{j[(\omega_d + \omega_s)t - (k_s + k_d) \cdot r]}$$ (2.14.133)

as the other terms cannot satisfy the phase matching condition given by

$$\omega_d + \omega_s = \omega_i \quad , \quad \text{and}$$ (2.14.134)

$$k_d + k_s = k_i \quad .$$ (2.14.135)

Similarly for the diffracted field equation

$$(\Delta P)_i^d \propto E_i(r_i) e^{j[(\omega_i - \omega_s)t - (k_i - k_s) \cdot r]} \quad .$$ (2.14.136)

From (2.14.130), one obtains

$$\nabla^2 E_i = \frac{1}{2}\left[k_i^2 E_i + 2jk_i \frac{dE_i}{dr_i} + \nabla^2 E_i \right] e^{j(\omega_i t - k_i \cdot r)} \quad .$$ (2.14.137)

Assuming $\nabla^2 E_i \ll k_i dE_i/dr_i$, i.e, that the coupling is small so that the variation in E_i is also small, we obtain from (2.14.126)

$$k_i \frac{dE_i}{dr_i} = j\mu \left[\frac{\partial^2}{\partial t^2}(\Delta P)_i^i \right] e^{j(\omega_i t - k_i \cdot r)}$$

$$= j\mu(\Delta\varepsilon)_i E_d(r_d)\omega_i^2 \quad , \quad \text{or}$$ (2.14.138)

$$\frac{dE_i}{dr_i} = j\eta_i E_d \quad , \quad \text{where}$$ (2.14.139)

$$\eta_i = \frac{\mu(\Delta\varepsilon)_i \omega_i^2}{k_i} = \omega_i \frac{\mu}{\sqrt{\mu\varepsilon_i}}(\Delta\varepsilon)_i = \frac{1}{2} \frac{\omega_i \Delta n}{c_0} \quad .$$ (2.14.140)

Similarly, one can derive

$$\frac{dE_d}{dr_d} = j\eta_d E_i \quad , \quad \text{where}$$ (2.14.141)

$$\eta_d = \omega_d \frac{\mu}{\sqrt{\mu\varepsilon_d}}(\Delta\varepsilon)_i = \frac{1}{2} \frac{\omega_d \Delta n}{c_0} = \frac{\pi \Delta n}{\lambda} \quad .$$ (2.14.142)

As

$$\omega_d \approx \omega_i$$ (2.14.143)

one has

$$\eta_d \approx \eta_i = \eta \quad .$$ (2.14.144)

Thus

$$\frac{dE_i}{dr_i} = -j\eta E_d \quad , \quad \text{and}$$ (2.14.145)

$$\frac{dE_d}{dr_d} = -j\eta E_i \quad .$$ (2.14.146)

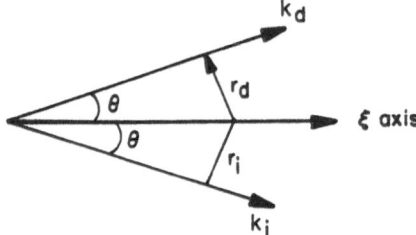

Fig. 2.79. Definition of the new coordinate system

Note that the above equations are very similar to those derived earlier (2.14.58, 59). If we define a new coordinate system along the bisector of the angle formed by k_i and k_d, given by ξ as shown in Fig. 2.79, we can write

$$r_i = \xi \cos \theta = r_d \qquad (2.14.147)$$

Substituting (2.14.147) into (2.14.145) and into (2.14.146) one obtains

$$\frac{dE_i}{d\xi} = -j\eta E_d \cos \theta \quad \text{and} \qquad (2.14.148)$$

$$\frac{dE_d}{d\xi} = -j\eta E_i \cos \theta \quad . \qquad (2.14.149)$$

The solutions for these equations are given by

$$E_i(\xi) = E_i(0) \cos (\eta \xi \cos \theta) \quad , \quad \text{and} \qquad (2.14.150)$$

$$E_d(\xi) = -jE_i(0) \sin (\eta \xi \cos \theta) \quad , \qquad (2.14.151)$$

where the incident field is given by $E_i(0)$. Substituting (2.14.147) for r_i and r_d, one can rewrite

$$E_i(r_i) = E_i(0) \cos (\eta r_i) \quad , \quad \text{and} \qquad (2.14.152)$$

$$E_d(r_d) = -jE_i(0) \sin (\eta r_d) \quad . \qquad (2.14.153)$$

The diffracted intensity is given by

$$I_d/I_i = \frac{E_d^2(L)}{E_i^2(0)} = \sin^2 \left(\frac{\pi L}{\lambda} \Delta n \right) \qquad (2.14.154)$$

which is the same expression as obtained previously (2.14.64). However, in the present formulation we can see the coupling between different terms given by the $(\Delta P)_i$ term, by expanding the expression for $(\Delta P)_i$ as follows:

$$(\Delta P)_i = \Delta \varepsilon_{ij} E_j \propto \Delta \left(\frac{1}{n^2} \right)_{ij} E_j \quad . \qquad (2.14.155)$$

For example,

$$(\Delta P)_x \propto \Delta \left(\frac{1}{n^2} \right)_{xx} E_x + \Delta \left(\frac{1}{n^2} \right)_{xy} E_y + \Delta \left(\frac{1}{n^2} \right)_{xz} E_z$$

$$= \Delta\left(\frac{1}{n^2}\right)_1 E_x + \Delta\left(\frac{1}{n^2}\right)_6 E_y + \Delta\left(\frac{1}{n^2}\right)_5 E_z \quad . \tag{2.14.156}$$

In the above equation we have neglected some constant terms related to $(-n_{i0}^3/2)$ because we are more interested in finding the coupling between the polarization of incident and diffracted fields. Similarly one can write

$$(\Delta P)_y = \Delta\left(\frac{1}{n^2}\right)_6 E_x + \Delta\left(\frac{1}{n^2}\right)_2 E_y + \Delta\left(\frac{1}{n^2}\right)_4 E_z \tag{2.14.157}$$

$$(\Delta P)_z = \Delta\left(\frac{1}{n^2}\right)_5 E_x + \Delta\left(\frac{1}{n^2}\right)_4 E_y + \Delta\left(\frac{1}{n^2}\right)_3 E_z \quad . \tag{2.14.158}$$

The corresponding Δn to be used in (2.14.154) for coupling to one electric field only is obtained by multiplying the relevant $\Delta(1/n^2)$ with the constant term. For the use of an isotropic solid with the compressional wave along the x direction we have

$$(\Delta P)_x \propto \Delta\left(\frac{1}{n^2}\right)_1 E_x \quad , \tag{2.14.159}$$

$$(\Delta P)_y \propto \Delta\left(\frac{1}{n^2}\right)_2 E_y \quad , \tag{2.14.160}$$

$$(\Delta P)_z \propto \Delta\left(\frac{1}{n^2}\right)_3 E_z \quad . \tag{2.14.161}$$

Thus the incident electric field and diffracted light field must be polarized in the same direction. However, for x polarized light

$$(\Delta n)_x = -\frac{n_0^3}{2} p_{11} S_1 \tag{2.14.162}$$

whereas for y or z polarized light

$$(\Delta n)_y = (\Delta n)_z = -\frac{n_0^3}{2} p_{12} S_1 \quad . \tag{2.14.163}$$

For the shear wave propagating along the x direction and particle velocity along the y direction, we have

$$(\Delta P)_x = \left(\frac{1}{n^2}\right)_6 E_y \tag{2.14.164}$$

$$(\Delta P)_y = \left(\frac{1}{n^2}\right)_6 E_x \tag{2.14.165}$$

$$(\Delta P)_z = 0 \quad . \tag{2.14.166}$$

Thus for this case only E_x and E_y couple with each other. If the incident light is

149

E_x, the diffracted light will be E_y and vice versa. However, E_z does not couple to E_x or E_y. The corresponding Δn to be used in (2.14.154) is given by

$$\Delta n = -\frac{n^3}{2}\Delta\left(\frac{1}{n^2}\right)_6 \quad .$$

If the incident wave is circularly polarized, the diffracted light will also be circularly polarized but of opposite rotation.

General Procedure. First determine θ_i and θ_d from the refractive index ellipsoid. Thus k_i and k_d are known. Find the two orthogonal polarizations for this k_i and k_d. Find the component of (ΔP) along the polarization directions which couple. Then determine the equivalent applicable Δn for (2.14.154).

2.14.6 Bragg Diffraction: Anisotropic Case

The electromagnetic wave with angular frequency ω and wave-vector K can also be considered to consist of photons with energy and momentum given by

$$\text{energy} = \hbar\omega = hf \tag{2.14.167}$$

$$\text{momentum} = \hbar k = \frac{h}{\lambda}i_k \tag{2.14.168}$$

where h is Planck's constant.

Similarly, for the acoustic wave, one has phonons with energy and momentum given by

$$\text{energy} = \hbar\omega_s = hf_s \quad , \tag{2.14.169}$$

$$\text{momentum} = \hbar k_s = (h/\lambda_s)i_{k_s} \quad . \tag{2.14.170}$$

If one regards acousto-optic interaction as a photon-phonon interaction, then due to the conservation of energy, one has

$$\hbar\omega_i = \hbar\omega_d - \hbar\omega_s \quad , \quad \text{or} \tag{2.14.171}$$

$$f_i = f_d - f_s \quad . \tag{2.14.172}$$

In general $f_s \ll f_d$, f_i. Similarly, to satisfy the momentum conservation law,

$$\hbar k_i = \hbar k_d - \hbar k_s \quad , \quad \text{or} \tag{2.14.173}$$

$$k_i = k_d - k_s \quad . \tag{2.14.174}$$

Note that

$$|k_i| = \frac{2\pi}{\lambda_0}n_i \quad ; \quad |k_d| = \frac{2\pi}{\lambda_0}n_d\left(\frac{f_i + f_s}{f_i}\right) \approx \frac{2\pi}{\lambda_0}n_d \quad , \quad \text{and} \tag{2.14.175}$$

$$|k_s| = \frac{2\pi f_s}{v_s} \tag{2.14.176}$$

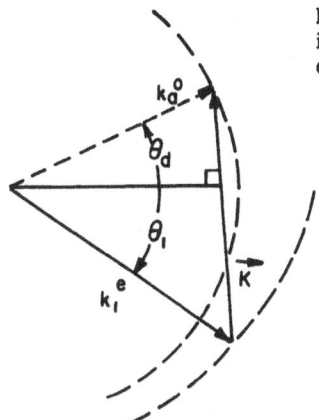

Fig. 2.80. Wave-vector construction describing Bragg diffraction in a positive uniaxial crystal when the incident optical wave is extraordinary polarized

where n_i and n_d are the refractive indices corresponding to the incident and diffracted waves respectively and v_s is the velocity of sound. Thus (2.14.174) becomes

$$n_i i_{k_i} = n_d i_{k_d} - \frac{f_s \lambda_0}{v_s} i_{k_s} \quad . \tag{2.14.177}$$

If we choose the angle between the normal to i_{k_s} and i_{k_i} as θ_i and similarly for i_{k_d}, θ_d (as shown in Fig. 2.80), then equating the parallel and normal components, we have

$$n_i \sin \theta_i = n_d \sin \theta_d - \frac{f_s \lambda_0}{v_s} \quad , \tag{2.14.178}$$

$$n_i \cos \theta_i = n_d \cos \theta_d \quad . \tag{2.14.179}$$

Note that k_i, k_d and k_s have to be coplanar.

Solving for $\sin \theta_i$ and $\sin \theta_d$ from (2.14.178) and (2.14.179) one obtains

$$\sin \theta_i = \frac{1}{2n_i} \frac{\lambda_0 f_s}{v_s} \left[1 + \left(\frac{v_s}{\lambda_0 f} \right)^2 (n_i^2 - n_d^2) \right] \quad , \tag{2.14.180}$$

$$\sin \theta_d = \frac{1}{2n_d} \frac{\lambda_0 f_s}{v_s} \left[1 - \left(\frac{v_s}{\lambda_0 f} \right)^2 (n_i^2 - n_d^2) \right] \quad . \tag{2.14.181}$$

For the isotropic case when $n_i = n_d$, we have

$$\sin \theta_B = \sin \theta_i = \sin \theta_d = \frac{\lambda_l}{2\lambda_s} \tag{2.14.182}$$

where $\lambda_l = \lambda_0/n$ is the wavelength of the light in the material itself. The vector diagram for this case is shown in Fig. 2.81. For uniaxial crystals, for diffraction in a plane perpendicular to the optical axis, we have

$$n = n_o \quad \text{or} \quad n_e$$

and the wave-vector surface is shown in Fig. 2.62. For this case, if k_i corre-

Fig. 2.81. Normal Bragg diffraction geometry

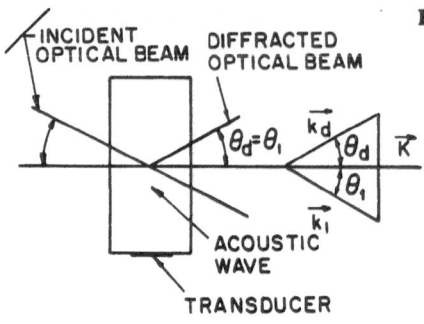

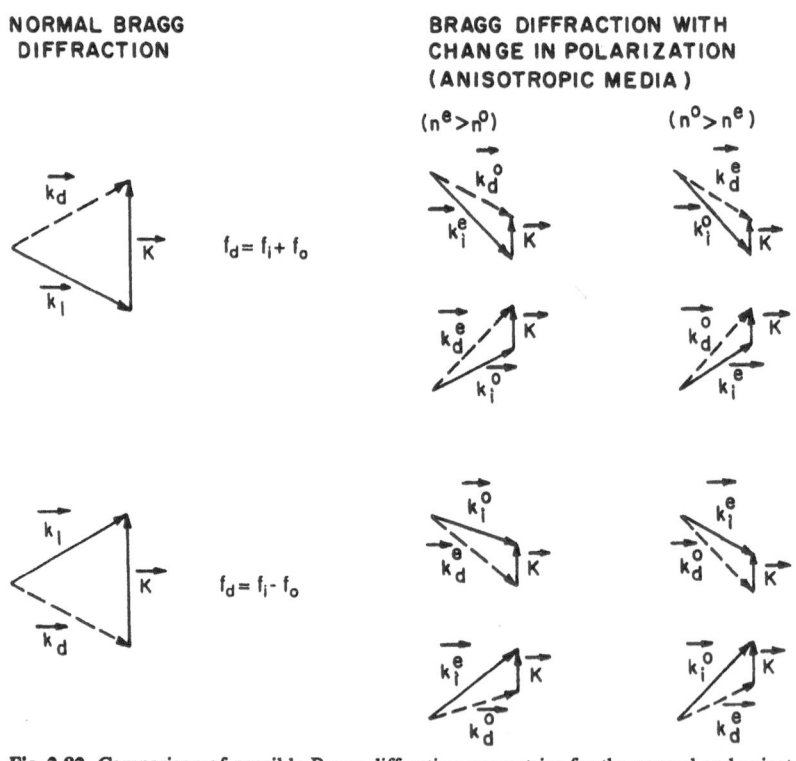

Fig. 2.82. Comparison of possible Bragg diffraction geometries for the normal and anisotropic cases. (Bragg diffraction in anisotropic crystals is "normal" unless the incident and diffracted polarizations differ). (From [2.19])

sponds to $n_e > n_o$ and k_d corresponds to n_o, then the vector diagram is shown in Fig. 2.82. The angles θ_i and θ_d obey (2.14.180, 181). A plot of θ_i and θ_d versus f_s is shown in Fig. 2.83 for a particular set of parameters $n_0 = 1.9$, $n_e = 2$, $v_s = 10^5$ cm/s and $\lambda_0 = 1\ \mu$m. We note that for $f \to f_{\min}$, $\theta_i \to +90°$ and $\theta_d \to -90°$. Also, as $f \to \infty$, $\theta_i \to +90°$ and $\theta_d \to 90°$, $f_{\min}$ is determined by (2.14.180) with $\theta_i = 90°$, i.e.,

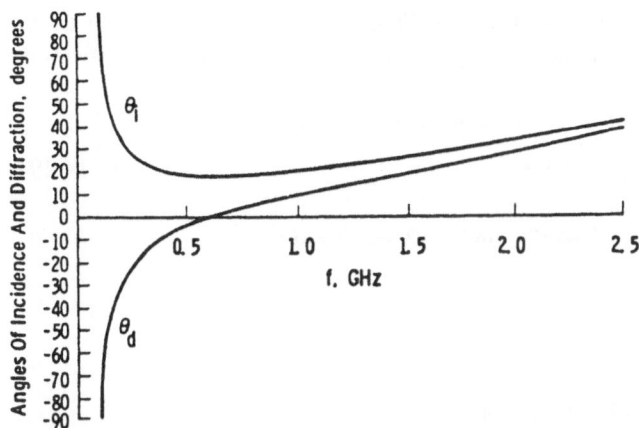

Fig. 2.83. Angles of incidence and diffraction for an anisotropic medium in which $n_i = 2$, $n_d = 1.9$, $v = 10^5$ cm/s and $\lambda = 1$ μm. (From [2.4])

$$1 = \frac{1}{2n_i} \frac{\lambda_0 f_{min}}{v_s} \left[1 + \left(\frac{v_s}{\lambda_0 f_{min}} \right)^2 (n_i^2 - n_d^2) \right] \quad , \tag{2.14.183}$$

from which we obtain

$$f_{min} = \frac{v_s(n_i - n_d)}{\lambda_0} \quad , \quad \text{and} \tag{2.14.184}$$

$$f_{max} = \frac{v_s}{\lambda_0}(n_i + n_d) \quad . \tag{2.14.185}$$

Note that for the case of f_{min}, k_s, k_d and k_i are collinear and thus obey the relationship

$$|k_s| + |k_i| = |k_d| \quad . \tag{2.14.186}$$

Using the above equation one can also easily obtain the expression for f_{min} given by (2.14.184).

Note that for f_{max},

$$|k_s| = |k_d| + |k_i| \quad . \tag{2.14.187}$$

To obtain the minimum angle, θ_i^{min} and the corresponding f_s, we use

$$\frac{\partial \theta_i}{\partial f} = 0 \quad . \tag{2.14.188}$$

Using (2.14.180) and differentiating with respect to θ_i, we obtain

$$\sin \theta_i^{min} = \frac{\lambda f_s}{n_i v_s} \tag{2.14.189}$$

and this occurs at

$$f_s = \frac{v_s}{\lambda}(n_i^2 - n_d^2)^{1/2} \quad . \tag{2.14.190}$$

153

Note that $\theta_d = 0$ for $\theta_i = \theta_i$ min. Because $\partial\theta_i/\partial f = 0$ at θ_i min, one can obtain a large bandwidth for which acousto-optic interaction can take place efficiently near f_s given by (2.14.190).

Optical activity from (2.15.38) can be written as

$$(n^2 - n_0^2)^2 = G^2 \quad , \quad \text{where} \tag{2.14.191}$$

$$n = \sqrt{\varepsilon_{\text{eff}}} \quad , \quad \text{the refractive index to be determined} \quad ,$$

$$n_0 = \sqrt{\varepsilon_{11}} \quad , \quad \text{and}$$

$$G = \varepsilon_{12} \quad .$$

For an anisotropic crystal, this becomes

$$(n^2 - n_1^2)(n^2 - n_2^2) = G^2 \quad . \tag{2.14.192}$$

In general G can be written as

$$G = g_{ij} l_i l_j \quad , \tag{2.14.193}$$

where l_i are the direction cosines of i_k. Note that $g_{ij} = g_{ji}$. For a positive ($n_e > n_0$) birefringent crystal, near the optic axis one has

$$\frac{n_1^2(\theta)\cos^2\theta}{n_0^2} + \frac{n_1^2(\theta)\sin^2\theta}{n_e^2} = 1 \tag{2.14.194}$$

and either

$$\frac{n_2^2(\theta)\cos^2\theta}{n_0^2} + \frac{n_2^2(\theta)\sin^2\theta}{n_e^2} = 1 \quad , \quad \text{or} \quad n_2(\theta) = n_0 \tag{2.14.195}$$

where θ is the angle from the optic axis.

In the presence of optical activity these equations become

$$\frac{n_1^2(\theta)\cos^2\theta}{n_0^2(1+\delta)^2} + \frac{n_1^2(\theta)\sin^2\theta}{n_e^2} = 1 \quad , \quad \text{and} \tag{2.14.196}$$

$$\frac{n_2^2(\theta)\cos^2\theta}{n_0^2(1-\delta)^2} + \frac{n_2^2(\theta)\sin^2\theta}{n_e^2} = 1 \tag{2.14.197}$$

where n_1 and n_2 refer to the circularly polarized light and δ is related to G. For $\theta = 0$,

$$n_1 = n_0(1 - \delta) \quad \text{and} \tag{2.14.198}$$

$$n_2 = n_0(1 + \delta) \tag{2.14.199}$$

as expected. For small θ, we have

$$n_1^2 = n_0^2\left(1 + 2\delta\,\cos^2\theta + \frac{n_e^2 - n_0^2}{n_e^2}\,\sin^2\theta\right) \quad \text{and} \tag{2.14.200}$$

$$n_2^2 = n_0^2(1 - 2\delta \cos^2 \theta) \ . \tag{2.14.201}$$

Substituting these values of n_1 and n_2 for n_i and n_d respectively in (2.14.180 and 181), we obtain for θ_i and θ_d, assuming $\delta \ll 1$,

$$\sin \theta_i \approx \frac{\lambda_0 v_s}{2n_0 f_s} \left[1 + \frac{4n_0^2 f_s^2}{\lambda^2 v_s^2} \delta + \frac{\sin^2 \theta_i f_s^2 n_0^2}{\lambda^2 v_s^2} \left(\frac{n_e^2 - n_0^2}{n_e^2} \right) \right] \tag{2.14.202}$$

$$\sin \theta_d \approx \frac{\lambda_0 v_s}{2n_0 f_s} \left[1 - \frac{4n_0^2 f_s^2}{\lambda^2 v_s^2} \delta - \frac{\sin^2 \theta_i}{\lambda^2 v_s^2} f_s^2 n_0^2 \left(\frac{n_e^2 - n_0^2}{n_e^2} \right) \right] \ . \tag{2.14.203}$$

Tellurium dioxide (TeO_2) is an optically active anisotropic acousto-optic crystal. Thus (2.14.196–203) are valid for TeO_2, which is also of practical importance as it has one of the largest acousto-optic figures of merit.

2.15 Magneto-optics

In dealing with magneto-optics, the polarization of the electromagnetic wave plays the most important role. To represent the light polarization as it passes through optical elements, it is convenient to introduce the Jones matrix formalism. This is considered in Sect. 2.15.1, which is followed by a description of opticcal activity and then by a review of magneto-optics: Faraday effect, the Voigt effect and the Kerr effect.

2.15.1 Polarization and the Jones Matrix

For light propagating in the z direction, the electric field in the unbounded medium can be written as

$$E = (i_x E_{0x} e^{j\phi_x} + i_y E_{0y} e^{j\phi_y}) e^{j(\omega t - kz)} \tag{2.15.1}$$

where E_{0x} and ϕ_x are the amplitude and phase of the x polarized component. Redefining the phase with respect to ϕ_x we have

$$E = (i_x E_{0x} + i_y E_{0y} e^{j\phi}) e^{j(\omega t - kz)} \tag{2.15.2}$$

where ϕ is the phase difference between the x and y polarized components of light. Equation (2.15.2) can be rewritten in terms of matrices as

$$E = \begin{bmatrix} E_{0x} \\ E_{0y} e^{j\phi} \end{bmatrix} [i_x i_y] e^{j(\omega t - kz)} \tag{2.15.3}$$

The first (2×1) matrix

$$J = \begin{bmatrix} E_{0x} \\ E_{0y} e^{j\phi} \end{bmatrix} \tag{2.15.4}$$

is known as the Jones polarization vector. Generally, J is given by

$$J = \begin{bmatrix} E_{0x}e^{j\phi_x} \\ E_{0y}e^{j\phi_y} \end{bmatrix} \quad . \tag{2.15.5}$$

It is of interest to look at the following special cases of Jones vectors:

a) x polarized light $\qquad\qquad\qquad J = \begin{bmatrix} E_{0x} \\ 0 \end{bmatrix} = E_{0x}\begin{bmatrix} 1 \\ 0 \end{bmatrix}; \qquad (2.15.6)$

b) y polarized light $\qquad\qquad\qquad J = \begin{bmatrix} E_{0y} \\ 0 \end{bmatrix} = E_{0y}\begin{bmatrix} 0 \\ 1 \end{bmatrix}; \qquad (2.15.7)$

c) circularly polarized light (right) $\qquad J = \begin{bmatrix} 1 \\ +j \end{bmatrix}; \qquad\qquad\qquad (2.15.8)$

d) circularly polarized light (left) $\qquad J = \begin{bmatrix} 1 \\ -j \end{bmatrix}; \qquad\qquad\qquad (2.15.9)$

When adding two waves with Jones vectors J_1 and J_2, the resultant Jones vector J is given by

$$J = J_1 + J_2 \quad . \tag{2.15.10}$$

If J_1 and J_2 are orthogonal to each other, then they satisfy the relation

$$[J_1][J_2]^T = 0 \quad . \tag{2.15.11}$$

Any optical element can be represented by a (2×2) Jones matrix M^J given by

$$M^J = \begin{bmatrix} a & b \\ c & d \end{bmatrix} \quad , \tag{2.15.12}$$

where the values of a, b, c and d depend on the type of the optical element. For example, consider the Jones matrices for the following optical elements:

i) *Linear Polarizers*

a) Polarization parallel to x

$$M^J = \begin{bmatrix} 1 & 0 \\ 0 & 0 \end{bmatrix} \tag{2.15.13}$$

b) Polarization parallel to y

$$M^J = \begin{bmatrix} 0 & 0 \\ 0 & 1 \end{bmatrix} \tag{2.15.14}$$

c) Polarization at an angle of $45°$ with respect to the x axis

$$M^J = \frac{1}{2}\begin{bmatrix} 1 & 1 \\ 1 & 1 \end{bmatrix} \tag{2.15.15}$$

ii) *Quarter wave plate*

$$M^J = \begin{bmatrix} 1 & 0 \\ 0 & j \end{bmatrix} \quad , \quad \text{or} \tag{2.15.16}$$

$$M^J = \begin{bmatrix} 1 & 0 \\ 0 & -j \end{bmatrix}$$

iii) *Isotropic phase retarder*

$$M^J = \begin{bmatrix} e^{j\phi} & 0 \\ 0 & e^{j\phi} \end{bmatrix} \tag{2.15.17}$$

iv) *Relative phase changer*

$$M^J = \begin{bmatrix} e^{j\phi_x} & 0 \\ 0 & e^{j\phi_y} \end{bmatrix} \tag{2.15.18}$$

v) *Right circular polarizer*

$$M^J = \frac{1}{2} \begin{bmatrix} 1 & j \\ -j & 1 \end{bmatrix} \tag{2.15.19}$$

vi) *Left circular polarizer*

$$M^J = \frac{1}{2} \begin{bmatrix} 1 & -j \\ j & 1 \end{bmatrix} \quad . \tag{2.15.20}$$

If light with Jones vector J_1 is incident on an optical element with Jones matrix M^J then the output Jones vector J_2 is given by

$$J_2 = M^J J_1 \quad . \tag{2.15.21}$$

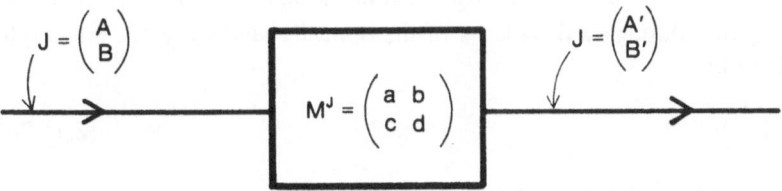

Fig. 2.84. Single optical element Jones matrix

For the case shown in Fig. 2.84

$$\begin{pmatrix} A' \\ B' \end{pmatrix} = \begin{pmatrix} a & b \\ c & d \end{pmatrix} \begin{pmatrix} A \\ B \end{pmatrix} \quad . \tag{2.15.22}$$

If more than one optical element is involved, then the resultant Jones matrix is the multiplication of the individual Jones matrices. For a three-element system shown in Fig. 2.85 we have

$$\begin{aligned} J_2 &= M^J J_1 \\ &= M_1^J M_2^J M_3^J J_1 \quad . \end{aligned} \tag{2.15.23}$$

157

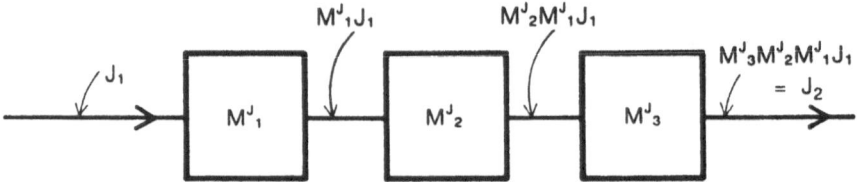

Fig. 2.85. Building up the Jones matrix for three optical elements

2.15.2 Optical Activity

Many crystals (e.g., quartz) have the property that they change the plane of polarization for incident linearly polarized light. This is known as optical activity. In our matrix notation, this implies that a crystal with optical activity has a Jones matrix given by

$$M^J = e^{j\psi} \begin{pmatrix} \cos\theta & \sin\theta \\ \sin\theta & \cos\theta \end{pmatrix} \quad , \tag{2.15.24}$$

where ψ is a constant phase angle and θ is the angle by which the plane of polarization is rotated. For example, if the incident Jones vector corresponds to x polarized light, as shown in Fig. 2.86, the output Jones vector is given by

$$J_2 = M^J J_1 = e^{j\psi} \begin{pmatrix} \cos\theta & \cos\theta \\ \sin\theta & \cos\theta \end{pmatrix} \begin{pmatrix} 1 \\ 0 \end{pmatrix} = e^{j\psi} \begin{pmatrix} \cos\theta \\ \sin\theta \end{pmatrix} \quad . \tag{2.15.25}$$

Optical activity can be explained in physical terms by considering that, in an optically active crystal, the right circularly polarized light and left circularly polarized light have different velocities and wave numbers given by v_r, k_r and v_l and k_l respectively. Thus, for the right-handed case, the input is $J_1 = \begin{pmatrix} 1 \\ j \end{pmatrix}$ and, after propagation through a distance z of the optically active crystal, the output will be given by

$$J_2 = \begin{pmatrix} 1 \\ j \end{pmatrix} e^{-jk_r z} \quad . \tag{2.15.26}$$

Similarly, for the left-handed case

$$J_2 = \begin{pmatrix} 1 \\ -j \end{pmatrix} e^{-jk_l z} \quad . \tag{2.15.27}$$

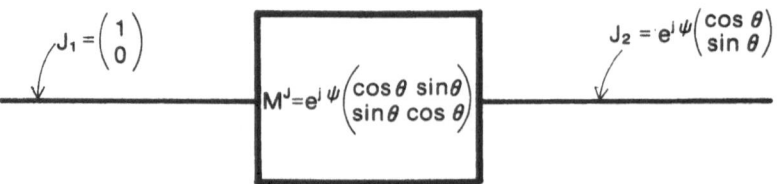

Fig. 2.86. The Jones matrix for optical activity. The incident Jones vector corresponds to x polarized light

158

To prove the result given by (2.15.25) we observe that the incident x polarized light can be decomposed into two, equal, right and left circularly polarized lights as follows

$$\begin{pmatrix} 1 \\ 0 \end{pmatrix} = \frac{1}{2}\begin{pmatrix} 1 \\ +j \end{pmatrix} + \frac{1}{2}\begin{pmatrix} 1 \\ -j \end{pmatrix} \quad . \tag{2.15.28}$$

Thus, using (2.15.26) and (2.15.27) the output for a length l of the crystal is given by

$$J_2 = \frac{1}{2}\begin{pmatrix} 1 \\ j \end{pmatrix} e^{-jk_r l} + \frac{1}{2}\begin{pmatrix} 1 \\ -j \end{pmatrix} e^{-jk_l l} \quad . \tag{2.15.29}$$

Defining

$$\psi = \tfrac{1}{2}(k_r + k_l)l \quad , \quad \text{and} \tag{2.15.30}$$

$$\theta = \tfrac{1}{2}(k_r - k_l)l \quad , \tag{2.15.31}$$

(2.15.29) can be rewritten as

$$J_2 = e^{j\psi}\begin{pmatrix} \cos\theta \\ \sin\theta \end{pmatrix} \quad . \tag{2.15.32}$$

We note that the polarization is rotated by an angle θ with respect to the x axis, θ being given by

$$\theta = (n_r - n_l)\frac{\pi l}{\lambda} = \delta l \tag{2.15.33}$$

where n_r and n_l are the respective refractive indices for the right- and left-handed circularly polarized light and δ is the specific rotary power.

The optical activity of a crystal corresponds to an effective dielectric tensor given by

$$\varepsilon = \varepsilon_0 \begin{bmatrix} \varepsilon_{11} & j\varepsilon_{12} & 0 \\ -j\varepsilon_{12} & \varepsilon_{11} & 0 \\ 0 & 0 & \varepsilon_{33} \end{bmatrix} \quad . \tag{2.15.34}$$

To relate ε_{11}, ε_{12} to n_r and n_l we use the wave equation given by (2.12.127) for the case $k_x = k_y = 0$.

$$-k^2 E + \omega^2 \mu ||\varepsilon|| E = 0 \quad . \tag{2.15.35}$$

As the electric field only has components E_x and E_y we obtain from (2.15.35) that

$$-k^2 E_x + \omega^2 \mu(\varepsilon_{11} E_x + j\varepsilon_{12} E_y) = 0 \quad , \quad \text{and} \tag{2.15.36}$$
$$-k^2 E_y + \omega^2 \mu(-j\varepsilon_{12} E_x + \varepsilon_{11} E_y) = 0 \quad .$$

Equations (2.15.36) have a solution if the following determinant is zero:

$$\begin{vmatrix} -k^2 + \omega^2 \mu \varepsilon_{11} & j\varepsilon_{12}\omega^2 \mu \\ -j\varepsilon_{12}\omega^2 \mu & -k^2 + \omega^2 \mu \varepsilon_{11} \end{vmatrix} = 0 \quad , \quad \text{or} \tag{2.15.37}$$

159

$$(k^2 - \omega^2 \mu \varepsilon_{11})(k^2 - \omega^2 \mu \varepsilon_{11}) = \omega^4 \mu^2 \varepsilon_{12}^2 \quad . \tag{2.15.38}$$

From (2.15.38) we obtain

$$k = \omega \sqrt{\mu \varepsilon_0} \sqrt{\varepsilon_{11} \pm \varepsilon_{12}} \quad . \tag{2.15.39}$$

By substituting the value of k in (2.15.36 or 37) we obtain

$$E_x = \pm j E_y \quad . \tag{2.15.40}$$

This yields

$$n_{\mathrm{r}} = \frac{k_{\mathrm{r}}}{\omega \sqrt{\mu \varepsilon_0}} = \sqrt{\varepsilon_{11} + \varepsilon_{12}} \quad \text{and} \tag{2.15.41}$$

$$n_{\mathrm{l}} = \sqrt{\varepsilon_{11} - \varepsilon_{12}} \quad . \tag{2.15.42}$$

Assuming $\varepsilon_{12} \ll \varepsilon_{11}$, we obtain

$$n_{\mathrm{r}} - n_{\mathrm{l}} \approx \frac{\varepsilon_{12}}{\sqrt{\varepsilon_{11}}} = \frac{\varepsilon_{12}}{n_0} \tag{2.15.43}$$

where n_0 is the ordinary index of refraction. The specific rotary power, δ, for this case is given by

$$\delta = \frac{\varepsilon_{12} \pi}{n_0 \lambda} \quad . \tag{2.15.44}$$

2.15.3 Magneto-optics: The Faraday Effect, the Voigt Effect and the Kerr Effect

Application of a magnetic field to a material affects the dielectric tensor and this, in turn, changes the electromagnetic wave propagation. Basically this magneto-optic process can be divided into three parts: the Faraday Effect, the Voigt Effect and the Kerr Effect.

Application of a magnetic field to a crystal sometimes makes it optically active. By activity we mean that for plane-polarized light incident on the crystal, the output light is polarized in a different plane. This is generally known as the Faraday effect. For applied magnetic induction B, and a crystal thickness l, the amount of rotation of the plane of polarization, θ, is

$$\theta = V B l \quad , \tag{2.15.45}$$

where V is a constant known as the Verdet constant that depends on the material used. For this case, the specific rotary power is given by

$$\delta = V B \quad . \tag{2.15.46}$$

The Faraday effect can easily be explained by the model of the movement of electronic charge with the resultant change in the displacement vector, which can be written as

$$D = \varepsilon_0 E + P = \bar{\bar{\varepsilon}} E \quad , \tag{2.15.47}$$

where P is the polarization vector associated with the material itself. The polarization P is induced by the applied electric field and thus can be written in the linear approximation as

$$P = \varepsilon_0 \overline{\overline{\chi}} E \quad , \tag{2.15.48}$$

where $\overline{\overline{\chi}}$ is the electrical polarizability tensor. Using (2.15.48 and 47) we obtain

$$\overline{\overline{\varepsilon}} = \varepsilon_0 (I + \overline{\overline{\chi}}) \quad . \tag{2.15.49}$$

If we model the isotropic material to consist of N electrons bound to the core, we have

$$P = -Nq\mathbf{r} \quad , \tag{2.15.50}$$

where $\mathbf{r}$ is the displacement of the electron from the equilibrium position due to the application of the electric field E. The vector $\mathbf{r}$ is also given by the force balance equation of the electron, namely

$$m \frac{d^2 \mathbf{r}}{dt^2} + m\gamma \frac{d\mathbf{r}}{dt} + K\mathbf{r} = -qE \quad , \tag{2.15.51}$$

where m is the mass, γ the damping constant, and K the force constant. As the time dependence of the E field is given by $e^{j\omega t}$, the steady state solution of (2.15.51) is expressed by

$$\mathbf{r} = \frac{-qE}{-m\omega^2 + j\omega m\gamma + K} \quad . \tag{2.15.52}$$

Thus using (2.15.51) we obtain

$$P = \frac{Nq^2 E}{-m\omega^2 + j\omega m\gamma + K} \quad . \tag{2.15.53}$$

The effective electric field seen by the electrons, as used in (2.15.51) differs from the actual applied field:

$$E_{\text{eff}} = E + \frac{1}{3\varepsilon_0} P \quad . \tag{2.15.54}$$

Thus (2.15.53) is modified to

$$P = \frac{Nq^2}{-m\omega^2 + j\omega m\gamma + K} \left(E + \frac{1}{3\varepsilon_0} P \right) \quad . \tag{2.15.55}$$

From (2.15.55) we obtain

$$\chi = \frac{Nq^2/m}{\omega_0^2 - \omega^2 + j\omega\gamma} \quad , \quad \text{where} \tag{2.15.56}$$

$$\omega_0 = \sqrt{\frac{K}{m} - \frac{Nq^2}{3\varepsilon_0 m}} \quad . \tag{2.15.57}$$

This is known as the effective resonance frequency. In the presence of a magnetic field, (2.15.51) becomes

$$m\frac{d^2\boldsymbol{r}}{dt^2} + m\gamma\frac{d\boldsymbol{r}}{dt} + K\boldsymbol{r} = -q\boldsymbol{E} - q\left(\frac{d\boldsymbol{r}}{dt} \times \boldsymbol{B}\right) \quad . \tag{2.15.58}$$

If we assume that $\boldsymbol{B} = \boldsymbol{i}_z B$ and, for simplification, neglect γ in (2.15.58), we obtain

$$\chi = \begin{pmatrix} \chi_{11} & j\chi_{12} & 0 \\ -j\chi_{12} & \chi_{11} & 0 \\ 0 & 0 & \chi_{33} \end{pmatrix} \quad , \quad \text{where} \tag{2.15.59}$$

$$\chi_{11} = \frac{Nq^2}{m\varepsilon_0} \frac{\omega_0^2 - \omega^2}{[(\omega_0^2 - \omega^2)^2 - \omega^2\omega_c^2]} \quad , \tag{2.15.60}$$

$$\chi_{323} = \frac{Nq^2}{m\varepsilon_0} \frac{1}{(\omega_0^2 - \omega^2)} \quad , \tag{2.15.61}$$

$$\chi_{12} = \frac{Nq^2}{m_0} \frac{\omega\omega_c}{[(\omega_0^2 - \omega^2)^2 - \omega^2\omega_c^2]} \quad , \tag{2.15.62}$$

$$\omega_0 = \sqrt{\frac{K}{m}} \quad , \quad \text{and} \tag{2.15.63}$$

$$\omega_c = \frac{qB}{m} \quad . \tag{2.15.64}$$

From (2.15.59 and 49), we note that an isotropic material in the presence of a magnetic field has an effective dielectric tensor similar to that for optical activity given by (2.15.34). Thus the specific rotary power for this case will be given by

$$\delta \approx \frac{\pi Nq^2}{\lambda m\varepsilon_0} \frac{\omega\omega_c}{(\omega_0^2 - \omega^2)^2} = \frac{Nq^3}{\lambda m^2\varepsilon_0} \frac{\omega B}{(\omega_0^2 - \omega^2)^2} \tag{2.15.65}$$

where the condition $\omega\omega_c \ll |\omega_0^2 - \omega^2|$ is assumed. We therefore note that, for this model, the Verdet constant is given by

$$V = \frac{Nq^3}{m^2\varepsilon_0} \frac{\omega}{(\omega_0^2 - \omega^2)^2} \quad . \tag{2.15.66}$$

So far we have not considered absorption in the medium. In the presence of absorption we can modify

$$n_r \to n_r - ja_r \quad \text{and}$$

$$n_l \to n_l - ja_l \quad ; \tag{2.15.67}$$

a_r and a_l are real quantities. The relationship between these quantities and the absorption coefficients α_r and α_l is given by

$$\alpha_r = \frac{4\pi a_r}{\lambda_0} \quad \text{and}$$

$$\alpha_l = \frac{4\pi a_l}{\lambda_0} \quad . \tag{2.15.68}$$

In the presence of absorption, a linearly polarized light propagating through a distance l will be elliptically polarized so that the major axis of the ellipse is rotated by θ_F from the incident plane of polarization, where

$$\theta_F = \text{Re}\left\{\frac{\pi l}{\lambda_0}(n_r - n_l)\right\} \tag{2.15.69}$$

where $\text{Re}\{v\}$ denotes the real part of v.

The ellipticity is expressed by

$$\varepsilon_F = -\tanh\left[\text{Im}\left\{\frac{\pi l}{\lambda_0}(n_r - n_l)\right\}\right] \quad . \tag{2.15.70}$$

To understand the Voigt effect analytically, we note that the permittivity tensor $\bar{\bar{\varepsilon}}$ for this case is given by

$$\bar{\bar{\varepsilon}} = \begin{pmatrix} \varepsilon_0 + \chi_{11} & j\chi_{12} & 0 \\ -j\chi_{12} & \varepsilon_0 + \chi_{11} & 0 \\ 0 & 0 & \varepsilon_0 + \chi_{33} \end{pmatrix} \quad . \tag{2.15.71}$$

Substituting this in (2.12.130), we obtain

$$n^4[(\varepsilon_0 + \chi_{11}) + (\chi_{33} - \chi_{11})\cos\theta']$$
$$-n^2\{[(\varepsilon_0 + \chi_{11})^2 - \chi_{12}^2 + (\varepsilon_0 + \chi_{11})(\varepsilon_0 + \chi_{33})]$$
$$-[(\varepsilon_0 + \chi_{11})^2 - \chi_{12}^2 - (\varepsilon_0 + \chi_{11})(\varepsilon_0 + \chi_{33})]\cos\theta'\}$$
$$+(\varepsilon_0 + \chi_{33})[(\varepsilon_0 + \chi_{11})^2 + \chi_{12}^2] = 0 \quad , \tag{2.15.72}$$

where

$$\cos\theta' = k_z/k \quad . \tag{2.15.73}$$

For the Voigt effect, light propagates in a direction perpendicular to the z direction. Thus

$$\cos\theta' = 0 \quad .$$

Substituting this value of $\cos\theta'$, and solving for n we obtain

$$n_\perp^2 = \frac{(\varepsilon_0 + \chi_{11})^2 - \chi_{12}^2}{\varepsilon_0 + \chi_{11}} \quad \text{and} \tag{2.15.74}$$

$$n_\parallel^2 = \varepsilon_0 + \chi_{33} \quad . \tag{2.15.75}$$

Substituting these values of n in (2.12.129) and considering propagation in the y direction, we obtain for $n_\perp$,

$$E_x = \frac{j\chi_{12}}{\varepsilon_0 + \chi_{11}} E_y \quad , \tag{2.15.76}$$

163

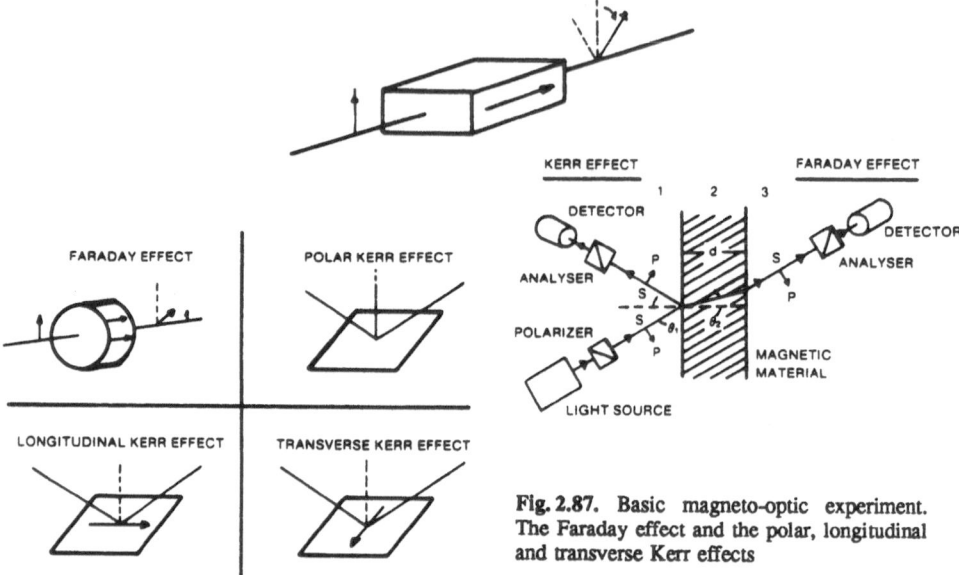

Fig. 2.87. Basic magneto-optic experiment. The Faraday effect and the polar, longitudinal and transverse Kerr effects

$$E_z = 0 \quad ,$$

and for $n_{\parallel}$

$$E_x = 0 \quad ,$$
$$E_y = 0 \quad .$$

Thus $n_{\perp}$ corresponds to the polarization perpendicular to the z axis, which is the direction of the applied magnetic field. Here $n_{\parallel}$ denotes the polarization parallel to the applied magnetic field.

The Kerr effect refers to the magneto-optic effect in reflection. This can be subdivided into three categories depending on the magnetic field directions as shown in Fig. 2.87. These are the polar Kerr effect, the longitudinal Kerr effect, and the equatorial (or transverse) Kerr effect. To solve the Kerr effect problem analytically using boundary conditions at the interface is a formidable task. In the following we shall just discuss certain important facts.

Let us consider the polar Kerr effect, where the magnetic field is perpendicular to the plane of incidence. For this case, the normal modes inside the magneto-optic medium are right and left circularly polarized light, as discussed in connection with the Faraday effect. As the refractive indices are different, the reflection coefficient given by

$$r = \frac{n-1}{n+1} \tag{2.15.77}$$

will also be different. If the incident light is linearly polarized, it can be resolved into two equal, right and left, circularly polarized components. However, the reflected light will be ellipticity polarized as the circular components will have different magnitudes and the major axis will be rotated from the direction of polarization of the incident light, due to the phase difference introduced. The rotation angle, θ_K, and the ellipticity ε_K is given by

$$\theta_K = -\tfrac{1}{2}(\phi_+ - \phi_-) \quad , \quad \text{and} \qquad (2.15.78)$$

$$\varepsilon_K = -\frac{|r_+| - |r_-|}{|r_+| + |r_-|} \quad , \quad \text{where} \qquad (2.15.79)$$

$$r_+ = |r_+|e^{j\phi_+} \quad ,$$
$$r_- = |r_-|e^{j\phi_-} \quad , \qquad (2.15.80)$$

and r_+ and r_- correspond to complex reflection coefficients for the right and left circularly polarized light respectively. If the incident light is not normal, then the reflected light is no longer a simple mode but a combination of the normal modes and is generally represented as follows:

$$\begin{pmatrix} E_p^r \\ E_s^r \end{pmatrix} = \begin{pmatrix} r_{pp} & r_{ps} \\ r_{sp} & r_{ss} \end{pmatrix} \begin{pmatrix} E_p^i \\ E_s^i \end{pmatrix} \qquad (2.15.81)$$

where the subscripts p and s refer to the state of polarization (p parallel and s perpendicular to the plane of incidence). The superscripts r and i refer to reflected and incident waves, while r_{pp}, r_{ps}, r_{sp} and r_{ss} refer to the reflection coefficients. For the incidence angle θ_0, and for the approximations

$$\varepsilon_0 + \chi_{11} \approx \varepsilon_0 + \chi_{33} = n_0^2 \quad , \quad \text{and}$$

$$\chi_{12} \ll \chi_{33} \quad ,$$

we list without derivation (see [2.8] for details) the above reflection coefficients, which take the following form:

Polar Kerr Effect

$$r_{pp} = \frac{n_0 r - r'}{n_0 r + r'}$$

$$r_{ss} = \frac{r - n_0 r'}{r + n_0 r'}$$

$$r_{sp} = r_{ps} = \frac{j\chi_{12}/n_0}{(r + n_0 r')(n_0 r + r')} \quad , \quad \text{where} \qquad (2.15.82)$$

$$r = \cos \theta_0$$

$$r' = \left(1 - \frac{1}{n_0^2} \sin^2 \theta_0 \right)^{1/2}$$

Longitudinal Kerr Effect

$$r_{pp} = \frac{n_0 r - r'}{n_0 r + r'}$$

$$r_{ss} = \frac{r - n_0 r'}{r + n_0 r'}$$

(2.15.83)

$$r_{ps} = -r_{sp} = \frac{j r \chi_{12}/n_0^2}{r'(n_{or} r + r')(r + n_0 r')}$$

Equatorial Kerr Effect

$$r_{tot} \approx \frac{n_0 \cos\theta_0 - 1}{n_0 \cos\theta_0 + 1}\left(1 + j\frac{(\chi_{12}/n_0^2)\sin 2\theta_0}{n_0^2 \cos^2\theta_0 - 1}\right)$$

$$r_{ss} \approx \frac{\cos\theta_0 - n_0}{\cos\theta_0 + n_0}$$

(2.15.84)

$$r_{sp} = r_{ps} = 0 \quad .$$

It is of interest to consider a basic magneto-optic experiment which uses a parallel plate of magnetic material. Polarized light is incident on the first surface. The transmitted light has a different polarization due to the Faraday effect. Similarly, the reflected light has properties dictated by the Kerr effect. For the simplified case, consider light propagation along the z axis, which is the axis of the applied magnetic field (polar Kerr effect situation). If one includes an absorptive medium, the following relationships can be derived for the Faraday effect:

$$\chi_{12} = \varepsilon_1' + j\varepsilon_1''$$

$$\Delta a = a_r - a_l$$

$$\Delta n = n_r - n_l$$

$$\varepsilon_1' = \frac{\lambda_0}{2\pi}\left(n\theta + \frac{1}{4}n\Delta a\right)$$

(2.15.85)

$$\varepsilon_1'' = \frac{\lambda_0}{\pi}\left(\theta a - \frac{1}{4}n\Delta a\right)$$

$$n = \tfrac{1}{2}(n_r + n_l)$$

$$a = \tfrac{1}{2}(a_r + a_l)$$

and θ is the Faraday rotation per unit length. Thus, by measuring θ and Δa, one can obtain $\chi_{\|}$. Similarly, for the Kerr effect, one measures the Kerr rotation ϕ_K

and Kerr ellipticity ε_K given by

$$\phi_K = \frac{(n^2 - a^2 - 1)\Delta a - 2na\Delta n}{(n^2 - a^2 - 1)^2 + 4n^2 k^2} \quad , \tag{2.15.86}$$

$$\varepsilon_K = \frac{(1 - n^2 + k^2)\Delta a - 2na\Delta n}{(n^2 - a^2 - 1)^2 + 4n^2 k^2} \quad . \tag{2.15.87}$$

To obtain $\chi_\parallel$ one uses the relationships

$$\varepsilon_1' = \phi_K(3n^2 a - a^2 - a) - \varepsilon_K(n^3 - 3n^2 a - n) \quad ,$$

$$\varepsilon_1'' = \varepsilon_K(3n^2 a - a^2 - a) - \phi_K(n^3 - 3n^2 a - n) \quad . \tag{2.15.88}$$

Note that $\Delta a = a_r - a_l$, is often referred to as magnetic circular dichroism.

2.16 Wave Equation with Source and Boundary

In this section, we will discuss the diffraction problem using the linear systems approach. This is followed by the solution of the scalar wave equation including source and boundary, leading to Kirchhoff's formulation of the diffraction problem. The vector electromagnetic wave equation with source and boundary is discussed next. The results of this section are applied to Fourier optics in the next section (Sect. 2.17). However, for the reader not familiar with Fourier optics, it is advisable to commence with Sect. 2.17.

2.16.1 Diffraction

Let us consider plane wave propagation. The wave is represented by

$$e^{j(\omega t - \mathbf{k} \cdot \mathbf{r})} \quad , \quad \text{where}$$

$$\begin{aligned} \mathbf{k} \cdot \mathbf{r} &= k_x x + k_y y + k_z z \\ &= 2\pi f_x x + 2\pi f_y y + 2\pi f_z z \quad . \end{aligned} \tag{2.16.1}$$

Note that $f_x = k_x/2\pi$ has the dimensions of m^{-1} and is called the spatial frequency in the x direction. If we consider only the space-dependent part, the phase front at any point x, y and z is represented by[6]

$$e^{j2\pi(f_x x + f_y y + f_z z)} \quad .$$

Considering propagation through a distance z as a spatial filter, as shown in Fig. 2.88, we note that the input at $z = 0$ is given by $\exp[j2\pi(f_x x + f_y y)]$. The output at z is given by $\exp[j2\pi(f_x x + f_y y)]\exp[jk_z z]$. Since

---

[6] For simplification in this section we represent the phase with a negative sign.

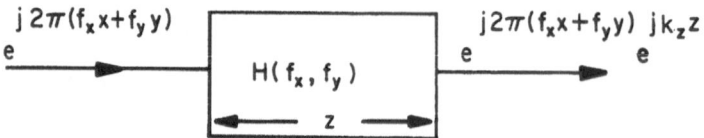

Fig. 2.88. Propagation through distance z, considered as a spatial filter

$$k_x^2 + k_y^2 + k_z^2 = k^2 = \left(\frac{2\pi}{\lambda}\right)^2 \quad , \tag{2.16.2}$$

we have

$$k_z^2 = \left(\frac{2\pi}{\lambda}\right)^2 - k_x^2 - k_y^2 = \left(\frac{2\pi}{\lambda}\right)^2 - (2\pi f_x)^2 - (2\pi f_y)^2 \quad .$$

Thus

$$k_z^2 = k^2[1 - (\lambda f_x)^2 - (\lambda f_y)^2] \quad . \tag{2.16.3}$$

Using this, we obtain that the output at z is

$$\exp\left[j2\pi(f_x x + f_y y)\right] \exp\left[jkz\sqrt{1 - (\lambda f_x)^2 - (\lambda f_y)^2}\right] \quad .$$

The frequency response of the filter which represents propagation through a free space of distance z is given by

$$H(f_x, f_y) = \exp\left[jkz\sqrt{1 - (\lambda f_x)^2 - (\lambda f_y)^2}\right] \quad . \tag{2.16.4}$$

The spatial impulse response, $h(x, y)$, is given by

$$h(x, y) = \mathcal{J}^{-1}[H(f_x, f_y)] \quad . \tag{2.16.5}$$

We shall show that in the far field approximation

$$h(x, y) = \frac{1}{j\lambda} \frac{e^{jkr}}{r} \quad , \quad \text{where} \tag{2.16.6}$$

$$r = (x^2 + y^2 + z^2)^{1/2} \quad . \tag{2.16.7}$$

However, first consider the simpler case of the Fresnel approximation. Expanding the right-hand side of (2.16.4) we obtain

$$H(f_x, f_y) = e^{jkz}\{1 - \tfrac{1}{2}[(\lambda f_x)^2 + (\lambda f_y)^2] + \ldots\} \quad . \tag{2.16.8}$$

Neglecting terms higher than the second in the expansion, we obtain the Fresnel approximation denoted by a subscript F. Thus

$$H_F(f_x, f_y) = e^{jkz}e^{-j(kz\lambda^2/2)(f_x^2 + f_y^2)}$$

$$= e^{jkz}e^{-j\pi z\lambda(f_x^2 + f_y^2)} \quad . \tag{2.16.9}$$

We can obtain $h_F(x, y)$ as

$$h_F(x, y) = e^{jkz} \mathcal{F}[e^{-j\pi z\lambda(f_x^2 + f_y^2)}] \quad . \tag{2.16.10}$$

Using $a^2 = jz\lambda$, we note

$$\mathcal{F}[e^{-j\pi z\lambda(f_x^2 + f_y^2)}] = e^{-\pi[(af_x)^2 + (af_y)^2]}$$

$$= \frac{1}{a^2}\exp\left[-\pi\left(\frac{x^2}{a^2} + \frac{y^2}{a^2}\right)\right] \quad . \tag{2.16.11}$$

The last expression is obtained using Fourier transform tables.

Using (2.16.11) in (2.16.10), one obtains

$$h_F(x, y) = \frac{e^{jkz}}{j\lambda z}e^{j(\pi/z\lambda)(x^2 + y^2)} \quad . \tag{2.16.12}$$

Thus, in the Fresnel approximation, an incident electric field, $E(x, y, 0)$ after propagation through a distance z becomes $E(x, y, z)$ given by

$$E(x, y, z) = E(x, y, 0) * h_F(x, y) \tag{2.16.13}$$

where * denotes the convolution operation (Sect. 3.1.1). The exact expression is

$$h(x, y) = \mathcal{F}^{-1}[H(f_x, f_y)]$$

$$= \int\!\!\!\int_{-\infty}^{+\infty} e^{jkz(1 - \lambda^2 f_x^2 - \lambda^2 f_x^2)^{1/2}} e^{j2\pi(f_x x + f_y y)} df_x df_y \quad . \tag{2.16.14}$$

Defining

$$f_x = f \cos \alpha, \quad f_y = f \sin \alpha \quad , \tag{2.16.15}$$

one obtains

$$h(x, y) = \int_0^\infty e^{jkz(1 - \lambda^2 f^2)^{1/2}} f \, df \int_0^{2\pi} e^{j2\pi f(x \cos \alpha + y \sin \alpha)} d\alpha \quad . \tag{2.16.16}$$

Substituting $x = r' \cos \phi$ and $y = r' \sin \phi$, the second integration becomes

$$\int_0^{2\pi} e^{j2\pi f(x \cos \alpha + y \sin \alpha)} d\alpha = \int_0^{2\pi} e^{j2\pi fr \cos(\alpha - \phi)} d\alpha$$

$$= 2\pi J_0(2\pi r' f) \quad . \tag{2.16.17}$$

The last expression is obtained using (B.8.29) and Appendix B. Substituting (2.16.17) into (2.16.16) yields

$$h(x, y) = 2\pi \int_0^\infty e^{jkz(1 - \lambda^2 f^2)^{1/2}} J_0(2\pi r' f) f \, df \quad . \tag{2.16.18}$$

Writing $s = 2\pi f$, we have

$$h(x, y) = \frac{1}{2\pi} \int_0^\infty e^{z(s^2 - k^2)^{1/2}} J_0(r' s) s \, ds \quad . \tag{2.16.19}$$

169

The following integral equation involving Bessel functions is useful:

$$\int_0^\infty J_0(bt)e^{-a(t^2-y^2)^{1/2}}(t^2-y^2)^{-1/2}t\,dt$$

$$= e^{-jy(a^2+b^2)^{1/2}}(a^2+b^2)^{1/2} \quad . \tag{2.16.20}$$

Differentiating both sides of the above equation with respect to a, one obtains

$$\int_0^\infty J_0(bt)e^{-a(t^2-y^2)^{1/2}}t\,dt = e^{-jy(a^2+b^2)}$$

$$\times \frac{a}{a^2+b^2}\left(\frac{1}{(a^2+b^2)^{1/2}}+jy\right) \quad . \tag{2.16.21}$$

Thus we see that

$$h(x,y) = \frac{1}{2\pi}\left[e^{jk(z^2+r'^2)^{1/2}}\frac{z}{z^2+r'^2}\left(\frac{1}{(z^2+r'^2)^{1/2}}+jk\right)\right]$$

$$= \frac{1}{j\lambda}\frac{e^{jkr}}{r}\frac{z}{r}\left(1+\frac{1}{jkr}\right) \quad . \tag{2.16.22}$$

If we make the far field approximation, i.e., $kr \gg 1$ and $z/r \approx 1$, and noting that $r'^2 + z^2 = r^2$, we finally obtain

$$h(x,y) = \frac{1}{j\lambda}\frac{e^{jkr}}{r} \quad . \tag{2.16.23}$$

Thus in the far-field approximation, an input electric field $E(x,y,0)$ at $z = 0$, after propagation through a distance z, becomes $E(x,y,z)$ given by

$$E(x,y,z) = E(x,y,0)*h(x,y)$$

$$= \frac{1}{j\lambda}\iint E(x',y',0)\frac{e^{jk|r-r'|}}{|r-r'|}dx'\,dy' \quad , \quad \text{where} \tag{2.16.24}$$

$$r = (x^2+y^2+z^2)^{1/2} \quad ,$$

$$r' = \sqrt{x'^2+y'^2} \quad , \quad \text{and}$$

$$|r-r'| = [(x-x')^2+(y-y')^2+z^2]^{1/2} \quad . \tag{2.16.25}$$

For the Fresnel approximation, one expands

$$|r-r'| = z\left(1+\frac{1}{2z^2}[(x-x')^2+(y-y')^2]\right)+\dots \quad . \tag{2.16.26}$$

and keeps only two terms in the expansion. Thus

$$E(x,y,z) \approx \frac{1}{j\lambda z}\iint E(x',y',0)e^{(jk/2z)[(x-x')^2+(y-y')^2]}dx'\,dy'$$

$$= E(x', y', 0) * \frac{1}{j\lambda z} e^{jkz} e^{(jk/2z)(x^2+y^2)} \quad . \tag{2.16.27}$$

This is identical to expression (2.16.13) derived directly. In this book the Fresnel approximation is assumed unless otherwise stated.

2.16.2 Solution of the Scalar Wave Equation with Source and Boundary

The scalar wave equation originating from a source term $s(x, y, z, t)$ is given by

$$\nabla^2 u - \frac{1}{c^2} \frac{\partial^2 u}{\partial t^2} = s(x, y, z, t) \quad , \tag{2.16.28}$$

where u is considered scalar and thus can be only a component of the electric field. We are interested in a solution of $u(x, y, z, t)$ in all the space within a closed volume V with a boundary surface S in terms of the source function and the given values of u at the boundary. We shall consider the region V to be simply connected. However, the solution derived will also be valid for multiply connected surfaces [2.13].

To obtain the solution, one needs Green's theorem. This states that if two functions ψ and ϕ have continuous first and second derivatives, then

$$\int\int\int_V (\psi \nabla^2 \phi - \phi \nabla^2 \psi) dV = \int\int_S \left(-\psi \frac{\partial \phi}{\partial n} + \phi \frac{\partial \psi}{\partial n} \right) dS \quad . \tag{2.16.29}$$

The proof is given in Appendix D. Let us choose ψ to be given by

$$\psi(x, y, z, t) = \frac{\omega(t - r/c)}{r} \quad , \tag{2.16.30}$$

where $\omega(t)$ is any twice differentiable function and r is given by

$$r^2 = (x - x_0)^2 + (y - y_0)^2 + (z - z_0)^2 \tag{2.16.31}$$

and x_0, y_0, z_0 is a point outside the volume V. We note that ψ as defined is a solution of the source-free wave equation, i.e.,

$$\nabla^2 \psi - \frac{1}{c^2} \frac{\partial^2 \psi}{\partial t^2} = 0 \tag{2.16.32}$$

excluding the point $r = 0$. Choosing $\phi = u$ we obtain from (2.16.30),

$$\psi \nabla^2 u - u \nabla^2 \psi = \psi \left(\frac{1}{c^2} \frac{\partial^2 u}{\partial t^2} - s \right) - u \frac{1}{c^2} \frac{\partial^2 \psi}{\partial t^2} \quad . \tag{2.16.33}$$

Using (2.16.29 and 33), we obtain

$$\frac{1}{c^2} \int\int\int_V \left(\frac{\omega}{r} \frac{\partial^2 u}{\partial t^2} - \frac{u}{r} \frac{d^2 \omega}{dt^2} \right) dV$$

$$= \int\int_S \left[-\frac{\omega}{r} \frac{\partial u}{\partial n} + u \frac{\partial (\omega/r)}{\partial n} \right] dS$$

$$+ \iiint\limits_{V} \frac{\omega s}{r} dV \quad , \tag{2.16.34}$$

where ω and $d^2\omega/dt^2$ are evaluated at

$$t' = t + r/c \quad . \tag{2.16.35}$$

Let us define $\omega(t)$ such that

$$\omega(t) = 0 \quad \text{for} \quad |t| \geq \frac{2r_{\mathrm{m}}}{c} \quad , \tag{2.16.36}$$

where r_{m} is the maximum distance between r_0 and any point in V. Then we have

$$\omega(t')|_{\pm T} = 0 \tag{2.16.37}$$

for $T = 3r_{\mathrm{m}}/c$.

Using integration by parts we obtain

$$-\int\limits_{-T}^{+T} \left[\omega(t') \frac{\partial^2 u}{\delta t^2} - u \frac{d^2\omega}{dt^2} \bigg|_{t=t'} \right] dt$$

$$= \left(\omega(t') - u \frac{d\omega}{dt} \bigg|_{t=t'} \right) \bigg|_{-T}^{T} = 0 \quad . \tag{2.16.38}$$

Integrating both sides of (2.16.34) from $-T$ to $+T$, we thus obtain

$$\iint\limits_{S} \int\limits_{-T}^{+T} \left(-\frac{\omega}{r} + \frac{\partial u}{\partial n} \frac{\partial(\omega/r)}{\partial n} \right) dt\, dS + \iiint\limits_{V} \int\limits_{-T}^{+T} \frac{\omega s}{r} dt\, dV = 0 \quad . \tag{2.16.39}$$

Substituting

$$\frac{\partial}{\partial} \frac{\omega(t')}{r} = \frac{\partial(1/r)}{\partial n} \omega(t) + \frac{1}{rc} \frac{\partial r}{\partial n} \frac{d\omega}{dt} \tag{2.16.40}$$

and

$$\int\limits_{-T}^{T} u \frac{d\omega}{dt} \bigg|_{t=t'} dt = [\omega(t')u]_{-T}^{+T} - \int\limits_{-T}^{T} \omega \frac{\partial u}{\partial t} dt = \int\limits_{-T}^{T} \omega \frac{\partial u}{\partial t} dt \tag{2.16.41}$$

into (2.16.39), we have

$$\iint\limits_{S} dS \int\limits_{-T+r/c}^{T+r/c} \omega(t') \left(-\frac{1}{r} \left[\frac{\partial u}{\partial n} \right]_{\mathrm{r}} + [u]_{\mathrm{r}} \frac{\partial(1/r)}{\partial n} - \frac{1}{rc} \frac{\partial r}{\partial n} \left[\frac{\partial u}{\partial n} \right]_{\mathrm{r}} \right) dt'$$

$$+ \iiint\limits_{V} dV \int\limits_{-T+r/c}^{T+r/c} \omega(t') \frac{[s]_{\mathrm{r}}}{r} dt' = 0 \quad , \tag{2.16.42}$$

where $[u]_{\mathrm{r}}$ means retardation, i.e., $[u]_{\mathrm{r}} = u(x, y, z, t - r/c)$. Using (2.16.36), we

172

can replace the limits $T + r/c$ and $-T + r/c$ by T and $-T$. Interchanging the order of integration, we thus obtain

$$\int_{-T}^{+T} \omega(t')\left\{ \iint_S \left(-\frac{1}{r}\left[\frac{\partial u}{\partial n}\right]_r + [u]_r \frac{\partial(1/r)}{\partial n} - \frac{1}{rc}\frac{\partial r}{\partial n}\left[\frac{\partial u}{\partial n}\right]_r \right) dS \right.$$

$$\left. + \iiint_V \frac{[s]_r}{r} dV \right\} dt' = 0 \quad . \tag{2.16.43}$$

Since $\omega(t)$ and the time origin are arbitrary, (2.16.44) is true only when

$$\iint_S \left(-\frac{1}{r}\left[\frac{\partial u}{\partial n}\right]_r + [u]_r \frac{\partial(1/r)}{\partial n} - \frac{1}{rc}\frac{\partial r}{\partial n}\left[\frac{\partial u}{\partial t}\right]_r \right) dS$$

$$+ \iiint_V \frac{[s]_r}{r} dV = 0 \quad . \tag{2.16.44}$$

Assuming that the point r_0 is as shown in Fig. 2.89 and the surface S consists of two surfaces S_0 and S_1 (S_1 being a sphere of radius a centered at r_0), then for integration over S_1, we have

$$\frac{\partial r}{\partial n} = 1 \quad \text{and} \quad \frac{\partial(1/r)}{\partial n} = -\frac{1}{a^2} \quad .$$

Thus

$$\lim_{a \to 0} \iint_{S_1} \left(-\frac{1}{r}\left[\frac{\partial u}{\partial n}\right]_r + [u]_r - \frac{1}{rc}\frac{\partial r}{\partial n}\left[\frac{\partial u}{\partial t}\right]_r \right) dS$$

$$= \lim_{a \to 0} \iint_{S_1} \left(-\frac{1}{a}\left[\frac{\partial u}{\partial n}\right]_r + \frac{[u]_r}{a^2} - \frac{1}{ac}\left[\frac{\partial u}{\partial t}\right]_r \right) dS$$

$$= -4\pi u(r_0, t) \quad . \tag{2.16.45}$$

Finally we obtain

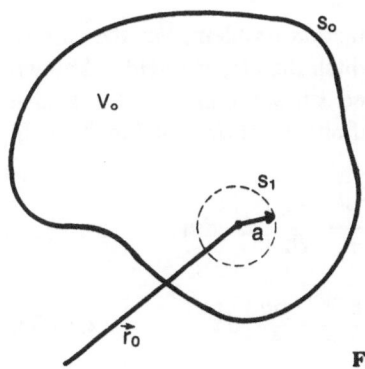

Fig. 2.89. Geometry used for the integration of (2.16.44)

173

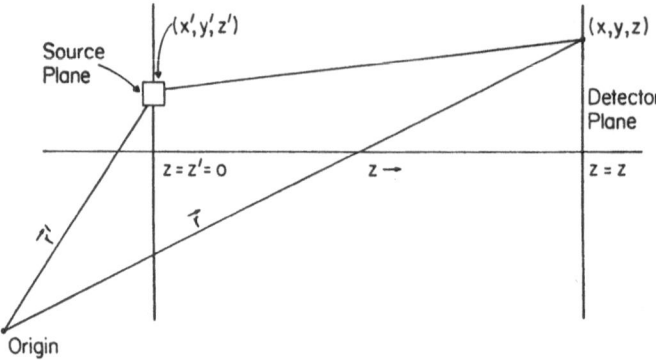

Fig. 2.90. Schematic diagram for the radiation problem showing the source and detector planes

$$u(\boldsymbol{r}_0, t) = u(x_0, y_0, z_0, t) = \frac{1}{4\pi} \int\!\!\int_{V_0}\!\!\int \frac{1}{r}[s]_r dV$$

$$+ \frac{1}{4\pi} \int\!\!\int_{S_0} \left(\frac{\partial(1/r)}{\partial n}[u]_r - \frac{1}{cr}\frac{\partial r}{\partial n}\left[\frac{\partial u}{\partial t}\right]_r - \frac{1}{r}\left[\frac{\partial u}{\partial n}\right]_r \right) dS \quad ,$$
(2.16.46)

where V_0 is a region of space containing the observation point ($\boldsymbol{r}_0$) and is bounded by the closed surface S_0.

For harmonic time dependence, i.e.,

$$u(x, y, z, t) = u(x, y, z)e^{j\omega t} \quad ,$$
(2.16.47)

(2.16.46) can be written as

$$u(\boldsymbol{r}_0) = \frac{1}{4\pi} \int\!\!\int_{S_0} \left[u\frac{\partial}{\partial n}\left(\frac{e^{-jkr}}{r}\right) - \frac{e^{-jkr}}{r}\frac{\partial u}{\partial n} \right] dS$$
(2.16.48)

where S_0 does not contain any sources. If we had started with the equation

$$\nabla^2 u + k^2 u = -s \quad ,$$
(2.16.49)

we could have obtained (2.16.48) directly.

Using (2.16.48), let us reformulate the diffraction problem. For the case of Fig. 2.90, where the source plane (the plane at which the electric field is known) and the observation plane are shown, the closed surface consists of a source plane and a hemisphere of radius tending to infinity as shown in Fig. 2.91. In this situation we have

$$u(\boldsymbol{r}_0) = \frac{1}{4\pi} \int\!\!\int_{S_0} \left[u(x_1, y_1)\frac{\partial}{\partial z}\left(\frac{e^{-jkr}}{r}\right) - \frac{e^{-jkr}}{r}\frac{\partial u}{\partial z} \right] dx_1 dy_1$$

$$+ \lim_{R \to \infty} \frac{1}{4\pi} \int\!\!\int_{S_\infty} \left[u\frac{\partial}{\partial r}\left(\frac{e^{-jkr}}{r}\right) - \frac{e^{-jkr}}{r}\frac{\partial u}{\partial r} \right] dS \quad .$$
(2.16.50)

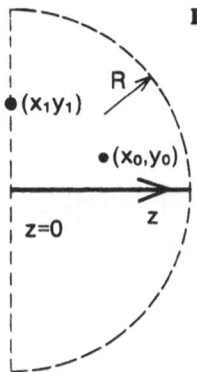

Fig. 2.91. Surface used for integration for the case of Fig. 2.90

The integration over S_∞ can be written as

$$\lim_{R \to \infty} \frac{1}{4\pi} \int \left[\left(-jk - \frac{1}{R}\right) \frac{e^{-jkr}}{R} u(R) - \frac{e^{-jkr}}{R} \frac{\partial u}{\partial r}\bigg|_{r=R} \right] R^2 d\Omega \quad .(2.16.51)$$

The above integral has a zero value provided the following radiation condition is satisfied:

$$\lim_{R \to \infty} R\left(\frac{\partial u}{\partial n} + jku\right) = 0 \quad . \tag{2.16.52}$$

For this case we have the diffraction formula given by

$$u(r_0) = \frac{1}{4\pi} \iint_{S_0} \left[u(x_1, y_1) \frac{\partial}{\partial z}\left(\frac{e^{-jkr}}{r}\right) - \frac{e^{-jkr}}{r} \frac{\partial u}{\partial z} \right] dx_1 dy_1 \quad . \tag{2.16.53}$$

If the electric field at $z = 0$ is such that it is a plane wave propagating in the z direction, then

$$\frac{\partial u}{\partial z} = -jku \quad . \tag{2.16.54}$$

Equation (2.16.53) for this case becomes

$$u(r_0) = \frac{j}{\lambda} \frac{z}{r} \iint u(x_1, y_1) \frac{e^{-jkr}}{r}\left(1 + \frac{1}{2jkr}\right) dx_1 dy_1 \quad . \tag{2.16.55}$$

For $kr \gg 1$, we obtain

$$u(r_0) = \frac{j}{\lambda} \iint u(x_1, y_1) \frac{e^{-jkr}}{r} dx_1 dy_1 \tag{2.16.56}$$

where

$$r = [(x_1 - x_0)^2 + (y_1 - y_0)^2 + z^2]^{1/2} \tag{2.16.57}$$

and we have assumed $r \approx z$. Equation (2.16.57) is the same as (2.16.64).

Note that $u(x_1 y_1)$ is not known in practical cases. What *is* known is that some wave is incident from the left on a screen or transparency at $z = 0$, the source plane. In general, it is assumed that $u(x_1 y_1)$ is obtained by multiplying the transmission function with the incident electric field. Although this does not always apply, it is valid for our purpose and for most other cases of practical interest.

2.16.3 Solution of the Vector Wave Equation with Source and Boundary

Assuming a harmonic time dependence, Maxwell's equations are given by

$$\nabla \times E = -j\omega B \quad ,$$
$$\nabla \times H = J_s + j\omega D \quad ,$$
$$\nabla \cdot E = \varrho_s/\varepsilon_0 \quad ,$$
$$\nabla \cdot B = 0 \quad .$$

$$(2.16.58)$$

The continuity equation is given by

$$\nabla \cdot J_s = -j\omega \varrho_s \quad , \tag{2.16.59}$$

where J_s and ϱ_s denote the source current and charge density respectively. Eliminating B and E, one has

$$\nabla \times (\nabla \times E) - k^2 E = -j\omega\mu J_s \quad , \quad \text{and}$$
$$\nabla \times (\nabla \times B) - k^2 B = \nabla \times (\mu J_s) \quad .$$

$$(2.16.60)$$

For this case, we need to use the vector Green's theorem given by

$$\int\int\int_V (F \cdot \nabla \times (\nabla \times G) - G \cdot \nabla \times (\nabla \times F) dV$$
$$= -\int\int_S (G \times (\nabla \times F) - F \times (\nabla \times G)) \cdot i_n dS \quad . \tag{2.16.61}$$

As with (2.16.29), F and G are vector functions of position in V with closed surface S. Also, F and G are assumed to have continuous first and second derivatives in V and on S. Let us choose $F = E$ and G to be given by

$$G = \frac{e^{-jkr}}{r}\alpha = \psi\alpha \quad , \tag{2.16.62}$$

where r is defined by (2.16.31) and α is an arbitrary constant vector. Note that G satisfies all the necessary conditions except at $r = 0$. Defining the surface S_1 which surrounds $r = r_0$ as shown in Fig. 2.89, it can be shown that

$$\alpha \cdot \int\int\int_V \left(j\omega\psi\mu J_s - \frac{\varrho_s}{\varepsilon_0}\nabla_s\psi \right) dV - \alpha \int\int_{S_0,S_1} (i_n \cdot E)\nabla_s\psi dS$$
$$= \alpha \int\int_{S_0,S_1} [j\omega\psi(i_n \times B) - (i_n \times E) \times \nabla_s\psi] dS \quad . \tag{2.16.63}$$

As α is arbitrary, equating both sides yields

$$\iint\limits_{S_1} [(\mathbf{i}_n \cdot \mathbf{E})\nabla_s\psi + (\mathbf{i}_n \times \mathbf{E}) \times \nabla_s\psi - j\omega\psi(\mathbf{i}_n \times \mathbf{B})]dS$$

$$= \int\int\limits_{V}\int \left(j\omega\psi\mu\mathbf{J}_s - \frac{\varrho_s}{\varepsilon_0}\nabla_s\psi\right)dV$$

$$- \iint\limits_{S_0} [(\mathbf{i}_n \cdot \mathbf{E})\nabla_s\psi + (\mathbf{i}_n \times \mathbf{E}) \times \nabla_s\psi - j\omega\psi(\mathbf{i}_n \times \mathbf{B})]dS \quad . \quad (2.16.64)$$

Performing the integration on S_1 and going to the limit $a \to 0$, we obtain

$$E(x_0, y_0, z_0) = \frac{1}{4\pi} \int\int\limits_{V}\int \left(\frac{\varrho_s}{\varepsilon_0}\nabla_s\psi - j\omega\psi\mu\mathbf{J}_s\right)dV$$

$$+ \frac{1}{4\pi} \iint\limits_{S_0} [(\mathbf{i}_n \cdot \mathbf{E})\nabla_s\psi + (\mathbf{i}_n \times \mathbf{E})$$

$$\times \nabla_s\psi - j\omega\psi(\mathbf{i}_n \times \mathbf{B})]dS \quad . \quad (2.16.65)$$

Substituting for ϱ_s from (2.16.59) we have

$$E(x_0, y_0, z_0) = \frac{1}{4\pi} \int\int\limits_{V}\int \frac{1}{j\omega t_0}[(\nabla \cdot \mathbf{J}_s)\nabla_s\psi + k \cdot \psi\mathbf{J}_s]dV$$

$$+ \frac{1}{4\pi} \iint\limits_{S_0} [(\mathbf{i}_n \cdot \mathbf{E})\nabla_s\psi + (\mathbf{i}_n \times \mathbf{E})$$

$$\times \nabla_s\psi - j\omega\psi(\mathbf{i}_n \times \mathbf{B})]dS \quad . \quad (2.16.66)$$

Similarly one can obtain for B,

$$B(x_0, y_0, z_0) = \frac{1}{4\pi} \int\int\limits_{V}\int \mu\mathbf{J}_s \times \nabla_s\psi \, dV$$

$$+ \frac{1}{4\pi} \iint\limits_{S_0} \left[\frac{j\omega\psi}{c^2}(\mathbf{i}_n \times \mathbf{E})\right.$$

$$\left. + (\mathbf{i}_n \times \mathbf{B})\nabla_s\psi + (\mathbf{i}_n \cdot \mathbf{B})\nabla_{gs}\psi\right]dS \quad . \quad (2.16.67)$$

If we consider Fig. 2.91 and consider the case where one surface is at infinity, we obtain the radiation conditions [equivalent to (2.16.52)], namely, that the surface integral must be zero, thus

$$\lim_{R \to \infty} R[\mathbf{i}_R \times \mathbf{E} - C\mathbf{B}] = 0 \quad \text{and}$$

$$\lim_{R \to \infty} R\left[\mathbf{i}_n \times \mathbf{B} + \frac{1}{c}\mathbf{E}\right] = 0 \quad . \quad (2.16.68)$$

We also need the following two finiteness conditions given by

$$\lim_{R \to \infty} R\mathbf{B} \quad \text{is finite} \quad \text{and}$$

$$\lim_{R \to \infty} R\mathbf{E} \quad \text{is finite} \quad . \quad (2.16.69)$$

Under these conditions in the absence of any source, we have that

$$E(r_0) = \frac{1}{4\pi} \iint\limits_{S_0} [(i_n \cdot E)\nabla_s \psi$$

$$+ (i_n \times E) \times \nabla_s \psi - j\omega \psi (i_n \times B)]dS \quad . \tag{2.16.70}$$

The above result is equivalent to (2.16.53) for the scalar case.

In the absence of any other surface boundaries (except at infinity), we have

$$E(x_0, y_0, z_0) = \frac{1}{4\pi} \int\!\!\int\limits_{V}\!\!\int \frac{1}{j\omega\varepsilon_0} [(J_s \cdot \nabla_s)\nabla_s \psi + k^2 \psi J_s]dV \tag{2.16.71}$$

and

$$B(x_0, y_0, z_0) = \frac{1}{4\pi} \int\!\!\int\limits_{V}\!\!\int \mu J_s \nabla_s \psi \, dV \quad . \tag{2.16.72}$$

The above two equations are fundamental to the electromagnetic radiation problem from sources such as antennas.

2.17 Fourier Optics

In this section we review some important results of physical optics, wave optics or Fourier optics. For full details, see [2.2, 9].

The general diffraction problem is shown in Fig. 2.90. The electric field at $r = (x, y, z)$ in the detector plane is given by

$$E(x, y, z) = \iint \frac{E(x', y')}{j\lambda |r - r'|} e^{j(\omega t - k|r - r'|)} dx' dy' \tag{2.17.1}$$

where $r = (x, y, 0)$ and the electric field at the source plane is denoted by $E(r')$.

Far-Field Approximation:

$$E(x, y, z) = \frac{e^{j\omega t}}{j\lambda z} \iint E(x', y', 0)e^{-jk|r - r'|} dx' dy' \tag{2.17.2}$$

where $z \gg x$, x', y and y'.

Fresnel Approximation:

$$z^3 \gg \frac{\pi}{4\lambda}[(x - x')^2 + (y - y')^2]_{\max}^2$$

$$E(x, y, z) = \frac{e^{j(\omega t - kz)}}{j\lambda z} \iint E(x', y', 0)e^{-(jk/2z)[(x-x')^2+(y-y')^2]} dx' dy'$$

$$= \frac{e^{j(\omega t - kz)}}{j\lambda z} E(x', y', 0) * e^{-(jk/2z)(x'^2 + y'^2)}$$

$$= \frac{e^{j(\omega t - kz)}}{j\lambda z} e^{-(j\pi/\lambda z)(x^2+y^2)} \iint E(x',y',0)e^{-(j\pi/\lambda z)(x'^2+y'^2)}$$

$$\times\ e^{j2\pi(f_x x' + f_y y')} dx'\, dy'$$

$$= \frac{e^{j(\omega t - kz)}}{j\lambda z} e^{-(j\pi/\lambda z)(x^2+y^2)} \mathcal{F}[E(x',y',0)e^{-(j\pi/\lambda z)(x'^2+y'^2)}] \quad . \tag{2.17.3}$$

$$f_x = \frac{x}{\lambda z}$$

$$f_y = \frac{y}{\lambda z}$$

Fraunhofer Approximation:

$$z \gg \frac{\pi}{\lambda}(x'^2 + y'^2)_{\text{max}}$$

$$E(x,y,z) = \frac{e^{j(\omega t - kz)}}{j\lambda z} e^{-(j\pi/\lambda z)(x^2+y^2)} \iint E(x',y',0)e^{(jk/2z)(xx'+yy')} dx'\, dy'$$

$$= \frac{e^{j(\omega t - kz)}}{j\lambda z} e^{-(j\pi/\lambda z)(x^2+y^2)} \mathcal{F}[E(x',y',0)]_{\substack{f_x = x/\lambda z \\ f_y = y/\lambda z}}$$

$$= \alpha_{\text{con}}(z)\mathcal{F}[E(x',y',0)]_{\substack{f_x = x/\lambda z \\ f_y = y/\lambda z}} \quad . \tag{2.17.4}$$

In many cases we have a transmitting mask in the source plane. The transmission function is defined as

$$T(x',y') = \frac{E_{\text{trans}}(x',y',0^-)}{E_{\text{inc}}(x',y',0^-)} \tag{2.17.5}$$

Examples of Fraunhofer diffraction with different masks:

a) One-dimensional rectangular aperture with uniform light propagating in the z direction.

$$T(x') = \begin{cases} 1 & \text{for} \quad -L_x/2 \le x' < L_x/2 \\ 0 & \text{otherwise} \end{cases} \tag{2.17.6}$$

$$E(x) = \alpha_{\text{con}} E_0 L_x \,\text{sinc}\left(\frac{x L_x}{\lambda z}\right) \tag{2.17.7}$$

$$I(x) = \frac{E_0^2}{\lambda^2 z^2} L_x^2 \,\text{sinc}^2\left(\frac{x L_x}{\lambda z}\right) \tag{2.17.8}$$

b) One-dimensional aperture centered at $x = x_0$:

$$T(x') = \begin{cases} 1 & x_0 - L_x/2 \le x' < x_0 + L_x/2 \\ 0 & \text{otherwise} \end{cases} \tag{2.17.9}$$

$$E(x) = \alpha_{\text{con}}(D) E_0 e^{-j2\pi f_x x_0} L_x \,\text{sinc}\,(f_x L_x) \tag{2.17.10}$$

$$I(x) = \frac{E_0^2}{\lambda^2 D^2}\left[L_x^2 \operatorname{sinc}\left(\frac{xL_x}{\lambda z}\right)\right]^2 \tag{2.17.11}$$

c) One-dimensional aperture with uniform light shining at an angle θ with respect to the optical axis:

$$E(x',0) = E_0 e^{-j2\pi f_{x0}x'} \quad , \quad \text{where} \tag{2.17.12}$$

$$f_{x0} = \frac{\sin\theta}{\lambda} \tag{2.17.13}$$

$$\begin{aligned} E(x) &= \alpha_{\text{con}}(D)\mathcal{F}[E_0 e^{-j2\pi f_{x0}x'}\operatorname{Rect}(L_x)] \\ &= \alpha_{\text{con}}(d)E_0 L_x \operatorname{sinc}[(f_x - f_{x0})L_x] \end{aligned} \tag{2.17.14}$$

$$I(x) = \frac{E_0^2}{\lambda^2 D^2}L_x^2 \operatorname{sinc}^2[(f_x - f_{x0})L_x] \tag{2.17.15}$$

d) Diffraction grating:

$$\begin{aligned} T(x') &= f(x') + f(x' - x_0) + f(x' - 2x_0) + \cdots \\ &\quad + f(x' - (N-1)x_0) \quad . \end{aligned} \tag{2.17.16}$$

$$E_{\text{tot}} = \left\{ \begin{array}{l} \text{Diffraction due to} \\ \text{single element } f(x') \end{array} \right\} \times \{\text{Interference term}\} \tag{2.17.17}$$

$$\text{Interference Term} = I_n = \frac{\sin(N\pi f_x x_0)}{\sin(\pi f_x x_0)} \quad . \tag{2.17.18}$$

For $N \to$ large, I_n has maxima at

$$f_x = \frac{m}{x_0} = f_m \quad . \tag{2.17.19}$$

Near f_m,

$$I_n \to N \operatorname{sinc}[N(f_x - f_m)x_0] \quad . \tag{2.17.20}$$

e) Phase grating:

$$T(x') = e^{-j\phi_0}e^{j\varrho/2\sin f_{x0}x'}\operatorname{Rect}(L_x) \quad . \tag{2.17.21}$$

$$E(x) = \alpha_{\text{con}}(D)\sum_{q=-\infty}^{+\infty} J_q\left(\frac{\varrho}{2}\right)\operatorname{sinc}\left[\frac{L_x}{\lambda z}(x - qf_{x0}\lambda z)\right] \quad . \tag{2.17.22}$$

$$I(x) = \frac{1}{\lambda^2 D^2}\left\{\sum_{q=-\infty}^{+\infty} J_q\left(\frac{\varrho}{2}\right)\operatorname{sinc}\left[\frac{L_x}{\lambda z}(x - qf_{x0}\lambda z)\right]\right\}^2 \quad . \tag{2.17.23}$$

f) Amplitude grating:

$$T(x,y) = \left(\frac{1}{2} + \frac{\varrho}{2}\cos 2\pi f_{x0}x\right)\operatorname{Rect}(L_x) \quad . \tag{2.17.24}$$

$$E(x) = E_0 \alpha_{con}(D) \left\{ \mathrm{sinc}\left(\frac{L_x x}{\lambda z}\right) + \frac{\varrho}{2} \mathrm{sinc}\left[\frac{L_x}{\lambda z}(x + f_{x0}\lambda z)\right] \right.$$
$$\left. + \frac{\varrho}{2} \mathrm{sinc}\left[\frac{L_x}{\lambda z}(x - f_{x0}\lambda z)\right] \right\} \quad . \tag{2.17.25}$$

$$I(x) = \frac{E_0^2}{\lambda^2 D^2} \left\{ \mathrm{sinc}^2\left(\frac{L_x x}{\lambda z}\right) + \frac{\varrho^2}{3} \mathrm{sinc}^2\left[\frac{L_x}{\lambda z}(x + f_{x0}\lambda z)\right] \right.$$
$$\left. + \frac{\varrho^2}{4} \mathrm{sinc}^2\left[\frac{L_x}{\lambda z}(x - f_{x0}\lambda z)\right] \right\} \quad . \tag{2.17.26}$$

The transmission function of a lens is given by

$$T(x', y') = e^{-j\phi_0} e^{j\pi(x'^2 + y'^2)/\lambda f} \tag{2.17.27}$$

where f is the focal length. Consider a situation where we have placed a lens with a focal length f at the aperture as shown in Fig. 2.92. If we assume that the aperture has a transmission function $T(x', y')$, the total transmission function $T'(x, y)$ to be used in the equations for Fresnel's diffraction formula is given by

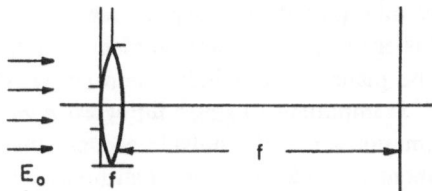

Fig. 2.92. Fresnel diffraction with a lens

$$T'(x, y) = T(x', y')e^{-j\phi_0} e^{j\pi(x'^2 + y'^2)/\lambda f} \quad . \tag{2.17.28}$$

For this case, we have

$$E(x, y, f) = \alpha_{con} e^{-j\phi_0} \mathcal{F}[E_{inc}(x', y')T(x', y')] \quad . \tag{2.17.29}$$

For

$$T(x', y') = T(r') = \begin{cases} 1 & \text{for } r' \leq \varrho \\ 0 & \text{otherwise} \end{cases}$$

we have

$$E(r, f) = E_0 e^{j(\omega t - kf)} e^{-jkr^2/2f} \frac{k\varrho^2}{j2f} \left(2\frac{J_1(k\varrho r/2f)}{k\varrho r/2f}\right) \quad . \tag{2.17.30}$$

$$I(r, f) = E_0^2 \left(\frac{k\varrho^2}{2f}\right)^2 \left(2\frac{J_1(k\varrho r/2f)}{k\varrho r/2f}\right)^2 \quad . \tag{2.17.31}$$

2.17.1 Holography

Holography was discovered by Dennis Gabor in 1947 in connection with three-dimensional viewing of X-ray images. Due to the recent availability of laser light and to the improvements made by many researchers (especially *Leith* and *Upatnieks* [2.11]), holography has become very useful and popular. To the layman, holography appears to be three-dimensional photography. However, other aspects of holography, such as the storage of optical information, are probably more important from the scientific point of view. Holography has been treated in many books and it is still being actively researched. The purpose of this section is to introduce the topic in terms of the Fourier optics developed in this book and discuss some of its applications. For a more detailed discussion, see [2.2, 9–12].

a) Photography

In order to examine holography, it is worthwhile to review the subject of photography. Generally, a light-sensitive silver compound based film is used in photography. This film has the property of recording the square of the incident light amplitude. The typical information we are interested in recording can be written as

$$E(x, y) = a(x, y)e^{-j\phi(x,y)} \quad . \tag{2.17.32}$$

For example, the view through a window may provide scenery of trees, birds and mountains. The reason we can see this scenery is because an electric field wavefront defined in (2.17.32) exists in the plane of the window and conveys complete information about the scenery. It is important to stress the word *complete*. At a particular instant, all the information about the outside scenery that can be obtained is present in that two-dimensional electric field distribution. It has both an amplitude part and a phase part. The phase part conveys some of the three-dimensional aspects of the scenery. For example, if there were a cat behind the tree and we looked directly at the tree, we would not see the cat. However, if we moved a little, as shown in Fig. 2.93, so that the cat were not obstructed by the tree, we would see it. This information, i.e., that looking along different lines of sight we see different parts of the scenery, is an aspect of so-called three-dimensional photography.

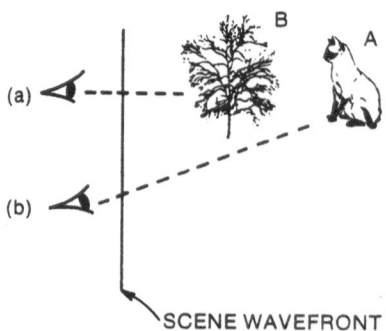

Fig. 2.93. The three-dimensional aspect of a scene to be recorded

In ordinary photography, we place the film near the window and record the square of the electric field (in fact, by using a camera lens we project the electric field at the window on the film). This is done by first exposing the film to the incident radiation for a specified length of time. This radiation performs a photochemical reaction which is proportional to the square of the electric field. After exposure, the film is chemically processed so as to produce a transmission function for the film given by

$$
\begin{aligned}
t(x,y) &= t_0 + \beta|E|^2 \\
&= t_0 + \beta[aa^*] \\
&= t_0 + \beta[(a(x,y)]^2 \quad .
\end{aligned}
\tag{2.17.33}
$$

Hence the transmission function of the exposed and developed film contains the information $[(a(x,y)]^2$. Of course, this film can be printed or viewed by shining light through it to obtain the picture of the scenery.

Although our goal was to obtain the information $a(x,y)\exp[-j\phi(x,y)]$, or to recreate the wavefront itself, what we have obtained by using photography is $[(a(x,y)]^2$. We have lost the phase information. Consequently, if we did take the picture in which the cat were behind the tree, conventional photography would never reveal it. To obtain that information, we must reproduce the E-field of the window itself, with all of its amplitude and phase variations, i.e., the whole wavefront must be reconstructed. This wavefront reconstruction can be performed by holography.

How can one record the phase of a light beam? A clue to this question can be found in the discussion on interference. As shown in Fig. 2.94, two point sources are denoted by A and B. If we take a photograph at the detector plane (in the Fraunhofer zone) we record a uniform light distribution for each point A or B. However, if both are present simultaneously, we obtain interference fringes. In that case the detected light intensity is given by

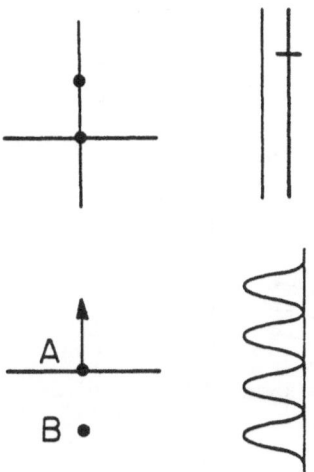

Fig. 2.94. The use of interference to obtain phase information

$$I(x) \propto E_A^2 + E_B^2 + 2E_A E_b \cos \alpha_0 \qquad (2.17.34)$$

where α_0 is the phase difference between the two electric fields at the detector point. We can therefore obtain the phase information with respect to another source. For example, the period of the recording in the detector plane gives information about the separation between the two sources, while the amplitude on the optical axis gives information about the phase difference between these sources.

By recording the square of the E-field in the presence of a reference source, the phase information can be recorded. Of course, this assumes that both sources are coherent. We can therefore infer that holography requires both coherent sources and a reference beam; as discussed in the next section.

b) Making a Hologram

As shown in Fig. 2.95 we are interested in the imaging of an object illuminated with a coherent source. The light rays carrying the information about the object have an electric field at the recording plane denoted by

$$E(x, y) = a(x, y)e^{-j\phi(x,y)} = a \quad . \qquad (2.17.35)$$

We shall denote this total electric field by a. A reference beam is also incident on the recording medium at the same time. This electric field is denoted by

$$E_{\text{ref}}(x, y) = A(x, y)e^{-j\psi(x,y)} = A \quad . \qquad (2.17.36)$$

As shown in Fig. 2.95, the light shining on the object and the reference beam may be produced by the same laser source to keep them coherent and they may thus interfere over larger areas. This has been performed in Fig. 2.95 by using a beam splitter and a prism.

One of the simplest cases arises when the reference beam is parallel, uniform and incident at an angle θ on the recording medium. In that case

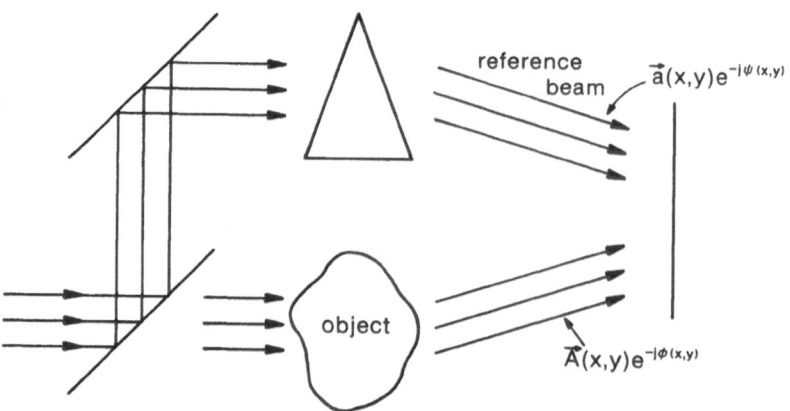

Fig. 2.95. The making of a hologram

$$A = A_0 e^{j2\pi f_0 x} \qquad\qquad\qquad\qquad (2.17.37)$$

where $f_0 = (\sin\theta)/\lambda$. It is obvious that this reference beam has a more or less single spatial frequency (due to finite aperture) which is determined by the incident angle θ. This reference beam frequency is called the carrier frequency, in analogy with radio-engineering terminology. The significance of this will be discussed in Sect. 2.17.1d.

The total incident electric field on the recording plane is given by

$$E_{tot} = a + A \quad . \qquad\qquad\qquad\qquad (2.17.38)$$

Hence the transmission factor of the correctly processed recording film will be given by

$$\begin{aligned}
t(x,y) &= t_0 + \beta(E_{tot}E_{tot}^*) \\
&= t_0 + \beta(aa^* + AA^* + aA^* + a^*A) \\
&= t_0 + \beta\{|a|^2 + |A|^2 + 2A(x,y)a(x,y)\cos\left[\phi(x,y) - \psi(x,y)\right]\} \quad .
\end{aligned}$$
$$(2.17.39)$$

Thus we see that, although we have recorded the square of the electric field magnitude, we have also been able to retain the phase information of the object, $\phi(x,y)$, due to the presence of the reference beam which interferes with the beam scattered by the object.

We recall that, in photography, the recording process would have been the same except that the reference beam is absent. We could then have immediately looked at this transparency and would have observed some resemblance to the object. This is not the case in holography. What we see in holograms is a multiplicity of different interference fringes which have absolutely no resemblance to the object at all. This is shown in Fig. 2.96. Thus, in a hologram, the viewing or reconstruction of the holographic field is a separate and essential process.

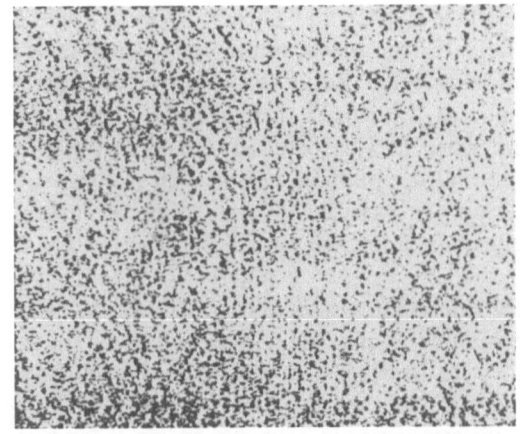

Fig. 2.96. Picture of an actual hologram

c) Reconstruction of the Hologram

To view the hologram, another coherent source must illuminate it. For the sake of simplicity, let us assume that the wavelengths of light for the reference beam and viewing beam are identical. The effect of their being different will be discussed later.

If the viewing beam electric field at the hologram plane is denoted by B, we obtain the electric field of the light emerging from the hologram:

$$E_{out} = t_0 B + \beta(|a|^2 + |A|^2)B + \beta(aA^* + a^* A)B \quad . \tag{2.17.40}$$

The output light thus consists of five individual terms. Again, let us assume for simplicity's sake that the viewing and the reference beams are identical in spatial frequency, i.e.,

$$B = B_0 e^{j2\pi f_0 x} \quad . \tag{2.17.41}$$

For this viewing field, we can denote the output E-field consisting of five terms given by

$$E_{tot} = E_1 + E_2 + E_3 + E_4 + E_5 \quad , \quad \text{where} \tag{2.17.42}$$

$$E_1 = t_0 B_0 e^{j2\pi f_0 x} \quad ,$$

$$E_2 = \beta |a|^2 B_0 e^{j2\pi f_0 x} \quad ,$$

$$E_3 = \beta |A_0|^2 B_0 e^{j2\pi f_0 x} \quad ,$$

$$E_4 = \beta A_0 B_0 a \quad , \quad \text{and}$$

$$E_5 = \beta A_0 B_0 a^* e^{-j2\pi f_0 x} \quad .$$

We immediately notice that the five beams emerge at three different angles without any overlap only if the spatial bandwidth of the field a is not too large. These are shown in Fig. 2.97. We see that E_1, E_2 and E_3 emerge at an angle θ with respect to the hologram. The field E_4 can be viewed directly, whereas E_5 appears at an angle which is approximately 2θ. Beam 4, except for a constant term $\beta A_0 B_0$, is an exact reproduction of the electric field associated with the object at the plane of the hologram when it was recorded. It contains all the amplitude and phase information, i.e., all the possible information it contained when the hologram was made. Thus we can view it as if the entire object were recreated. Beam 5 also contains the information about the object but it is a phase conjugate picture.

In the above discussion, a very important assumption was made, namely, that beams 4 and 5 do not overlap with each other or with beams 1, 2, and 3. We can immediately see that, if they do overlap, then while looking directly, for example, we will see not only a but also the other beams. This separation of the beams or the use of the carrier frequency to have the beams emerge in

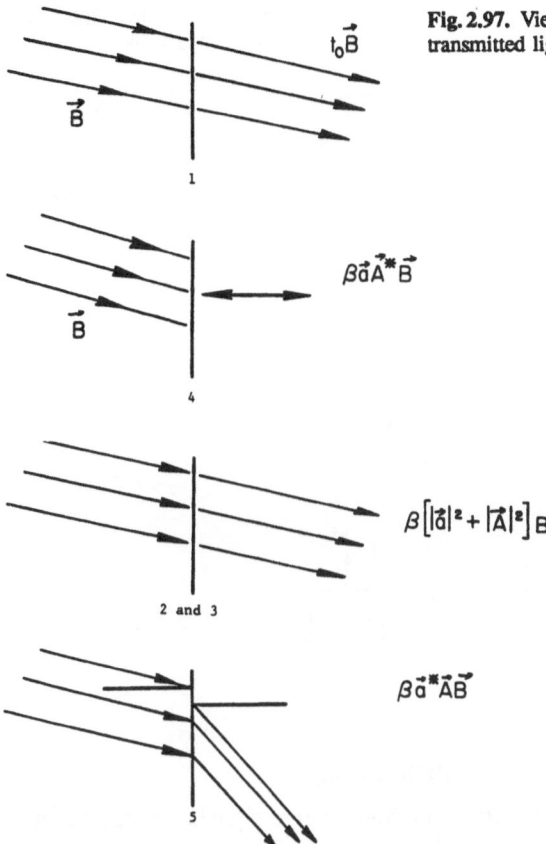

Fig. 2.97. Viewing of a hologram and its different transmitted light components

$t_0 \vec{B}$

$\vec{B}$

1

$\vec{B}$

$\beta \vec{a} \vec{A}^* \vec{B}$

4

$\beta \left[|\vec{a}|^2 + |\vec{A}|^2 \right] B$

2 and 3

$\beta \vec{a}^* \vec{A} \vec{B}$

5

different directions was the contribution of *Leith* and *Upatnieks* in their classic paper [2.11]. The Gabor hologram did not have this separation and it is therefore of much poorer quality.

Let us assume that the spatial bandwidth associated with the object is Δf_x. For simplification let us consider one dimension only. This means that the frequency components of a are limited to the frequency spread

$$f_0 - \frac{\Delta f_x}{2} < f < f_0 + \frac{\Delta f_x}{2} \quad .$$

For the electric field of any object, we can then identify an associated spatial bandwidth which denotes the rapidity of the spatial variations in amplitude and phase. In that spatial frequency plane we can then plot all the output beams, E_1-E_5, as shown in Fig. 2.98. We see that, if the beams must not overlap, we must satisfy

$$f_{x0} > \tfrac{3}{2}(\Delta f_x) \quad . \tag{2.17.43}$$

187

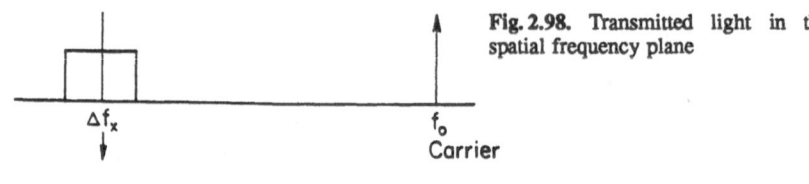

Fig. 2.98. Transmitted light in the spatial frequency plane

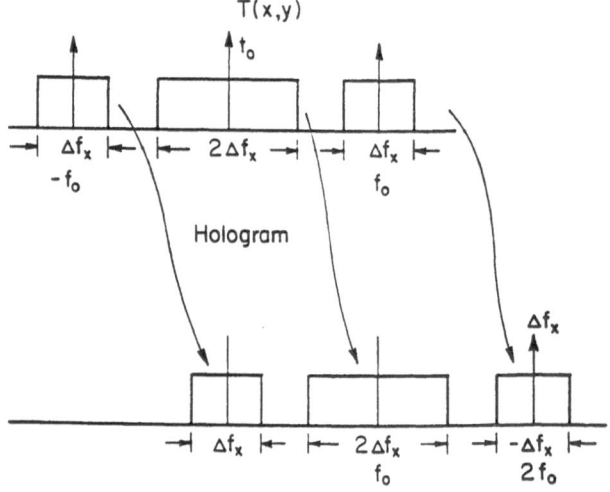

d) Analogy with Radio and Information Storage

In radio engineering, the aim is to transmit and receive signals that are functions of time. For example, those signals may be the music of an orchestra performance. An illustrative signal with bandwidth Δf is shown in Fig. 2.99. Most radio signals have an information bandwidth $\Delta f \approx 30 \, \text{kHz}$. To send such a radio signal directly via an electric wire or through a medium by radiation, a separate wire or medium is needed for each signal. For example, the signals of two orchestras cannot be sent via the wire directly because at the output the sum of the two time signals was received and there is no way to separate them. However, if we

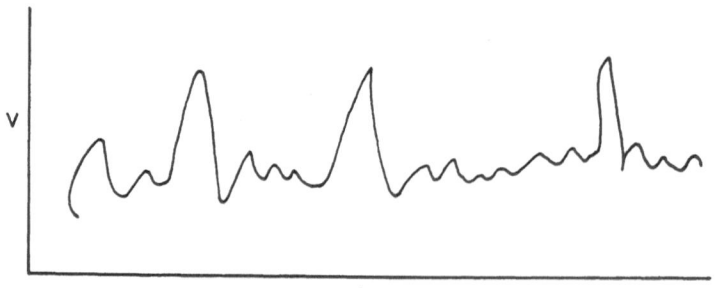

Fig. 2.99. Typical radio wave form

modulate these signals with different carrier frequencies $f_1, f_2 \gg \Delta f$ then, even if they pass through the same wire or medium, we can separate them by frequency filters and by remixing with carrier frequencies. This frequency multiplexing and demultiplexing is essential for the operation of radio and TV.

With regard to holography, we still have spatial frequency, albeit two-dimensional. We also have a bandwidth of the signal, which is now the E-field due to the object, a. If we do not add any carrier frequency, we have a situation somewhat similar to photography. However, if we multiplex with different carrier frequencies, we can then store and view many objects or pictures via the same hologram, just as if different audio signals were passing through the same wire.

To implement this arrangement let us assume that the bandwidth of each picture is Δf_x. We can then record one page or picture with a carrier frequency f_{x0}, the next picture or page with a carrier frequency $f_{x0} + 2\Delta f_x$ and so on, as shown in Fig. 2.100. Notice that all the pictures are recorded in the same film. However, when viewing it with different carrier lights, or light incident at different angles, we see the different pictures wihtout any overlap or distortion. Hence, we recognize that enormous amounts of information can be stored in one piece of film such as a whole book in just one square area of film, provided the resolution of the film is high enough.

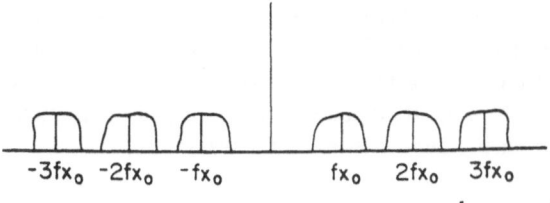

$$-3fx_0 \quad -2fx_0 \quad -fx_0 \qquad fx_0 \quad 2fx_0 \quad 3fx_0$$

$\longrightarrow fx$ \qquad **Fig. 2.100.** Multiplexing in holograms

e) Some Features of Holograms

i) If in (2.17.40), $B \neq B_0 \exp(j2\pi f_0 x)$, i.e. has a different spatial frequency f_B given by $B = B_0 \exp(j2\pi f_B x)$, then the output E-field will emerge from the hologram at angles θ_-, θ_B and θ_+, given respectively by

$$\frac{\sin\theta_B}{\lambda_B} = f_B \quad ; \quad \frac{\sin\theta_-}{\lambda_B} = f_B - f_0 \quad ; \quad \frac{\sin\theta_+}{\lambda_B} = f_B + f_0 \quad ,$$

where λ_B is the wavelength of the viewing light.

ii) We have not discussed the brightness of the image when viewed through the holographic process. It transpires that the diffraction efficiency of thin-film holograms is not very large, so that the image may not be bright. However, one can increase the diffraction efficiency by using a thick-film hologram. In thick-film holograms, the efficiency may be great enough so that instead of needing a laser beam, a simple white light might suffice for viewing purposes. The different colors are then diffracted at different angles and if the separation between them is sufficiently large, one can view what is known as a white light hologram.

iii) Another form of white light hologram has also been developed. In this kind, no thick film is used; instead, the diffraction efficiency is increased at the cost of the perspective in one direction. To construct such a hologram, one must first produce a regular hologram of the object. This is then illuminated with a viewing light and another reference beam is used to make a second hologram by using the real image of the first. This second hologram, when viewed with white light produces bright single color images at different angles.

iv) So far we have implied that there is no difference between the holographic recording film and that used in photography. From the photochemical point of view, there is no difference. However, the photographic film needs to record only a bandwidth of Δf_x whereas the holographic film must also be able to record the carrier frequency. Thus the holographic film must generally have a much higher resolution. Because the higher resolution films generally need to be exposed for a longer time for the same intensity of incident light, the light sources must be coherent over this recording time period. While this might be a restriction that is difficult to overcome one can still use a high power laser source so that the exposure time is shorter.

Problems

2.1 An electromagnetic plane wave in air passes through a dielectric sheet (normal to the direction of travel). The sheet is 0.58 wavelength thick. The plane wave continues on to infinity after passing through the sheet. If the relative dielectric constant of the sheet is 2, find the reflected electric field and the intensity of the wave which goes on, for an incident intensity of $1 \, \text{W/m}^2$.

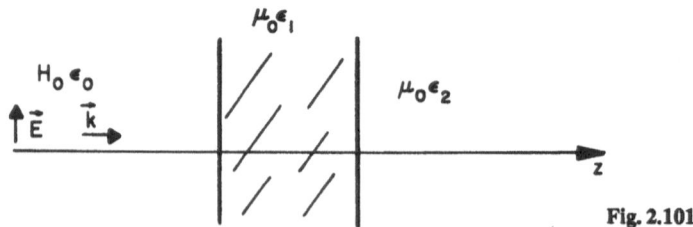

Fig. 2.101

2.2

a) Consider the situation shown in Fig. 2.101. If $\varepsilon_2 = 6.25$ and the wavelength of the incident wave $\lambda_0 = 5 \times 10^{-7} \, \text{m} = 5000 \, \text{Å}$, calculate ε_1 and d so that there is no reflected wave in the air.

b) Find an expression for the transmission coefficient if the angle of incidence is θ in place of $0°$. ($E \perp$ plane of incidence.)

2.3

a) Both sides of a flat pane of glass having refractive index 1.5 are coated with a dielectric material having refractive index 2. If the glass thickness is 6 mm, calculate the reflected and transmitted power for 1 W laser light ($\lambda = 6000$ Å) incident normally. Dielectric material thickness: $\lambda/4$.

b) Calculate the electric field at the focus if the transmitted beam has a diameter of 5 mm and it is focused with a lens of focal length 5 mm.

2.4

a) Consider 1 mW electromagnetic radiation of wavelength 1.0 μm incident at an angle of 56.3° from air on a glass plate (Fig. 2.102). If the polarization of the electric field is in the plane of incidence and the glass is 1 mm thick, calculate the transmitted power. The refractive index of glass is 1.5.

b) If the glass is coated on the back with a dielectric layer having refractive index 2 and thickness 1 μm, calculate the transmitted power.

c) Now the wavelength of the radiation is changed to 0.5 μm. Repeat part (b) only.

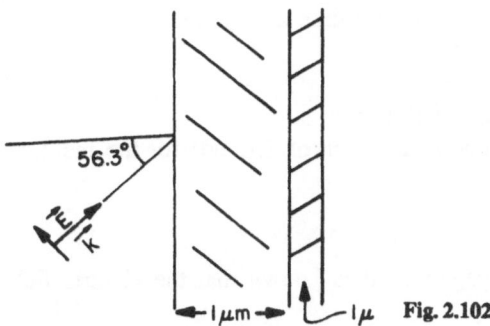

56.3°

$\leftarrow$ lμm $\rightarrow$ $-$lμ **Fig. 2.102**

2.5

a) Calculate the Brewster angle θ_B for an air/glass interface. The refractive index of glass is 1.5.

b) 1 W of unpolarized laser light is incident on the interface at an angle θ_B. Calculate the incident and the transmitted electric field, both magnitude and polarization.

c) Repeat part (b) for an incident angle of $\theta_B + 1°$.

2.6

a) Light of wavelength $\lambda = 0.5$ μm is incident on a glass medium from air as shown in Fig. 2.103. It is found that for an incidence angle of 60°, no reflected beam is observed. 1) Find the incident electric vector. 2) Find the refractive index of the glass medium.

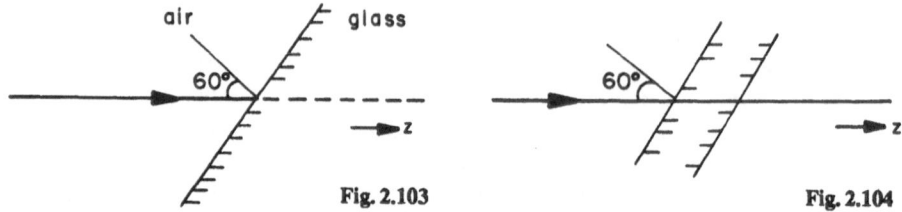

air   glass

60°

Fig. 2.103   Fig. 2.104

b) This part of the problem is the same as part (a) except a flat glass plate of thickness 0.2 mm is used (Fig. 2.104). Calculate the transmitted power density in the air.

c) For case (b), is there any change in transmitted power if the incident E vector is rotated by 90°? If the answer is yes, explain how you would calculate it. If not, explain why.

2.7 A plane electromagnetic wave ($\lambda = 1.0\,\mu$m) is incident from air to glass as shown in Fig. 2.105. The total incident power is $1\,\text{W/m}^2$. It is known that the y component of the incident electric field is 15 V/m.

a) Calculate the incident E-field and H-field. Give a numerical answer showing all the components and their variation with time and space.

b) Do the same for the reflected field.

c) Do the same for the transmitted field.

d) Calculate the reflected and transmitted power.

e) Determine the plane of polarization (direction of E-field) for the incident, reflected and transmitted light.

2.8

a) Let us consider free space propagation. It is known that the electric field ($\lambda = 1.0\,\mu$m) at $z = 0$ is given by

$$E(x, y, z = 0) = \frac{x}{10^{-4}} \left(\frac{2y^2}{10^{-8}} - 1 \right) \exp\left(\frac{x^2 + y^2}{2 \times 10^{-8}} \right) \,,$$

where MKS units are used.

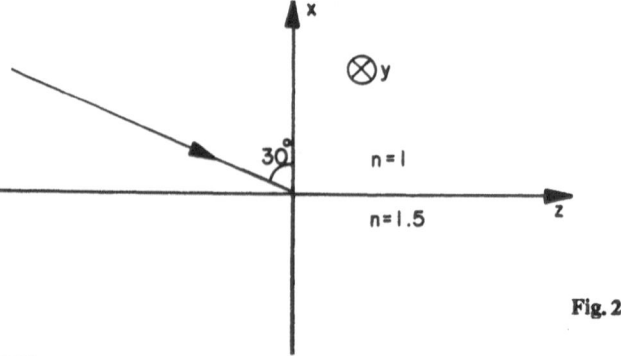

Fig. 2.105

i) Write an exact expression for the electric field at z.
ii) Plot the magnitude of the electric field at $z = 1\,\text{km}$.

b) It is known that the electric field ($\lambda = 1.0\,\mu\text{m}$) propagating inside a graded optical fiber at $z = 0$ is given by the same expression as in (a).

i)   Determine the refractive index of the fiber as a function of its radius.
ii) Calculate the electric field at a distance z.
iii) Calculate the group velocity if $dn(\lambda)/d\lambda = -10^{-10}\,\text{m}^{-1}$.

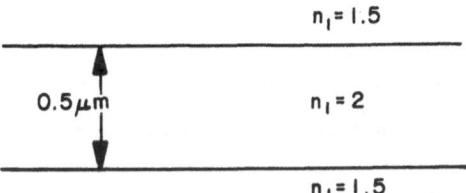

$n_1 = 1.5$

$0.5\,\mu\text{m}$ $n_1 = 2$

Fig. 2.106

$n_1 = 1.5$

2.9 A slab waveguide is constructed with the dimensions shown in Fig. 2.106.

a) If we want single-mode propagation, to what light frequency range should we restrict ourselves?
b) Calculate the group velocity of the wave at the center of the frequency band calculated in (a).
c) Calculate the smallest pulse width that can propagate if we restrict ourselves to the upper half of the frequency band. Also, if this pulse is propagated along 1 km of the filter, what will be the pulse shape at the output end?

2.10

a) The electric field for a propagating wave ($\lambda = 1.0\,\mu\text{m}$) at $z = 0$ is given by

$$E(x, y, z = 0) = \sqrt{\frac{2}{\pi}}\exp\left(-\frac{x^2 + y^2}{10^{-8}}\right) \quad \text{(MKS units)} \quad .$$

Calculate the spatial frequency components of this electric field.
b) Calculate the electric field in the *spatial frequency domain* at $z = f$ where the wave propagates through free space.
c) A lens is placed at $z = f$. The focal length of the lens is $f = 10\,\text{cm}$. Calculate the electric field in the spatial frequency domain of the wave after it passes through the lens.
d) Calculate the electric field at $z = 2f$.

2.11

a) A certain commercial helium/neon laser is advertised to have a far-field divergence angle of 2 millirad at $\lambda = 632.8\,\text{nm}$. What is the spot size w_0?
b) The power in this laser is 10 mW. What is the peak electric field in V/cm at $r = z = 0$?

c) Electromagnetic energy can only come in packages of hf. If one thousand more photons per second were emitted by this laser, what would be the new power specification?

2.12 Consider a 1-W TEM$_{0,0}$ beam of $\lambda = 514.5$ nm from an argon-ion laser with a minimum spot size $w_0 = 3$ mm located at $z = 0$.

a) How far will this beam propagate before the spot size is 1 cm?
b) What is the radius of curvature of the phase front at this distance?
c) What is the amplitude of the electric field at $r = 0$?

2.13 Suppose that a Gaussian beam with $w = 2$ cm and a planar wavefront impinges upon a lens of focal length $f = 4$ cm ($\lambda = 1.0\,\mu$m).

a) If $z = 0$ is the location of the lens, where does the output beam reach its minimum spot size?
b) What is the far-field divergence angle?

2.14 A typical fiber with a quadratic variation of index $n(r) = n_0 - \Delta n(r/a)^2$ has the following parameters: $n_0 = 1.5$, $\Delta n = 8 \times 10^{-3}$, and $a = 20\,\mu$m.

a) Assume that all modes with numbers $(m, p \leq 20)$ are excited more-or-less uniformly at $z = 0$ and with a common phase. Calculate the phase shift of the (m, p) mode relative to the $(0,0)$ mode at $z = 1$ km. Estimate the distribution of phases of the various modes relative to $m = p = 0$.
b) Calculate the range of values of the group velocities for the various modes.
c) A GaAs semiconductor laser at $\lambda = 8420$ Å excites this fiber with a 1-ns pulse. Estimate the degree of pulse spreading due to the fact that the group velocities of the various modes are all different.
d) Use the results of part (c) to estimate the spread in the pulse width after a distance of 1 km.

2.15

a) Calculate the numerical aperture, R, and cutoff conditions for the TE modes of the slab waveguide shown in Fig. 2.107. ($n_1 = 1.60$, $n_2 = 1.46$, $d = 0.6\,\mu$m, and $\lambda_0 = 1\,\mu$m.)
b) What happens to the number of propagating modes if d is increased to 3 μm?
c) What happens to the number of propagating modes if λ_0 is increased to 1.3 μm?

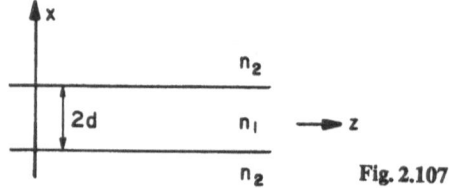

Fig. 2.107

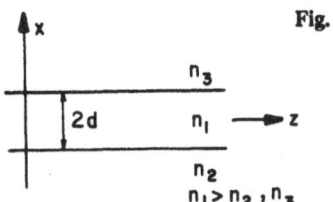

Fig. 2.108

2.16 For the asymmetric slab waveguide shown in Fig. 2.108, derive the characteristic equations for the TE and TM modes. Show that the TE mode equation reduces to that for the symmetric case.

2.17 $1 \, W/m^2$ light ($\lambda = 0.5 \, \mu m$) is incident at a vacuum/solid ($n = 1.5$) interface at an angle of $40°$ with respect to the normal. The incident wave is unpolarized. Calculate the polarization of the wave inside the solid and the electric field vector for each polarization.

2.18 Consider the optical system consisting of two lenses shown in Fig. 2.109. The incident light wave at $z = 0$ is given by

$$E(x, y, 0) = A \exp\left[-\frac{x^2 + y^2}{w^2(0)}\right] \exp(j\omega t)$$

where $w(0) = 1 \, cm$ and $\lambda = 0.5 \, \mu m$. Calculate the electric field at $z = 30$ and plot it as a function of x and y.

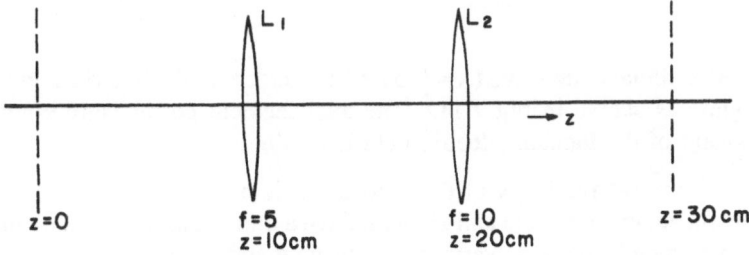

Fig. 2.109

2.19 For an integrated optics experiment a symmetrical single-mode slab waveguide is designed as shown in Fig. 2.110.

a)   Calculate the cutoff frequency of the next TE mode and thus determine the approximate bandwidth of the waveguide.
b)   Plot the mode-shape of the propagating mode and the next TE mode.
c)   Calculate the penetration depth of the propagating mode in the $n = 1.1$ region.
d)   Estimate the smallest pulse one can propagate through this guide.

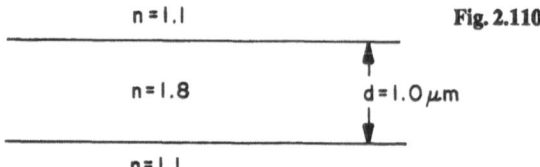

Fig. 2.110

2.20 The D vs E relationship for an anisotropic crystal is given by

$$\begin{pmatrix} D_x \\ D_y \\ D_z \end{pmatrix} = \begin{pmatrix} 2\varepsilon_0 & 0 & 0 \\ 0 & 2\varepsilon_0 & 0 \\ 0 & 0 & 3\varepsilon_0 \end{pmatrix} \begin{pmatrix} E_x \\ E_y \\ E_z \end{pmatrix} .$$

a) Plot the index ellipsoid.
b) If it is known that the electric field is propagating along the x-direction, can you determine the possible E-, D-, B- and H-fields for this wave?
c) What are the phase and group velocities (both magnitude and direction)?

2.21 For the system shown in Fig. 2.111 with impulse response given by $h(x)$, calculate the output.

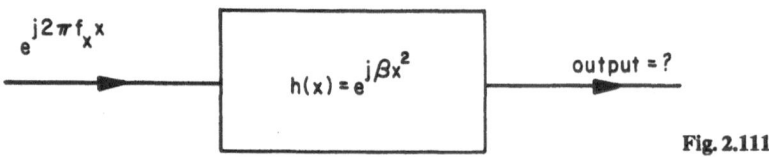

Fig. 2.111

2.22 A plane electromagnetic wave ($\lambda = 0.5\ \mu m$) is incident at the Brewster angle from air to glass as shown in Fig. 2.112. The total incident power is $10\ W/m^2$. The y component of the incident electric field is 60 V/m.

a) Calculate the refractive index of the second medium.
b) Calculate the incident E-field and H-field. Give a numerical answer showing all the components and their variations with time and space.
c) Do the same for the reflected field.

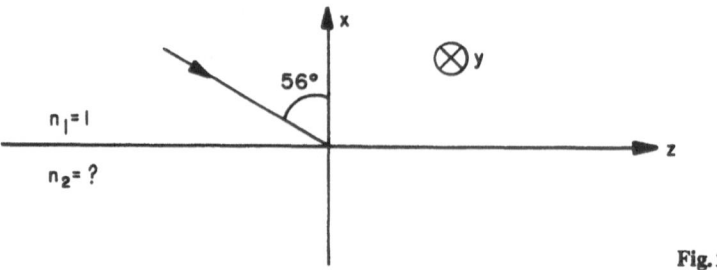

Fig. 2.112

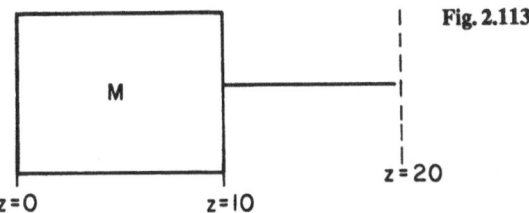

Fig. 2.113

z = 0   z = 10

d) Calculate the transmitted power.

e) Determine the plane of polarization (direction of E-field) for the reflected light.

2.23 It is known that the electric field ($\lambda = 0.5\,\mu$m) at $z = 0$ is given by

$$E(x, y, z) = \frac{x}{10^{-4}}\left(8\frac{y^3}{10^{-12}} - 12\frac{y}{10^{-4}}\right)\exp\left(-\frac{x^2 + y^2}{10^{-8}}\right) \; ,$$

where MKS units are used. Calculate the E-field at $z = 20$ when an optical system with matrix M occupies from $z = 0$ to $z = 10$, see Fig. 2.113.

2.24 A typical fiber with a quadratic variation of index $n(r) = n_0 - \Delta n(r/b)^2$ has the following parameters: $n_0 = 1.7$, $\Delta n = 10^{-2}$ and $b = 40\,\mu$m. $\lambda = 1.0\,\mu$m. It is known that only the (1,1) mode is propagating.

a) Write an expression for the E-field as a function of x, y, z and t.

b) Calculate the phase velocity.

c) Calculate the group velocity assuming no material dispersion.

2.25 For an integrated optics experiment, a symmetrical slab waveguide is designed as shown in Fig. 2.114.

a) Calculate the cutoff frequencies of the first two TE modes.

b) Plot the mode-shape of the second propagating mode.

c) Calculate the penetration depth of the second propagating mode in the outer cladding.

d) Calculate the numerical aperture of the guide.

2.26 Consider the optical system shown in Fig. 2.115. A lens with focal length f and size D is placed at $z = 0$; the output is at $z = f$.

a) Calculate the impulse response of the system.

b) A mask is placed at $z = 0^-$ and light parallel to the z axis is incident on it.

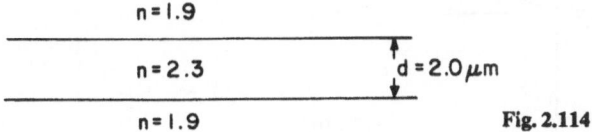

n = 1.9

n = 2.3 $d = 2.0\,\mu$m

n = 1.9   Fig. 2.114

197

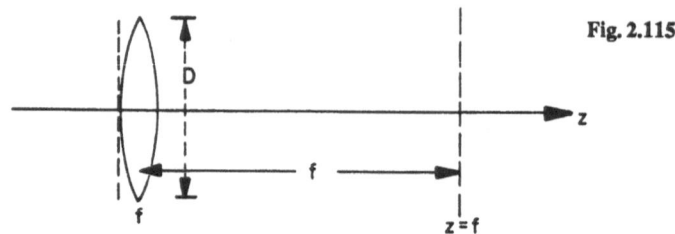

Fig. 2.115

The transmission function of the mask is given by $T(x, y) = \cos(2\pi f_0 x)$.
Find the output of the optical system.

c) Repeat (b) for light incident at an angle θ to the z axis in the xz plane.

2.27 The transmission function of a Fabry-Perot interferometer is given by

$$T = \left| \frac{E_t}{E_i} \right|^2 = \frac{T_{max}}{1 + F \sin^2 \frac{\delta}{2}} \quad \text{where}$$

$$F = \frac{4R_1}{(1 - R_1)^2} \quad ; \quad P_2 : \text{ reflection coefficient}$$

$$P_2 = p_0 e^{j\theta} \quad ; \quad R_1 = p_0^2 \quad ; \quad \delta = \Delta - 2\theta$$

$$\Delta = \frac{4\pi n d \cos r}{\lambda} \quad .$$

Use the transmission line model and Fig. 2.116 to derive the equation for T.

2.28

a) Design an antireflection coating for light of $\lambda = 5000\,\text{Å}$ normally incident
   on glass ($n = 1.8$).
b) Determine the effectiveness of this coating for incident light of different
   wavelength by plotting reflection coefficient vs λ (3500 Å–8500 Å).

2.29 A plane electromagnetic wave ($\lambda = 0.7\,\mu\text{m}$) is incident from air to a
parallel plate of glass ($n = 1.35$). It is observed that for a certain plane of polarization, all the incident light (1 W/m^2) is transmitted and there is no reflected
component.

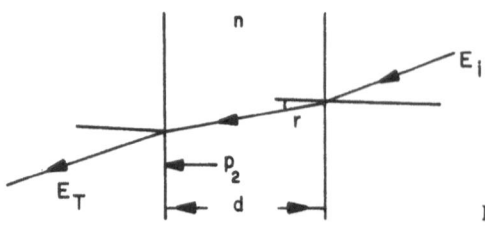

Fig. 2.116

a) Calculate the angle of incidence.
b) Write an expression for the reflection coefficient (R) if the polarization used is perpendicular to that for the case of no reflected component. The thickness of the plate is d.
c) Calculate d for the case of $R = 0$ as discussed in (b).
d) Calculate the magnitude of H-field in the glass for the no-reflection case.

2.30

a) Let us consider free-space propagation. It is known that the electric field ($\lambda = 0.5\,\mu$m) is given by

$$E(x, y, z = 0) = \left(\frac{2x^2}{10^{-6}} - 1\right)\left(\frac{2y^2}{10^{-6}} - 1\right)\exp\left(-\frac{x^2 + y^2}{2 \times 10^{-6}}\right) \quad,$$

where MKS units are used.

i) Write an expression for the E-field at z.
ii) Write an expression for the phase velocity.

b) This time the propagation is inside a graded optical fiber and the same expression for the E-field at $z = 0$ is to be considered.

i) Determine the refractive index of the fiber as a function of its radius.
ii) Write an expression for the E-field at z.

2.31 A slab waveguide is constructed with the dimensions shown in Fig. 2.117.

a) If we want single-mode propagation, to what light frequency range should we restrict ourselves?
b) Write expressions for all the components of the E- and H-fields for this single mode inside the guide.
c) Repeat (b) for outside the guide.

2.32 The D vs E relationship for an anisotropic crystal is given by

$$\begin{bmatrix} D_x \\ D_y \\ D_z \end{bmatrix} = \begin{bmatrix} 1.5\varepsilon_0 & 0 & 0 \\ 0 & 1.5\varepsilon_0 & 0 \\ 0 & 0 & 2\varepsilon_0 \end{bmatrix} \begin{bmatrix} E_x \\ E_y \\ E_z \end{bmatrix} \quad.$$

It is known that light is propagating at 45° to both the y and z axes in the yz plane. It is also known that the D-field is in the x direction. Write expressions for the D-field of this light as a function of x, y, z and t.

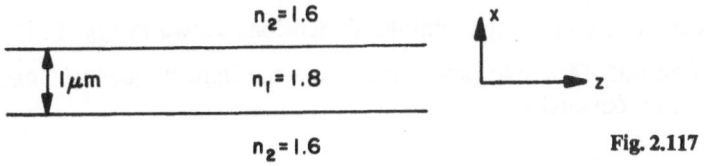

Fig. 2.117

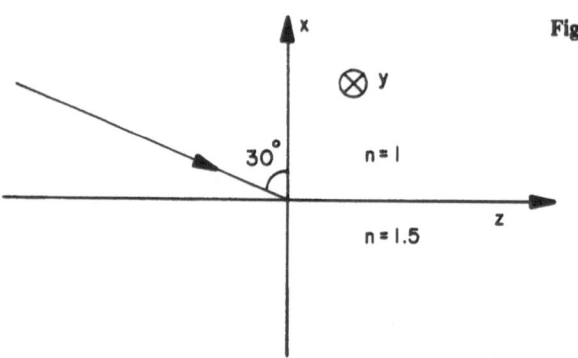

Fig. 2.118

2.33 A plane electromagnetic wave ($\lambda = 0.5\,\mu$m) is incident from air to glass as shown in Fig. 2.118. The total incident power is $10\,\text{W/m}^2$. It is known that the y component of the incident electric field is $50\,\text{V/m}$.

a) Calculate the incident E-field. Give a numerical answer showing all the components and their variation with time and space.
b) Do the same for the reflected field.
c) Do the same for the transmitted field.
d) Calculate the reflected and transmitted power.
e) Determine the plane of polarization (direction of E-field) for the reflected light.

2.34

a) Let us consider free-space propagation. It is known that the electric field at $z = 0$ ($\lambda = 0.5\,\mu$m) is given by

$$E(x, y, z = 0) = xy \exp\left(-\frac{x^2 + y^2}{10^{-6}}\right) \quad,$$

where MKS units are used.

i) Write an expression for the E-field at z. Include all space and time dependence.
ii) Plot the constant phase wavefront near z.

b) This time the propagation is inside a graded optical fiber and the same expression for the E-field at $z = 0$ is to be considered.

i) Determine the refractive index of the fiber as a function of its radius.
ii) Write an expression for the E-field at z.

2.35 A slab waveguide is constructed with the dimensions shown in Fig. 2.119.

a) If we want first odd TE mode propagation, to what light frequency range should we restrict ourselves?

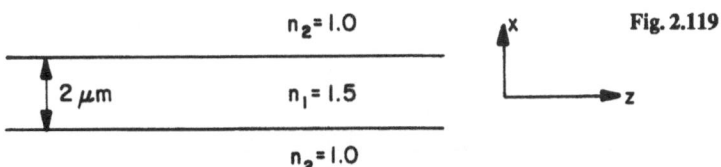

Fig. 2.119

b) Write expressions for all the components of the E- and H-fields for this mode inside the guide.

c) Repeat (b) for outside the guide.

2.36 The D vs E relationship for an anisotropic crystal is given by

$$
\begin{bmatrix} D_x \\ D_y \\ D_z \end{bmatrix} = \begin{bmatrix} 1.5\varepsilon_0 & 0 & 0 \\ 0 & 2\varepsilon_0 & \varepsilon_0 \\ 0 & \varepsilon_0 & 2\varepsilon_0 \end{bmatrix} \begin{bmatrix} E_x \\ E_y \\ E_z \end{bmatrix} .
$$

a) Plot the index ellipsoid.

b) It is known that the wave is propagating along the x direction. Determine the possible E- and D-fields for the wave.

c) What are the phase and group velocities (both magnitude and direction)?

2.37 An acousto-optic (A-O) spectrum analyzer is to be designed. The following components are available.

1. A-O device with center frequency 200 MHz, bandwidth 100 MHz and time duration 20 μs. The velocity of sound in the A-O device is 3×10^3 m/s.

2. Lenses of different focal lengths and aperture sizes.

3. Photodetector arrays of different numbers of elements and different element sizes.

4. A collimated laser source with $\lambda = 0.6\ \mu$m.

    a) For Raman-Nath operation, show with a diagram how the spectrum analyzer is to be designed. Give the angle of incidence, focal length and aperture of the lens, photodetector array size and number. If $(\Delta n)_0/n_0 = 10^{-5}$, $n_0 = 1.5$ and $L/\lambda = 10$, calculate the diffraction efficiency.

    b) Repeat (a) for the Bragg diffraction case.

2.38

a) Calculate and plot the Fraunhofer diffraction pattern (intensity) of 50 identical slits as a function of spatial frequency. The slits are 1 μm $\times$ 1 cm and their separation is 2 μm. The detector plane is 5 m away from the source plane. The light ($\lambda = 0.5\ \mu$m) is incident parallel to the optical axis.

b) A photodetector is to be used to measure the diffraction pattern. Estimate suitable dimensions for the photodetector.

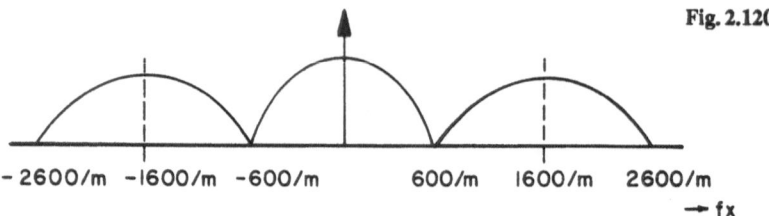

Fig. 2.120

$$-2600/m \quad -1600/m \quad -600/m \qquad 600/m \quad 1600/m \quad 2600/m$$
$$\longrightarrow f_x$$

2.39 A hologram is made using a reference beam ($\lambda = 0.6\,\mu$m) incident at an angle of $6°$ with respect to the optical axis (z) and the xz plane. The spatial frequency components of the object are shown in Fig. 2.120.

a) Plot the transmission function of the hologram as a function of spatial frequency f_x. Include numerical values.

b) To view the hologram, light ($\lambda = 0.7\,\mu$m) is incident on the hologram plate at an angle of $15°$. Plot the transmitted light as a function of f_x.

c) What is the smallest angle for the reference light that we can use to make the hologram for clear viewing?

2.40 Collimated laser light ($6\,\text{W/m}^2$ and $\lambda = 0.6\,\mu$m) is incident at an angle of $5°$ with respect to the optical axis (z) on a lens of focal length $100\,$cm. Very near the lens, different masks can be placed and the radiation pattern can be observed at the focal plane. Consider no variation along the y axis.

a) Write an expression for the E-field associated with the incident light. Include numerical values.

b) Write an expression for the transmission function of the lens. Include numerical values.

c) If the transmission function of the mask is as shown in Fig. 2.121, write an expression for the intensity at the focal plane.

d) Plot the intensity at the focal plane for the mask shown in Fig. 2.122. Give numerical values.

2.41 A hologram is made using a reference beam ($\lambda = 0.5\,\mu$m) incident at an angle of $8°$ with respect to the optical axis (z) in the xz plane. The light from the object is incident at an angle of $-5°$ and has an angular spread of $1°$.

a) Plot the transmission function of the hologram as a function of spatial frequency f_x. Include numerical values.

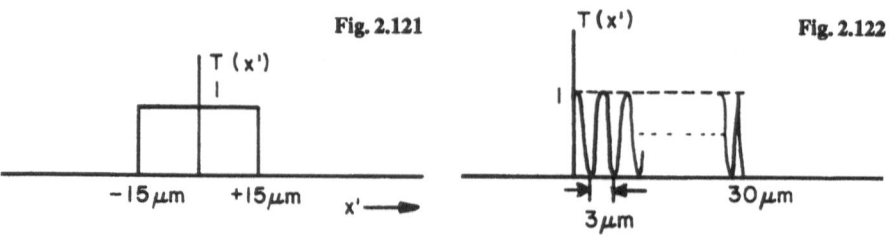

Fig. 2.121

Fig. 2.122

b) We would like to view the hologram at an angle of 15° with light of wavelength $\lambda = 0.7\,\mu$m due to a design constraint. Write an expression for the electric field associated with the light which one should use to illuminate the hologram for viewing.

c) Plot the transmitted light as a function of angle from the optical axis for the case (b).

d) If the size of the hologram is $20\,\text{cm} \times 20\,\text{cm}$ and the reference beam is known to have an angular spread of 0.1°, how much angular spread is allowed in the illuminating beam for viewing?

e) Can you estimate the coherence time of the reference beam needed for the making of the hologram? Justify.

2.42 100 rectangular apertures, each of dimension $1\,\mu$m $(x') \times 100\,\mu$m (y') with a separation between them of $20\,\mu$m along x', are placed in the path of uniform light ($\lambda = 5000\,\text{Å}$) incident parallel to the optical axis. A lens of 50 cm focal length is used to observe the Fraunhofer diffraction at the focal plane.

a) Give an analytical expression for the electric field at the focal plane.

b) Plot the intensity in the focal plane as a function of spatial frequency.

c) What is the minimum size of the detector required to collect most of the light intensity [as found in part (b)]?

d) What approximation did you use for part (a) – Fresnel or Fraunofer?

e) Repeat part (a) for the incident light 5° off axis (in the $x'z$ plane).

2.43 An experiment is to be designed to evaporate a pellet of size $1\,\mu$m. A laser ($\lambda = 3000\,\text{Å}$) with a beam diameter of 1 cm is used in the experiment.

a) Find the focal length of the lens if the pellet is placed at the focal plane.

b) It is observed that when the laser power is 100 kW the pellet at the focal plane evaporates. Estimate the peak electric field of the light at the pellet under these conditions.

c) Write an expression for the intensity at the focal plane (assume a one-dimensional problem).

2.44 A hologram is constructed with a reference beam incident at an angle of 8° with respect to the optical axis (z) in the xz plane. In place of an object, another beam is used incident at an angle of −5° with respect to the optical axis in the xz plane. The angular spread of the "object" beam is 0.5°. ($\lambda = 0.5\,\mu$m).

a) Plot the electric field of the object beam as a function of spatial frequency. Include numerical values.

b) Write an expression for the electric field of the reference beam including x and z dependence.

c) Plot the transmission function of the hologram as a function of spatial frequency. Include numerical values.

d) The hologram is illuminated with light with $\lambda = 0.8\,\mu$m at an angle of 2°

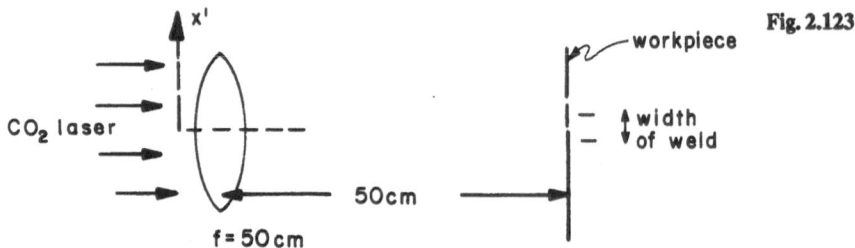

Fig. 2.123

with respect to the optical axis. Plot the transmitted light as a function of angle from the optical axis. Include numerical values.

2.45 A CO_2 laser ($\lambda = 10\,\mu m$) welder is to be designed. The output of the laser is considered to be uniform and parallel to the optical axis. A lens of focal length 50 cm and diameter 1 cm is used to focus the light on the workpiece, see Fig. 2.123. Assume it to be a one-dimensional problem. Estimate the width of the weld.

2.46

a) i) A CO_2 laser beam has 1 cm diameter. It is plane polarized and has a peak power of 1 GW. Estimate the electric field associated with it in free space.
   ii) If the propagation direction is z and the electric field is polarized in the x direction, determine the magnetic field vector.

b) In a diffraction experiment with light of $0.5\,\mu m$ wavelength, the distance between the source plane and the detector plane is 40 km.

   i) Estimate the dimension of the effective sources in the source plane, using the Fraunhofer approximation.
   ii) Repeat for the Fresnel approximation.

2.47 A hologram is made using a reference beam ($\lambda = 0.6\,\mu m$) incident at an angle of 6° with respect to the optical axis (z) in the xz plane. The light from the object is incident at an angle of $-2°$ and has an angular spread of 0.5°.

a) Plot the transmission function of the hologram as a function of spatial frequency f_x. Include numerical values.
b) We would like to view the hologram at an angle of 10° with light of $\lambda = 0.8\,\mu m$, due to a design constraint. Write an expression for the electric field associated with the light which one should use to illuminate the hologram for viewing.
c) Plot the transmitted light as a function of angle from the optical axis for the case (b).

d) If the size of the hologram is 1 cm × 1 cm and the reference beam is known to have an angular spread of 0.1°, how much angular spread is allowed in the illuminating beam for viewing?

e) Can you estimate the coherence time of the reference beam needed for the making of the hologram? Justify.

2.48 An optical swith is to be designed which operates by producing a spark at the focus of the lens. It requires an electric field of 1 kV/mm to produce the spark and the light is incident on a lens of size 2 cm × 2 cm and focal length 10 cm (Fig. 2.124). Calculate the minimum incident power in watts for $\lambda = 1.0\,\mu$m.

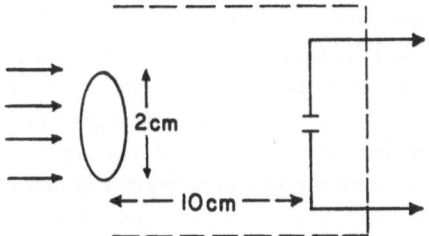

Fig. 2.124

3. Signal Processing Fundamentals

In this chapter, we shall review material relevant to the understanding, design and application of optical and other devices discussed in this book. Signals can be analog, discrete or digital. For the analog case, the time or space variable is continuous. For the discrete case, the time axis is sampled at a fixed interval or for discrete values, even though the amplitude remains analog or continuous. Of course, for a digital system, the time is discrete and the amplitude is represented by a digital number. The spatial signal can also be two- or multidimensional.

Sections 3.1 and 3.2 discuss analog and discrete one-dimensional systems, respectively. The concepts of Fourier and z-transform, impulse and delta response, sampling theorem and aliasing are introduced. Noise and stochastic processes are dealt with in Sect. 3.3, which also includes the matched filter. Transversal and recursive filters are discussed in Sect. 3.4. They are followed by Sect. 3.5 on adaptive filters, which includes least mean squares estimation, the Wiener-Hopf filter, the Widrow-Hoff algorithm and lattice filter. The next two sections discuss power spectra estimation and Kalman filtering. Section 3.8 points out the complexity introduced by the two-dimensional systems. The review ends with a discussion of the ambiguity function, the Wigner distribution function and triple correlation.

3.1 Analog Signals and Systems

It is of interest to define the following analog signals at the outset. These functions or signals will be used often in this text and are shown in Fig. 3.1; they are defined as follows:

Step function
$$u(t) \begin{cases} = 0 & t < 0 \\ = 1 & t \geq 0 \end{cases}$$

Sign function
$$\text{sgn}(t) \begin{cases} = +1 & t > 0 \\ = -1 & t < 0 \end{cases}$$

Rectangular function
$$P(t) \begin{cases} = 0 & t < -\frac{1}{2} \\ = 1 & -\frac{1}{2} \leq t \leq \frac{1}{2} \\ = 0 & t > \frac{1}{2} \end{cases}$$

Delta function
$$\delta(t) \begin{cases} = 1 & \text{for } t = 0 \\ = 0 & \text{otherwise} \end{cases}$$

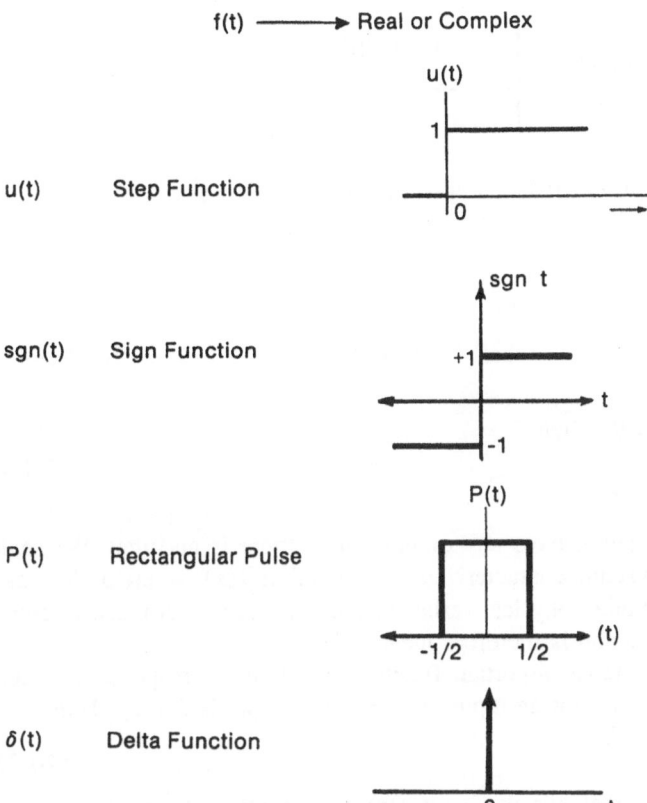

f(t) ⟶ Real or Complex

u(t)   Step Function

sgn(t)   Sign Function

P(t)   Rectangular Pulse

δ(t)   Delta Function

Fig. 3.1. Sketches of analog signals (functions)

with

$$\int\limits_{-\infty}^{+\infty} \delta(t)dt = 1$$

3.1.1 Linear Systems

Let us consider a so-called black box with input $f(t)$ and output $g(t)$ as shown in Fig. 3.2. The output $g(t)$ can be represented by a one-to-one mapping function represented by L:

$$g(t) = L[f(t)] \quad . \tag{3.1.1}$$

For a linear system

$$L[A_1 f_1(t) + A_2 f_2(t)] = A_1 L[f_1(t)] + A_2 L[f_2(t)] \quad , \tag{3.1.2}$$

where A_1 and A_2 are complex numbers and f_1 and f_2 any two functions that satisfy (3.1.1). Also, most of the time we shall deal with time-invariant systems,

207

Fig. 3.2. A "black box" with input $f(t)$ and output $g(t)$

i.e.,

$$L[f(t - t_0)] = g(t - t_0)$$ (3.1.3)

where t_0 is a constant. As we shall be dealing with causal systems, we know that if

$$f(t) = 0 \quad \text{for } t < 0 \quad \text{then}$$

(3.1.4)

$$g(t) = 0 \quad \text{for } t < 0 \quad .$$

This means that we cannot have any output unless there is an input. We shall also consider some systems characterized by a function $f(x)$, where x does not denote time, but some other physical variable such as position. For those systems, causality does not hold and is therefore meaningless.

Let us define the most important function, the impulse response function $h(t, \tau)$ of a linear system. For an input $\delta(t - \tau)$, the output is $h(t, \tau)$. Thus

$$h(t, \tau) = L[\delta(t - \tau)] \quad .$$ (3.1.5)

We immediately prove that, for any input $f(t)$, the output is given by

$$g(t) = \int_{-\infty}^{+\infty} f(\tau) h(t, \tau) d\tau \quad .$$ (3.1.6)

Proof. We note that

$$f(t) = \int_{-\infty}^{+\infty} f(\tau) \delta(t - \tau) d\tau \quad .$$ (3.1.7)

Thus, we can consider the input as a sum of a multiplicity of delta functions $\delta(t - \tau)$ with amplitude $f(\tau)$ as shown in Fig. 3.3. For each input $f(\tau)\delta(t - \tau)$, the output is given by $f(\tau)h(t, \tau)$. Hence the total output is given by (3.1.6). $\square$

The linear system described by (3.1.5 and 6) is generally known as a time-variant system. Much simplification results if one considers time-invariant systems, for which

$$h(t, \tau) = h(t - \tau) \quad .$$

The impulse response then shifts only with the time of the application of the impulse, without any other change. This is shown in Fig. 3.4 for both the time-

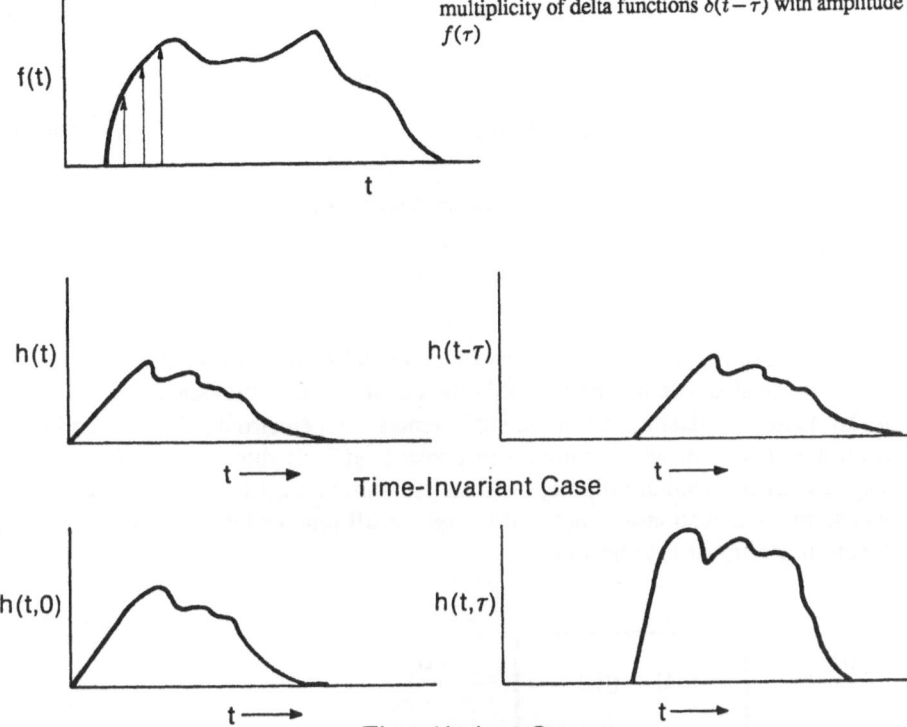

Fig. 3.3. A function $f(t)$ may be considered as a multiplicity of delta functions $\delta(t-\tau)$ with amplitude $f(\tau)$

f(t)

t

h(t)   h(t-τ)

t ⟶   t ⟶

Time-Invariant Case

h(t,0)   h(t,τ)

t ⟶   t ⟶

Time-Variant Case

Fig. 3.4. In the time-invariant case (*top*) the response is simply shifted in time when the impulse is applied at $t = \tau$ instead of at $t = 0$. In the time variant case (*bottom*) the shape of the response changes

invariant and time-variant cases. Because of the shift property in the time-invariant case, the impulse response can also be defined as the response to an impulse at the origin, $\delta(t)$, i.e., for an input $\delta(t)$ the output is $h(t)$. Thus

$$h(t) = L\delta(t) \quad .$$

Equation (3.1.6) for this case becomes

$$g(t) = \int\limits_{-\infty}^{+\infty} f(\tau)h(t-\tau)d\tau \quad .$$

The quantity on the right-hand side is also the definition of convolution of two functions f and h and is denoted symbolically by

$$f*h = \int\limits_{-\infty}^{+\infty} f(\tau)h(t-\tau)d\tau \quad . \tag{3.1.8}$$

If we substitute $t - \tau = \tau'$, we obtain

$$f*h = \int\limits_{-\infty}^{+\infty} h(\tau')f(t-\tau')d\tau \quad \text{or}$$

$$f*h = \int\limits_{-\infty}^{+\infty} h(\tau)f(t-\tau)d\tau = h*f \quad . \tag{3.1.9}$$

For a causal system (3.1.8 and 9) can be rewritten as

$$f*h = h*f = \int\limits_{0}^{t} h(\tau)f(t-\tau)d\tau = \int\limits_{0}^{t} f(\tau)h(t-\tau)d\tau \quad .$$

The results obtained above are fundamental to the understanding of many devices. We shall often need to obtain the convolution of two signals, e.g., f and g. To obtain the desired result we shall generate an electronic signal f and pass it through a system whose impulse response is $g(t)$, as shown in Fig. 3.5. As we shall see in the second volume, one can use simple masks to construct a device whose impulse response is $g(t)$. Of course, as all practical devices are finite, the function $g(t)$ must also be finite.

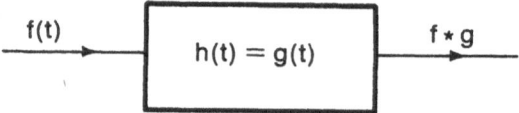

Fig. 3.5. An electronic signal $f(t)$ is passed through a system with the impulse response $g(t)$

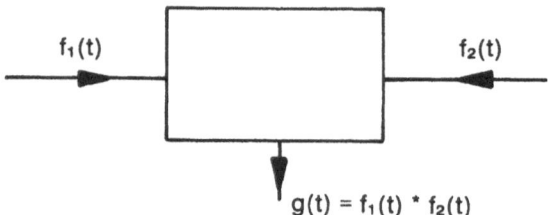

Fig. 3.6. The convolver is a three-terminal device

Another way to perform the convolution of two signals is to use a so-called convolver. It is a three port device (Fig. 3.6) using surface acoustic wave (SAW) or other wave interactions and is discussed in detail in the second volume (Chap. 2).

Again the functions $f(t)$ and $g(t)$ have to be finite. An example illustrates how the convolution is obtained. To obtain the convolution at a particular time t we note that

$$[f*g](t) = \int f(\tau)g(t-\tau)d\tau \quad . \tag{3.1.10}$$

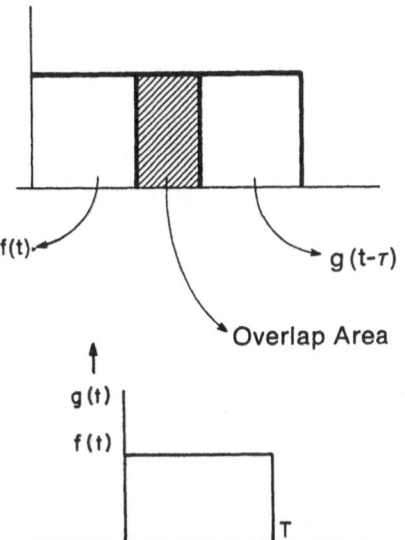

Fig. 3.7. (a) To convolve f and g, g is folded backwards and shifted along the time axis, then f and g are multiplied together and integrated. (b) Convolution of two identical rectangular pulses (*left*) gives a triangular form (*right*)

f(t)

g (t-τ)

Overlap Area   (a)

g(t)

f(t)

T

t →

f*g

t →   2T   (b)

This means that one has to multiply the two functions f and g, with g folded backwards and shifted in the time axis by t, and integrate. This is illustrated in Fig. 3.7a. Figure 3.7b demonstrates the convolution of two rectangular pulse signals, f and g.

3.1.2 Fourier Transforms and Frequency Response

The Fourier transforms[1] $G(f)$ of a function $g(t)$ is defined as

$$G(f) = \int_{-\infty}^{+\infty} g(t)e^{-j\omega t}dt \quad , \tag{3.1.11}$$

where $\omega = 2\pi f$ and $g(t)$ is bounded and goes to zero asymptotically as $t \to 0$. Physically $G(f_0)$ represents the magnitude and phase of the component of $g(t)$ having frequency f_0.

Multiplying both sides of (3.1.11) by exp $(j\omega t)$ and integrating with respect to f from $-\infty$ to ∞, one obtains

$$\int_{-\infty}^{+\infty} G(f)e^{j\omega t}df = \int e^{j2\pi ft} \int_{-\infty}^{+\infty} g(t')e^{-j2\pi ft'}dt' \, df \quad . \tag{3.1.12}$$

Interchanging the order of integration on the right-hand side and noting that

---

[1] We shall always use the corresponding capital letter to denote the Fourier transform of a quantity.

$$\int_{-\infty}^{+\infty} e^{j2\pi f(t-t')} df = \delta(t - t') \quad , \tag{3.1.13}$$

one obtains the inversion formula

$$g(t) = \int_{-\infty}^{+\infty} G(f) e^{j\omega t} df \quad . \tag{3.1.14}$$

The following six properties of Fourier transforms can easily be derived:[2]

Symmetry property $\quad G(t) \leftrightarrow g(-f)$

Conjugate functions $\quad g^*(t) \leftrightarrow G^*(-f)$

Scaling $\qquad\qquad\quad g(at) \leftrightarrow \dfrac{1}{|a|} G\left(\dfrac{f}{a}\right)$

Shifting $\qquad\qquad\quad g(t - a) \leftrightarrow e^{-j a \omega} G(f)$

Convolution $\qquad\quad\ g_1(t) * f_2(t) \leftrightarrow G_1(f) F_2(f)$

Parseval's theorem

$$\int_{-\infty}^{+\infty} f_1(t) f_2^*(t) dt = \int_{-\infty}^{+\infty} F_1(f) F_2(f) df \quad \text{or}$$

$$\int_{-\infty}^{+\infty} |f(t)|^2 dt = \int |F(f)|^2 df \quad \text{when}$$

$$f_1(t) = f_2(t) = f(t) \quad .$$

Parseval's theorem is related to the conservation of energy. Some important and frequently used Fourier transform pairs are tabulated in Table 3.1. The sinc function, plotted in Fig. 3.8, is a very important function and will often be used. The function $\Lambda(t)$ is shown in Fig. 3.9; the comb function is defined in (3.2.23).

The frequency response of a linear system is defined to be $H(f)$, if the output of the system is $g(t) = H(f) \exp(j\omega t)$ for an input of $\exp(j\omega t)$. From Fig. 3.10 we thus obtain, by using the results of (3.1.9), that

$$g(t) = \int e^{j\omega(t-\tau)} h(\tau) d\tau$$

$$= e^{j\omega t} \int h(\tau) e^{j\omega \tau} d\tau \quad , \quad \text{or}$$

$$g(t) = e^{j\omega t} H(f) \quad , \tag{3.1.15}$$

where $h(t)$ is the impulse response of the system and $H(f)$ is the Fourier transform of $h(t)$. Using the definition of frequency response we obtain that $H(f)$ is the frequency response of the linear system whose impulse response is $h(t)$.

---

[2] The notation $\leftrightarrow$ denotes a Fourier transform pair.

Table 3.1. List of Fourier transform pairs

Function	Fourier transform
$\delta(t)$	1
$\text{sgn}(t)$	$\frac{2}{j\omega}$
$u(t) = \frac{1}{2} + \frac{1}{2}\,\text{sgn}(t)$	$\pi\delta(\omega) + \frac{1}{j\omega}$
$P(t)$	$\text{sinc}(f)$
$\cos(\omega_0 t)$	$\pi\delta(\omega - \omega_0) + \pi\delta(\omega + \omega_0)$
$e^{-\pi t^2}$	$e^{-\pi f^2}$
$\Lambda(t)$	$\text{sinc}^2(f)$
$\text{comb}(t)$	$\text{comb}(f)$
$\text{sech}(\pi t)$	$\text{sech}(\pi f)$
$\cos\pi(t^2 - \frac{1}{8})$	$\cos\pi(f^2 - \frac{1}{8})$
$\sin\pi(t^2 - \frac{1}{8})$	$\sin\pi(f^2 - \frac{1}{8})$
$\exp(\pm j\pi t^2)$	$\exp(\pm j\pi/4)\exp(\pm j\pi f^2)$
$\exp[-\pi t^2/(a+jc)]$	$\sqrt{a+jc}\,\exp[-\pi(a+jc)f^2]$ a and c real and $a \geq 0$

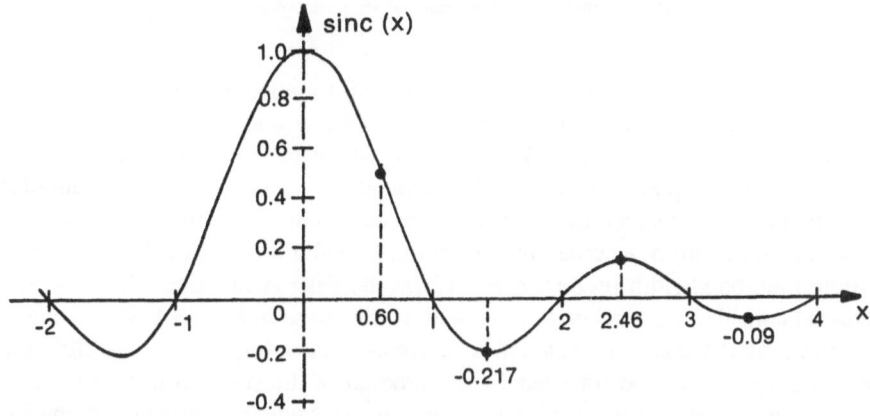

Fig. 3.8. Sinc function : $\text{sinc}\,x = (\sin\pi x)/\pi x$

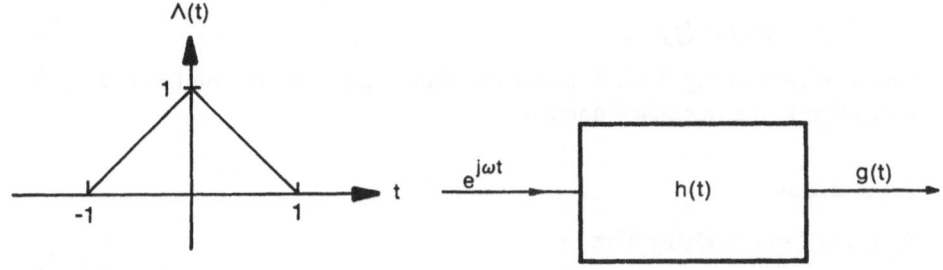

Fig. 3.9. The function $\Lambda(t)$

Fig. 3.10. Linear time-invariant system with single-frequency input

213

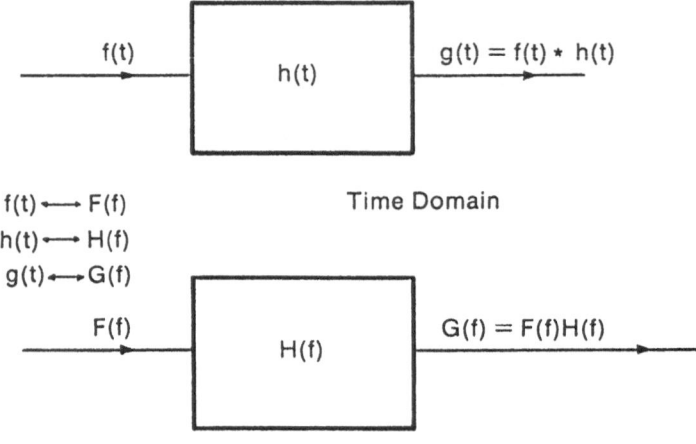

Fig. 3.11. Relationship between analyses in the time and frequency domains

This leads to an interesting observation. All impulse responses of physical devices are finite or time limited. However, we know that the Fourier transform of a rectangular function is a sinc function, which theoretically goes to $\pm\infty$. The frequency response must therefore extend to infinity for a finite impulse response device. However, as the energy contained in most practical cases beyond a certain bandwidth or a certain time duration is negligible, practical devices with a finite time-bandwidth product can perform the desired functions. Some of the practical considerations related to this will be discussed in Sect. 3.4. We also note that some of the main characteristics of signal processing devices are the time duration, bandwidth and time-bandwidth product of the signal it can process.

A linear system can also be analyzed in the frequency domain. If the input and output to the system are $F(f)$ and $G(f)$ respectively, in the frequency domain, we then obtain, using the convolution theorem from Table 3.1, that

$$G(f) = H(f)F(f) \quad . \tag{3.1.16}$$

This is shown in Fig. 3.11. Thus convolution in the time domain becomes multiplication in the frequency domain.

3.1.3 Examples

a) Delay Line of Delay Time τ

$$h(t) = \delta(t - \tau) \quad ,$$
$$H(f) = e^{-j2\pi f\tau} \quad . \tag{3.1.17}$$

214

b) Chirp Impulse Response

$$h(t) = e^{j\beta t^2} \quad \text{for an up-chirp} ,$$
$$h(t) = e^{-j\beta t^2} \quad \text{for a down-chirp} , \tag{3.1.18}$$

where β is a constant known as the chirp rate. The impulse response is plotted in Fig. 3.12 and we observe that, for the instantaneous frequency $f(t)$, (3.1.18) can be written as

$$h(t) = e^{j\theta(t)} .$$

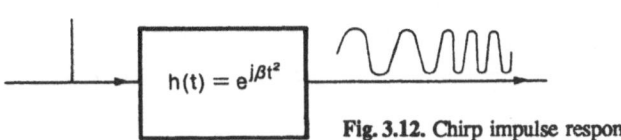

Fig. 3.12. Chirp impulse response

From this we obtain

$$f(t) = \frac{1}{2\pi} \frac{d\theta(t)}{dt} = \frac{\beta t}{\pi} . \tag{3.1.19}$$

Hence the instantaneous frequency increases linearly with time for an up-chirp and decreases linearly for a down-chirp. As the chirping of birds is of the same nature, this term was selected for the signals. Taking the Fourier transform of (3.1.18) one obtains, using the scaling theorem,

$$H(f) = \left(\frac{\pi}{\beta}\right)^{1/2} e^{-j\pi/4} e^{j\pi f^2/\beta} = \left(\frac{\pi}{\beta}\right)^{1/2} e^{-j\pi/4} e^{-j2\pi f \tau(f)} \tag{3.1.20}$$

where

$$\tau(f) = \frac{\pi^2}{2\beta} f . \tag{3.1.21}$$

Thus, if we have a dispersive delay line whose delay time is a linear function of frequency given by (3.1.21), its impulse response will be a chirp signal. For a delay line we recall that $\tau = l/v$, where v is the velocity of the wave used to make the delay line and l is the distance between the transducers.

If v is dispersive or dependent on frequency, then the delay line will also be dispersive. As mentioned in Sect. 4.1, the Lamb wave and other waves are dispersive and they can therefore be used to make dispersive delay lines, but the magnitude of the dispersion cannot easily be controlled. An easier and better way to implement chirp devices is by using SAW tapped delay lines and this will be discussed in Sect. 4.2.3.

c) Real-Time Fourier Transformation Using a Chirp Algorithm

Consider the system shown in Fig. 3.13. The signal $f(t)$ to be Fourier transformed is multiplied with a down-chirp and used as an input to a device that has an impulse response which is an up-chirp. After multiplication with another

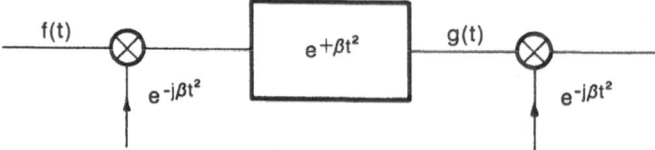

Fig. 3.13. Real-time Fourier transformer

down-chirp, the output of the device is the real-time Fourier transform of the input signal.

Proof

$$g(t) = \int f(\tau)e^{-j\beta\tau^2}e^{j\beta(t-\tau)^2}\,d\tau$$

$$= e^{j\beta t^2}\int f(\tau)e^{-j2\beta t\tau}\,d\tau$$

$$= e^{j\beta t^2}\int f(\tau)e^{-j2\pi f\tau}\,d\tau \qquad (3.1.22)$$

$$= e^{j\beta t^2}F(f) \quad \text{where} \quad f = \frac{\beta t}{\pi}\ . \qquad (3.1.23)$$

Thus the final output is the Fourier transform with a frequency given by $\frac{\beta t}{\pi}$. ☐

Hence a chirp transform maps the Fourier transform of the input signal into the time domain. The chirp transform algorithm is symbolically given by

$$F\left(f = \frac{\beta t}{\pi}\right) = \{[f(t)e^{-j\beta t^2}]*e^{j\beta t^2}\}e^{-j\beta t^2}\ . \qquad (3.1.24)$$

The same result is also obtained if the input and output of the chirp device are multiplied by an up-chirp and the device itself is a down-chirp device. This real-time Fourier transformation can be implemented using different devices in different configurations and will be discussed in detail in Sect. 4.6 and the second volume.

d) Pulse Compression

Let us consider the signal $f(t)$ in Fig. 3.14, which is a rectangular pulse of duration T. Thus the input to the chirp device is a down-chirp signal of finite duration. In practice a negative frequency centered around zero frequency cannot

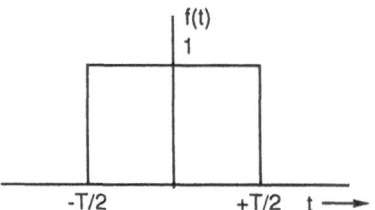

Fig. 3.14. The signal $f(t)$ considered for pulse compression

216

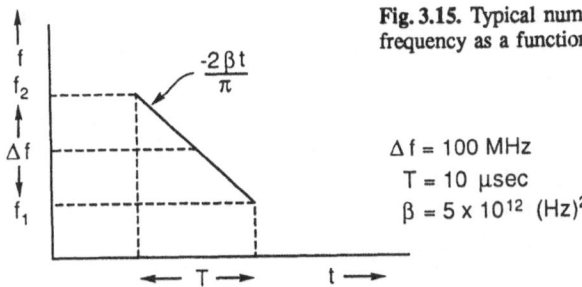

Fig. 3.15. Typical numerical values for the chirp signal; frequency as a function of time

$$\Delta f = 100 \text{ MHz}$$
$$T = 10 \text{ μsec}$$
$$\beta = 5 \times 10^{12} \text{ (Hz)}^2$$

be obtained. However, as we shall see later, SAW devices work around a center frequency, so negative frequencies with respect to this reference frequency can be obtained. This signal is shown in Fig. 3.15, where it is assumed that $T = 10 \,\mu s$, $\beta = 5 \times 10^{12} \text{ (Hz)}^2$, $f_2 - f_1 = \Delta f = 100 \text{ MHz}$. The output of the up-chirp device after multiplication with another down-chirp will be

$$F(t) = \int_{-T/2}^{T/2} e^{-j2\pi ft} dt = T \,\text{sinc}\,(fT) = T \frac{\sin(\beta T t)}{\beta T t} \tag{3.1.25}$$

as f is given by (3.1.23). For the function $F(f)$, the output is plotted in Fig. 3.16 and we see that the 3 dB width of the pulse is given by

$$\Delta t \sim \frac{1}{2\beta T} = \frac{1}{\Delta f} \quad . \tag{3.1.26}$$

Hence the original pulse of duration T has been compressed to Δt, the compression ratio being given by

$$\frac{T}{\Delta t} = T \Delta f \quad ,$$

which is equal to the time-bandwidth (TB) product. Thus in our numerical example the TB product is 1000. The concept of pulse compression for radar pulses was originally proposed by *Darlington* [3.1] to improve the detected signal-to-noise ratio for a finite amplitude pulse. Because of constraints imposed by the

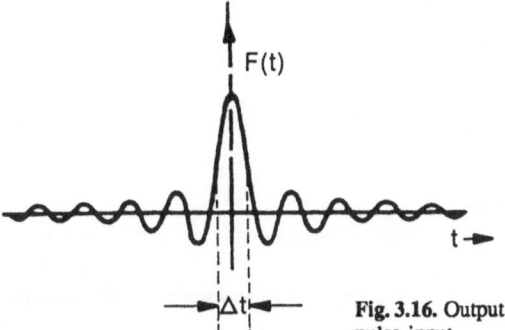

Fig. 3.16. Output of the real-time transformer for rectangular pulse input

217

magnetron tubes that generate the radar pulse, one cannot increase the power indefinitely. However, by using pulse compression, one obtains the effect of higher power without losing resolution. As we shall see later, the arrangement described here also acts as a matched filter for the chirp signal. For radar use, the detected pulse compressed output contains not only the main lobe but also side lobes, the largest of which is only 12 dB down. This is often unacceptable. However, improvement of side lobes' performance can be obtained by shaping the chirp pulse, and this will be discussed in Sect. 3.4.

3.1.4 Hilbert Transform and Causality

The Hilbert transform $\hat{f}(t)$ of a signal $f(t)$ is defined as

$$\hat{f}(t) = f(t) * \frac{1}{\pi t} = \frac{1}{\pi} \int_{-\infty}^{+\infty} \frac{f(\tau)}{t - \tau} d\tau \quad . \tag{3.1.27}$$

Fourier transforming both sides, we obtain

$$\mathcal{F}[f(t)] = F(f)\{-j \operatorname{sgn}(f)\} \quad . \tag{3.1.28}$$

Taking the inverse transform of (3.1.28), we obtain

$$\hat{f}(t) = \mathcal{F}^{-1}[F(f)\{-j \operatorname{sgn}(f)\}] \quad . \tag{3.1.29}$$

Using (3.1.29), Fig. 3.17 shows how one can obtain a real-time Hilbert transformer using two real-time Fourier transformers discussed previously.

To obtain the restriction imposed on the linear system by causality, we note that the output $g(t)$ is given by

$$g(t) = 0 \quad \text{for } t < 0 \quad \text{if}$$
$$f(t) = 0 \quad \text{for } t < 0 \quad .$$

Thus

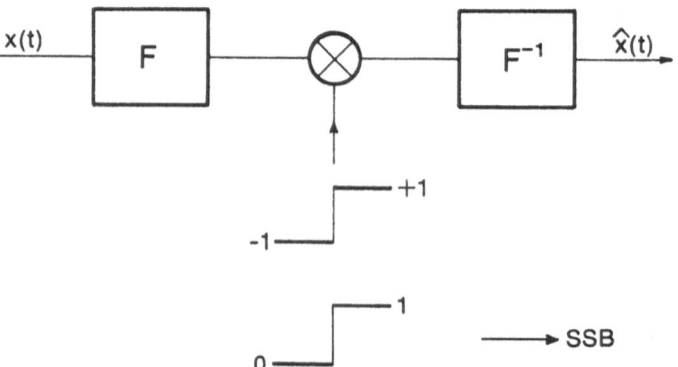

Fig. 3.17. A real-time Hilbert transformer using two real-time Fourier transformers (SSB: single side band)

$$h(t) = 0 \quad \text{for } t < 0 \quad \text{or}$$
$$h(t) = h(t)u(t) \quad .$$

(3.1.30)

Taking the Fourier transform of both sides we obtain

$$H(f) = H(f) * \left(\frac{1}{2}\delta(f) + \frac{1}{j2\pi f} \right) \quad .$$

Denoting the real and imaginary parts of the frequency response of the device by $R(f)$ and $X(f)$, respectively, $H(f)$ can be written as

$$H(f) = R(f) + jX(f) \quad \text{or}$$

(3.1.31)

$$R(f) + jX(f) = \frac{1}{2}R(f) + \frac{X(f) * f^{-1}}{2\pi}$$
$$+ j\left(\frac{1}{2}X(f) - \frac{1}{2}R(f) * f^{-1} \right) \quad .$$

(3.1.32)

Equating the real and imaginary parts of the equation, we obtain

$$R(f) = \frac{1}{\pi}\left(X(f) * \frac{1}{f} \right) = \frac{1}{\pi} \int\limits_{-\infty}^{\infty} \frac{X(f')}{f - f'} df' \quad \text{and}$$

$$X(f) = -\frac{1}{\pi}\left(R(f) * \frac{1}{f} \right) = -\frac{1}{\pi} \int\limits_{-\infty}^{+\infty} \frac{R(f')}{f - f'} df' \quad .$$

(3.1.33)

Thus the real and imaginary parts of the frequency response of any physical system are related to each other by a Hilbert transform. This relationship is also known in physics textbooks as the Kramers-Kronig relationship. Equation (3.1.33) can be shown (Problem 3.17) to be given by

$$R(f) = -\frac{2}{\pi} \int\limits_{0}^{\infty}\int\limits_{0}^{\infty} X(f') \sin f't \cos 2\pi ft \, df' dt$$

$$= \frac{2}{\pi} \int\limits_{0}^{\infty} \frac{f'X(f')}{4\pi^2 f^2 - f'^2} df' \quad ,$$

$$X(f) = -\frac{2}{\pi} \int\limits_{0}^{\infty}\int\limits_{0}^{\infty} R(f') \cos f't \sin 2\pi ft \, df' dt$$

$$= -4f \int\limits_{0}^{\infty} \frac{R(f')}{4\pi^2 f^2 - f'^2} df' \quad .$$

3.1.5 Time-Variant Systems

A time-variant system has an impulse response $h(t, t')$ which is obtained as the output when the input to the system is $\delta(t - t')$. If $h(t, t')$ can be written as

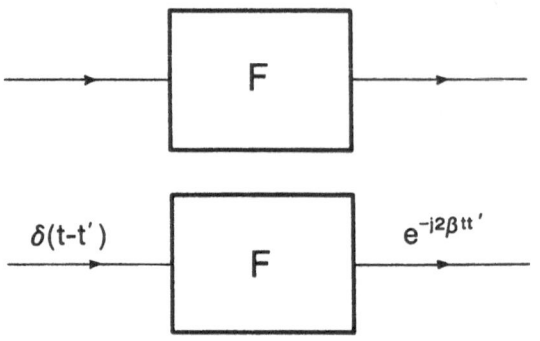

Fig. 3.18. Example of a time-variant system: a real-time Fourier transformer

$h(t - t')$, then the system is time invariant. Thus the response of a time-variant system depends on when the impulse is applied. The best example of a time-variant system is the real-time Fourier transformer discussed in Sect. 3.1.3. For this system, as shown in Fig. 3.18, we have

$$h(t, t') = e^{-j2\beta tt'} \quad . \tag{3.1.34}$$

For a time-variant system we generally have

$$g(t) = \int\limits_0^\infty h(t, t') f(t') dt' \tag{3.1.35}$$

for a causal system.

For a noncausal system we have

$$g(x) = \int\limits_{-\infty}^{+\infty} h(x, x') f(x') dx' \quad . \tag{3.1.36}$$

Thus the convolution relationship given by (3.1.9) does not apply.

It is of interest to note the impulse responses for the following time-variant systems:

i) Multiplication by a function $m(t)$

$$g(t) = m(t) f(t) \tag{3.1.37}$$

$$h(t, t') = m(t') \delta(t - t') \tag{3.1.38}$$

ii) Cosine transform

$$g(t) = \int \cos (2\pi tt') f(t') dt' \tag{3.1.39}$$

$$h(t, t') = \cos (2\pi tt') \tag{3.1.40}$$

iii) Laplace transform

$$g(t) = \int e^{-tt'} f(t')dt' \qquad (3.1.41)$$

$$h(t, t') = e^{-tt'} \quad . \qquad (3.1.42)$$

iv) Hankel transform

$$g(t) = 2\pi \int t' J_0(2\pi tt') f(t')dt' \qquad (3.1.43)$$

$$h(t, t') = 2\pi t' J_0(2\pi tt') \qquad (3.1.44)$$

3.2 Discrete Systems

The relationship between a continuous signal $f(t)$ and its counterpart $f(n)$ in a discrete system or sampled data system is shown in Fig. 3.19. Here n is an integer that runs from zero to infinity and replaces the continuous variable t. The sampling time interval Δt is related to the bandwidth of the system.

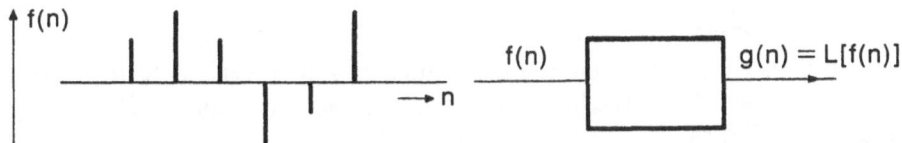

Fig. 3.19. The signal $f(n)$, counterpart to the continuous signal $f(t)$, in a discrete or sampled data system

Fig. 3.20. The input and output of a discrete system

Similar to the continuous case, one can define important signals for the discrete case, given by

Step sequence $\quad u(n) = \begin{cases} 1 & n \geq 0, \\ 0 & n < 0, \end{cases}$

Delta sequence $\quad \delta(n) = \begin{cases} 1 & n = 0, \\ 0 & n \neq 0. \end{cases}$ $\qquad (3.2.1)$

As shown in Fig. 3.20 the output $g(n)$ of a discrete system is some mapping of the input $f(n)$, the mapping being denoted symbolically as

$$g(n) = L[f(n)] \quad . \qquad (3.2.2)$$

For a linear system

$$L[a_1 f_1 + a_2 f_2] = a_1 L(f_1) + a_2 L(f_2) \quad , \qquad (3.2.3)$$

where a_1 and a_2 are arbitrary complex numbers and f_1 and f_2 are two arbitrary inputs. The time invariance for the discrete case is given by the relationship

$$L[f(n - k)] = g(n - k) \quad . \tag{3.2.4}$$

Similarly to the impulse response, we define what is known as the delta response $h(n)$ as

$$h(n) = L[\delta(n)] \quad . \tag{3.2.5}$$

For an input $f(n)$, the output is obtained by discrete convolution. Thus, as

$$f(n) = \sum_{k=-\infty}^{+\infty} f(k)\delta(n - k) \quad , \tag{3.2.6}$$

we obtain

$$\begin{aligned} g(n) &= \sum_{k=-\infty}^{+\infty} f(k)h(n - k) \\ &= \sum f(n - k)h(k) \quad . \end{aligned} \tag{3.2.7}$$

So far we have assumed a one-to-one relationship between the continuous time case and the discrete time case. However, the equivalent of the Fourier transform for the discrete system is known as the z-transform and is defined as

$$F(z) = \sum_{n=-\infty}^{+\infty} f(n)z^{-n} \tag{3.2.8}$$

provided that the series converges. Equation (3.2.8) is valid only for certain ranges of values for z. The inversion formula for the z-transform is obtained by expanding $F(z)$ as

$$F(z) = \sum C(n)z^{-n} \quad . \tag{3.2.9}$$

If an expression for $F(z)$ given by (3.2.9) can be obtained, then

$$f(n) = C(n) \quad . \tag{3.2.10}$$

Similarly to the frequency response, one can obtain the z-response of a system, $H(z)$, as shown in Fig. 3.21. If we apply an input z^{-n}, then the output of the system is $H(z)z^{-n}$, where $H(z)$ is known as the z-response. To obtain the relationship between $h(n)$ and $H(z)$ we note from (3.2.7) that

$$H(z)z^n = \sum_{k=-\infty}^{+\infty} z^n h(n - k)$$

$$= \sum_{k=-\infty}^{+\infty} z^{n-k} h(k) = z^n \sum_{k=-\infty}^{+\infty} z^{-k} h(k) \quad . \tag{3.2.11}$$

Hence $H(z)$ is the z-transform of the delta response.

Similarly to the convolution theorem for the Fourier transform, one can easily derive the so-called discrete convolution theorem given as follows:

z^n ⟶ $H(z)z^n$

Fig. 3.21. The z-response of a system

If

$$\sum_{k=-\infty}^{+\infty} f_1(k)f_2(k-h) = f_3(h) \quad,$$

then

$$F_1(z)F_2(z) = F_3(z) \quad. \qquad (3.2.12)$$

For any general input $f(n)$ to a system with an impulse response $h(n)$, we obtain by using the discrete convolution theorem that

$$G(z) = H(z)F(z) \quad.$$

3.2.1 Examples

a) Delay Line

$$\begin{aligned} h(n) &= \delta(n-1) \\ H(z) &= z^{-1} \end{aligned} \qquad (3.2.13)$$

Proof. For a delay line with input $f(n)$

$$g(n) = \sum_{k=-\infty}^{+\infty} f(n-k)\delta(n-1) = f(n-1) \quad.$$

For

$$\begin{aligned} f(n) &= z^n \quad, \\ g(n) &= z^{n-1} = z^{-1}z^n \quad. \end{aligned}$$

Thus

$$H(z) = z^{-1} \quad. \qquad \qquad \square$$

b) Electronic Circuit

Consider the system in Fig. 3.22. Such a system can easily be fabricated using charge coupled devices (CCDs). We are interested in finding the z-response of the device. As there are feedbacks, the calculation is not straightforward. We assume that the input is z^n. The output is given by

$$g(n) = z^n H(z) = a_0 H_1(z)z^n - a_1 H_1(z)z^{n-1} \quad, \qquad (3.2.14)$$

where $H_1(z)$ is the z-response of the system at the point 1, shown in the figure. Notice that each delay introduces a term z^{-1}. The following relationship involving $H_1(z)$ can be written:

223

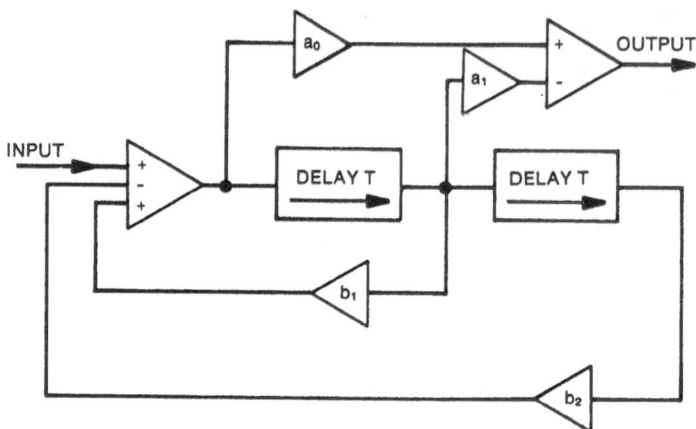

Fig. 3.22. Electronic circuit used as an example (two-pole filter)

$$H_1(z)z^n = z^n + b_1 H_1(z)z^{n-1} - b_2 H_1(z)z^{n-2} \quad \text{or}$$

$$H_1(z) = (1 - b_1 z^{-1} + b_2 z^{-2})^{-1} \quad . \tag{3.2.15}$$

Substituting the value of $H_1(z)$ into (3.2.15) we obtain

$$H(z) = H_1(z)(a_0 - a_1 z^{-1}) \quad \text{or}$$

$$H(z) = \frac{a_0 z^2 - a_1 z}{z^2 - b_1 z + b_2} \quad . \tag{3.2.16}$$

We therefore note that all the feedback terms are in the denominator and the other terms are in the numerator. It is easy to visualize that the more complex and general case shown in Fig. 3.23 can easily be solved to obtain

$$H(z) = \frac{\displaystyle\sum_{k=0}^{N} a_k z^{-k}}{1 - \displaystyle\sum_{k=1}^{N} b_k z^{-k}} \quad . \tag{3.2.17}$$

c) Effect of Finite Transfer Inefficiency in a CCD Signal Processing Device

An N-stage CCD is represented in Fig. 3.24. There is a clock in the CCD and at each cycle the signal is transferred from the left-hand bin to the next bin. Ideally all the electrons in the single bin representing the signal should be transferred. However, due to physical limitations to be discussed later, a fraction, $\bar{\varepsilon}$, is left per cell. The clocks in the CCD are generally multiphase, i.e., for each transfer of signal, the signal charges may be moved through p cells (where p denotes the multiplicity of the multiphase clock). For example, $p = 4$ for a 4-phase clock.

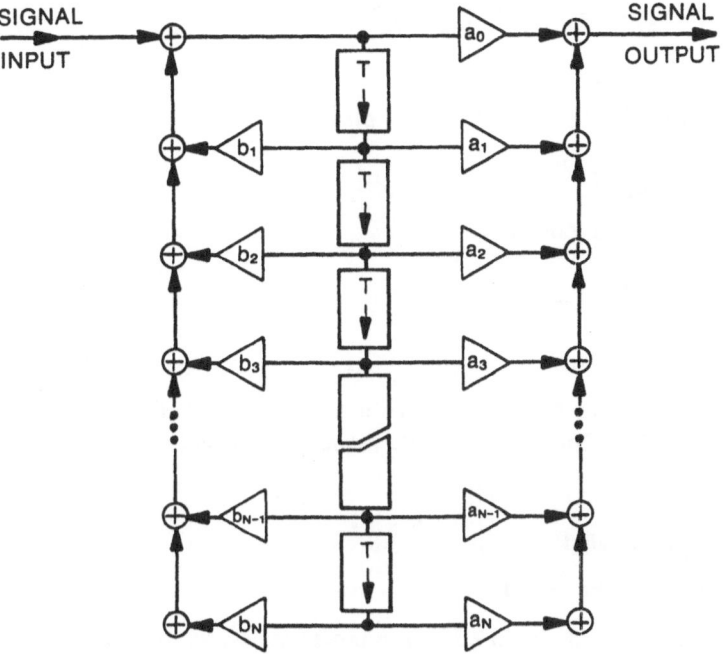

Fig. 3.23. General form of the recursive filter with both feedback and feedforward of the delayed signal

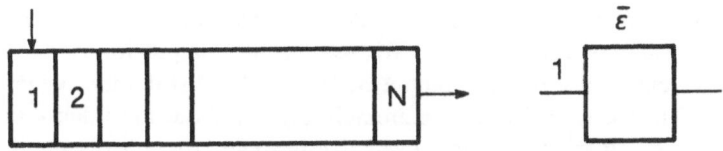

Fig. 3.24. Schematic N-stage CCD

Thus the effective transfer inefficiency per stage ε is given by

$$\varepsilon = p\bar{\varepsilon} \quad . \tag{3.2.18}$$

Let us consider the impulse response of a single stage. Ideally it should be

$$0, 1, 0, 0, 0 \quad ,$$

because the delta input is at $n = 0$ and the ideal output is 1 at a delay of 1 time unit. However, in an actual device, the first two outputs are obviously 0 and $1 - \varepsilon$. To calculate the next output, we note that when the output was $1 - \varepsilon$, an amount of charge ε was left over. Of this amount, a fraction $(1 - \varepsilon)$ will appear in the next clock cycle. The third output will therefore be $\varepsilon(1 - \varepsilon)$. We can similarly calculate the outputs for the fourth, fifth, etc., clock cycles to be given by $\varepsilon^2(1 - \varepsilon)$, $\varepsilon^3(1 - \varepsilon)$, ..., respectively. Hence the output or the impulse response is given by

225

$$0, \quad 1 - \varepsilon, \quad (1 - \varepsilon)\varepsilon, \quad (1 - \varepsilon)\varepsilon^2, \quad (1 - \varepsilon)\varepsilon^3, \ldots \quad .$$

The z-response is given by

$$H(z) = (1 - \varepsilon)z^{-1} + (1 - \varepsilon)\varepsilon z^{-2} + (1 - \varepsilon)\varepsilon^2 z^{-3} + \ldots$$
$$= z^{-1}\frac{1 - \varepsilon}{1 - \varepsilon z^{-1}} \quad . \tag{3.2.19}$$

For N complete stages, we obtain

$$H_N(z) = \left(z^{-1}\frac{1 - \varepsilon}{1 - \varepsilon z^{-1}}\right)^N$$
$$= z^{-N}\exp\left[N\ln(1 - \varepsilon) - \ln(1 - \varepsilon z^{-1})\right] \quad . \tag{3.2.20}$$

In general the value of ε for practical devices is small, e.g., less then 10^{-3} and sometimes around 10^{-6}. Thus we can safely assume $\varepsilon \ll 1$ and obtain

$$H_N(z) \simeq z^{-N}\exp\left[N\varepsilon(z^{-1} - 1)\right] \quad , \tag{3.2.21}$$

where we have approximated

$$\ln(1 - x) \approx -x \quad \text{for } x \ll 1 \quad .$$

Expanding the right-hand side of (3.2.21) in a power series, we obtain

$$H_N(z) \approx z^{-N}\exp(-N\varepsilon)\sum_{k=1}^{\infty}z^{-k}\frac{(N\varepsilon)^k}{k!} \quad . \tag{3.2.22}$$

This shows that the z-response therefore is a Poisson distribution as shown in Fig. 3.25 for representative values of $N\varepsilon$. We see that the output for $N\varepsilon = 4$ certainly does not represent a delayed output but a Poisson distribution, which looks like noise. One must therefore be extremely careful about the fidelity of the filter for N stages where N is large, even if ε is small. As an example, if

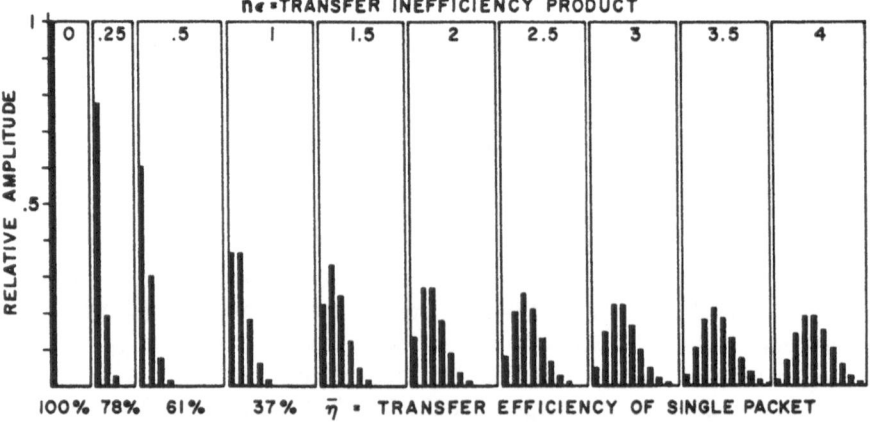

Fig. 3.25. Appearance of output signals upon injection of a single charge packet and transfer through CCDs for several different values of total transfer inefficiency products. (From [3.4])

$N = 1000$ is chosen for $\varepsilon = 10^{-3}$, the device will not perform as expected. However, for $\varepsilon = 10^{-5}$, $N = 1000$ is acceptable.

3.2.2 Sampling Theorem and Aliasing

All signals in the physical world are continuous and we can discretize them according to a sampling rate $1/T$, where T is the sampling time. Such a discrete signal can then be processed through a device like a CCD. Sometimes, the output signal of a CCD must again be converted into a continuous time signal for our observation. In this section, we shall discuss any restrictions on T that must be imposed due to the finite bandwidth of the input continuous time signal. We shall also see that the necessary evil of sampling is aliasing and that, in general, two anti-aliasing filters will be needed for CCD filters.

Let us define the comb function by

$$\text{comb}\,(t) = \sum_{n=-\infty}^{+\infty} \delta(t - n) \quad .$$ (3.2.23)

It can easily be shown that

$$\mathcal{F}[\text{comb}\,(t)] = \text{comb}\,(f) \quad .$$ (3.2.24)

This important function is shown in Fig. 3.26.

For a function $g(t)$ that is band limited we note that

$$G(f) = 0 \quad \text{for } f \geq \sigma \quad ,$$ (3.2.25)

where σ is the highest frequency component for this signal, see Fig. 3.27. The sampled version $g_\text{s}(t)$ of $g(t)$ is shown in Fig. 3.28 [along with $g(t)$] and can be written as

$$g_\text{s}(t) = \text{comb}\,(t/T)g(t) \quad ,$$ (3.2.26)

where T is the sampling interval. We also note that

$$G_\text{s}(f) = T \,\text{comb}\,(fT) * G(f)$$

$$= \sum_{n=-\infty}^{\infty} \delta\left(f - \frac{n}{T}\right) * G(f) \quad , \quad \text{or}$$

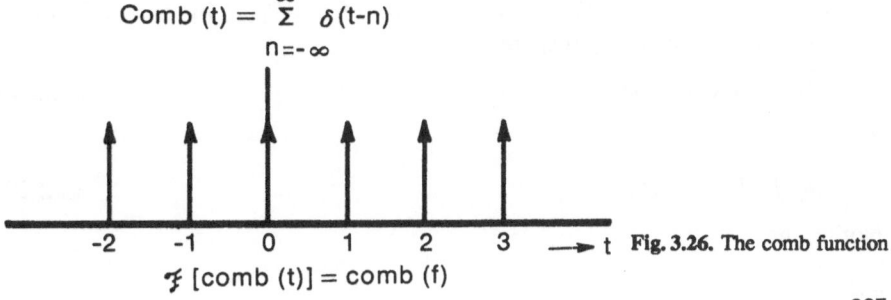

Comb (t) $= \sum\limits_{n=-\infty}^{\infty} \delta\,(t-n)$

$\mathcal{F}\,[\text{comb}\,(t)] = \text{comb}\,(f)$

Fig. 3.26. The comb function

227

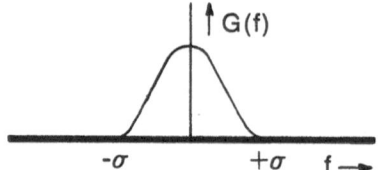

Fig. 3.27. The Fourier transform $G(f)$ of a band-limited function $g(t)$

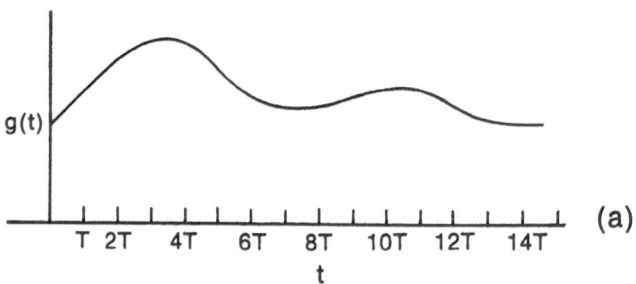

(a)

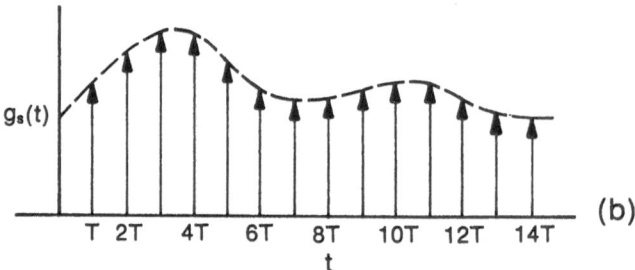

(b)

Fig. 3.28. (a) A function $g(t)$ and (b) its sampled version $g_s(t)$ with a sampling interval T

$$G_s(f) = \sum_{n=-\infty}^{\infty} G\left(f\frac{n}{T}\right) \ . \tag{3.2.27}$$

Thus, the frequency components of the sampled signal are given by an infinite summation of $G(f)$ shifted in frequency by nf_0 where $n = 0, \pm 1, \pm 2 \ldots$ and $f_0 = 1/T$ is the sampling frequency. This infinite folding and mingling of frequencies for the sampled data is known as aliasing. Two representative forms of $G_s(f)$ for two different sampling rates are shown in Fig. 3.29. In the first case, we observe that all the spectra are completely mixed up and cannot be separated if the sampling frequency $f_0 < 2\sigma$. However, $f_0 > 2\sigma$ for the second case, and the spectra are separated. We can recover $G(f)$ from $G_s(f)$ by using a low pass filter having the frequency response

$$H(f) = \mathrm{rect}\,(f/2\sigma) \ . \tag{3.2.28}$$

Thus

$$G_S(f)H(f) = G(f) \ . \tag{3.2.29}$$

Noting that

228

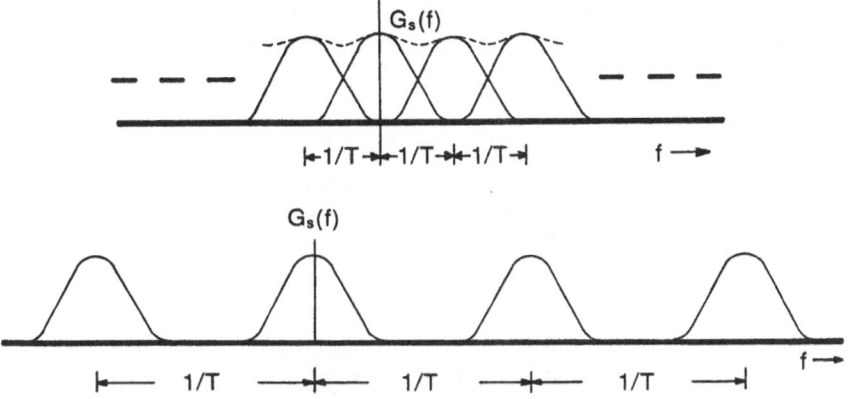

Fig. 3.29. $G_s(f)$ for two different sampling rates $1/T$

$$h(t) = 2\pi \operatorname{sinc}(2\sigma t) = \mathcal{F}^{-1}[H(f)] \quad, \tag{3.2.30}$$

we find that $g(t)$ can also be written as

$$\begin{aligned}
g(t) &= h(t)*g_s(t) \\
&= h(t)*T\sum g(nT)\delta(n-nT) \\
&= \sum 2\sigma T g(nT) \operatorname{sinc}[2\sigma(t-nT)] \\
&= \sum g\left(\frac{n}{2\sigma}\right) \operatorname{sinc}\left[2\sigma\left(t-\frac{n}{2\sigma}\right)\right] \quad.
\end{aligned} \tag{3.2.31}$$

This expression is known as the sampling theorem. We also note that we must sample the signal at a rate

$$f_0 > 2\sigma \tag{3.2.32}$$

to recover it later by filtering. This is known as the Nyquist criterion.

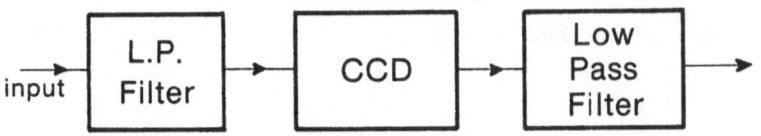

Fig. 3.30. Arrangement for a CCD to be used with a continuous time input-output signal

Figure 3.30 shows how a CCD can be used with a continuous time input-output signal. It must have two anti-aliasing filters to ensure that the input to the device is limited to the bandwidth σ. Of course, an anti-aliasing filter is simply a low pass filter. One might argue that for the input one really does not require the anti-aliasing filter because $g(t)$ is band limited. However, there is always noise present having frequency components that extend beyond σ and this will corrupt the signals. The clock rate in the device determines the sampling frequency and this must be higher than the Nyquist rate.

3.2.3 Frequency Response of a Discrete Time Filter

In Sect. 3.2.1, we obtained different examples of the z-response of some filters. If we use these filters with proper anti-aliasing filters we know that we obtain continuous signal input-output. The next topic of interest is how to obtain the frequency response from the z-response.

If we consider a time- and band-limited signal then we know that the Fourier transform is related to the discrete Fourier series. Remembering the comb function and its transform denoted by $\bar{\delta}(t)$ nad $\bar{\delta}(\omega)$ (Sect. 3.2.2), we have

$$\bar{\delta}(t) = \sum_{-\infty}^{+\infty} \delta(t + nT) \quad , \quad \text{and} \tag{3.2.33}$$

$$\bar{\delta}(\omega) = \sum_{-\infty}^{+\infty} \delta(\omega - n\omega_0) \quad . \tag{3.2.34}$$

Hence $\bar{\delta}(t) \leftrightarrow \omega_0 \bar{\delta}(\omega)$ where $\omega_0 = 2\pi/T$.

Consider a periodic function with period T, as shown in Fig. 3.31. If we define a complete period of the function $f(t)$ as

$$f_0(t) = \begin{cases} f(t) & |t| < T/2 \\ 0 & |t| > T/2 \end{cases} \tag{3.2.35}$$

we see that $f(t)$ can be written as

$$f(t) = \sum_{-\infty}^{+\infty} f_0(t + nT) = \bar{\delta}(t) * f_0(t) \quad . \tag{3.2.36}$$

Thus

$$\begin{aligned} F(\omega) &= \omega_0 F_0(\omega) * \bar{\delta}(\omega) \\ &= \omega_0 F_0(\omega) \sum \delta(\omega - n\omega_0) \\ &= \omega_0 \sum F_0(n\omega_0)\delta(\omega - n\omega_0) \quad . \end{aligned} \tag{3.2.37}$$

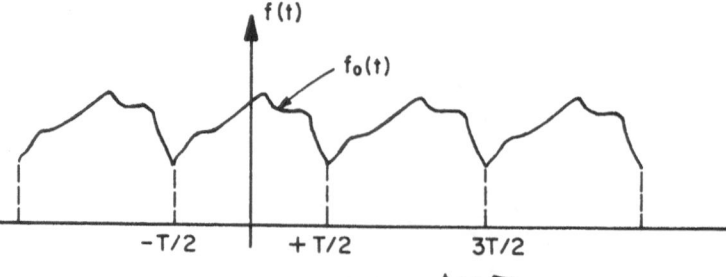

Fig. 3.31. Periodic function with period T

Recalling that

$$F_0(n\omega_0) = \int_{-T/2}^{+T/2} f(t)e^{-jn\omega_0 t}\,dt \qquad (3.2.38)$$

and defining $a_n = (1/T)F_0(n\omega_0)$, we obtain from (3.2.37) the familiar Fourier series given by

$$F(\omega) = a_n\delta(\omega - n\omega_0) \quad , \quad \text{or}$$

$$f(t) = \sum a_n e^{jn\omega_0 t} \quad . \qquad (3.2.39)$$

It is of interest to note that the comb function is periodic with period T and can thus be expanded in a Fourier series given by

$$\overline{\delta}(t) = \sum a_n e^{jn\omega_0 t} \quad , \quad \text{where}$$

$$a_n = \frac{1}{T} \int_{-T/2}^{+T/2} \delta(t)e^{-jn\omega_0 t}\,dt = \frac{1}{T} \quad ,$$

because $f_0(t) = \delta(t)$ for this case. Substituting the value of a_n we obtain

$$\overline{\delta}(t) = \frac{1}{T}\sum e^{jn\omega_0 t} \quad . \qquad (3.2.40)$$

If we consider a system with an arbitrary impulse response given by $f(t)$ and input $\overline{\delta}(t)$, then the output $g(t)$ is given by

$$\overline{f}(t) = \sum_{h=-\infty}^{+\infty} f(t + nT) \quad . \qquad (3.2.41)$$

However, if instead of a $\overline{\delta}(t)$ input we apply the equivalent input given on the right-hand side of (3.2.40), then once more we obtain

$$g(t) = \frac{1}{T}\sum F(n\omega_0)e^{jn\omega_0 t} \quad , \qquad (3.2.42)$$

where $F(\omega)$ is the Fourier transform of $f(t)$. Equating (3.2.41) and (3.2.42), we obtain

$$f(t + nT) = \frac{1}{T}\sum F(n\omega_0)e^{jn\omega_0 T} \quad . \qquad (3.2.43)$$

The above formula is known as Poisson's sum formula.

To numerically evaluate the Fourier transform of a function $f(t)$, one must replace the infinite limits, approximate the resulting integral with a sum and evaluate the sum for a discrete set of values of ω. This form of the Fourier integral is known as the discrete Fourier series.

From (3.2.38) we note that

$$F(n\omega_0) = \int_{-T/2}^{+T/2} \overline{f}(t)e^{-jn\omega_0 t}\,dt \quad , \quad \text{where} \qquad (3.2.44)$$

231

$$\overline{f}(t) = \sum_{n=-\infty}^{+\infty} f(t + nT) \quad . \tag{3.2.45}$$

Thus, if $f(t)$ is known for all values of t, $F(n\omega_0)$ can be evaluated numerically exactly. However, if the duration of $f(t)$ is less than T, then

$$\overline{f}(t) = f(t) \quad \text{for} \quad |t| < T/2 \quad .$$

Thus, for time-limited signals, we have

$$F(n\omega_0) = \int_{-T/2}^{+T/2} 2(t) e^{-jn\omega_0 t} \, dt \quad . \tag{3.2.46}$$

In order to obtain the discrete Fourier series we note that (3.2.44) can be rewritten as

$$\overline{f}(mT_1) = \frac{1}{T} \sum_{n=-\infty}^{+\infty} F(n\omega_0) e^{jn\omega_0 mT_1} \quad , \tag{3.2.47}$$

where T_1 is the sampling period and we have substituted $T = mT_1$, where m is any integer from 0 to $N - 1$, $T = NT_1$, $\omega_1 = 2\pi/T_1$ and $\omega_0 = \omega_1/N$. Thus (3.2.47) becomes

$$\begin{aligned} \overline{f}(mT_1) &= \frac{1}{T} \sum_{n=-\infty}^{+\infty} F(n\omega_0) e^{j2\pi mn/N} \\ &= \frac{1}{T} \sum_{n=-\infty}^{+\infty} F(n\omega_0) W_N^{mn} \quad , \quad \text{where} \end{aligned} \tag{3.2.48}$$

$W_N = e^{j2\omega/N}$ and, by definition, $W_N^N = 1$.

We next rewrite $n = k + \gamma N$ and (3.2.48) becomes

$$\begin{aligned} \overline{f}(mT_1) &= \frac{1}{T} \sum_{k=0}^{N-1} W_N^{km} \sum_{\gamma=-\infty}^{+\infty} F((k + \gamma N)\omega_0) W_N^{km} \\ &= \frac{1}{T} \sum_{k=0}^{N-1} W_N^{km} \sum_{\gamma=-\infty}^{+\infty} F(k\omega_0 + \gamma\omega_1) \quad , \quad \text{or} \end{aligned} \tag{3.2.49}$$

$$\overline{f}(mT_1) = \frac{1}{T} \sum_{k=0}^{N-1} \overline{F}(k\omega_0) W_N^{km} \quad , \quad \text{where} \tag{3.2.50}$$

$$\overline{F}(k\omega_0) = \sum_{\gamma=-\infty}^{+\infty} F(k\omega_0 + \gamma\omega_1) \quad . \tag{3.2.51}$$

If the sampling rate is higher than the Nyquist rate or the total bandwidth of the signal $f(t)$, 2σ, is less than ω_1, then

$$\overline{F}(\omega) = \begin{cases} F(\omega) & \text{for } |\omega| < \sigma \quad , \\ 0 & \text{for } |\omega| > \sigma \text{ and } \omega_1 > 2\sigma \quad . \end{cases}$$

Also, if the signal is time-limited to T then in the time interval 0 to T

$$\bar{f}(t) = f(t) \quad \text{for } t < T \quad .$$

We can rewrite (3.2.50) as

$$f(mT_1) = \sum_{k=0}^{N-1} F(k\omega_0)W_N^{km} \quad , \quad m = 0, \ldots, N-1 \quad . \tag{3.2.52}$$

The above equation represents the discrete Fourier series. By multiplying both sides of (3.2.52) with W_N^{-mn} and summing over all possible values, we obtain

$$\sum_{m=0}^{N-1} f(mT_1)W_N^{-mn} = \sum_{m=0}^{N-1} W_N^{-mn} \sum_{k=0}^{N-1} W_N^{mk} F(k\omega_0)$$

$$= \sum_{k=0}^{N-1} F(k\omega_0) \sum_{k=0}^{N-1} W_N^{m(k-n)} \quad . \tag{3.2.53}$$

Now

$$\sum_{N=0}^{N-1} W_N^{m(k-n)} = \frac{W_N^{N(k-n)} - 1}{W_N^{k-n} - 1} = \begin{cases} 0 & \text{for } k \neq n \quad , \\ N & \text{for } k = n \quad . \end{cases}$$

We thus obtain

$$F(n\omega_0) = \sum_{m=0}^{N-1} f(mT_1)W_N^{-mn} \quad . \tag{3.2.54}$$

There is a one-to-one correspondence between $F(n\omega_0)$ and the components $f(m, t_1)$ of a discrete Fourier series. Here $F(n\omega_0)$ and $f(mT_1)$ are two periodic series with period N. To calculate all the frequency components $F(n\omega_0)$ numerically from the $f(mT_1)$ values, we note that we need $N-1$ multiplications excluding $n = 0$ for each value of $F(n\omega_0)$. The total number of multiplications required is thus $(N-1)^2$. When computers perform numerical calculations the multiplication operation is much more time consuming than subtraction and addition, thus it is of great interest to find an algorithm which reduces this number of multiplications. This is achieved by using what are known as Fast Fourier Transform (FFT) algorithms.

A FFT algorithm is based on the fact that if N is even, it is convenient to break up (3.2.54) in the form

$$F(n\omega_0) = B(n\omega_0) + W_N^n C(n\omega_0) \quad , \tag{3.2.55}$$

where $B(n\omega_0)$ and $C(n\omega_0)$ are the discrete Fourier components corresponding to the even and odd values of $f(mT_1)$, respectively. Thus, to calculate $B(n\omega_0)$ from even values of m, we need to multiply $(N/2 - 1)^2$ times. Hence the total number of multiplications needed to evaluate (3.2.55) is

$$2\left(\frac{N}{2} - 1\right)^2 + (N - 1) = \frac{N^2}{2} - N + 1 \quad , \tag{3.2.56}$$

which is smaller than $(N-1)^2 = N^2 - 2N + 1$ for N a positive integer. Actually, the saving in number of multiplications is a factor of 2 for a large value of N. If $N = 2^s$, the process can be repeated many times starting with $N = 2$ and it can then be shown that the total number of multiplications needed is $(N/2) \log_2 N$. Thus using FFT for $N = 1064$, we need to multiply 50 times, instead of $\sim 250\,000$ multiplications required in the ordinary way. The FFT has therefore played a very important role in the use of digital techniques for signal processing.

a) Frequency Response from z-Response for Discrete Time Systems

To obtain a relationship between $H(z)$ and $H(f)$ we use the relationship between $f(t)$, $g(t)$ and $f(nT) = f(n)$ and $g(nT) = g(n)$ where T is the proper sampling time. For the continuous case, as shown in Fig. 3.10 the frequency response $H_c(\omega)$ is obtained by applying an input $\exp(j\omega t)$ so that the output will be given by

$$g(t) = H_c(\omega)e^{j\omega t} \quad .$$

For the equivalent sampled data case, we know for an input z^n that the output is

$$g(n) = H(z)z^n \quad .$$

If $z = \exp(j\omega t)$ or the input is $\exp(jn\omega T)$, then the output is given by

$$g(nT) = H(e^{j\omega T})e^{jn\omega T} \quad . \tag{3.2.57}$$

Thus $H_c(\omega)$ is equivalent to $H(\exp(j\omega t))$. However, $H(\exp(j\omega t))$ is a periodic function whereas $H_c(\omega)$ is not. If $f(t)$ is band limited then

$$F(\omega) = 0 \quad \text{for } |\omega| > \sigma \quad \text{where } \sigma = \frac{\pi}{T}$$

so that

$$H_c(\omega) = H(e^{j\omega T}) \quad \text{for } |\omega| < \sigma \quad .$$

Thus, the procedure to obtain the frequency response from the z-response is to replace z by $e^{j\omega t}$, i.e.,

$$H(f) = H(z = e^{j\omega t}) \quad \text{for } |\omega| < \sigma \quad . \tag{3.2.58}$$

Using (3.2.58) and (3.3.16), we obtain the frequency response of the two pole filter shown in Fig. 3.22 as

$$H(z) = \frac{a_0 z^2 - a_1 z}{(z - \gamma_1)(z - \gamma_2)} \quad , \tag{3.2.59}$$

where γ_1 and γ_2 are poles and are given by

$$\gamma_{1,2} = \frac{b_1}{2} \pm \sqrt{\frac{b_1^2}{4} - b_2} = e^{\pm j\alpha} \quad .$$

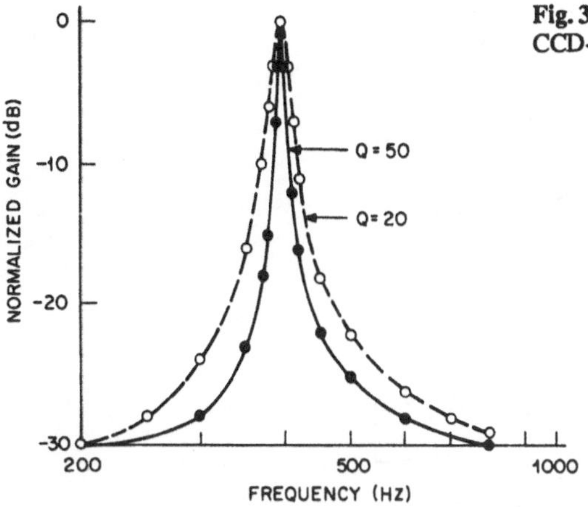

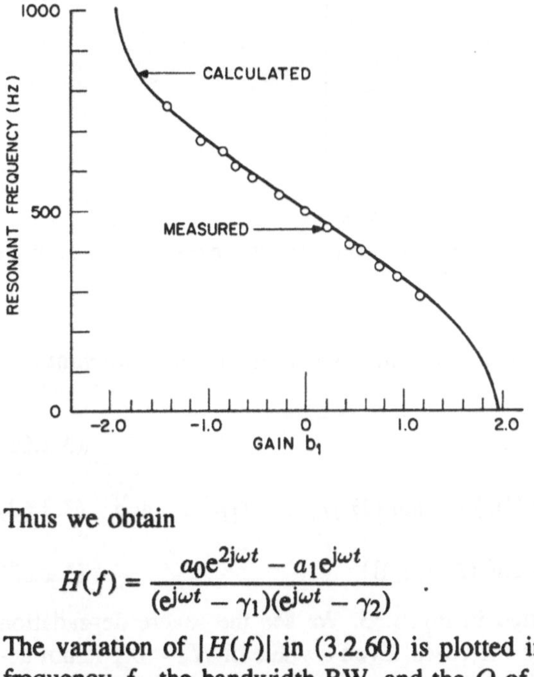

Fig. 3.32. Frequency response of a two-pole CCD-implemented filter. (From [3.5])

Thus we obtain

$$H(f) = \frac{a_0 e^{2j\omega t} - a_1 e^{j\omega t}}{(e^{j\omega t} - \gamma_1)(e^{j\omega t} - \gamma_2)} \quad . \tag{3.2.60}$$

The variation of $|H(f)|$ in (3.2.60) is plotted in Fig. 3.32 where the resonance frequency f_r, the bandwidth BW, and the Q of the two-pole filter are given by

$$f_r = \frac{1}{2\pi T} \cos^{-1}(b_1/2\sqrt{b_2}) \quad ,$$

$$BW = \frac{1}{\pi T} |\ln \sqrt{b_2}| \quad ,$$

$$Q = \frac{\cos^{-1}(b_1/2\sqrt{b_2})}{|\ln b_2|} \quad .$$

The above equations are derived assuming that $b_2 \to 1$, i.e., a high Q condition.

235

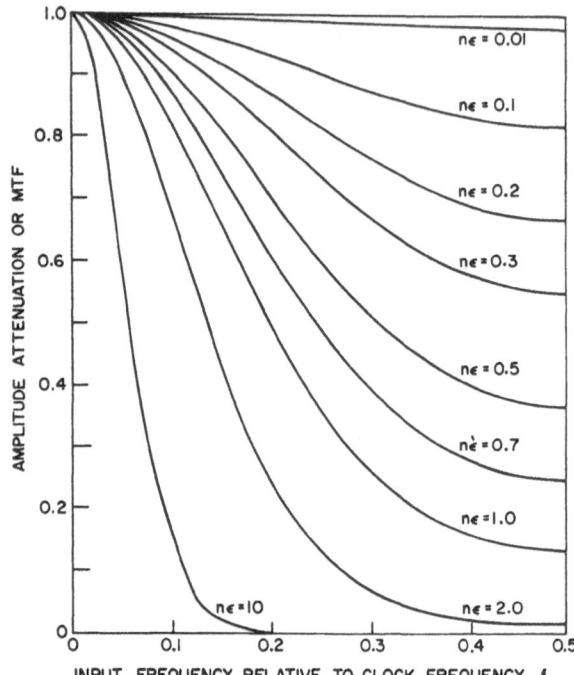

Fig. 3.33. Frequency dependence of the transfer inefficiency. Note the severe degradation of the device performance as $N\varepsilon$ increases. (From [3.4])

For the frequency dependence effect of the transfer inefficiency we similarly use (3.2.21) to obtain

$$H(z) \propto \exp\left[N\varepsilon(z^{-1} - 1)\right] \quad, \tag{3.2.61}$$

$$H_N(f) \propto \exp\left\{N\varepsilon[\cos(2\pi f/f_c) - j\sin(2\pi f/f_c) - 1]\right\} \quad, \quad \text{and} \quad \tag{3.2.62}$$

$$|H_N(f)| \propto \exp\left\{-N\varepsilon[1 - \cos(2\pi f/f_c)]\right\} \quad. \tag{3.2.63}$$

Some typical responses are plotted in Fig. 3.33. We see the severe degradation of the device performance as $N\varepsilon$ increases. Even a value of $N\varepsilon = 0.1$ is not acceptable if one insists on using the whole acceptable frequency range determined by the Nyquist criterion.

b) Chirp z Algorithm for Discrete Time Signals

In Sect. 3.1.3, we worked out an example showing how one can obtain a real-time Fourier transform using chirp devices. Similar real-time discrete Fourier transforms can be performed using sampled versions of chirp devices. However, before discussing this algorithm, we note that the actual signal is complex, having both amplitude and phase. As the sampled data processing is performed in the base band, the signal has real and imaginary components and each component

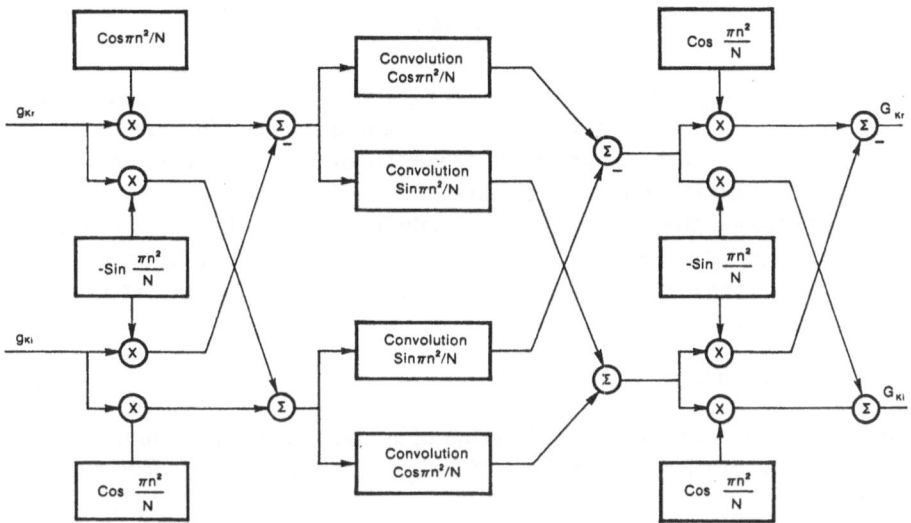

Fig. 3.34. Implementation of (3.2.65) in the sampled data case

should be dealt with separately. For processing using SAW devices, this is carried out around a center frequency, and complex operations can thus be performed. The signal $g(n)$ can be written as

$$g(n) = g_r(n) + jg_{im}(n) \quad , \tag{3.2.64}$$

where the subscripts r and im represent the real and imaginary parts respectively. The chirp transform algorithm given by (3.1.24) and reproduced below

$$F\left(t = \frac{\beta t}{\pi}\right) = \{[f(t)e^{-j\beta t^2}] * e^{j\beta t^2}\}e^{-j\beta t^2}$$

yields, after substitution of real and imaginary parts, the expression

$$\begin{aligned}
F_r(f) + jF_{im}(f) = {} & [g_r \cos \beta t^2 + g_{im} \sin \beta t^2] * \cos \beta t^2 \\
& - (g_{im} \cos \beta t^2 - g_r \sin \beta t^2) * \sin \beta t^2 \\
& + j(g_r \cos \beta t^2 + g_{im} \sin \beta t^2) * \sin \beta t^2 \\
& + j(g_{im} \cos \beta t^2 - g_r \sin \beta t^2) * \cos \beta t^2 \\
& \times (\cos \beta t^2 - j \sin \beta t^2) \quad .
\end{aligned} \tag{3.2.65}$$

The implementation of the above equation in the sampled data case is shown in Fig. 3.34, where t is replaced by n and the sampling time is normalized to unity. We observe that in the base band case, four chirp devices are needed for convolution, whereas only one is needed for the continuous radio frequency (rf) centered case.

3.3 Noise and Stochastic Processes

One of the main purposes of signal processing is to improve the signal-to-noise ratio in a system. It is therefore very important to understand the physics behind noise, which is a stochastic process and therefore different from the deterministic signals discussed in the last few sections.

A stochastic process $x(t)$ is defined only in some probability space. We can never know exactly what the value will be at a certain instant. However, if the process is repeated many many times, we can determine the probability of it having a certain value. The expected mean value is defined as

$$\eta(t) = E\{x(t)\} = \int xP(x)dx \quad , \tag{3.3.1}$$

where $E\{u\}$ denotes the expected value of u and P is the probability density function appropriate for the problem. If the system is ergodic[3], the expected value varies as

$$\eta(t) \to \frac{1}{2T} \int\limits_{-T}^{+T} x(t - \tau)d\tau \tag{3.3.2}$$

as T goes to infinity. As nearly all physical processes are ergodic, the approximation where the expected value is replaced by the time average is justified.

The auto- and cross-correlations between two stochastic processes are defined, respectively, by

$$R_{xx}(t_1, t_2) = E\{x(t_1)x^*(t_2)\} \quad \text{and} \tag{3.3.3}$$

$$R_{xy}(t_1, t_2) = E\{x(t_1)y^*(t_2)\} \quad . \tag{3.3.4}$$

It is occasionally convenient to take out the mean value or the bias in the above definitions. Thus we obtain the auto- and cross-covariances given by

$$C_{xx}(t_1, t_2) = E\{[x(t_1) - \eta_x(t_1)][x^*(t_2) - \eta_x^*(t_2)]\} \tag{3.3.5}$$

and

$$C_{xy}(t_1, t_2) = E\{[x(t_1) - \eta_x(t_1)][y^*(t_2) - \eta_y^*(t_2)]\} \quad . \tag{3.3.6}$$

If the statistical properties are invariant to a shift in t, the process is defined as a strict sense stationary process. However, if

$$\eta(t) = \eta_0 \quad \text{and} \tag{3.3.7}$$

$$R(t_1, t_2) = R(\tau) \quad , \quad \text{where} \quad t_2 - t_1 = \tau \quad , \tag{3.3.8}$$

the process is defined as a wide sense stationary process. For an ergodic system which is wide sense stationary, we obtain for (3.3.2–6) that

---

[3] For an ergodic system, the time average can be considered as the expected value. Note that for most physical systems this is true if a long enough time is allowed for averaging.

$$\eta_0 = \frac{1}{2T} \int\limits_{-T}^{+T} x(\tau)d\tau \quad , \quad T \to \infty \quad , \tag{3.3.9}$$

$$R_{xx}(\tau) = \int\limits_{-\infty}^{+\infty} x(\tau)x^*(t+\tau)d\tau = x(t) \boxplus x(t) \quad ,$$

$$R_{xy}(\tau) = \int\limits_{-\infty}^{+\infty} x(\tau)y^*(t+\tau)d\tau = x(t) \boxplus y(t) \quad ,$$

$$C_{xx}(\tau) = \int\limits_{-\infty}^{+\infty} [x(\tau) - \eta_0][x^*(t+\tau) - \eta_0]d\tau \quad , \quad \text{and}$$

$$C_{xy}(\tau) = \int\limits_{-\infty}^{+\infty} [x(\tau) - \eta_{x0}][y^*(t+\tau) - \eta_{y0}]d\tau \quad ,$$

where the symbol $\boxplus$ denotes correlation.

White noise is a good example of a stationary process. It is given by

$$R(\tau) = I_0\delta(\tau) \quad \text{or} \tag{3.3.10}$$

$$R(t_1, t_2) = I(t_1)\delta(t_1 - t_2) \quad .$$

For a stationary process, one defines the power spectrum $S(\omega)$, given by

$$S(\omega) = \int\limits_{-\infty}^{+\infty} R(\tau)e^{j\omega\tau}d\tau \quad . \tag{3.3.11}$$

For white noise, as expected,

$$S(\omega) = I_0 = \text{const} \quad . \tag{3.3.12}$$

Any other form of noise is referred to as colored noise. Some typical examples of colored noises in different cases are listed below.

For many physical processes, the autocorrelation $R(\tau)$ is given by

$$R(\tau) = I_0 e^{-|\tau|/\tau_0} \quad , \tag{3.3.13}$$

where τ_0 is a characteristic lifetime and I_0 is a constant given by

$$I_0 = E\{x(t)x^*(t)\} \quad . \tag{3.3.14}$$

Thus I_0 is the variance of the random process. The corresponding $S(\omega)$ is

$$S(\omega) = \frac{I_0}{1 + (\omega\tau_0)^2} \quad . \tag{3.3.15}$$

For Johnson noise in a semiconductor or conductor, τ_0 is the momentum relaxation time associated with different scattering mechanism of the carriers and is on the order of 10^{-12} s. Thus for most of the frequency range of interest,

$$S(\omega) \approx I_0 \quad \text{as} \quad (\omega\tau_0)^2 \ll 1 \tag{3.3.16}$$

and the Johnson noise behaves like white noise.

There is a different interpretation of τ_0, which is probably more useful when we discuss light incoherence. Then τ_0 is related to coherence time and (3.3.15) implies that the random process $x(t)$ is correlated approximately to $x(t+\tau)$ for $0 \le |\tau| \le \tau_0$. For example, consider a coherent signal

$$x(t) = e^{j\omega_0 t} \quad . \tag{3.3.17}$$

For this case $R(\tau) = e^{j\omega_0 \tau}$ and

$$S(\omega) = \delta(\omega - \omega_0) \quad . \tag{3.3.18}$$

However, if there is some phase incoherence involved in $x(t)$, then

$$R(\tau) = e^{j\omega_0 \tau} e^{-|\tau|/\tau_0} \quad \text{and} \tag{3.3.19}$$

$$S(\omega) = \frac{1}{1 + (\omega - \omega_0)^2 \tau_0^2} \quad . \tag{3.3.20}$$

Thus we see that, for a coherence time τ_0, the power spectrum is not a delta function at $\omega = \omega_0$ but has a spread Δf given by

$$\Delta f \sim \frac{1}{\tau_0} \quad . \tag{3.3.21}$$

It is of interest to note the following important properties of the power spectra and autocorrelation:

a) $\quad R(-\tau) = R^*(\tau)$. $\hspace{5cm}$ (3.3.22)

b) $\quad S(\omega)$ is real. $\hspace{5.5cm}$ (3.3.23)

c) $\quad$ If $x(t)$ is real then

$$R(-\tau) = R(\tau) \tag{3.3.24}$$

$\quad$ and $S(\omega)$ is even and real.

d) $\quad E\{|x(t)|^2\} = R(0) = \int\limits_{-\infty}^{+\infty} S(f)df$. $\hspace{2.5cm}$ (3.3.25)

e) $\quad S(\omega) \ge 0$, thus $R(\tau)$ is positive-definite. $\hspace{1.7cm}$ (3.3.26)

f) $\quad R(\tau) \le R(0)$. $\hspace{5cm}$ (3.3.27)

3.3.1 Linear Systems with Stochastic Input

Let us consider a linear system with impulse response $h(t)$. If the input is a stochastic input $x(t)$ the output is also a stochastic process $y(t)$ given by

$$y(t) = \int x(t - \alpha)h(\alpha)d\alpha \quad . \tag{3.3.28}$$

The expected value of $y(t)$ is given by

$$E\{y(t)\} = \int E\{x(t - \alpha)\}h(\alpha)d\alpha \quad , \tag{3.3.29}$$

where the integration can be performed later after taking the expected value of $x(t)$, as $h(t)$ is deterministic.

To calculate the autocorrelation of the output signal, we note that

$$y(t_1)y^*(t_2) = \iint x(t_1 - \alpha)x^*(t_2 - \alpha)h(\alpha)h^*(\alpha)d\alpha \, d\beta \tag{3.3.30}$$

and

$$
\begin{aligned}
R_{yy}(t_1, t_2) &= E\{y(t_1)y^*(t_2)\} \\
&= \int E\{x(t_1 - \alpha)y^*(t_2)\}h(\alpha)d\alpha \\
&= \int R_{xy}(t_1 - \alpha, t_2)h(\alpha)d\alpha \\
&= \iint R_{xx}(t_1 - \alpha, t_2 - \beta)h(\alpha)h^*(\beta)d\alpha \, d\beta \quad .
\end{aligned}
\tag{3.3.31}
$$

In the above expression, we have used

$$R_{xx}(t_1, t_2) = E\{x(t_1)x^*(t_2)\} \quad \text{and} \tag{3.3.32}$$

$$
\begin{aligned}
R_{xy}(t_1, t_2) &= E\{x(t_1)y^*(t_2)\} \\
&= \int E\{x(t_1)x^*(t_2 - \beta)\}h^*(\beta)d\beta \\
&= \int R_{xx}(t_1, t_2 - \beta)h^*(\beta)d\beta \quad .
\end{aligned}
$$

Again, if the processes are stationary, the above equations simplify to

$$R_{xx}(t_1, t_2) = R_{xx}(\tau) \quad , \quad \text{where} \quad t_2 - t_1 = \tau \quad , \tag{3.3.33}$$

$$
\begin{aligned}
R_{xy}(\tau) &= \int R_{xx}(\tau + \beta)h^*(\beta)d\beta \\
&= R_{xx}(\tau)*h^*(-\tau) \quad ,
\end{aligned}
\tag{3.3.34}
$$

$$
\begin{aligned}
R_{yy}(\tau) &= \int R_{xy}(\tau - \alpha)h(\alpha)d\alpha \\
&= R_{xy}(\tau)*h(\tau) \\
&= [R_{xx}(\tau)*h^*(-\tau)]*h(\tau) \quad .
\end{aligned}
\tag{3.3.35}
$$

We note that

$$E\{|y(t)|^2\} = R_{yy}(0) \quad . \tag{3.3.36}$$

Taking Fourier transforms of the relevant quantities as $h(\tau) \to H(\omega)$, $h^*(-\tau) = H^*(\omega)$, $R_{xx}(\tau) \to S_{xx}(\omega)$, $R_{xy}(\tau) \to S_{xy}(\omega)$ and $R_{yy}(\tau) \to S_{yy}(\omega)$, we obtain

$$S_{xy}(\omega) = S_{xx}(\omega)H^*(\omega) \quad \text{and} \tag{3.3.37}$$

$$\begin{aligned} S_{yy}(\omega) &= S_{xy}(\omega)H(\omega) \\ &= S_{xx}(\omega)|H(\omega)|^2 \quad . \end{aligned} \tag{3.3.38}$$

3.3.2 Matched Filters

Let us consider an input to a filter with a signal component $f(t)$ and a noise component $n(t)$. The signal $f(t)$ is considered to be deterministic. The input is

$$x(t) = f(t) + n(t) \quad . \tag{3.3.39}$$

If we pass this input through a filter with impulse response $h(t)$, the output is given by

$$\begin{aligned} y(t) &= x(t)*h(t) \\ &= y_f(t) + y_n(t) \quad , \quad \text{where} \end{aligned} \tag{3.3.40}$$

$$y_f(t) = f(t)*h(t) \quad \text{and}$$

$$y_n(t) = h(t)*n(t) \quad .$$

In the matched filtering problem, we are interested in maximizing the signal-to-noise ratio S/N, where S and N are signal and noise power, respectively. The particular $h(t)$ which performs this maximization is known as the matched filter for the signal.

If the power spectrum for the noise is given by $N(\omega)$, we obtain

$$N = E\{y_n^2(t)\} = \frac{1}{2\pi} \int\limits_{-\infty}^{+\infty} N(\omega)|H(\omega)|^2 d\omega \quad . \tag{3.3.41}$$

Similarly, at $t = t_0$, the output due to the signal component can be written as

$$\begin{aligned} y_f(t_0) &= \frac{1}{2\pi} \int\limits_{-\infty}^{+\infty} F(\omega)H(\omega)e^{j\omega t_0} d\omega \\ &= \int\limits_{-\infty}^{+\infty} \frac{F(\omega)}{\sqrt{N(\omega)}} H(\omega)\sqrt{N(\omega)}e^{j\omega t_0} d\omega \quad . \end{aligned} \tag{3.3.42}$$

We note the following Schwarz inequality:

$$\left| \int f_1(x)f_2(x)dx \right| \le \int |f_1(x)|^2 dx \int |f_2(x)|^2 dx \quad , \tag{3.3.43}$$

where the equality holds when

242

$$f_1(x) = K f_2^*(x) \tag{3.3.44}$$

where K is a constant. Using

$$f_1(x) = \frac{F(\omega)}{\sqrt{N(\omega)}} \quad \text{and} \quad f_2(x) = H(\omega)\sqrt{N(\omega)}$$

we note that

$$S = |y_f(t_0)|^2 \le \int \frac{|F(\omega)|}{N(\omega)} d\omega \int |H(\omega)|^2 N(\omega) d\omega \quad , \tag{3.3.45}$$

which yields

$$\frac{S}{N} = \frac{|y_f(t_0)|^2}{E\{|y_n(t)|\}^2} \le \int \frac{|F(\omega)|^2}{N(\omega)} d\omega \quad . \tag{3.3.46}$$

The right-hand side is a constant. For maximum S/N, we obtain

$$\left(\frac{S}{N}\right)_{\max} = \frac{|\dot{y}_f(t_0)|^2}{E\{|y_n(t)|\}^2} = \int \frac{|F(\omega)|^2}{N(\omega)} d\omega \quad \text{or} \tag{3.3.47}$$

$$\frac{F(\omega)}{\sqrt{N(\omega)}} = K H^*(\omega)\sqrt{N(\omega)} e^{-j\omega t_0} \quad \text{or}$$

$$H^*(\omega) = K \frac{F(\omega)}{N(\omega)} e^{-j\omega t_0} \quad . \tag{3.3.48}$$

For white noise, we have $N(\omega) = N_0$, which leads to a matched filter frequency response (note that causal solution requires that t_0 be greater than a minimum value)

$$H(\omega) = \frac{K}{N_0} F^*(\omega) e^{-j\omega t_0} \quad \text{or} \tag{3.3.49}$$

$$h(t) = \frac{K}{N} f^*(t_0 - t) \quad \text{and} \tag{3.3.50}$$

$$\left(\frac{S}{N}\right)_{\max} = \frac{S}{N_0} \quad . \tag{3.3.51}$$

The output signal for this case is given by

$$y_f(t) = \frac{K}{N_0} \int\limits_{-\infty}^{+\infty} f(\tau) f^*(t_0 - t + \tau) d\tau \quad \text{or}$$

$$y_f(t_0) = \frac{K}{N_0} \int\limits_{-\infty}^{+\infty} f(t) f^*(t_0 - t) dt = \frac{K}{N_0} f(t) \boxplus f(t) \quad , \tag{3.3.52}$$

which is simply the autocorrelation of the signal. Thus the matched filter is also called a correlator for the case of white noise. For colored noise, if $N(\omega)$ is

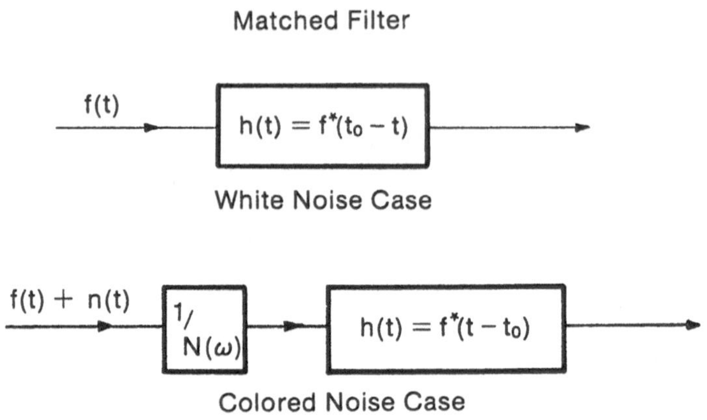

Matched Filter

Fig. 3.35. Block diagrams of matched filters for the white noise (*top*) and colored noise (*bottom*) cases

known, then one can pre-whiten the signal and noise through division with $N(\omega)$ by using a device in the Fourier domain. In that case, the block diagram of the matched filter is as shown in the lower half of Fig. 3.35.

a) Cauchy-Schwarz Inequality

If

$$I(y) = Ay^2 + 2By + C$$
$$= (y - B + \sqrt{B^2 - AC})(y - B - \sqrt{B^2 - AC}) \geq 0$$

for all real values of y, then we note that $B^2 - AC \leq 0$ because otherwise $I(y) \leq 0$ for $y = B$. If $B^2 - AC = 0$, then $I(y) = 0$ for $y = B$.

b) Schwarz Inequality

Consider the following function $I(y)$ of y for all real values of y:

$$I(y) = \int_a^b [z(x) - yw(x)]^2 dx \geq 0 \quad,$$

where $z(x)$ and $w(x)$ are real functions of x. Then

$$I(y) = \int_a^b z(x)dx - 2y \int_a^b z(x)w(x)dx + y^2 \int_a^b w^2(x)dx \quad.$$

Using the Cauchy-Schwarz inequality, we get

$$\left| \int_a^b z(x)w(x)dx \right|^2 \leq \int_a^b z^2(x)dx \int_a^b w^2(x)dx \quad. \tag{3.3.53}$$

The above inequality is known as the Schwarz inequality.

We note that if

$$\left| \int_a^b z(x)w(x)dx \right|^2 = \int_a^b z^2(x)dx \int_a^b w^2(x)dx \tag{3.3.54}$$

then for some value $y = k$, $I(k) = 0$. Thus

$$I(k) = \int_a^b [z(x) - kw(x)]^2 dx = 0 \quad \text{or}$$

$$z(x) = kw(x) \quad . \tag{3.3.55}$$

The proof can be extended for a complex function of x by considering

$$I(y) = \int_a^b |z(x) - ye^{j\theta}w^*(x)|^2 dx \quad , \quad \text{where}$$

$$\int_a^b z(x)w(x)dx = Be^{j\theta} \quad .$$

For the complex case, one obtains in place of (3.3.55) the equation

$$z(x) = kw^*(x) \quad . \tag{3.3.56}$$

3.3.3 Matched Filters from the Point of View of Maximum Output

A matched filter can also be examined from a different point of view: If the input to the linear system is $f(t)$, we wish to determine the impulse response of the system if the output $g(t)$ is to be maximum at a time $t = t_0$ (Fig. 3.36). We then have

$$g(t_0) = \int_{-\infty}^{+\infty} f(\tau)h(t_0 - \tau)d\tau = \int_{-\infty}^{\infty} F(f)H(f)e^{j2\pi f t_0}df \quad . \tag{3.3.57}$$

We note that the energy E of the signal is given by

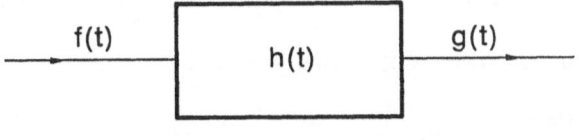

$$g(t_0) = \int_{-\infty}^{+\infty} f(\tau)\, h(t_0 - \tau)\, d\tau = \int_{-\infty}^{+\infty} F[f]\, H[f]\, e^{j2\pi f t_0}\, df$$

Fig. 3.36 Block diagram of a filter with input $f(t)$, impulse response $h(t)$ and output $g(t)$

245

$$E = \int\limits_{-\infty}^{+\infty} |f(t)|^2 dt = \int\limits_{-\infty}^{+\infty} |F(f)|^2 df = \text{const} \quad . \tag{3.3.58}$$

Using the Schwarz inequality, we have

$$|g(t_0)|^2 = \left| \int\limits_{-\infty}^{+\infty} F(f)H(f)e^{j\omega_0 t} d\omega \right|^2$$

$$\leq \int\limits_{-\infty}^{+\infty} |F(f)|^2 df \int\limits_{-\infty}^{+\infty} |H(f)e^{j2\pi f_0 t}|^2 df \quad . \tag{3.3.59}$$

The maximum value of $|g(t_0)|$ occurs for

$$F(f) = KH^*(f)e^{-j2\pi f t_0}$$

and we obtain

$$H(f) = \frac{1}{K} F^*(\omega)e^{-j2\pi f t_0} \quad \text{or}$$

$$h(t) = \frac{1}{K} f^*(t_0 - t) \quad . \tag{3.3.60}$$

Hence the matched filter also gives the largest output for a given input.

3.3.4 Matched Filtering of Stochastic Signals

In this section we discuss the problem of obtaining a matched filter for a signal $f(t)$ that is not deterministic but a stochastic process by itself. We shall therefore be dealing with the expected mean value and variance of f. The signal power S for this case is given by

$$S = |E\{y_f\}|^2 \quad , \tag{3.3.61}$$

where y_f is the output of the filter when only the signal f is at the input. We then have

$$y_f = f * h \quad . \tag{3.3.62}$$

and (3.3.61) can be written as

$$S = \left| \int E\{F(\omega)H(\omega)e^{-j\omega t_0}\} d\omega \right|^2 \quad , \tag{3.3.63}$$

where t_0 is a constant. The noise power N is similarly given by

$$N = E\{y_n^2(t)\} = \int S_{yy}(\omega) d\omega \quad . \tag{3.3.64}$$

Note that $y = (f + n) * h$, so that

$$N = \int S_{(f+n)(f+n)}(\omega) |H(\omega)|^2 d\omega \quad . \tag{3.3.65}$$

246

The signal-to-noise ratio to be maximized is

$$\frac{S}{N} = \frac{|\int E\{F(\omega)H(\omega)e^{-j\omega t_0}\}d\omega|^2}{\int S_{(f+n)(f+n)}(\omega)|H(\omega)|^2 d\omega} \quad . \tag{3.3.66}$$

Following the argument discussed in Sect. 3.3.2a, we obtain that the optimum filter is given by

$$H(\omega) = \frac{F^*(\omega)e^{-j\omega t_0}}{S_{(f+n)(f+n)}(\omega)} \quad . \tag{3.3.67}$$

The power spectrum or Fourier transform of the variance of the processes f and n is given by

$$S_{(f+n)(f+n)} = S_{ff}(\omega) + S_{nn}(\omega) + S_{nf}(\omega) \quad . \tag{3.3.68}$$

If f and n are uncorrelated, then

$$S_{nf}(\omega) = 0$$

and (3.3.67) simplifies to

$$H(\omega) = \frac{F^*(\omega)e^{-j\omega t_0}}{S_{ff}(\omega) + S_{nn}(\omega)} \quad . \tag{3.3.69}$$

In (3.3.69) we must properly define $\mathcal{F}^*(\omega)$ of a stochastic process. It is given by

$$F^*(\omega) = \mathcal{F}[E\{f\}] = \mathcal{F}[\eta_f(t)] \quad , \tag{3.3.70}$$

where $\eta_f = E\{f\}$ is the expected value of the stochastic process f. As $\eta_f(t)$ is not a constant, the f process is not stationary. However, we can define the covariance $C_{ff}(\tau)$ by

$$C_{ff}(\tau) = E\{f(t)f^*(t+\tau) - \eta(t)\eta^*(t+\tau)\} \quad . \tag{3.3.71}$$

Note that we assume (3.3.71) to be a function of τ only, in which case

$$S_{ff} = \mathcal{F}[C_{ff}(\tau)] \quad . \tag{3.3.72}$$

To understand (3.3.69) we consider the case when $f(t)$ tends to the deterministic case, so that

$$F^*(\omega) \rightarrow \mathcal{F}[f(t)] \quad \text{and}$$

$$S_{ff}(\omega) \rightarrow 0 \quad .$$

Thus (3.3.69) becomes

$$H(\omega) = \frac{F^*(\omega)e^{-j\omega t_0}}{S_{nn}(\omega)} \quad , \tag{3.3.73}$$

which is identical to the deterministic case given by (3.3.49).

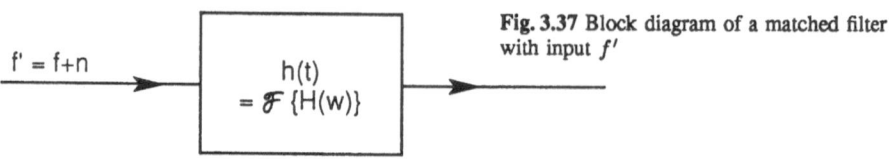

Fig. 3.37 Block diagram of a matched filter with input f'

$$f' = f+n \qquad \boxed{\begin{array}{c} h(t) \\ = \mathcal{F}\{H(w)\} \end{array}}$$

We can look at another interesting variation of the deterministic case by considering the noise term with f. Thus our input to the filter is only $f' = f + n$ as shown in Fig. 3.37. For this case, the matched filter equation given by (3.3.69) becomes

$$H(\omega) = \frac{F'^*(\omega)e^{-j\omega t_0}}{S_{f'f'}(\omega)} \quad , \tag{3.3.74}$$

where we note that

$$F'^*(\omega) = (\mathcal{F}[E\{f + n\}])^* = (\mathcal{F}[f])^* = F^*(\omega) \quad . \tag{3.3.75}$$

In (3.3.75) we have used the property that the expected value of n is zero and the expected value of f is f. Similarly,

$$S_{f'f'}(\omega) = S_{nn}(\omega) \quad . \tag{3.3.76}$$

Because the process is deterministic, its variance tends to zero. Thus for this special case (3.3.74) becomes

$$H(\omega) = \frac{F^*(\omega)e^{-j\omega t_0}}{S_{nn}(\omega)} \quad , \tag{3.3.77}$$

which is identical to (3.3.73).

The main difference between stochastic and deterministic signals is that, in the former case, the signal itself has some variance, which must be included as noise power to obtain the best possible filter to maximize the output signal-to-noise ratio.

3.3.5 Noise and Stochastic Processes: Discrete

In this section, we develop the mathematical representation of noise and stochastic processes as discrete time systems. Corresponding to the analog stochastic process $x(t)$, we have a discrete process $x(n)$, which is a sequence of real or complex random variables. Here, n is an integer number from zero to infinity which replaces the continuous variable t. As discussed in Sect. 3.2, the sampling time interval Δt and the bandwidth of the process are related by the sampling theorem.

For the discrete stochastic process, we can define the following quantities:

Mean $\qquad \eta(n) = E\{x(n)\}.$ $\qquad\qquad$ (3.3.78)

Autocorrelation $\quad R(n_1, n_2) = E\{x(n_1)x^*(n_2)\}.$ $\qquad$ (3.3.79)

Autocovariance $\quad C(n_1, n_2) = E\{[x(n_1) - \eta(n_1)][x(n_2) - \eta(n_2)]^*\}$

$$= R(n_1, n_2) - \eta(n_1)\eta^*(n_2). \qquad (3.3.80)$$

Cross-correlation $R_{xy}(n_1, n_2) = E\{x(n_1)y^*(n_2)\}.$ $\qquad (3.3.81)$

Cross-covariance $C_{xy}(n_1, n_2) = R_{xy}(n_1, n_2) - \eta_x(n_1)\eta_y^*(n_2).$ $\qquad (3.3.82)$

If the statistical properties are invariant under a shift in n, the process is known as a strict sense stationary process. However, if

$$\eta(n) = \eta_0 \quad \text{and}$$

$$R(n_1, n_2) = R(n_2 - n_1) = R(m) \quad \text{where} \quad n_2 = n_1 + m \quad , \qquad (3.3.83)$$

the process is called wide sense stationary. Note that

$$R(-m) = R^*(m) \quad . \qquad (3.3.84)$$

The z-transform $S(z)$ of $R(m)$ is given by

$$S(z) = \sum_{m=-\alpha}^{+\alpha} R(m)z^{-m} \quad . \qquad (3.3.85)$$

If $x(n)$ is obtained by sampling an analog process with sampling period T, then the power spectrum $S(\omega)$ can be written as

$$S(\omega) = S(z = e^{j\omega T}) \quad \text{for} \quad |\omega| < \sigma \qquad (3.3.86)$$

and $S(\omega)$ is always real because of (3.3.84). For this case, one can write

$$R(m) = \frac{1}{2\sigma} \int_{-\infty}^{+\sigma} S(\omega)e^{jm\omega T} d\omega \quad . \qquad (3.3.87)$$

Thus

$$E\{|x(n)|^2\} = R(0) = \frac{1}{2\sigma} \int_{-\infty}^{+\infty} S(\omega)d\omega \quad . \qquad (3.3.88)$$

For white noise

$$R[n_1 N_2] = I(n_1)\delta(n_1 - n_2) \quad . \qquad (3.3.89)$$

For stationary white noise

$$R(m) = I\delta(m) \quad , \qquad (3.3.90)$$

where I is a constant and $S(\omega) = I$.

Let us consider a linear system with impulse response $h(n)$. Then, if the input $x(n)$ is stochastic, the output $y(n)$ is also a stochastic process given by

$$y(n) = \sum_{k=-\infty}^{+\infty} x(n - k)h(k) \quad . \qquad (3.3.91)$$

The expected value of $y(n)$ is given by

$$E\{y(n)\} = \sum_{k=-\infty}^{+\infty} E\{x(n-k)\}h(k) \quad . \tag{3.3.92}$$

The autocorrelation, $R_{yy}(n_1, n_2)$ is given by

$$R_{yy}(n_1, n_2) = \sum_{k=-\infty}^{+\infty} R_{xy}(n_1 - k, n_2)h(k) \tag{3.3.93}$$

where

$$R_{xy}(n_1, n_2) = \sum_{k=-\infty}^{+\infty} R_{xx}(n_1, n_2 - k)h^*(k) \quad . \tag{3.3.94}$$

If the process $x(n)$ is stationary, then the output $y(n)$ is also stationary and we obtain

$$R_{xy}(m) = \sum_{k=-\infty}^{+\infty} R_{xx}(m+k)h^*(k) \quad \text{and} \tag{3.3.95}$$

$$R_{yy}(m) = \sum_{k=-\infty}^{+\infty} R_{xy}(m-k)h(k) \quad : \tag{3.3.96}$$

Substituting $R_{xy}(m)$ in the expression for $R_{yy}(m)$ and rearranging the terms, one obtains

$$R_{yy}(m) = R_{xx}(m)\hbar(m) \quad , \quad \text{where} \tag{3.3.97}$$

$$\hbar(m) = \sum_{k=-\infty}^{+\infty} h(m+k)h^*(k) \quad . \tag{3.3.98}$$

Using the z-transform properties, we can write

$$S_{xy}(z) = S_{xx}(z)H^*(z) \quad , \tag{3.3.99}$$

$$S_{yy}(z) = S_{xy}(z)H(z) \quad , \tag{3.3.100}$$

$$S_{yy}(z) = S_{xx}|H(z)|^2 \quad , \tag{3.3.101}$$

where we have used the notation

$$S_{ij}(z) = \sum_{k=-\infty}^{+\infty} R_{ij}(m)z^{-m} \quad . \tag{3.3.102}$$

As expected, the analog signal equations (3.3.37) and (3.3.38) correspond to the discrete time signal equations (3.3.99) and (3.3.101).

3.3.6 Matrix Methods[4]

It is sometimes convenient to represent discrete signals, in terms of a matrix of dimension $N \times 1$, where N represents the length of the signal, considered finite. This method is fundamental to the manipulation of a signal by digital computers. However, as we shall see in the second volume, optical matrix processors are fast becoming a reality. The matrix method is a prerequisite for the appreciation of the benefits of these matrix processors for signal processing functions such as adaptive filtering, matched filtering and deconvolution.

Suppose we want to rewrite the discrete convolution equation (3.2.7) in terms of matrices:

$$g(n) = \sum_{k=-\infty}^{+\infty} f(n-k)h(k) \quad . \tag{3.2.7}$$

First we note that the input $f(n)$ can be represented as an $N \times 1$ matrix

$$F = \begin{bmatrix} f(0) \\ \vdots \\ f(N-1) \end{bmatrix} \quad . \tag{3.3.103}$$

Similarly, the output $g(n)$ can be represented by an $N' \times 1$ matrix

$$G = \begin{bmatrix} g(0) \\ \vdots \\ g(N'-1) \end{bmatrix} \quad . \tag{3.3.104}$$

Equation (3.2.7) represents the following set of equations:

$$g(0) = f(0)h(0)$$

$$g(1) = f(1)h(0) + h(1)f(0)$$

$$\vdots \tag{3.3.105}$$

where we have assumed $f(n)$ and $h(n)$ are zero for $n < 0$. We define an H-matrix $N \times (N+M)$ given by

$$H = \begin{bmatrix} h(0) & 0 & 0 & \cdots & 0 \\ h(1) & h(0) & 0 & \cdots & 0 \\ h(2) & h(1) & h(0) & \cdots & 0 \\ \vdots & \vdots & \vdots & & \vdots \\ h(M) & h(M-1) & h(M-2) & \cdots & 0 \\ 0 & h(M) & h(M-1) & \cdots & 0 \\ \vdots & \vdots & \vdots & & \vdots \\ 0 & 0 & 0 & \cdots & h(M) \end{bmatrix} \quad . \tag{3.3.106}$$

It is assumed that the impulse response is time limited and thus M denotes the

---

[4] See Appendix A for an elementary discussion of the matrix properties needed in this book.

extent of $h(n)$ beyond which it is zero. We note that (3.3.105) can be written as

$$G = HF \quad . \tag{3.3.107}$$

For this case H is a lower triangular and Toeplitz matrix. Thus we need to know only the first column, having M elements. Also

$$N' = N + M \quad . \tag{3.3.108}$$

In many cases $M \ll N$. For this case H is a banded Toeplitz matrix, as shown below. We shall discuss the case of linear variant systems where the convolution theorem does not apply in a later section. It will be shown that the H-matrix is not Toeplitz for that case. Thus the Toeplitz property is directly related to the invariance property of the linear system.

Let us consider the important example of a tapped delay line, as shown in Fig. 3.38a. The tap weights are labeled by W_i as shown. The H-matrix for this case is banded with p elements, where p is the number of taps:

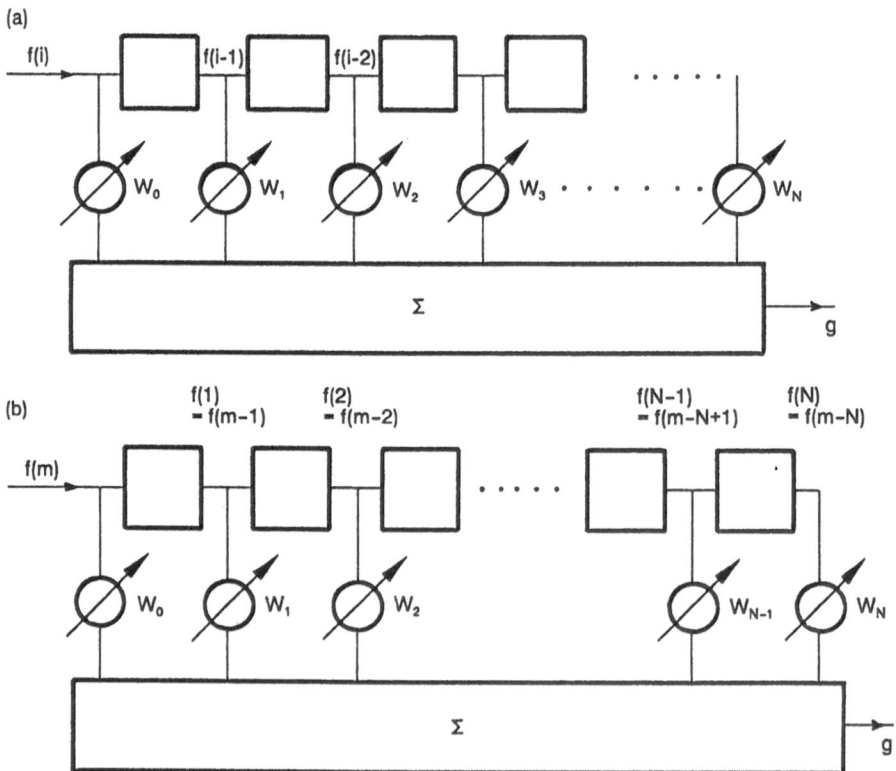

Fig. 3.38a,b. A tapped delay line with two ways of labeling the tap weights

$$H = \begin{vmatrix} W_0 & 0 & 0 & \cdots \\ W_1 & W_0 & 0 & \\ W_2 & W_1 & W_0 & \\ \vdots & \vdots & \vdots & \\ W_p & W_{p-1} & W_{p-2} & \\ 0 & W_p & W_{p-1} & \\ 0 & 0 & W_p & \\ \vdots & \vdots & & \\ 0 & 0 & 0 & \end{vmatrix} \quad . \tag{3.3.109}$$

Excluding the initial p elements of the G matrix, we see that one can write

$$g(i) = \sum_{k=0}^{p} W_k f(i - k) \quad . \tag{3.3.110}$$

It is interesting to note that all banded H-matrices yield an equation like (3.3.110). We shall see in Sect. 3.5 on adaptive filters that the tap weights can also be changed in an iterative fashion. For that case, the H-matrix is no longer Toeplitz, However, if we redefine W and f at the mth iteration as

$$W(m) = \begin{pmatrix} W_1 \\ W_2 \\ W_3 \\ \vdots \\ W_p \end{pmatrix} = \begin{pmatrix} W_1(m) \\ W_2(m) \\ W_3(m) \\ \vdots \\ W_p(m) \end{pmatrix}$$

and

$$f(m) = F = \begin{pmatrix} f_1 \\ f_2 \\ \vdots \\ f_p \end{pmatrix} = \begin{pmatrix} f(m - 1) \\ f(m - 2) \\ \vdots \\ f(m - p) \end{pmatrix} \quad , \tag{3.3.111}$$

(3.3.110) can be rewritten as

$$Y(m) = Y = W^{\mathrm{T}} F \quad . \tag{3.3.112}$$

Note that, in the F matrix, we have rearranged the labeling as shown in Fig. 3.38b, so that we have $g(i) = \sum W_k f_k$ rather than (3.3.110).

For stochastic signals one can also define the following elements of the correlation matrix:

$$R_{ij} = E[X_i X_j] \quad . \tag{3.3.113}$$

For example, the correlation matrix for white noise is diagonal:

$$R_{ij} \text{ (white noise)} = \begin{vmatrix} I_0 & & \\ & \ddots & \\ & & I_0 \\ & & & \ddots \end{vmatrix} \quad . \tag{3.3.114}$$

The correlation matrix of a colored noise with coherence time τ_0 given by

$$\tau_0 = p\Delta t \quad , \tag{3.3.115}$$

where Δt is the sampling time, can be written as

$$R_{ij} \text{ (colored)} = \begin{vmatrix} I_0 & I_1 & I_2 \cdots & I_p \cdots \\ I_1 & & & \\ \vdots & & & \\ I_p & & & \\ \vdots & & & \end{vmatrix} . \tag{3.3.116}$$

This is a banded matrix with bandwidth $2p - 1$. Note that for a coherent signal the correlation matrix is a full matrix with few nonzero elements.

3.3.7 Matched Filters: Discrete Case

It is of interest to consider the matched filter for the discrete case. We shall use the matrix notation as a prelude to the adaptive filter to be discussed in the next section also using matrix notation. The matched filter problem is defined in Fig. 3.38b, where the filter with the delta response $W(n)$ is to be optimized for an input X. The quantity $W(n)$ is given by

$$W(n) = W_0\delta(n) + W_1\delta(n-1) + W_2\delta(n-2) + \dots W_N\delta(n-N) \tag{3.3.117}$$

where W_i are the weights of the discrete time system filter. As discussed in the last section, if we represent $W(n)$ by a $N \times 1$ matrix denoted by W, we can write

$$Y = W^T X \quad , \quad \text{where} \quad X = f + n \quad . \tag{3.3.118}$$

In matrix notation the signal power S for this case is given by

$$S = (W^T f)^2 \quad . \tag{3.3.119}$$

The noise power N is given by

$$\begin{aligned} N &= E\{(W^T n)(W^T n)^T\} \\ &= E\{W^T nn^T W\} \quad . \end{aligned} \tag{3.3.120}$$

The last equation follows because $(W^T)^T = W$.

Defining the correlation matrix for the noise process n as R_{nn}, we note

$$R_{nn} = nn^T \quad . \tag{3.3.121}$$

Thus

$$N = W^T R_{nn} W \quad . \tag{3.3.122}$$

The signal-to-noise ratio S/N to be maximized is given by

$$\frac{S}{N} = \frac{(W^T f)^2}{W^T R_{nn} W} \quad . \tag{3.3.123}$$

Differentiating with respect to W we obtain

$$\frac{\partial}{\partial W}\left(\frac{S}{N}\right) = \frac{2(W^T f)f}{W^T R_{nn} W} - \frac{(W^T f)^2}{(W^T R_{nn} W)^2}[W^T R_{nn} - R_{nn} W] \quad (3.3.124)$$

Equating (3.3.124) to zero, we obtain for maximum signal-to-noise ratio that

$$W = \left(\frac{W^T R_{nn} W}{W^T f}\right) R_{nn}^{-1} f \quad . \quad (3.3.125)$$

The quantity in parentheses in (3.3.125) is a scalar and can be normalized to unity. Then the matched filter weight matrix W is given by

$$W = R_{nn}^{-1} f. \quad (3.3.126)$$

It is of interest to note that, if the signal f is the only input, the matched filter output is given by

$$Y_f = W^T f = (R_{nn}^{-1} f)^T f \quad . \quad (3.3.127)$$

If the noise process is white, then

$$R_{nn} = N_0 I \quad , \quad (3.3.128)$$

where N_0 is a constant and I is the identity matrix. For this case

$$(R_{nn}^{-1})^T = R_{nn}^{-1} = \frac{1}{N_0} I \quad (3.3.129)$$

and (3.3.127) can be written as

$$Y_f = \frac{1}{N_0} I f^T f \quad . \quad (3.3.130)$$

Equation (3.3.128) is the discrete correlation equation. If R_{nn} is not diagonal as in (3.3.116) we can always factorize R_{nn} as follows:

$$R_{nn} = Q \Lambda Q^T \quad (3.3.131)$$

where Λ is a diagonal matrix and Q is the matrix associated with the eigenvectors. Equation (3.3.131) can be written as

$$\begin{aligned} R_{nn} &= (Q \Lambda^{+1/2} \Lambda^{1/2} Q^T) \\ &= (Q \Lambda^{1/2})(\Lambda^{1/2} Q)^T \\ &= R R^T \quad \text{where} \end{aligned} \quad (3.3.132)$$

$$R = Q \Lambda^{1/2} \quad \text{and} \quad (3.3.133)$$

$$R^T = (Q \Lambda^{1/2})^T = \Lambda^{1/2} Q^T \quad . \quad (3.3.134)$$

The output signal, y, for this case with input f and n will be given by

$$Y = (R_{nn}^{-1}f)^{\mathrm{T}}(f+n)$$
$$= [(RR^{\mathrm{T}})^{-1}f]^{\mathrm{T}}(f+n)$$
$$= (R^{-1}f)^{\mathrm{T}}[R^{-1}(f+n)] \quad . \tag{3.3.135}$$

Thus $R^{-1}(f+n)$ is equivalent to prewhitening and it is very similar to the case discussed in connection with analog signals. This is obvious if one notes that for the analog case

$$Y(\omega) = H(\omega)[F(\omega) + n(\omega)]$$
$$= \frac{F^*(\omega)e^{-j\omega t_0}}{N(\omega)}[F(\omega) + n(\omega)]$$
$$= \frac{F^*(\omega)e^{-j\omega t_0}}{\sqrt{N(\omega)}} \frac{F(\omega) + n(\omega)}{\sqrt{N(\omega)}} \quad . \tag{3.3.136}$$

The similarity between (3.3.135) and (3.3.136) is then apparent.

3.4 Filters

In signal processing, we very often need a device, known as a filter, whose frequency response or impulse response is prescribed. In this section, we shall discuss how a filter with a prescribed frequency can be obtained by means of various implementations, which can be subdivided into the following categories:

1. Conventional filters using R, L, C and including active elements, i.e., amplifiers.
2. Transversal filters using a tapped delay line or an equivalent structure.
3. Recursive filters.

Another way of subdividing them is into passive, active and digital, etc.

A typical transversal filter is shown in Fig. 3.39. Here the unit delay time T, can be obtained by optical fibers, SAW propagation and interdigital transducers or CCDs with clock time related to T or a shift register. Details of the exact implementations will be discussed in Chap. 4. We are interested in certain fundamental properties that are common to all of them. For a finite N-element transversal filter, the impulse response is given by

$$h(t) = \sum_{n=0}^{N} a_n \delta(t - nT) \quad , \tag{3.4.1}$$

where a_n is the complex weight for the nth tap. As the frequency response is the Fourier transform of $h(t)$, we have

$$H(\omega) = \sum_{n=0}^{N} a_n e^{jn\omega t} \quad . \tag{3.4.2}$$

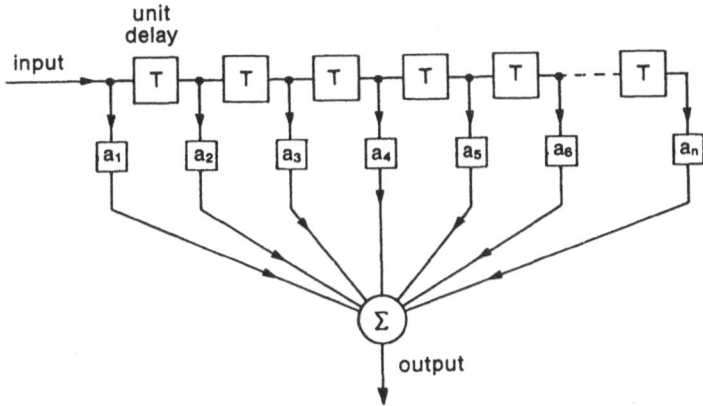

Fig. 3.39. Schematic diagram of a transversal filter

Sometimes, however, a prescribed $H(\omega)$ is given and we need to design the transversal filter by choosing the value of N and proper values of a_n. In other words, $H(\omega)$ is given and we take a Fourier transform to obtain the impulse response. Actually, for the tapped delay line, the discrete Fourier series is a better approach and we can use the fast Fourier transform.

The sampling rate or the delay time will be determined by the Nyquist rate, and the number of taps N will be determined by the acceptable distortion in the actual frequency response compared to that desired. To clarify this point, let us consider the following simple example.

We are interested in designing the bandpass filter whose frequency response is shown in Fig. 3.40. As the response desired is ideally rectangular, we know that the impulse response is a sinc function, which extends from $-\infty$ to $+\infty$. However, we cannot make a physical tapped delay line whose length is infinite. Hence we decide on a particular length denoted by the weighting function $W(t)$ which is also rectangular. The implemented impulse response is thus given as a finite impulse response and its actual frequency response is obtained by convolving the Fourier transform of the weighting function (also a sinc function) and the desired frequency response. The implemented frequency response thus has ripples in both the pass band and the rejection band, and the transition from pass band to rejection band is not sharp. The ripples or oscillations in the resulting frequency response are a well-known phenomenon, called Gibbs' oscillations, which occur whenever a sharp transition is involved. These ripples can be reduced at the expense of the sharpness of the transition by using weighting functions $w(t)$ that are not rectangular. This is illustrated in Fig. 3.41. We recall that if $w(t)$ extends to infinity, its Fourier transform becomes a delta function and the implemented filter becomes the desired filter.

Over the years, many weighting functions have been used. Some of these are discussed below:

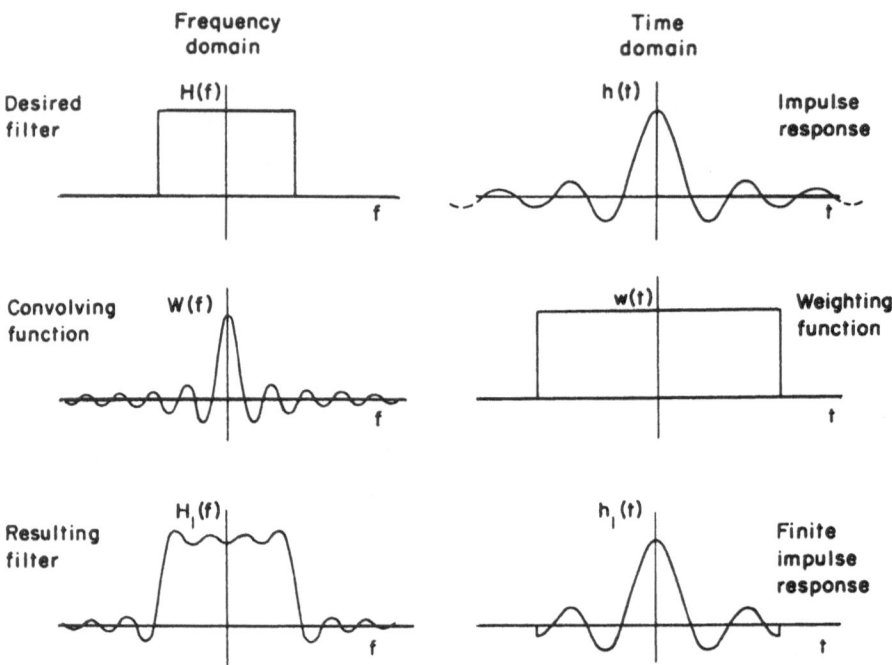

Fig. 3.40. Filter design using truncations of an infinite impulse response. (From [3.6])

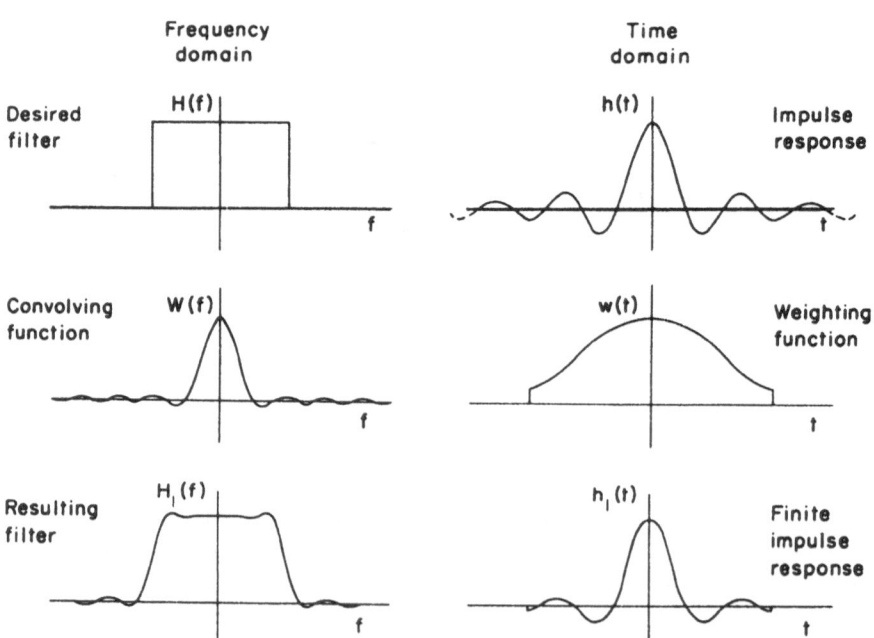

Fig. 3.41. Filter design using weighting (window) functions to reduce Gibbs' oscillations. (From [3.6])

i) Fejer weighting: For a symmetric array of $2N + 1$ elements, each element is weighted as

$$w(n) = \frac{N + 1 - |n|}{N + 1} \quad . \tag{3.4.3}$$

This weighting is a linear taper which decreases to zero at the end. It has been used in antenna arrays and it eliminates sidelobes but almost doubles the transition bandwidth compared to the unweighted design.

ii) Cesaro in summation weighting:

$$w(n, m) = \begin{cases} 1 & \text{for } n = 0, \\ \dfrac{N(N - 1) \dots (N - |n| + 1)}{(N + m)(N + m - 1) \dots (N + m - |n| + 1)} & \text{for } n \neq 0. \end{cases} \tag{3.4.4}$$

The m signifies the degree of weighting, with $m = 0$ signifying no weighting and $m = 1$ signifying the weighting equivalent to that of Fejer.

iii) Lanczos weighting:

$$w(n) = \frac{\sin\left[n\pi/(N + 1)\right]}{n\pi/(N + 1)} \quad . \tag{3.4.5}$$

This reduces the sidelobes in the frequency response to $-38.5\,\text{dB}$ compared to $-21\,\text{dB}$ for the unweighted one.

iv) Hamming weighting: For a delay line length of 2τ, the Hamming weighting is given by

$$w(t) = \begin{cases} 0.54 + 0.46 \cos\left(\dfrac{\pi t}{\tau}\right) & \text{for } |t| < \tau \quad , \\ 0 & \text{for } |t| > \tau \quad . \end{cases} \tag{3.4.6}$$

The first lobe of the Fourier transform of the Hamming function contains 96 %– 99 % of its energy.

v) Dolph-Chebyshev weighting: This window specifies the maximum sidelobes acceptable and then designs the best solution. It is given in terms of the Fourier transform coefficients $W(k)$ of the window function $W(n)$:

$$W(k) = (-1)^k \frac{\cos\left\{N \cos^{-1}\left[\beta \cos\left(\pi k/N\right)\right]\right\}}{\cosh\left[N \cosh^{-1}(\beta)\right]} \quad ,$$

$$0 \leq k \leq N - 1 \quad , \tag{3.4.7}$$

where

$$\beta = \cosh\left[\frac{1}{N} \cosh^{-1}(10^\alpha)\right] \quad \text{and}$$

$$\cos^{-1}(x) = \begin{cases} \frac{\pi}{2} - \tan^{-1}(x/\sqrt{1.0 - x^2}) & |x| \leq 1.0 \\ \ln(x + \sqrt{x - 1.0}) & |x| \geq 0 \end{cases}.$$

The value of α determines the sidelobe level. For $\alpha = 3$, the sidelobes will be 60 dB below the main lobe.

vi) *Prolate spheroidal weighting – optimum window:* For this case the weighting functions are prolate spheroidal functions which depend on the time-bandwidth product, $\sigma\tau$, and are the solution of the following integral equation with λ the maximum eigenvalue:

$$\int_{-\tau}^{+\tau} \theta(x) \frac{\sin[\sigma(t-x)]}{\pi(t-x)} dx = \lambda\theta(t) \quad \text{or} \tag{3.4.8}$$

$$\theta(t)p_\tau(t) * \frac{\sin \sigma t}{\pi t} = \lambda\theta(t) \quad.$$

In the Fourier domain the equation becomes

$$\mathcal{F}[\theta(t)p_\tau(t)]p_\sigma(\omega) = \lambda\phi(\omega) \quad. \tag{3.4.9}$$

It can be shown that these functions are optimum in the sense that, for a time-limited function, they have most of the energy within a band relative to the total energy.

vii) *Kaiser weighting:* Kaiser weighting is an approximation to prolate spheroidal weighting. It is given by

$$w(t) = \begin{cases} \dfrac{I_0(\omega_a\sqrt{\tau^2 - t^2})}{I_0(\omega_a\tau)} & \text{for } |t| < \tau \\ 0 & \text{for } |t| > \tau \end{cases} \tag{3.4.10}$$

where I_0 is the modified Bessel function of the first kind and zeroth order. It is found that the Kaiser function is very similar to the prolate spheroidal functions. The quantity ω_a specifies the trade-off between transition bandwidth and ripple. Some numerical values are shown in Table 3.2.

viii) *Bartlett window/triangle window:* The time function is a triangle

$$w(t) = q_\tau(t) \quad. \tag{3.4.11}$$

ix) *Tukey window:* The weighting function is a raised cosine given by

$$w(t) = \frac{1}{2}\left(1 + \cos\frac{\pi t}{\tau}\right)p_\tau(t) \quad. \tag{3.4.12}$$

The Hamming window is obtained by adding a constant to a raised cosine and normalizing.

Table 3.2. Transition width (TW) and ripple for Bessel weighting function of time duration 2τ. (From [3.6])

Parameter $\omega_0\tau$	TW at τ	Filter response	
		Peak-to-peak ripple at band edge [dB]	Peak out-of-band sidelobe [dB]
0	0.50	1.4	−21
1	0.56	1.0	−24
2	0.80	0.52	−30
3	1.03	0.21	−37
4	1.32	0.076	−45
5	1.64	0.026	−54
6	2.00	0.0094	−62
7	2.23	0.0040	−71

A generalized Tukey window is often defined for the discrete signals as follows:

$$\omega(n) = \begin{cases} 1\cdot 0, & 0 \leq |n| \leq \alpha\frac{N}{2}, \\ \frac{1}{2}\left\{1\cdot 0 + \cos\left[\pi\frac{n-\alpha N/2}{2(1-\alpha)N/2}\right]\right\}, & \alpha\frac{N}{2} \leq |n| \leq \frac{N}{2}, \end{cases}$$

where α is a parameter less than 1.

x) Specified zero crossing − optimum window: This is given by

$$w(t) = k(1 + 2\cos at)p_\tau(t) \quad, \tag{3.4.13}$$

$$\alpha = \frac{-2\sin a\tau}{2a\tau + \sin 2a\tau} \quad. \tag{3.4.14}$$

The Hamming window is a special case of this.

xi) Minimum amplitude moment:

$$w(t) = \frac{1}{\tau}\left|\sin\left(\frac{\pi t}{\tau}\right)\right| + \left(1 - \frac{|t|}{\tau}\right)\cos\left(\frac{\pi t}{\tau}\right)p_\tau(t) \quad. \tag{3.4.15}$$

xii) Parzen:

$$w(t) = \frac{3}{\tau}q_{\tau/2}(t)*q_{\tau/2}(t) \quad. \tag{3.4.16}$$

xiii) Hanning window/$\cos^\alpha$ window: The Hanning window is given by

$$w(n) = 0.5 + 0.5\cos\left(\frac{2n}{N}\pi\right) \quad,$$

$$n = -\frac{N}{2}, \ldots, -1, 0, 1, \ldots, \frac{N}{2} - 1 \quad, \tag{3.4.17}$$

261

for $\alpha = 1, 2, 3, 4$, etc. The $\cos^\alpha$ window is given by

$$w(n) = \cos^\alpha\left(\frac{n}{N}\pi\right) \quad , \quad n = -\frac{N}{2}, \ldots, 1-, 0, 1, \ldots, \frac{N}{2} \quad , \qquad (3.4.18)$$

where the most common values of α are the integers $1 - 4$. Note that $\alpha = 2$ corresponds to the Hanning window.

xiv) Blackman window/Blackman-Harris window: This is given by

$$w(n) = \sum_{m=0}^{N/2} a_m \cos\left(\frac{2\pi}{N}mn\right) \quad . \qquad (3.4.19)$$

Note that, for $a_0 \neq 0$, $a_1 \neq 0$ and $a_m = 0$, where m is 2, 3, $\ldots$, we obtain either a Hanning or a Hamming window. This window is therefore a more general one. The a_m values are chosen such that

$$\sum a_m = 1 \qquad (3.4.20)$$

and from $m - 1$ equations obtained by using the fact that the side lobes will be zero at specified points in the frequency axis. A particular case with three nonzero coefficients, generally referred to as a Blackman window, is given by

$$w(n) = 0.42 + 0.50 \cos\left(\frac{2\pi}{N}n\right) + 0.08 \cos\left(\frac{2\pi}{N}2n\right) \quad . \qquad (3.4.21)$$

The cases with more than three nonzero coefficients are referred to as Blackman-Harris windows.

xv) Riemann window: This is the central lobe of the sinc function given by

$$w(n) = \frac{\sin(2\pi n/N)}{(2\pi n/N)} \quad , \quad 0 \leq |n| \leq \frac{N}{2} \quad . \qquad (3.4.22)$$

xvi) Delavalle-Poussin window:

$$w(n) = \begin{cases} 1.0 - 6\left(\frac{n}{N/2}\right)^2\left[1.0 - \frac{|n|}{N/2}\right] & \text{for } 0 \leq |n| \leq \frac{N}{4} \quad , \\ 2\left[1.0 - \frac{|n|^3}{N/2}\right] & \text{for } \frac{N}{4} \leq |n| \leq \frac{N}{2} \quad . \end{cases} \qquad (3.4.23)$$

xvii) Bohman window:

$$w(n) = \left(1.0 - \frac{|n|}{N/2}\right) + \cos\left(\pi\frac{|n|}{N/2}\right) + \frac{1}{\pi}\sin\left(_{\shortparallel}\frac{|n|}{N/2}\right) \quad ,$$

$$0 \leq |n| \leq \frac{N}{2} \quad . \qquad (3.4.24)$$

xviii) Poisson window:

$$w(n) = \exp\left(-\alpha \frac{|n|}{N/2}\right) \quad , \quad 0 \le |n| \le \frac{N}{2} \quad , \tag{3.4.25}$$

where α is a parameter with values 2, 3, 4, etc.

xix) Hanning-Poisson window:

$$w(n) = 0.5\left[1.0 + \cos\left(\frac{\pi n}{N/2}\right)\right]\exp\left(-\alpha \frac{|n|}{N/2}\right) \quad . \tag{3.4.26}$$

xx) Gaussian window:

$$w(n) = \exp\left[-\frac{1}{2}\left(\alpha \frac{n}{N/2}\right)^2\right] \quad . \tag{3.4.27}$$

xxi) Cauchy window:

$$w(n) = \frac{1}{1.0 + [\alpha \frac{n}{N/2}]^2} \quad , \tag{3.4.28}$$

where $\alpha = 3, 4, 5$, etc.

xxii) Riesz window:

$$w(n) = 1.0 - \left|\frac{n}{N/2}\right|^2 \quad . \tag{3.4.29}$$

This is the simplest polynomial window.

xxiii) Barcilon-Temes window: This is defined in terms of Fourier coefficients $W(k)$ of the window function $W(n)$ and is obtained by minimizing the weighted sidelobe energy, i.e., by minimizing

$$\int_W^\infty |H(\omega)|^2 \frac{\omega}{\sqrt{\omega^2 - W^2}} d\omega \quad .$$

The window is given by

$$W(k) = (-1)^k \frac{A \cos[y(k)] + B\{\frac{y(k)}{C} \sin[y(k)]\}}{[C + AB]\{[\frac{y(k)}{C}]^2 + 1.0\}} \quad , \tag{3.4.30}$$

where

$$A = \sinh C = \sqrt{10^{2\alpha} - 1} \quad ,$$
$$B = \cosh C = 10^\alpha \quad ,$$
$$C = \cosh^{-1} 10^\alpha \quad ,$$
$$\beta = \cosh\frac{C}{N} \quad , \quad \text{and}$$

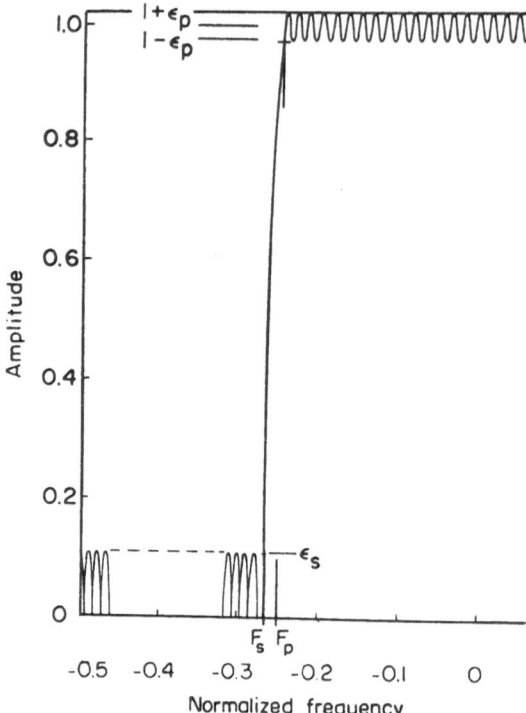

Fig. 3.42. Notation for stopband edge (F_s), passband edge (F_p), and ripples ε_p and ε_s. Frequency scale is $(f - f_0)/2f_0$ where f_0 is the center frequency. Only one-half of the frequency response is shown. (From [3.6])

$$y(k) = N \cos^{-1}\left(\beta \cos \pi \frac{k}{N}\right) \quad .$$

Here α is a parameter which determines the sidelobe level as in the Dolph-Chebyshev case.

There are some interesting rules of thumb that are rather helpful for filter design. For a filter of $2N + 1$ elements

$$N\left(\frac{\text{transition width}}{\text{center frequency}}\right) \simeq 2 \tag{3.4.31}$$

for ε_p and $\varepsilon_s \sim 0.01$. Here ε_p and ε_s are the maximum ripple magnitudes in the pass band and stop band, respectively, and are shown in Fig. 3.42. Another rule is given by

$$2N(F_s - F_p) \approx 0.55(1 + \log_{10} \varepsilon_p)\log_{10} \varepsilon_s$$
$$+ 0.9 \log_{10} \varepsilon_p + 2.7 \quad , \tag{3.4.32}$$

where F_p and F_s are the passband and stopband edges, respectively, see Fig. 3.42. Some numerical results are given in Table 3.3.

Table 3.3. Ripple and transition width for equiripple filter with $(2N + 1)$ elements. (From [3.6])

Peak-to-peak passband ripple [dB]	Stopband ripple [dB]	$N \times \left(\dfrac{\text{Transition width}}{\text{Center frequency}} \right)$	Approximate number of elements $(2N + 1)$ for $\left(TW/f_0 \right) = 0.04$
0.55	−40	1.9	96
0.55	−50	2.0	103
0.55	−60	2.2	110
0.17	−40	2.0	101
0.17	−50	2.2	115
0.17	−60	2.5	129
0.055	−40	2.1	106
0.055	−50	2.5	127
0.055	−60	2.9	147

The performances of the individual windows are shown in Table 3.4 where the highest sidelobe levels and sidelobe fall-off in dB/octave are given. The table also includes coherent gain, equivalent noise bandwidth, 3-dB bandwidth, scallop loss, worst case process loss and 6.0 dB bandwidth. These are defined and discussed below.

Equivalent noise bandwidth (ENBW) is given by

$$\text{ENBW} = \frac{\sum\limits_n w^2(n)}{\left[\sum\limits_n w(n) \right]^2} \ . \tag{3.4.33}$$

This represents the accumulated, normalized noise power for the detection problem with white noise when the window is considered as a filter. The coherent gain or processing gain (PG) is given by

$$\begin{aligned} \text{PG} &= \frac{\text{Output signal-to-noise ratio}}{\text{Input signal-to-noise ratio}} \\ &= \frac{\left[\sum\limits_n w(n) \right]^2}{\sum\limits_n w^2(n)} \ . \end{aligned} \tag{3.4.34}$$

This is the inverse of the equivalent noise bandwidth. Processing gain refers to the concept that DFT is a bank of matched filters and to the effect of the window in reducing the matched filtered output.

The 3.0 dB and 6.0 dB bandwidths refer to the bandwidths of the main lobe in terms of frequency bins. They are also related to the ability of the windowed output to distinguish two nearby frequencies of equal amplitude.

Table 3.4. Windows and figures of merit. (From [3.2]; BW: bandwidth)

Window		Highest side-lobe level [dB]	Side-lobe fall-off [dB/Oct]	Coherent gain	Equiv. noise BW [bins]	3.0-dB BW [bins]	Scallop loss [dB]	Worst case process loss [dB]	6.0-dB BW [bins]
Rectangle		−13	−6	1.00	1.00	0.89	3.92	3.92	1.21
Triangle		−27	−12	0.50	1.33	1.28	1.82	3.07	1.78
cos $\alpha \lvert x \rvert$	$\alpha = 1.0$	−23	−12	0.64	1.23	1.20	2.10	3.01	1.65
Hanning	$\alpha = 2.0$	−32	−18	0.50	1.50	1.44	1.42	3.18	2.00
	$\alpha = 3.0$	−39	−24	0.42	1.73	1.66	1.08	3.47	2.32
	$\alpha = 4.0$	−47	−30	0.38	1.94	1.86	0.86	3.75	2.59
Hamming		−43	−6	0.54	1.36	1.30	1.78	3.10	1.81
Riesz		−21	−12	0.67	1.20	1.16	2.22	3.01	1.59
Riemann		−26	−12	0.59	1.30	1.26	1.89	3.03	1.74
De la Valle-Poussin		−53	−24	0.38	1.92	1.82	0.90	3.72	2.55
Tukey	$\alpha = 0.25$	−14	−18	0.86	1.10	1.01	2.96	3.39	1.38
	$\alpha = 0.50$	−15	−18	0.75	1.22	1.15	2.24	3.11	1.57
	$\alpha = 0.75$	−19	−18	0.63	1.36	1.31	1.73	3.07	1.80
Bohman		−46	−24	0.41	1.79	1.71	1.02	3.54	2.38
Poisson	$\alpha = 2.0$	−19	6	0.44	1.30	1.21	2.09	3.23	1.09
	$\alpha = 3.0$	−24	−6	0.32	1.65	1.45	1.46	3.64	2.08
	$\alpha = 4.0$	−31	6	0.25	2.08	1.75	1.03	4.21	2.50
Hanning-	$\alpha = 0.5$	−35	−18	0.43	1.61	1.54	1.26	3.33	2.14
Poisson	$\alpha = 1.0$	−39	−18	0.38	1.73	1.64	1.11	3.50	2.30
	$\alpha = 2.0$	none	−18	0.29	2.02	1.87	0.87	3.94	2.65
Cauchy	$\alpha = 3.0$	−31	−6	0.42	1.48	1.34	1.71	3.40	1.90
	$\alpha = 4.0$	−35	−6	0.33	1.76	1.50	1.36	3.83	2.20
	$\alpha = 5.0$	−30	−6	0.28	2.06	1.68	1.13	4.28	2.53
Gaussian	$\alpha = 2.5$	−42	−6	0.51	1.39	1.33	1.69	3.14	1.86
	$\alpha = 3.0$	−55	−6	0.43	1.64	1.55	1.29	3.40	2.18
	$\alpha = 3.5$	−69	−6	0.37	1.90	1.79	0.94	3.73	2.52
Dolph-	$\alpha = 2.5$	−50	0	0.53	1.39	1.33	1.70	3.12	1.85
Chebyshev	$\alpha = 3.0$	−60	0	0.48	1.51	1.44	1.44	3.23	2.01
	$\alpha = 3.5$	−67	0	0.45	1.62	1.55	1.25	3.35	2.17
	$\alpha = 4.0$	−80	0	0.42	1.73	1.65	1.10	3.48	2.31
Kaiser-	$\omega_a \tau / \pi = 2.0$	−46	−6	0.49	1.50	1.43	1.46	3.20	1.99
Bessel	$\omega_a \tau / \pi = 2.5$	−57	−6	0.44	1.65	1.57	1.20	3.38	2.20
	$\omega_a \tau / \pi = 3.0$	−69	−6	0.40	1.80	1.71	1.02	3.56	2.39
	$\omega_a \tau / \pi = 3.5$	−82	−6	0.37	1.93	1.83	0.89	3.74	2.57

Table 3.4. (Continued)

Barcilon-	$\alpha = 3.0$	−53	−6	0.47	1.56	1.49	1.34	3.27	2.07
Temes	$\alpha = 3.5$	−58	−6	0.43	1.67	1.59	1.18	3.40	2.23
	$\alpha = 4.0$	−68	−6	0.41	1.77	1.69	1.05	3.52	2.36
Exact Blackman		−51	−6	0.46	1.57	1.52	1.33	3.29	2.13
Blackman		−58	−18	0.42	1.73	1.68	1.10	3.47	2.35
Minimum 3-sample Blackman-Harris		−67	−6	0.42	1.71	1.66	1.13	3.45	1.81
Minimum 4-sample Blackman-Harris		−92	−6	0.36	2.00	1.90	0.83	3.85	2.72
61 dB 3-sample Blackman-Harris		−61	−6	0.45	1.61	1.56	1.27	3.34	2.19
74 dB 4-sample Blackman-Harris		−74	−6	0.40	1.79	1.74	1.03	3.56	2.44

Scalloping loss is given by

$$\frac{|\sum_n w(n)e^{-j\pi n/N}|}{\sum_n w(n)} = \frac{|W(\pi/N)|}{W(0)}.$$ (3.4.35)

This represents the additional loss in processing gain when the signal frequency is midway between two bin frequencies (i.e., multiples of $1/N$).

The worst case processing loss is the sum of maximum scalloping loss and of processing loss.

3.5 Adaptive Filters

Adaptive filters are a class of filters whose impulse response is not fixed but changes or adapts according to some desired criterion. For example, consider the tapped delay line with weights W_i, as shown in Fig. 3.38. For an input X, the output is given by

Continuous case $\qquad\qquad Y(t) = W(t)*X(t),$ (3.5.1)

Discrete case $\qquad\qquad Y_i = \sum W_k X_{i-k},$ (3.5.2)

Matrix notation $\qquad\qquad Y = W^T X.$ (3.5.3)

As discussed in Sect. 3.3.6, for the case of matrix notation, the W_i's are rearranged. $W(t)$ denotes the impulse response of the tapped delay line in the continuous case. The tap weights or the impulse response of the filter is not fixed. The tap weights are changed such that the output Y becomes equal to D, the desired output. To achieve this, we need the arrangement shown in Fig. 3.43,

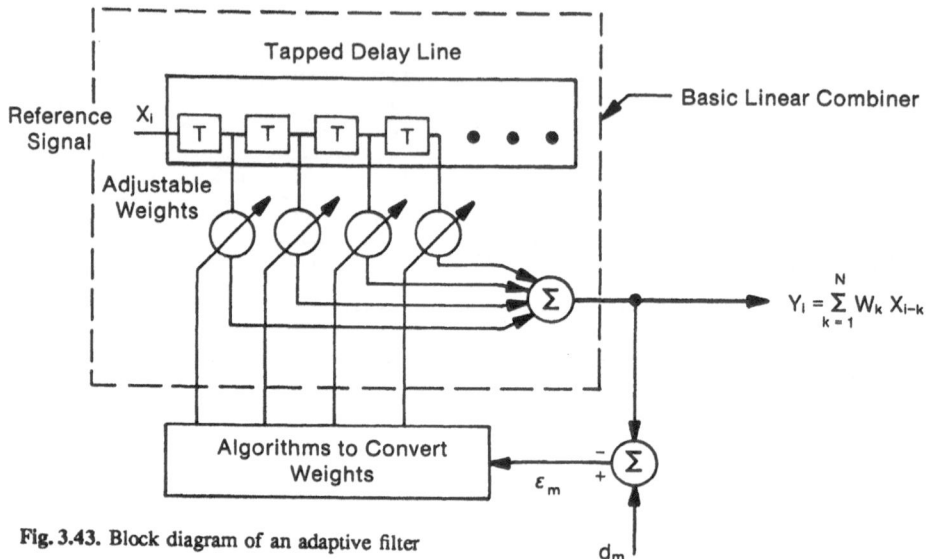

Fig. 3.43. Block diagram of an adaptive filter

which is an adaptive filter. This filter takes the output Y, calculates the error ε from the desired response D, and then, using an algorithm to be discussed in Sect. 3.5.1, tries to make the error tend to zero by adjusting the weights W. In general, a least mean square error is used as the criterion for achieving this goal. This is considered prior to the proper discussion of the adaptive filter.

3.5.1 Linear Mean Squares Estimation

Given a random variable x, we want to estimate another random variable y as a linear function of x. We also want to use the least mean square error criterion. If we denote the estimate of y by $\hat{y}$, then

$$\hat{y} = \text{estimate of } y = ax \quad , \tag{3.5.4}$$

where a is the unknown constant to be determined. The error of the estimate ε is given by

$$\varepsilon = y - \hat{y} = y - ax \quad . \tag{3.5.5}$$

The mean square error is

$$\overline{\varepsilon^2} = E\{(y - ax)^2\} \quad . \tag{3.5.6}$$

To obtain the minimum of $\overline{\varepsilon^2}$ we differentiate $\overline{\varepsilon^2}$ with respect to a and set the result to zero, namely

$$\frac{\partial \overline{\varepsilon^2}}{\partial a} = E\{2(y - ax)x\} = 0 \quad \text{or}$$

$$E\{\varepsilon x\} = 0 \quad . \tag{3.5.7}$$

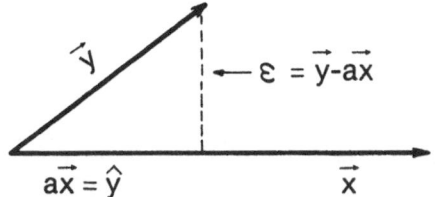

Fig. 3.44. The orthogonality principle

Also

$$a = \frac{E\{yx\}}{E\{x^2\}} \quad .$$ (3.5.8)

Equation (3.5.7) is known as the orthogonality principle and is shown graphically in Fig. 3.44. It implies that the error ε and the given random variable are orthogonal to each other for the best estimate in the least mean squares sense. As $E\{\varepsilon x\}$ also represents cross-correlation between ε and x, it tells us that the best estimate is when ε and x are not correlated at all.

We can easily extend these concepts to a more complex case for which y is a function of n given stochastic variables denoted by $x_1, \ldots, x_i, \ldots, x_n$:

$$y = f(x_1, x_2, \ldots, x_i, \ldots, x_n) \quad .$$ (3.5.9)

Let us denote the estimate $\hat{y}$ of y by

$$\hat{Y} = \sum W_i x_i \quad .$$ (3.5.10)

Here we have used W_i rather than a for the estimation constants. The main reason for this choice is that in the next section we want to apply these results to an adaptive filter using a tapped delay line. For that case, W_i is the ith tap weight (Fig. 3.41). In the present case ε is given by

$$\varepsilon = Y - \hat{Y} \quad .$$ (3.5.11)

Thus

$$\overline{\varepsilon^2} = E[(Y - \hat{Y})^2] = E[(y - \sum W_i x_i)^2] \quad .$$ (3.5.12)

For the least mean square estimation, we need

$$\frac{\partial \overline{\varepsilon^2}}{\partial W_i} = 0 \quad \text{or}$$

$$E\left[-2Y x_i + 2x_i \sum_k W_k X_k\right] = 0 \quad .$$ (3.5.13)

If we define the correlation matrix R_{ij} and the steering vector P_i as

$$R_{ij} = E\{x_i x_j\} \quad \text{and}$$ (3.5.14)

$$P_i = E\{y x_i\}$$ (3.5.15)

then (3.5.13) can be written as separate linear equations given by

$$R_{11}W_1 + R_{12}W_2 + \ldots + R_{1j}W_j + \ldots + R_{1n}W_n = P_1$$
$$R_{21}W_1 + R_{22}W_2 + \ldots \qquad\qquad\qquad\quad = P_2 \quad. \qquad (3.5.16)$$
$$\vdots \qquad\quad \vdots \qquad\qquad\qquad\qquad\qquad \vdots$$

To obtain the optimum values of W_i, we need to solve the above set of linear equations. If we use matrix notation, we can write

$$W = R^{-1}P \quad. \qquad (3.5.17)$$

It is of interest to derive the above equation using matrix notation from the beginning, namely

$$\hat{Y} = W^T X \quad, \qquad (3.5.18)$$

$$\varepsilon = y - \hat{y} = Y - W^T X \quad, \qquad (3.5.19)$$

$$\overline{\varepsilon^2} = E\{(Y - W^T X)^2\} \quad, \quad \text{or} \qquad (3.5.20)$$

$$\overline{\varepsilon^2} = E\{Y^2\} - 2E\{YX\}W + W^T E\{XX^T\}W$$
$$= E\{Y^2\} - 2PW + W^T RW \quad. \qquad (3.5.21)$$

To obtain the least mean squares estimate, we have

$$\nabla_W \overline{\varepsilon^2} = 0 \quad \text{or} \qquad (3.5.22)$$

$$-2P + 2RW = 0 \quad \text{or}$$

$$W = R^{-1}P \quad.$$

We also note that $E\{\varepsilon X\} = 0$, i.e., the orthogonality principle holds. The cross-correlation between the error and the input goes to zero.

Before we discuss the adaptive filter, it is of interest to derive (3.5.17) for the continuous case as shown in Fig. 3.45. The output y is given by

$$y = x * h = \int x(t - \tau)h(\tau)d\tau \quad. \qquad (3.5.23)$$

We pose the problem of optimizing $h(t)$ such that the stochastic signal input $x(t)$ causes the output $y(t)$ to behave like $d(t)$, which is the desired signal in the least mean squares sense. The error for this case is given by

$$\varepsilon(t) = d(t) - y(t) \quad. \qquad (3.5.24)$$

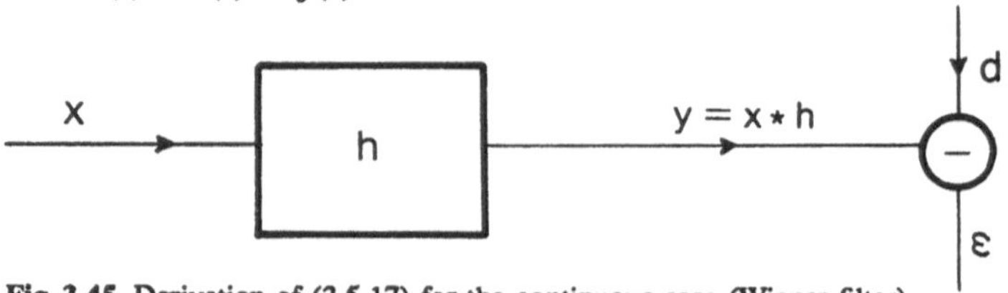

Fig. 3.45. Derivation of (3.5.17) for the continuous case (Wiener filter)

The mean square error $\overline{\varepsilon^2}$ is given by

$$
\begin{aligned}
\overline{\varepsilon^2} &= E\{\varepsilon^2\} \\
&= E\{(d - y)^2\} \\
&= E\{(d - x*h)^2\} \quad .
\end{aligned}
$$
(3.5.25)

From the orthogonality principle of least mean squares, we note that the expected values of ε and x must be zero. Thus

$$ E\{(d - x*h)x\} = 0 \quad \text{or} $$

$$ E\left\{ \left(d(t) - \int_{-\infty}^{+\infty} x(t - \alpha)h(\alpha)d\alpha \right) x(t - \tau) \right\} = 0 \quad . $$
(3.5.26)

Equation (3.5.26) must be valid for all values of τ, and can be rewritten in terms of correlation integrals:

$$ R_{dx}(\tau) = \int_{-\infty}^{+\infty} R_{xx}(\tau - \alpha)h(\alpha)d\alpha \quad , \quad \text{for all values of } \tau \quad . $$
(3.5.27)

Using power spectra, one obtains

$$ S_{dx}(\omega) = S_{xx}(\omega)H(\omega) \quad \text{or} $$
(3.5.28)

$$ H(\omega) = \frac{S_{dx}(\omega)}{S_{xx}(\omega)} \quad . $$
(3.5.29)

Equation (3.5.29) is the continuous version of the matrix equation given by (3.5.17) and is generally known as a Wiener filter. Note that the filter response given by (3.5.29) is not causal. A causal solution can be obtained by using the causal version of (3.5.28) given by

$$ R_{dx}(\tau) = \int_{-\infty}^{\tau} R_{xx}(\tau - \alpha)h(\alpha)d\alpha \quad . $$
(3.5.30)

The above equation is known as the Wiener-Hopf equation, whose solution is well known but will not be discussed here. Note that the solution for the discrete case given by (3.5.17) is causal because we start with a causal filter given by the tapped delay line with weights given by W.

3.5.2 Least Mean Squares Adaptive Filters

Returning to Fig. 3.43, we note that, for this case,

$$ Y = d \quad \text{and} $$
(3.5.31)

$$ P_i = E\{dX_i\} \quad . $$
(3.5.32)

Hence to obtain the adaptive filter we must adjust the weight vector to equal that

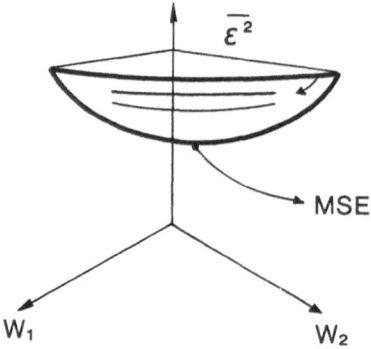

Fig. 3.46. Principle of the method of steepest descent

derived in (3.5.19). To obtain W_{opt}, we must find the correlation matrix and the steering vector and then solve (3.5.16). This is a very difficult task, especially if it has to be performed in real time. One particular method is by iteration using the method of steepest descent. Let us consider that the weight vector is $W(t)$ at a certain time. We then change $W(t)$ according to the equation

$$\frac{dW}{dt} = -\mu \nabla_W \overline{\varepsilon^2} \tag{3.5.33}$$

where μ is a constant to be chosen. This is illustrated in Fig. 3.46, where $\overline{\varepsilon^2}$ is plotted as a function of W_j in n-dimensional space and the objective is to reach the minimum point by following the steepest gradient. As we shall see, μ determines how fast we achieve the optimum value. If μ is too large, instability ensues and the result may diverge rather than converge to the actual solution.

Equation (3.5.33) can be rewritten as

$$\frac{dW}{dt} = 2\mu P - 2\mu RW \quad , \tag{3.5.34}$$

$$\frac{dW_k}{dt} = 2\mu P_k - 2\mu \sum_j R_{kj} W_j \quad \text{and} \tag{3.3.35}$$

$$W_k(m+1) - W_k(m) = 2\mu E\{\varepsilon(m)X(m-k)\} \quad . \tag{3.5.36}$$

In the last equation, m represents the number of iterations. Implementation of (3.5.36) is also difficult because one needs to calculate the expected value of the error and the input. It was Widrow and Hoff who first pointed out that, if one replaces the expected value by the present value, the implementation is rather simple. Furthermore, as we shall see shortly, the solution of the following Widrow-Hoff equation yields the optimum W as the asymptotic solution:

$$W_k(m+1) = W_k(m) + 2\mu\varepsilon(m)X(m-k) \quad . \tag{3.5.37}$$

Note that, for a continuous system, (3.5.35) can be rewritten as

$$\frac{d}{dt} W(t) = -2\mu\varepsilon(t)x(t) \quad \text{or} \tag{3.5.38}$$

272

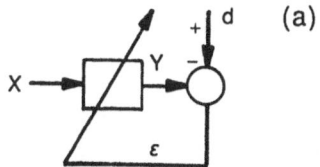

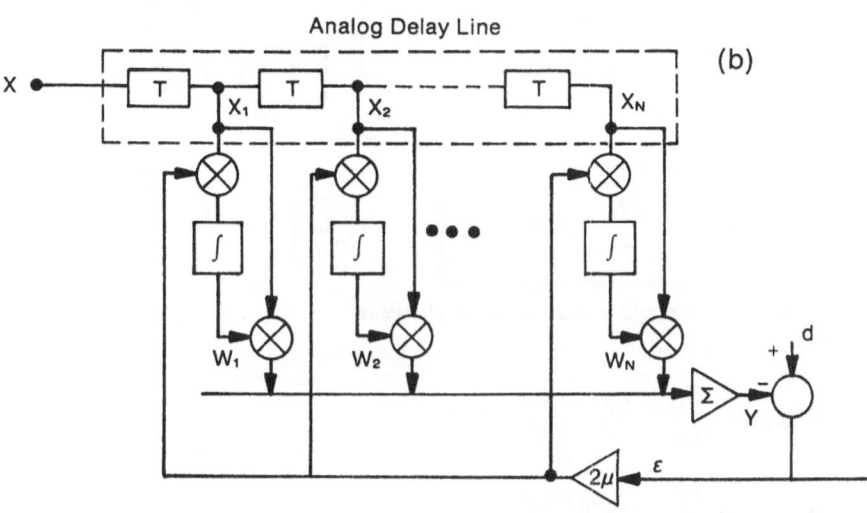

Fig. 3.47. Analog (a) and discrete (b) implementations of the Widrow-Hoff algorithm. (From [3.7])

$$W(t) = -2\mu \int_0^t \varepsilon(t')x(t')dt' \quad . \tag{3.5.39}$$

Figure 3.47 shows the analog and discrete implementations of the Widrow-Hoff algorithm.

Before we proceed further, it is illuminating to consider the simple example shown in Fig. 3.48. The input to the two-tap delay line is a signal at angular frequency ω with unknown amplitude and phase. We want the desired output to be a fixed amplitude d_0 and fixed phase ϕ. The delay time T is such that $\omega T = 90°$. Thus, we have the following relationships:

$$\begin{aligned}
d &= d_0 \cos(\omega t + \phi) \quad , \\
X_1 &= x_0 \cos \omega t + n_1(t) \quad , \\
X_2 &= x_0 \sin \omega t + n_2(t) \quad ,
\end{aligned} \tag{3.5.40}$$

where $n_1(t)$ and $n_2(t)$ are noise components. If we assume the input noise to be white Gaussian with a variance σ_n, we obtain

$$R_{11} = R_{22} = \frac{x_0^2}{2} + \sigma_n^2 \quad , \quad R_{12} = R_{21} = 0 \quad , \tag{3.5.41}$$

$$P_1 = \frac{x_0 d_0}{2} \cos \phi \quad , \quad P_2 = \frac{x_0 d_0}{2} \sin \phi \quad , \tag{3.5.42}$$

273

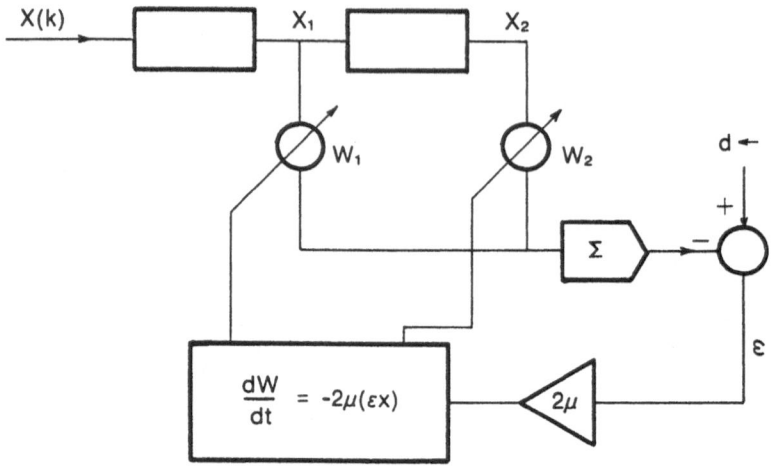

Fig. 3.48. Adaptive filter with a two-element weight vector

$$W_{1(\text{opt})} = \frac{P_1}{R_{11}} = \frac{x_0 d_0 \cos \phi}{x_0^2 + 2\sigma_n^2} \quad \text{and}$$

$$(3.5.43)$$

$$W_{2(\text{opt})} = \frac{P_2}{R_{22}} = \frac{x_0 d_0 \sin \phi}{x_0^2 + 2\sigma_n^2} \; .$$

The equation of steepest descent becomes

$$\frac{dW_1}{dt} = \mu x_0 d_0 \cos \phi - 2\mu \left(\frac{x_0^2}{2} + \sigma_n^2 \right) W_1 \quad \text{or}$$

$$W_1 = W_{1(\text{opt})} - \frac{1}{\mu(x_0^2 + 2\sigma_n^2)} e^{-t/\tau_1} \quad \text{where} \qquad (3.5.44)$$

$$\tau_1 = \frac{1}{\mu(x_0^2 + 2\sigma_n^2)} \; . \qquad (3.5.45)$$

Here τ_1 represents the time constant associated with the adjustment of the weight W_1. Similar equations can be solved for W_2. Note that the adaptive filter will follow d, i.e., the output will be $d_0 \cos(\omega t + \phi)$ even if we change the values of d_0 and ϕ. However, we must keep d_0 or ϕ constant for τ_1 or τ_2, whichever is greater. To show that in the limit $m \to \infty$ the weight vectors W_k approach the optimum solution given by (3.5.17), we take the expected value of (3.5.37), namely,

$$E\{W(m+1)\} = E\{W(m)\} - 2\mu E\{[d - W^T(m)X(m)]X(m)\}$$
$$= (I + 2\mu R)E\{W(m)\} - 2\mu P \; . \qquad (3.5.46)$$

The above equation holds only when the sampling period is long enough and the inputs $X(m)$ are stationary processes such that $X(m)$ and $X(m+1)$ are uncorrelated. Under these conditions, $W(m)$ is dependent on $W(m-1)$, $W(m-2)$, ..., $W(0)$ but independent of $X(m)$.

If the initial value of W at $m = 0$ is denoted by $W(0)$, then the $(m+1)$th iteration of the above equation yields

$$E\{W(m+1)\} = [1 + 2\mu R]^{m+1} W(0)$$
$$- 2\mu \sum_{i=0}^{m} [I + 2R]^i P \quad . \tag{3.5.47}$$

We diagonalize the correlation matrix R as follows:

$$R = Q^{-1} \Lambda Q \quad , \tag{3.5.48}$$

where Λ is the diagonal matrix with eigenvalues λ_i given by

$$\Lambda = \begin{bmatrix} \lambda_1 & & & \\ & \lambda_2 & & \\ & & \ddots & \\ & & & \lambda_n \end{bmatrix} \quad . \tag{3.5.49}$$

Note that R is symmetric and positive definite and thus all the eigenvalues are positive. Using (3.5.48) for R we obtain

$$E\{W(m+1)\} = (I + 2\mu Q^{-1} \Lambda Q)^{m+1} W(0) - 2\mu \sum_{i=0}^{m} (I + 2\mu Q^{-1} \Lambda Q)^i P$$
$$= Q^{-1} (I + 2\mu \Lambda)^{m+1} Q W(0)$$
$$- 2\mu Q^{-1} \sum_{i=0}^{m} (I + 2\mu \Lambda)^i Q P \quad . \tag{3.5.50}$$

If the diagonal terms of the matrix $(I + 2\mu\Lambda)$ are less than 1, then

$$\lim_{m \to \infty} (I + 2\mu\Lambda)^{m+1} \to 0 \quad . \tag{3.5.51}$$

Under this condition, we can write

$$E\{W(m+1)\} = -2\mu Q^{-1} \sum_{i=0}^{m} (I + 2\mu\Lambda)^i Q P \quad . \tag{3.5.52}$$

We note that

$$\lim_{m \to \infty} \sum_{i=0}^{m} (I + 2\mu\Lambda)^i = -\frac{1}{2\mu} \Lambda^{-1} \quad , \tag{3.5.53}$$

as each diagonal term can be summed using the geometric series summation formula as follows:

$$\sum_{i=0}^{\infty} (1 + 2\mu\lambda_i)^i = \frac{1}{1 - (1 + 2\mu\lambda_i)} = -\frac{1}{2\mu} (\lambda_i)^{-1} \quad . \tag{3.5.54}$$

Thus (3.5.50) becomes in the limit

$$\lim_{m \to \infty} E\{W(m+1)\} = Q^{-1}\Lambda^{-1}QP = R^{-1}P \quad . \tag{3.5.55}$$

If the condition given by (3.5.46) is satisfied, the Widrow-Hoff solution approaches the optimum solution given by (3.5.17). If λ_{max} is the maximum eigenvalue of the R matrix, then we obtain

$$-\frac{1}{\lambda_{max}} < \mu < 0 \quad . \tag{3.5.56}$$

The above convergence condition for μ must be satisfied for the adaptive filter to operate. Note that

$$\lambda_{max} \leq \operatorname{Tr}\{R\} \quad \text{and} \tag{3.5.57}$$

$$\operatorname{Tr}\{R\} = E\{X^{T}(m)X(m)\} \tag{3.5.58}$$

$$= \sum_{i=1}^{n} E\{X_i^2\} = \text{total input power} = P_{in} \quad . \tag{3.5.59}$$

The convergence condition given by (3.5.56) can be written as

$$-\frac{1}{P_{in}} < \mu < 0 \quad . \tag{3.5.60}$$

Although the proof of the convergence equation assumes that the independent successive input samples are uncorrelated, it is found that this condition can be relaxed.

It is of interest to derive the time dependence of the weight vector for the continuous case. From (3.5.35) we have

$$\frac{dW(t)}{dt} = 2\mu P - 2\mu RW(t) \quad . \tag{3.5.61}$$

Substituting R from (3.5.48) we obtain

$$\frac{dW(t)}{dt} = 2\mu P - 2\mu Q^{-1}\Lambda QW \quad . \tag{3.5.62}$$

Multiplication of both sides by Q yields

$$\frac{d}{dt}(QW) = 2\mu QP - 2\mu \Lambda QW \quad . \tag{3.5.63}$$

Solving for QW we obtain

$$QW = (QP)\Lambda^{-1} - \frac{1}{\mu}\Lambda^{-1}[E_1] \tag{3.5.64}$$

where E_1 is a $N \times 1$ matrix with elements given by

$$E_i = e^{-t/2\mu\lambda_i} \quad . \tag{3.5.65}$$

Thus the jth element of (3.5.65) is given by

$$(QW)_j = \frac{(QP)_j}{\lambda_j} - \frac{1}{\mu\lambda_j}e^{-t/2\mu\lambda_j}. \tag{3.5.66}$$

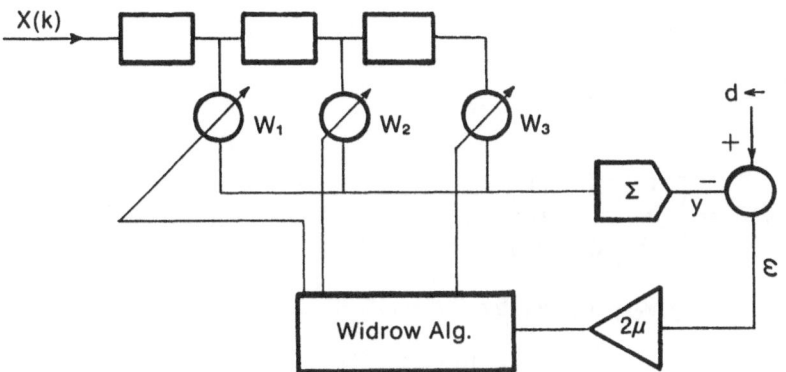

Fig. 3.49. Adaptive filter with a three-element weight vector

Note that (3.5.17) can be rewritten as

$$W_{\text{opt}} = R^{-1}P$$
$$= Q^{-1}\Lambda^{-1}QP \quad \text{or}$$

$$QW_{\text{opt}} = \lambda^{-1}QP \quad . \tag{3.5.67}$$

Using the above equation we can rewrite (3.5.63) as

$$\frac{d}{dt}(QW) = QW_{\text{opt}} - 2\mu\Lambda^{-1}[E_1] \quad . \tag{3.5.68}$$

It is enlightening to consider a numerical example, as shown in Fig. 3.49, which is the same as that considered in Fig. 3.48, except that now we have three elements of the weight vector. We can easily calculate the R matrix

$$R = \begin{bmatrix} x_0^2/2 + \sigma_n^2 & 0 & x_0^2/2 \\ 0 & x_0^2/2 + \sigma_n^2 & 0 \\ x_0^2/2 & 0 & x_0^2/2 + \sigma_n^2 \end{bmatrix} \quad . \tag{3.5.69}$$

The P matrix is

$$P = \begin{bmatrix} (d_0 x_0/2)\cos\phi \\ -(d_0 x_0/2)\sin\phi \\ (d_0 x_0/2)\cos\phi \end{bmatrix} \quad . \tag{3.5.70}$$

The eigenvalues of the R matrix are easily found to be

$$\lambda_1 = \sigma_n^2 \quad ,$$

$$\lambda_2 = \frac{x_0^2}{2} + \sigma_n^2 \quad \text{and} \tag{3.5.71}$$

$$\lambda_3 = x_0^2 + \sigma_n^2 \quad .$$

After normalization the Q matrix is given by

$$Q = \begin{bmatrix} \frac{1}{\sqrt{2}} & 0 & -\frac{1}{\sqrt{2}} \\ 0 & 1 & 0 \\ -\frac{1}{\sqrt{2}} & 0 & \frac{1}{\sqrt{2}} \end{bmatrix} . \tag{3.5.72}$$

Note that

$$Q^T = Q^{-1} = \begin{bmatrix} 1 & 0 & +1 \\ 0 & 1 & 0 \\ -1 & 0 & 1 \end{bmatrix} . \tag{3.5.73}$$

We also note that QW_{opt} is given by

$$QW_{\text{opt}} = \begin{bmatrix} 0 \\ -d_0 x_0 \sin\phi / x_0^2 + \sigma_n^2 \\ d_0 x_0 \cos\phi / x_0^2 + \sigma_n^2 \end{bmatrix} . \tag{3.5.74}$$

The time constants associated with convergence of the weight vectors are

$$\begin{aligned} \tau_1 &= 2\mu\sigma_n^2 , \\ \tau_2 &= 2\mu\left(\frac{x_0^2}{2} + \sigma_n^2\right) \quad \text{and} \\ \tau_3 &= 2\mu(x_0^2 + \sigma_n^2) . \end{aligned} \tag{3.5.75}$$

3.5.3 Lattice Filters

The adaptive filter shown in Fig. 3.43 can also be used to estimate the input signal if we make the desired signal and the reference signal the same. This case is shown in Fig. 3.50. Note that the filter cancels all the correlated signals and the error signal is mostly white. The adaptive filter in Fig. 3.50 can also be implemented using a lattice filter, which uses delay elements but allows feedback, instead of the transversal filter implementation discussed so far. A typical lattice filter is shown in Fig. 3.51. We note that, for an l-stage filter, we need to determine l coefficients, $K_0, K_1, \ldots, K_{l-1}$. As shown later, the lattice filter has two distinct advantages compared to the Widrow-Hoff filter. These are (i) faster convergence and (ii) the fact that for the determination of an $(N + 1)$ stage filter one need calculate only the $(N+1)$th stage, if an N-stage filter has already been calculated,

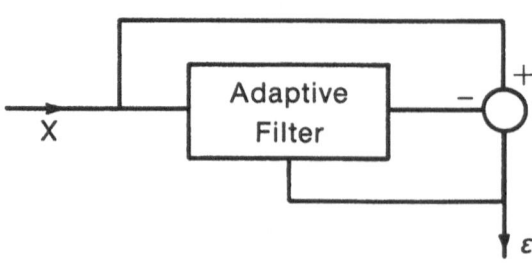

Fig. 3.50. An adaptive filter can be used to estimate the input signal if the desired signal is the reference signal

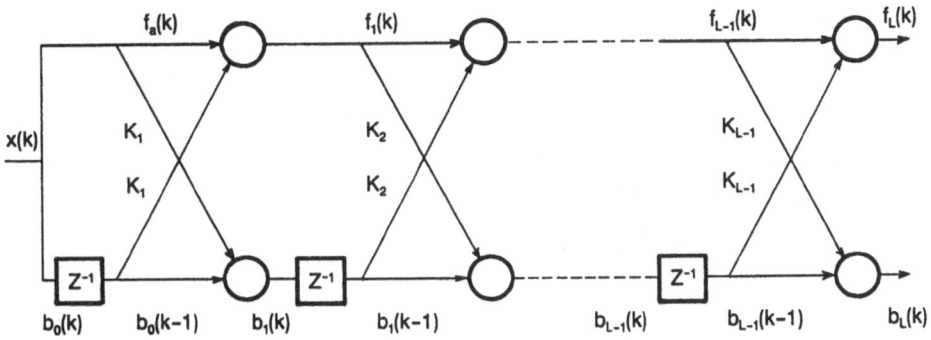

Fig. 3.51 Lattice filter

because the stage n ($n \leq N$) of any filter with length equal to or greater than N is always the same. Note that this is not true for a Widrow-Hoff filter. In Fig. 3.51 we denote the estimation error by $f_l(k)$ and $b_l(k)$, where $f_l(k)$ is the forward error sequence at the lth stage and kth time step. Similarly, $b_l(k)$ is called the backward error sequence. Note that

$$x(k) = \hat{x}(k) + f_l(k) \quad \text{and} \tag{3.5.76}$$

$$x(k - l) = \hat{x}(k - l) - b_l(k) \quad , \tag{3.5.77}$$

where $x(k)$ denotes the input signal to be estimated using its last l samples, $x(k - 1), \ldots, x(k - l)$, and $\hat{x}(k)$ denotes the estimated value. The feedback sequence $K_l(k)$ is called the reflection coefficients or weights. From Fig. 3.51 we note

$$f_0(k) = g_0(k) = x(k) \quad , \tag{3.5.78}$$

$$f_{l+1}(k) = f_l(k) - K_{l+1}(k)b_l(k - 1) \quad \text{and} \tag{3.5.79}$$

$$b_{l+1}(k) = b_l(k) - K_{l+1}(k)f_l(k) \quad . \tag{3.5.80}$$

If the reflection coefficients are fixed and not dependent on this discrete time coefficient, i.e., $k_l(k) = K_l$, we can write

$$f_0(k) = b_0(k) = x(k) \quad , \tag{3.5.81}$$

$$f_{l+1}(k) = f_l(k) + K_{l+1}b_l(k - 1) \quad \text{and} \tag{3.5.82}$$

$$b_{l+1}(k) = b_l(k - 1) + K_{l+1}f_l(k) \quad . \tag{3.5.83}$$

Using z-transforms, we obtain

$$F_0(z) = B_0(z) = X(z) \quad , \tag{3.5.84}$$

$$F_{l+1}(z) = F_l(z) + K_l z^{-1}B_l(z) \tag{3.5.85}$$

and

279

$$B_{l+1}(z) = K_{l+1}F_l(z) + B_l(z)z^{-1} \quad . \tag{3.5.86}$$

If we define the forward and backward transfer functions at stage l by $A_l(z)$ and $G_l(z)$, respectively, then

$$A_l(z) = \frac{F_l(z)}{x(z)} = \frac{F_l(z)}{F_0(z)} \quad \text{and} \tag{3.5.87}$$

$$G_l(z) = \frac{B_l(z)}{x(z)} = \frac{B_l(z)}{B_0(z)} \quad . \tag{3.5.88}$$

We thus obtain from (3.5.84–86)

$$A_0(z) = G_0(z) = 1 \quad , \tag{3.5.89}$$

$$A_l(z) = A_{l-1}(z) + K_l z^{-1}G_{l-1}(z) \quad \text{and} \tag{3.5.90}$$

$$G_l(z) = K_l A_{l-1}(z) + z^{-1}G_{l-1}(z) \quad . \tag{3.5.91}$$

Thus

$$G_l(z) = z^{-l}A_l(z^{-1}) \quad . \tag{3.5.92}$$

If we denote $A_l(z)$ by a polynomial given by

$$A_l(z) = \sum_{k=0}^{l} W_{lk}z^{-k} \tag{3.5.93}$$

then $W_l(k)$ are the polynomial coefficients for an l-stage lattice filter. It is obvious that one can consider W_{lk} as the kth weight of the l-stage transversal filter that is equivalent to the l-stage lattice filter. For this case one can also write

$$G_l(z) = \sum_{k=0}^{l} W_{l,l-k}z^{-k} \tag{3.5.94}$$

and $G_l(z)$ is the reverse polynomial corresponding to $A_l(z)$. We note that

$$W_{l,0} = 1 \quad \text{and} \tag{3.5.95}$$

$$W_{ll} = K_l \quad . \tag{3.5.96}$$

We also note that

$$A_{l-1}(z) = \frac{A_l(z) - K_l B_l(z)}{1 - K_l^2} = \frac{A_l(z) - K_l z^{-l}A_l(z^{-1})}{1 - K_l^2} \quad . \tag{3.5.97}$$

Thus, if $A_p(z)$ is known with $a_p(0) = 1$, one can generate all the polynomials $A_m(z)$, $m < p$, and the coefficients K_l using (3.5.89, 90 and 96). Note that $|k_l| \neq 1$ for a stable solution. It can be shown that, if

$$-1 < K_l < 1 \quad \text{with} \quad 1 \leq l \leq m \quad , \tag{3.5.98}$$

then all the zeros of $A_m(z)$ fall inside the unit circle and the filter is a minimum phase filter. Similarly, for the corresponding case, $B_m(z)$ is a maximum phase filter. The condition given by (3.5.98) justifies the labeling of K_l's as reflection coefficients.

We note that the z-transform of the residual error is given by

$$F_l(z) = A_l(z)X(z) = \sum_{k=0}^{l} W_{lk} z^{-k} x(z) \quad . \tag{3.5.99}$$

Thus

$$f_l(k) = \sum_{k=0}^{l} W_{lk} x(l-k) \quad . \tag{3.5.100}$$

The mean square error at the lth stage, E_l, is given by

$$
\begin{aligned}
E_l &= \langle f_l^2(k) \rangle \\
&= \sum_{k=0}^{l} \sum_{i=0}^{l} W_{lk} W_{li} \langle x(l-k) x(l-i) \rangle \\
&= \sum_{k=0}^{l} \sum_{i=0}^{l} W_{lk} W_{li} R(i-k) \quad , \tag{3.5.101}
\end{aligned}
$$

where

$$R(n) = \langle x(m) x(m-n) \rangle \tag{3.5.102}$$

and represents the autocorrelation of the signal $x(k)$.

Equation (3.5.101) can also be written as

$$E_l = \sum_{i=-m}^{m} u_m(i) R(i) \quad , \quad \text{where} \tag{3.5.103}$$

$$u_m(i) = \sum_{k=0}^{m-|i|} W_{mk} W_{m,k+|i|} \quad . \tag{3.5.104}$$

Thus, $u_m(i)$ is the autocorrelation of the impulse response of $A_m(z)$. Similarly, one can show that

$$E_l = \langle b_l^2(k) \rangle = \langle f_l^2(k) \rangle \quad . \tag{3.5.105}$$

One can also show that the recursive relation obeyed by the mean square error E_l is given by

$$
\begin{aligned}
E_l &= E_{l-1} + 2K_l \langle f_{l-1}(k) b_{l-1}(k-1) \rangle + K_l^2 E_{l-1} \\
&= (1 + 2r_{l-1} K_l + K_l^2) E_{l-1} \quad , \tag{3.5.106}
\end{aligned}
$$

where we have defined r_{l-1}, the correlation coefficient between $f(l-1, k)$ and $b(l-1, k-1)$, as

$$r_{l-1} = \frac{\langle f_{l-1}(k) b_{l-1}(k-1) \rangle}{\sqrt{\langle f_{l-1}^2(k) \rangle} \sqrt{\langle b_{l-1}^2(k-1) \rangle}} = \frac{\langle f_{l-1}(k) b_{l-1}(k) \rangle}{E_{l-1}} \quad . \tag{3.5.107}$$

281

Hence the residual energy at stage l is known in terms of the energy at stage $l-1$, the correlation coefficient r_{l-1}, and K_l. Using (3.5.106) one obtains by iteration

$$E_l = R(0) \prod_{i=1}^{l} (1 + 2r_{i-1}K_i + K_i^2)$$

(3.5.108)

where

$$E_0 = R(0) \quad .$$

(3.5.109)

For a lattice filter of p stages, one must determine the p unknown coefficients K_p. Once K_p are known, one can determine the values of W_{pk} by using the least mean squares algorithm or the minimization of E_p, as in Sect. 3.4. The optimum values are given by

$$R(0)W_{p1} + R(1)W_{p2} + \ldots + R(p-1)W_{pp} = -P_1 = -R(1)$$
$$R(1)W_{p1} + R(0)W_{p2} + \ldots + R(p-2)W_{pp} = -P_2 = -R(2)$$

$$\vdots \qquad \vdots \qquad \vdots \qquad \vdots \qquad \vdots$$

or

$$R_p W_p = P \quad \text{where}$$

(3.5.110)

$$P_i = \langle x(k)x(k+i) \rangle = R(i) \quad .$$

(3.5.111)

Note that the W_p column matrix is really the W matrix given in Sect. 3.5.2. Thus

$$W_p = -[R_p]^{-1} R_i \quad .$$

(3.5.112)

as in the case of the Wiener filter. For this optimum case, we note that the minimum mean square error is given by

$$E_p = \sum_{k=0}^{l} W_{pk} R(k) \quad .$$

(3.5.113)

Incorporating (3.5.110 and 113) into one matrix equation we can write

$$\begin{bmatrix} R(0) & R(1) & R(2) & \cdots & R(p) \\ R(1) & R(0) & R(1) & \cdots & R(p-1) \\ \vdots & \vdots & \vdots & & \vdots \\ R(p) & R(p-1) & R(p-2) & \cdots & R(0) \end{bmatrix} \begin{bmatrix} 1 \\ W_{p1} \\ \vdots \\ W_{p0} \end{bmatrix} = \begin{bmatrix} E_p \\ 0 \\ \vdots \\ 0 \end{bmatrix}$$

or

$$R_p W_p = e_p \quad .$$

(3.5.114)

Note that in the matrix W_p of (3.5.114) we have included $W_{p0} = 1$. Here R_p is the pth-order autocorrelation matrix of the signal $x(k)$. It is a symmetric and Toeplitz matrix. Let us define the following matrices:

$$C_p = \begin{bmatrix} 1 & & & & \\ W_{11} & 1 & & & \\ W_{22} & W_{21} & 1 & & \\ \vdots & & & & \\ W_{pp} & W_{p,p-1} & \cdots & & 1 \end{bmatrix} , \tag{3.5.115}$$

$$E_p = \begin{bmatrix} E_0 & & & & \\ & E_1 & & & \\ & & E_2 & & \\ & & & \ddots & \\ & & & & E_p \end{bmatrix} . \tag{3.5.116}$$

Note that C_p is a lower triangular and Toeplitz matrix and E_p is a diagonal matrix. The elements of C_p include all the optimum transversal filter weights for stages up to p. Thus the second column represents the weights for a $(p-1)$ stage filter. By direct multiplication, it can be shown that

$$C_p R_p C_p^T = E_p \quad . \tag{3.5.117}$$

Thus C_p diagonalizes the autocorrelation matrix by using the Gram-Schmidt orthogonalization as discussed in Appendix A.

If we define

$$x_p = [x(k), x(k-1), \ldots, x(k-p)]^T \tag{3.5.118}$$

and

$$b_p = [b_0(k), b_1(k), \ldots]^T \tag{3.5.119}$$

we obtain

$$b_p = C_p x_p \quad . \tag{3.5.120}$$

Note that x_p and b_p are $(p \times 1)$ column matrices. Using (3.5.120), we obtain the autocorrelation matrix for the backward errors, R_p^b, as

$$\begin{aligned} R_p^b &= \langle b_p b_p^T \rangle = C_p \langle x_p x_p^T \rangle C_p^T \\ &= C_p R_p C_p^T = E_p \quad . \end{aligned} \tag{3.5.121}$$

Thus the autocorrelation matrix for the backward errors is diagonal. Another interpretation of (3.5.121) is that the expected values of the backward errors are orthogonal to each other, as given by the relationship

$$\langle b(i,k)b(j,k) \rangle = \begin{cases} E_i & i = j \\ 0 & i \neq j \end{cases} , \tag{3.5.122}$$

This orthogonalization decouples the successive stages from each other. *Itakura* [3.3] recognized this decoupling and argued that the values of the reflection coefficients K_l can be obtained by minimizing the mean square error at each stage rather than the whole mean square error. One can show that the global minimization is actually equivalent to the sequence of local minimizations. Using the expression for E_l given by (3.5.106) and differentiating with reference to K_l, we obtain

$$\frac{\partial E_l}{\partial K_l} = (2r_{l-1} + 2K_l)E_{l-1} = 0 \quad \text{or} \tag{3.5.123}$$

$$K_l = -r_{l-1} = \frac{\langle f_{l-1}(k)b_{l-1}(k-1)\rangle}{E_{l-1}} \quad . \tag{3.5.124}$$

Note that the calculation of each reflection coefficient does not depend on any of the following reflection coefficients. This is the most important result of the lattice filter because it allows one to determine a $(N+1)$ stage filter by calculating just the $(N+1)$th stage if the other N stages are already known.

The minimum mean square error for the lth stage is given by

$$E_l = (1 - K_l^2)E_{l-1}$$
$$= R(0) \prod_{i=1}^{l} (1 - K_l^2) \quad . \tag{3.5.125}$$

The following properties of the forward and backward error sequences can be derived:

$$\langle f_m(n)x(n-i)\rangle = 0 \quad , \quad 1 \leq i \leq m \quad . \tag{3.5.126}$$

$$\langle b_m(n)x(n-i)\rangle = 0 \quad , \quad 0 \leq i \leq m-1 \quad . \tag{3.5.127}$$

$$\langle f_m(n)x(n)\rangle = E_m \quad . \tag{3.5.128}$$

$$\langle b_m(n)x(n)\rangle = E_m \quad . \tag{3.5.129}$$

$$\langle f_i(n)f_j(n)\rangle = E_{\max(i,j)} \quad . \tag{3.5.130}$$

$$\langle b_i(n)b_j(n)\rangle = E_i\delta_{ij} \quad . \tag{3.5.131}$$

$$\langle f_i(n)f_j(n-t)\rangle = 0 \quad \text{for} \quad \begin{cases} 0 \leq t \leq i-j-1, & i > j, \\ 0 \geq t \geq i-j+1, & i < j. \end{cases} \tag{3.5.132}$$

$$\langle f_i(n+i)f_j(n+j)\rangle = E_i\delta_{ij} \quad . \tag{3.5.133}$$

$$\langle b_i(n+i)b_j(n+j)\rangle = E_{\max(i,j)} \quad . \tag{3.5.134}$$

$$\langle f_i(n)b_j(n)\rangle = \begin{cases} K_jE_j, & i \geq j \\ 0, & i < j \end{cases} \quad i,j \geq 0 \quad . \tag{3.5.135}$$

$$\langle f_i(n)b_i(n-1)\rangle = -K_{i+1}E_i \quad . \tag{3.5.136}$$

$$\langle b_i(n-1)x(n)\rangle = -K_{i+1}E_i \quad . \tag{3.5.137}$$

284

$$\langle f_i(n+1)x(n-i)\rangle = -K_{i+1}E_i \quad . \tag{3.5.138}$$

$$\langle f_i(n)b_j(n-1)\rangle = \left\{ \begin{array}{ll} 0, & i>j \\ -K_{j+1}E_j, & i\leq j \end{array} \right\} \quad i,j\geq 0 \quad . \tag{3.5.139}$$

a) Different Forms of Lattice Structures

The lattice shown in Fig. 3.49 needs two multipliers and two adders. As multipliers are generally more expensive, the alternative forms that use only one multiplier are of practical interest.

To derive the one-multiplier lattice structure we obtain from (3.6.90, 91), by adding and subtracting, $S_m K_m A_{m-1}(z)$ and $S_m K_m z^{-1}G_{m-1}(z)$, respectively, where

$$A_m(z) = (1 - S_m K_m)A_{m-1}(z) + S_m K_m T_m(z) \tag{3.5.140}$$

and

$$G_m(z) = K_m T_m(z) + (1 - S_m K_m)z^{-1}G_{m-1}(z) \quad . \tag{3.5.141}$$

Here $S_m = \pm 1$ is a sign parameter and

$$T_m(z) = A_{m-1}(z) + S_m z^{-1}G_{m-1}(z) \quad . \tag{3.5.142}$$

If we use the lattice structure shown in Fig. 3.52, where each stage is multiplied by a multiplier M_m, $1\leq m \leq p$, for a p-stage filter, we can scale the lattice transfer function. The scaled transfer function is given by

$$\hat{A}_p(z) = P_p A_p(z) \quad \text{and} \tag{3.5.143}$$

$$\hat{G}_p(z) = P_p G_p(z) \quad , \tag{3.5.144}$$

where

$$P_p = \prod_{m=1}^{p} M_m \quad . \tag{3.5.145}$$

Writing the scaled version of the recursive equation, we have

$$\hat{A}_m(z) = M_m[(1 - S_m K_m)\hat{A}_{m-1}(z) + S_m K_m \hat{T}_m(z)] \quad , \tag{3.5.146}$$

$$\hat{G}_m(z) = M_m[K_m \hat{T}_m(z) + (1 - S_m K_m)z^{-1}\hat{G}_{m-1}(z)] \tag{3.5.147}$$

and

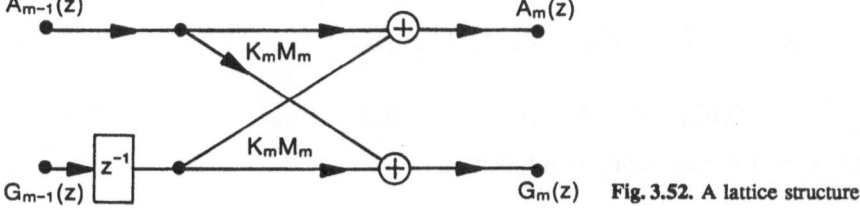

Fig. 3.52. A lattice structure

285

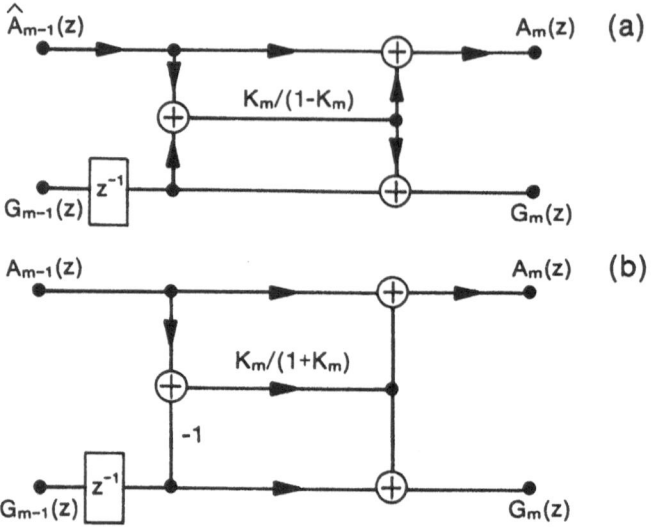

Fig. 3.53. Implementation of (3.5.150, 151) for (a) $S_m = 1$ and (b) $S_m = -1$

$$\hat{T}_m(z) = \hat{A}_{m-1}(z) + S_m z^{-1} \hat{G}_m(z) \quad . \tag{3.5.148}$$

If we set

$$M_m = \frac{1}{1 - S_m K_m} \quad , \tag{3.5.149}$$

we obtain

$$\hat{A}_m(z) = \hat{A}_{m-1}(z) + S_m \frac{K_m}{1 - S_m K_m} \hat{T}_m(z) \tag{3.5.150}$$

and

$$G_m(z) = \frac{K_m}{1 - S_m K_m} \hat{T}_m(z) + z^{-1} \hat{G}_{m-1}(z) \quad . \tag{3.5.151}$$

The implementation of the above two equations needs only one multiplier and three adders as shown in Fig. 3.53a for $S_m = 1$ and in Fig. 3.53b for $S_m = -1$. One can easily derive the minimum mean square error as

$$E_m = R(0) \prod_{i=1}^{m} \frac{1 + S_i K_i}{1 - S_i K_i} \quad . \tag{3.5.152}$$

Another structure that also uses only one multiplier (shown in Fig. 3.54) can be derived as follows. For this case, $S_m z^{-1} G_{m-1}(z)$ and $S_m A_{m-1}(z)$ are added to and subtracted from (3.5.90) and (3.5.91), respectively. We obtain

$$A_m(z) = T_m(z) + (K_m - S_m) z^{-1} G_{m-1}(z) \tag{3.5.153}$$

and

$$G_m(z) = (K_m - S_m) A_{m-1}(z) + S_m T_m(z) \quad . \tag{3.5.154}$$

In the scaled version using multipliers

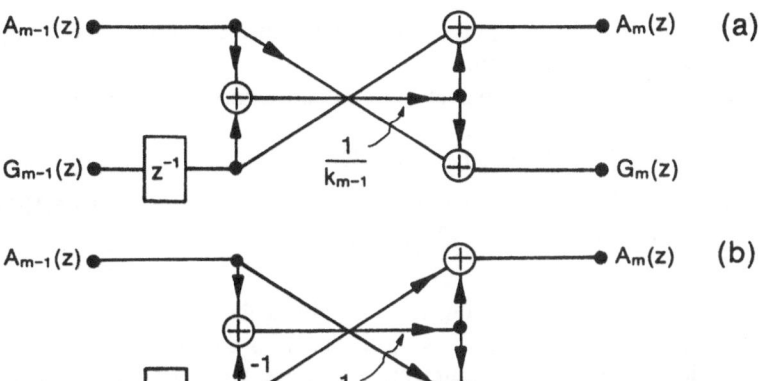

Fig. 3.54. Implementation of (3.5.156, 157) for (a) $S_m = 1$ and (b) $S_m = -1$

$$M_m = \frac{1}{K_m - S_m} \quad , \tag{3.5.155}$$

we obtain

$$\hat{A}_m(z) = \frac{1}{K_m - S_m}\hat{T}_m(z) + z^{-1}\hat{G}_{m-1}(z) \quad \text{and} \tag{3.5.156}$$

$$\hat{G}_m(z) = A_{m-1}(z) + S_m\frac{1}{K_m - S_m}\hat{T}_m(z) \quad . \tag{3.5.157}$$

b) Adaptive Lattice Filters

So far we have considered K_l to be constant, determined by (3.5.124), which assumes that we know the statistics of the signal and hence the error. As for the case of a Widrow-Hoff filter, in reality we want to find these coefficients by using only n known samples of the signal $x(k), \ldots, x(k-n)$, and we should be able to do it recursively. The gradient method consists of updating the reflection coefficients K_l as follows:

$$K_l(n+1) = K_l(n) + \frac{\beta}{\hat{E}_l(n) + \hat{E}_f(n)}$$
$$\times \left[f_l(n)b_{l+1}(n) + b_l(n-1)f_{l+1}(n) \right] \tag{3.5.158}$$

where

$$\hat{E}_b(n) = \gamma\hat{E}_b(n-1) + (1-\gamma)b_l^2(n-1) \quad , \tag{3.5.159}$$

$$\hat{E}_f(n) = \gamma\hat{E}_f(n-1) + (1-\gamma)f_l^2(n-1) \tag{3.5.160}$$

and β is a constant which controls the convergence rate. The form of (3.5.158) is very similar to the Widrow-Hoff transversal filter implementation. The constant γ $(0 < \gamma < 1)$ controls the time-averaging window length and is generally set to a value that approximates the adoption time of the filter. In many practical cases, the term $[\hat{E}_b(n) + \hat{E}_f(n)]$ is incorporated into β because division is generally expensive to obtain in hardware form.

Another method, which is similar to the gradient algorithm, is to choose

$$K_l(n+1) = -\frac{2\sum_{i=i_0}^{n} \beta^{n-1} f_{l-1}(i) b_{l-1}(i-1)}{\sum_{i=i_0}^{n} \beta^{n-1} f_{l-1}^2(i) + b_{l-1}^2(i-1)} . \qquad (3.5.161)$$

In the above equation, we are updating the lth gain at the $(n+1)$ clock cycle from the values up to the nth cycle. Here i_0 depends on the memory size used for the computation. If i_0 is a fixed integer, e.g., zero, then we have the case of a growing memory. For a fixed memory size, $i_0 = n - M + 1$, where M is the memory size. The quantity β lies between 0 and 1 and represents a fading memory for $\beta < 1$. If we write

$$K_l(n+1) = -\frac{C(n)}{D(n)} , \qquad (3.5.162)$$

the recursive relations for $C(n)$ and $D(n)$ for the fading, growing memory are given by

$$C(n) = \beta C(n-1) + 2 f_{l-1}(n) b_{l-1}(n-1) \quad \text{and} \qquad (3.5.163)$$

$$D(n) = \beta D(n-1) + f_{l-1}^2(n) + b_{l-1}^2(n-1) . \qquad (3.5.164)$$

Thus β effectively determines how fast the filter adapts to the changing signal. The closer β is to 1, the longer the adaption time.

For Fig. 3.51 one can also write

$$K_l(n+1) = K_l(n) - \frac{f_{l-1}(n) b_l(n) + b_{l-1}(n-1) f_l(n)}{D(n)} . \qquad (3.5.165)$$

From this equation, we note that, for the growing, nonfading memory, $D(n) \to \infty$ and K_l reaches a steady-state value for a stationary signal. Thus (3.5.165) is very similar to the steepest descent gradient. We have already noted the advantages of the adaptive lattice filter. Because of the orthogonalization and decoupling properties of this structure, the convergence is much faster than for the transversal filter. Also, the convergence time is almost independent of the eigenvalue spread of R.

3.6 Power Spectra Estimation

Power spectra estimation of a stochastic process is an important problem with many applications. The problem is also sometimes referred to as a time series problem or the finding of hidden periodicities. Typically, data are taken or available about a process as a set of contiguous data measurements taken at equispaced time intervals T and represented as

$$x(1), x(2), x(3), z(4), \ldots, x(N) \quad,$$

where N is the data length and the time interval is normalized to 1. The data consist of an actual process and some noise is involved in the measurement. Also, N is finite. The problem of power spectra estimation can be divided into two major parts: First, obtaining the autocorrelation of the process from the finite set of data and, secondly, finding the power spectra from the autocorrelation values. We shall tackle the second problem first. The three models to be discussed briefly are

i)   the MA (Moving Average) or all-zero model,
ii) the AR (Auto-Regressive) or maximum entropy spectral analysis or all-pole model,
iii) ARMA – a combination of the AR and MA models.

In many problems of stationary random processes, it is desired to model the process itself with the autocorrelation values known at discrete time intervals. These are denoted as

$$R_x(0), R_x(1) \quad, \ldots \quad \text{where}$$

$$R_x(m) = \langle x(n)x^*(n+m) \rangle \quad. \tag{3.6.1}$$

The power spectrum S_x is given by

$$S_x(\omega) = \sum_{n=-\infty}^{+\infty} R_x(n) e^{-jn\omega} \quad. \tag{3.6.2}$$

$$R_x(m) = \frac{1}{2\pi} \int_{-\pi}^{+\pi} S_x(e^{j\omega}) e^{-j\omega m} d\omega \quad. \tag{3.6.3a}$$

In the z-transform domain, we have

$$S_x(z) = \sum_{n=-\infty}^{+\infty} R_x(n) z^{-n} \quad. \tag{3.6.3b}$$

The most useful model is shown in Fig. 3.55, where the random process to be modeled is the output of the linear system with a transfer function to be determined with white noise as the input. If the transfer function is given by

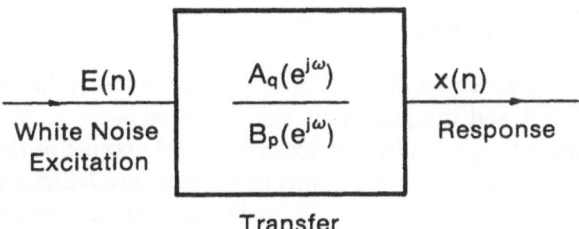

Fig. 3.55. Model of a stationary random process

$$\frac{A_q(\omega)}{B_p(\omega)} = \frac{a_0 + a_1 e^{-j\omega} + \ldots + a_q e^{-jq\omega}}{1 + b_1 e^{-j\omega} + \ldots + b_p e^{-jp\omega}} \quad , \tag{3.6.4}$$

the power spectrum is given by

$$S_x(e^{j\omega}) = \left| \frac{A_q(\omega)}{B_p(\omega)} \right|^2 \tag{3.6.5}$$

because the power spectrum of white noise is 1. The time series $x(n)$ for this process can be written as

$$x(n) + \sum_{k=1}^{p} b_k x(n-k) = \sum_{k=0}^{q} a_k \varepsilon(n-k) \quad , \tag{3.6.6}$$

where $\varepsilon(n)$ represents a sequence of zero mean unit variance, uncorrelated random variable, or normalized white noise. Multiplying both sides of (3.6.6) by $x(n-m)$ and taking the expected values, one obtains the so-called Yule-Walker equation

$$\sum_{k=0}^{p} b_k R_x(n-k) = \sum_{j=0}^{q} a_j h^*(j-n) \quad . \tag{3.6.7}$$

Here $h^*(n)$ is the delta response of the linear system represented by (3.6.6). The term $h(n)$ can also be obtained from the transfer function (3.6.4) by an inverse Fourier transform. For a causal system, $h(n) = 0$ for $n < 0$.

3.6.1 MA Model

For the MA model

$$B_p(\omega) = 1 \quad . \tag{3.6.8}$$

Thus (3.6.6) becomes

$$x(n) = \sum_{k=0}^{q} a_k \varepsilon(n-k) \quad . \tag{3.6.9}$$

Equation (3.6.7) simplifies to

$$R_x(n) = \begin{cases} \displaystyle\sum_{k=0}^{q} a_k a_{k-n}^*, & -q \leq n \leq q, \\ 0, & \text{otherwise} . \end{cases} \tag{3.6.10}$$

Note that $h(n) = a_n$ for $0 \leq n \leq q$. We see from (3.6.10) that the autocorrelation $R_x(n)$ is of finite length $2q + 1$ and the process is said to be of order q. The reason for calling it a moving average is obvious because $(2q + 1)$ points of the autocorrelation are used and, as more and more points of the autocorrelation are obtained, only $2q + 1$ of them are used each time to update the power spectra. To obtain the power spectra we note that

290

$$S_x(z) = \sum_{n=-q}^{+q} R_x(n)z^{-n}$$

$$= \sum_{n=-q}^{+q} \sum_{k=0}^{q} a_k a_{k-n}^* z^{-n}$$

$$= \sum_{k=0}^{q} a_k z^{-k} \sum_{m=0}^{q} a_m^* z^{-m} \quad . \tag{3.6.11}$$

The autocorrelation matrix $R_x(n)$ is complex conjugate symmetrical, i.e.,

$$R_x(-n) = R_x^*(n) \quad . \tag{3.6.12}$$

Hence the zeros of $S_x(z)$ must occur in reciprocal pairs and we can always write

$$S_x(z) = \alpha^2 \prod_{k=1}^{q} (1 - z_k z^{-1})(1 - z_k^* z^{-1}) \quad , \tag{3.6.13}$$

where α is a real scalar constant. Comparing (3.6.11) and (3.6.13) we obtain

$$\sum_{k=0}^{q} a_k z^{-1} = \alpha \prod_{k=1}^{q} (1 - z_k z^{-1}) \quad . \tag{3.6.14}$$

The a_k values can be obtained in terms of z_k from (3.6.14) by equating equal powers of z^{-1}. Note that the factorization given by (3.6.13) is unique. However, as the roots occur in reciprocal pairs, a term like $(1 - z_k z^{-1})$ can be replaced by $(1 - z_k^{-1} z^{-1})$ without altering (3.6.13) but altering instead the values of a_k obtained from (3.6.14). In general, there are two sets of a_k parameters that can be obtained. A minimum phase requirement is used to obtain the optimum set. Note that, if only the power spectrum is needed, without the model, one can directly derive

$$S_x(\omega) = \sum_{n=-q}^{+q} w(n) R_x(n) e^{-j\omega n} \quad , \tag{3.6.15}$$

where a window function $w(n)$ is added to reduce the sidelobes due to truncation.

3.6.2 AR Model

For the AR model

$$A_q(\omega) = a_0 \quad . \tag{3.6.16}$$

Thus (3.6.6) becomes

$$x(n) + \sum_{k=1}^{p} b_k x(n-k) = a_0 \varepsilon(n) \quad . \tag{3.6.17}$$

Equation (3.6.7) is rewritten for this model as

$$R_x(n) + \sum_{k=1}^{p} b_k R_x(n-k) = \begin{cases} |a_0|^2, & n = 0, \\ 0, & n \geq 0. \end{cases} \tag{3.6.18}$$

For this case, $h(0) = a_0$ and $h(n) = 0$ for $n < 0$. Equation (3.6.17) can be expressed in matrix form as

$$\begin{bmatrix} R_x(0) & R_x(-1) & \cdots & R_x(-p) \\ R_x(1) & R_x(0) & \cdots & R_x(-p+1) \\ \vdots & \vdots & & \vdots \\ R_x(p) & R_x(p-1) & \cdots & R_x(0) \end{bmatrix} \begin{bmatrix} 1 \\ b_1 \\ \vdots \\ b_p \end{bmatrix} = \begin{bmatrix} |a_0|^2 \\ 0 \\ \vdots \\ 0 \end{bmatrix}, \tag{3.6.19}$$

which in turn can be written as

$$RB = |a_0|^2 e_1 \quad, \tag{3.6.20}$$

where R is a $(p+1) \times (p+1)$ matrix, B is a $(p+1) \times 1$ matrix and e_1 is the $(p+1) \times 1$ matrix with p elements zero and the first one having a value of 1. The elements of matrix R are given by

$$R_{ij} = R_x(i-j) = R_{ji}^* \quad \begin{cases} 1 \leq i \leq p+1, \\ 1 \leq j \leq p+1. \end{cases} \tag{3.6.21}$$

One can solve for B from (3.6.19) to obtain

$$B = |a_0|^2 R^{-1} e_1 \quad. \tag{3.6.22}$$

Note that, to obtain the values of B, one can use the Levinson-Durbin algorithm. Equation (3.6.19) can also be obtained from the maximum entropy principle. For this purpose one maximizes the entropy measure given by

$$\int_{-\pi}^{+\pi} \log [S_x(\omega)] d\omega$$

subject to conditions of known $p+1$ autocorrelation values $R_x(n)$. The spectral density in the AR model is given by

$$\begin{aligned} S(\omega) &= \left| \frac{a_0}{1 + b_1 e^{-j\omega} + \ldots + b_p e^{-jp\omega}} \right|^2 \\ &= \frac{|a_0|^2}{|B_p(\omega)|^2} \\ &= \frac{|a_0|^2}{\displaystyle\prod_{k=1}^{p}(1 - p_k e^{-j\omega})(1 - p_k^* e^{j\omega})} \quad, \end{aligned} \tag{3.6.23}$$

where p_k are roots of $B_p(z)$. Thus, this model is ideal for spectra containing peaks.

3.6.3 ARMA Model

The MA model is ideal if the spectrum contains sharp notches (i.e., zeros) whereas the AR model is more effective if the spectrum contains peaks (i.e., poles). In general, it is expected that the spectrum may contain both peaks and notches and one must use a combination of both models, i.e., the ARMA model. For this general case, no simplification results in (3.6.4) and (3.6.6). The power spectra in the ARMA model are given by

$$
S(\omega) = \left| \frac{A_q(\omega)}{B_p(\omega)} \right|^2
$$

$$
= |b_0|^2 \frac{\displaystyle\prod_{k=1}^{q}(1 - z_k e^{-j\omega})(1 - z_k^* e^{-j\omega})}{\displaystyle\prod_{k=1}^{p}(1 - p_k e^{-j\omega})(1 - p_k^* e^{-j\omega})} . \tag{3.6.24}
$$

Thus the ARMA model has q zeros located at z_k and p poles located at p_k.

The solution of (3.6.7) to obtain a_k and b_k is very difficult because these parameters appear in a nonlinear fashion through the impulse response $h(n)$. However, considerable simplification is obtained if a_k and b_k are evaluated separately. Using this method, one obtains linear equations for all b_k. For $n > q$, (3.6.7) becomes

$$
\sum_{k=0}^{p} b_k R_x(n - k) = 0 \quad \text{for} \quad n \geq q + 1 . \tag{3.6.25}
$$

One can solve the above set of linear equations for b_k. However, to obtain b_k, only a small subset of autocorrelation values are used and it is better to use more than p R_x values for a better model. This overdetermined set is generally solved by a least squares fit to obtain the b_k's. If we choose $q + 1 \leq n \leq q + t$ where $t > p + 1$, then (3.6.7) can be written as

$$
\begin{bmatrix}
R_x(q+1) & R_x(q) & \cdots & R_x(q-p+1) \\
R_x(q+2) & R_x(q+1) & \cdots & R_x(q-p+2) \\
\vdots & \vdots & & \vdots \\
R_x(q+t) & R_x(q+t-1) & \cdots & R_x(q-p+t)
\end{bmatrix}
\begin{bmatrix}
1 \\
b_1 \\
\vdots \\
b_p
\end{bmatrix}
=
\begin{bmatrix}
0 \\
0 \\
\vdots \\
0
\end{bmatrix} \tag{3.6.26}
$$

or

$$
R_1 B = \theta ,
$$

where θ is the $t \times 1$ zero vector, R_1 is the $t \times (p+1)$ autocorrelation matrix and B is the $(p+1) \times 1$ AR parameter vector.

There are two methods of solving (3.6.26) for B. We shall discuss them without derivation.

Method 1: Find the orthonormal eigenvector x_1 of the positive definite Hermitian matrix $R_1^* R_1$ associated with its minimum eigenvalue. Then

293

$$B = \frac{1}{x_1(1)} \boldsymbol{x}_1 \qquad\qquad (3.6.27)$$

where $x_1(1)$ denotes the first component of $\boldsymbol{x}_1$. This method involves minimizing $B^* R_1^* R_1 B$ with the constraint $B^* B = I$.

Method 2: Solve the following equation for B:

$$R_1^* R_1 B = \alpha e_1 \quad , \qquad\qquad (3.6.28)$$

where the normalizing constant α is chosen such that $b_0 = 0$. In this method, one is minimizing $B^* R_1^* R_1 B$ with the constraint $b_0 = 1$.

3.7 Kalman Filtering

We have already discussed the matched, Wiener, adaptive Widrow-Hoff and lattice filters. The Kalman filter is quite different from these, mainly because its formulation is in terms of the state-space method. Actually, it can be regarded as a least mean squares estimation problem (discussed in Sect. 3.5) using this state-space formulation. Its practical success can also be attributed to the recursive solution of the problem.

3.7.1 State-Space Formulation

Let us consider the most general case of a linear system with random forcing functions $W(t)$ and deterministic inputs $U(t)$. The system state vector $\boldsymbol{x}(t)$ can be written in terms of the matrices $F(t)$, $G(t)$ and $L(t)$ as follows:

$$\dot{\boldsymbol{x}}(t) = \overline{\overline{F}}(t)\boldsymbol{x}(t) + \overline{\overline{G}}(t)\boldsymbol{w}(t) + \overline{\overline{L}}(t)\boldsymbol{u}(t) \quad . \qquad\qquad (3.7.1)$$

The output state vector $\boldsymbol{y}(t)$ is related to the state vector $\boldsymbol{x}(t)$ through the matrix $\overline{\overline{B}}(t)$:

$$\boldsymbol{y}(t) = \overline{\overline{B}}(t)\boldsymbol{x}(t) \quad . \qquad\qquad (3.7.2)$$

The important point to note in (3.7.1) is that it is a first-order vector-matrix differential equation. One might wonder how a process like that shown in Fig. 3.56 and discussed in connection with the ARMA model can be represented by state vectors. This is accomplished as follows: The differential equation satisfying the process is given by

$$[D^n + a_{n-1} D^{n-1} + \ldots a_1(t)D + a_0(t)]x(t)$$
$$= [b_{n-1} D^{n-1} + \ldots b_0]w(t) \quad . \qquad\qquad (3.7.3)$$

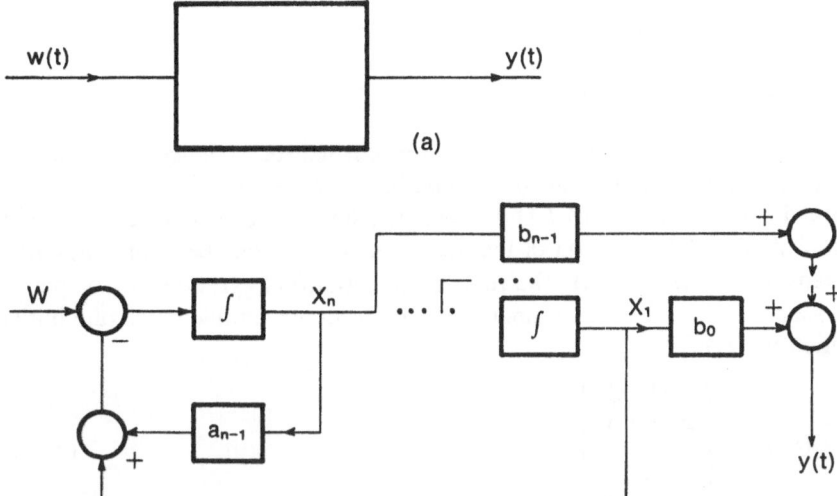

Fig. 3.56. (a) Block diagram of a system with input $\omega(t)$ and output $y(t)$. (b) State-space representation of the system in (3.8.5)

Defining

$$x_2 = \dot{x}_1 \quad ,$$
$$x_3 = \dot{x}_2 \quad ,$$
$$\vdots \qquad\qquad (3.7.4)$$
$$x_n = \dot{x}_{n-1} \quad ,$$

one finds (3.7.3) is equivalent to

$$\dot{x}_1 = x_2 \quad ,$$
$$\dot{x}_2 = x_3 \quad ,$$
$$\vdots \qquad\qquad (3.7.5)$$
$$\dot{x}_{n-1} = x_n \quad ,$$
$$\dot{x}_n = -a_0 x_1 + \ldots + w(t) \quad ,$$
$$y = b_0 x_1 + b_2 x_2 + \ldots + b_{n-1} x_n \quad .$$

Writing this in the form of (3.7.1 and 2), we have

$$\begin{bmatrix} \dot{x}_1 \\ \dot{x}_2 \\ \vdots \\ \dot{x}_n \end{bmatrix} = \begin{bmatrix} 0 & 1 & 0 & \cdots \\ 0 & 0 & 1 & \cdots \\ \vdots & \vdots & \vdots & \\ -a_0 & -a_1 & -a_2 & \cdots & -a_{n-1} \end{bmatrix} \begin{bmatrix} x_1 \\ x_2 \\ \vdots \\ x_n \end{bmatrix} + \begin{bmatrix} 0 \\ 0 \\ \vdots \\ w \end{bmatrix} \qquad (3.7.6)$$

295

and

$$y = [b_0 \ldots b_{n-1}] \begin{bmatrix} x_1 \\ \vdots \\ x_n \end{bmatrix} \quad . \tag{3.7.7}$$

In general, $\dot{x}$ is an $n \times 1$ matrix, F is an $n \times n$ matrix, G is an $n \times r$ matrix, W is an $r \times 1$ matrix, L is an $n \times s$ matrix and U is an $s \times 1$ matrix, where $r \leq n$ and $s \leq n$. Equation (3.7.1) consists of a homogeneous unforced part and two forced parts. The solution of the homogeneous part can be written in terms of a transition matrix $\phi(t, t_0)$. The ith column $\phi_i(t, t_0)$ of $\phi(t, t_0)$ is the output of the system at time t due to an input 1 at time t_0 to the ith integrator (the initial condition for $\dot{x}_i$)

$$\phi_i(t, t_0) = \begin{bmatrix} x_1(t, t_0)_i \\ x_2(t, t_0)_i \\ \vdots \\ x_n(t, t_0)_i \end{bmatrix} \quad . \tag{3.7.8}$$

Because of linearity, we see that the output due to $\bar{x}(t_0)$ at t_0 must be

$$\begin{bmatrix} x_1(t) \\ x_2(t) \\ \vdots \\ x_n(t) \end{bmatrix} = \begin{bmatrix} \phi_1(t, t_0) & \phi_2(t, t_0) & \cdots & \phi_n(t, t_0) \end{bmatrix} \begin{bmatrix} x_1(t_0) \\ x_2(t_0) \\ \vdots \\ x_n(t_0) \end{bmatrix} \tag{3.7.9}$$

or

$$x(t) = \overline{\overline{\Phi}}(t, t_0)x(t_0) \quad .$$

We see that the transition matrix must satisfy the following matrix equation:

$$\frac{d}{dt}\overline{\overline{\Phi}}(t, t_0) = \overline{\overline{F}}(t)\overline{\overline{\Phi}}(t, t_0) \quad , \quad \text{where} \tag{3.7.10}$$

$$\Phi(t, t) = I \quad . \tag{3.7.11}$$

We note the following relationships for any t_0, t_1, t_2:

$$\Phi(t_2, t_0) = \Phi(t_2, t_1)\Phi(t_1, t_0) \quad , \tag{3.7.12}$$

$$\Phi(t, t) = I = \Phi(t, t_0)\Phi(t_0, t) \quad ,$$
$$\Phi^{-1}(t, t_0) = \Phi(t_0, t) \quad , \tag{3.7.13}$$

$$|\Phi(t, t_0)| \neq 0 \quad .$$

For a stationary system

$$\Phi(t, t_0) = \Phi(t - t_0) \tag{3.7.14}$$

and from (3.7.10) we have

$$\Phi(t - t_0) = e^{F(t-t_0)} \quad . \tag{3.7.15}$$

Coming back to (3.7.1), we can now write the complete solution as

$$x(t) = \Phi(t, t_0)x(t_0) + \int\limits_{t_0}^{t} \Phi(t, \tau)G(\tau)\omega(\tau)d\tau + \int\limits_{t_0}^{t} \Phi(t, \tau)L(\tau)U(\tau)d\tau. \quad (3.7.16)$$

The above equation can be written in the discrete time system as

$$X_{k+1} = \Phi_k X_k + G_k W_k + L_k U_k \quad , \quad\quad\quad\quad\quad\quad (3.7.17)$$

where

$$X(k) = X(t_k) \quad , \quad\quad\quad\quad\quad\quad\quad\quad\quad\quad\quad (3.7.18)$$

$$\Phi(k) = \phi(t_{k+1}, t_k) \quad , \quad\quad\quad\quad\quad\quad\quad\quad\quad (3.7.19)$$

$$G_k W_k = \int\limits_{t_k}^{t_k+1} \Phi(t_{k+1}, \tau)G(\tau)\omega(\tau)d\tau \quad\quad\quad\quad\quad (3.7.20)$$

and

$$L_k U_k = \int\limits_{t_k}^{t_k+1} \Phi(t_{k+1}, \tau)L(\tau)U(\tau)d\tau \quad . \quad\quad\quad (3.7.21)$$

Note that $G_k W_k$ and $L_k U_k$ cannot be separated. Hence in many cases, in the absence of any deterministic input, we write

$$X_{k+1} = \Phi_k X_k + W_k \quad \text{and} \quad\quad\quad\quad\quad\quad (3.7.22)$$

$$Y_k = B_k X_k \quad . \quad\quad\quad\quad\quad\quad\quad\quad\quad\quad (3.7.23)$$

The state space formulation is not unique because one can always define a new $x'(t)$ given by

$$\boldsymbol{x}'(t) = \overline{\overline{A}}(t)x(t) \quad , \quad\quad\quad\quad\quad\quad\quad\quad (3.7.24)$$

where $\overline{\overline{A}}$ is not singular.

3.7.2 The Kalman Filter

Let us consider a problem which can be modeled as

$$X_{k+1} = \phi_k X_k + W_k \quad \text{and}$$

$$Y_k = B_k X_k \quad .$$

We shall assume that W_k is an uncorrelated zero mean function having covariance given by Q_k. Thus we have

$$E\{W_k W_j^T\} = \begin{cases} Q_k, & k = j \quad , \\ 0, & k \neq j \quad . \end{cases} \quad\quad\quad (3.7.25)$$

Let us also consider that a measurement is made at each discrete time to update the process as follows:

$$Z_k = H_k X_k + \nu_k \quad . \tag{3.7.26}$$

Here Z_k is the $m \times 1$ measurement vector, H_k is the $m \times n$ matrix giving the ideal connection between the measurement and the state vector at $t = t_k$ and ν_k is the measurement error, an $m \times 1$ matrix. We shall assume that the ν_k sequence is also white, zero mean, and that its covariance is given by

$$E\{\nu_k \nu_j^T\} = \begin{cases} R_k, & k = j \quad , \\ 0, & k \neq j \quad . \end{cases} \tag{3.7.27}$$

We assume that the cross-correlation between ν_k and W_k is zero. We shall discuss later how to deal with the situation when this is not the case. The Kalman filtering problem is to obtain the optimum updated estimate $\hat{X}_k(+)$ given the measurement Z_k and the previous estimate $\hat{X}_k(-)$. As a linear recursive solution is sought, we assume

$$\hat{X}_k(+) = K_k' \hat{X}_k(-) + K_k Z_k \quad , \tag{3.7.28}$$

where K_k' and K_k are unknown weighting matrices to be determined from the least mean squares error condition. Defining the errors by e_k, we have

$$e_k^+ = X_k - \hat{X}_k(+) \quad \text{and} \tag{3.7.29}$$

$$e_k^- = X_k - \hat{X}_k(-) \quad . \tag{3.7.30}$$

Using (3.7.22, 26 and 28) we have

$$e_k^+ = [K_k' + K_k H_k - I]\hat{X}_k + K_k' \hat{X}_k(-) + K_k \nu_k \quad . \tag{3.7.31}$$

By definition $E\{\nu_k\} = 0$.

If we assume that $E\{e_k^-\}$ is also zero, i.e., that we have a previous unbiased estimate, then to obtain an unbiased estimate, we must then have

$$E\{e_k^+\} = 0 \quad \text{or}$$

$$K_k' = I - K_k H_k \quad . \tag{3.7.32}$$

Thus (3.7.28) becomes

$$\hat{X}_k(+) = (I - K_k H_k)\hat{X}_k(-) + K_k Z_k \quad \text{or} \tag{3.7.33}$$

$$\hat{X}_k(+) = \hat{X}_k(-) + K_k[Z_k - H_k \hat{X}_k(-)] \quad .$$

The estimation error for this case is given by

$$e_k^+ = (I - K_k H_k)\hat{X}_k(-) + K_k \nu_k \quad . \tag{3.7.34}$$

Let us define error covariance by P_k. Thus

$$P_k^+ = E\{e_k^+ e_k^{+T}\}$$

$$= E\{(X_k - \hat{X}_k(+))(X_k - X_k(+))^{\mathrm{T}}\}$$
$$= E\{[(X_k - \hat{X}_k(-)) - K_k(H_k X_k + \nu_k - H_k \hat{X}_k(-))]\}$$

or

$$P_k^+ = (I - K_k H_k)P_k^-(I - K_k H_k)^{\mathrm{T}} + K_k R_k K_k^{\mathrm{T}} \quad . \tag{3.7.35}$$

We are interested in finding the gain constant K_k that minimizes P_k^+ in the least squares sense. We must therefore minimize the trace of P_k^+ with respect to K_k:

$$\frac{\partial}{\partial K_k}(\mathrm{Tr}\{P_k\}) = 0 \quad \text{or} \tag{3.7.36}$$

$$-2(I - K_k H_k)P_k^- H_k^{\mathrm{T}} + 2K_k R_k = 0 \quad . $$

Solving for K_k, we find the optimum gain

$$K_k = P_k^- - H_k^{\mathrm{T}}[H_k P_k^- H_k^{\mathrm{T}} + R_k]^{-1} \quad , \tag{3.7.37}$$

where K_k is generally referred to as the Kalman gain matrix. Substituting this value of K_k in (3.7.35) we obtain the optimized value

$$P_k^+ = (I - K_k H_k)P_k^- \quad . \tag{3.7.38}$$

To use these equations recursively, we need P_{k+1}^- and $X_{k+1}(-)$. These are obtained as follows:

$$X_{k+1}(-) = \Phi_k X_k(+) \quad \text{and} \tag{3.7.39}$$

$$P_{k+1}^- = \Phi_k P_k^+ \Phi_k^{\mathrm{T}} + Q_k \quad . \tag{3.7.40}$$

It can also be shown by straightforward manipulations that

$$K_k = P_k^+ H_k^{\mathrm{T}} R_k^{-1} \quad , \tag{3.7.41}$$

which is much simpler than (3.7.37). The fact that it is an optimal solution can easily be verified by performing the following calculation to show that the error and the estimate are uncorrelated and orthogonal to each other:

$$E\{e_k^+ X_k(+)\} = 0 \quad . \tag{3.7.42}$$

The block diagram and information flow diagram of the discrete Kalman filter equations are shown in Fig. 3.57.

We note the following:

1) If W_k and ν_k are Gaussian, then the filter is optimum. Even a nonlinear filter cannot do better.

2) If W_k and ν_k are not Gaussian, then the filter is an optimum linear filter.

3) For the continuous case, the error covariance P obeys the equation

$$\dot{P} = FP + PF^{\mathrm{T}} + GQG^{\mathrm{T}} - PG^{\mathrm{T}}R^{-1}HP \quad . \tag{3.7.43}$$

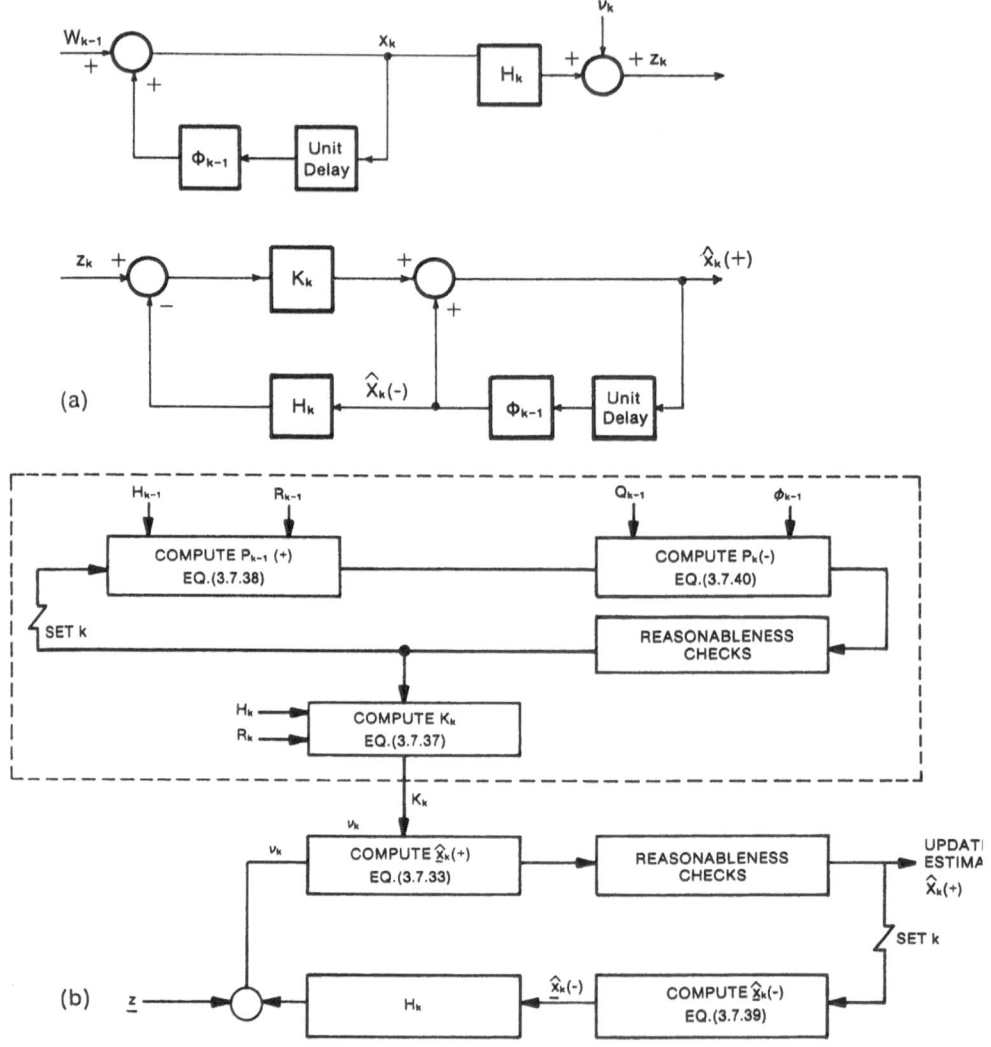

Fig. 3.57. (a) Information flow diagram and (b) the block diagram of the discrete Kalman filter equations

The Kalman filter solution is given by

$$\dot{\hat{X}} = F\hat{X} + PH^T R^{-1}[Z - H\hat{X}]$$
$$= F\hat{X} + K(t)[Z - H\hat{X}] \quad . \tag{3.7.44}$$

Physically, the rate of change of variance or uncertainty $\dot{P}$ has three components: $(FP + PF^T)$ is due to the homogeneous solution part, GQG^T is the increase of uncertainty due to process noise and the term $PH^T R^{-1}HP$ accounts for the decrease in uncertainty due to the measurement. Equation (3.7.43) is nonlinear and is known as the matrix Ricatti equation.

4) If $W(t)$ and $\nu(t)$ are correlated to each other, we have

$$E\{\omega(t)\nu^T(t)\} = c(t)\delta(t - \tau) \quad . \tag{3.7.45}$$

For this case one can make an equivalent problem given by

$$\dot{X} = (F - DH)X + DZ + (GW - D\nu) \tag{3.7.46}$$

and

$$Z = HX + \nu \quad , \tag{3.7.47}$$

where $D = GCR^{-1}$. Now $(GW - D\nu)$ is the process noise and DZ is the known input. For this case

$$E\{(GW - D\nu)\nu^T\} = 0 \quad . \tag{3.7.48}$$

Thus the Kalman gain for this case is given by

$$K = [PH^T + GC]R^{-1} \quad . \tag{3.7.49}$$

5) If the system model includes deterministic inputs, we must use the following equations for the discrete case:

$$X_k = \Phi_{k-1}X_{k-1} + W_{k-1} + L_{k-1}U_{k-1} \tag{3.7.50}$$

and

$$Z_k = H_k X_k + \nu_k \quad .$$

The Kalman filtering equations for this case are

$$\begin{aligned}
\hat{X}_{k+1} &= \Phi_{k-1}\hat{X}_{k-1}(+) + L_{k-1}U_{k-1} \\
&\quad + K_k[Z_k - H_k\Phi_{k-1}\hat{X}_{k-1}(+) - H_kL_{k-1}U_{k-1}] \quad .
\end{aligned} \tag{3.7.51}$$

The corresponding equations for the continuous case are

Model

$$\dot{X}(t) = F(t)X(t) + G(t)W(t) + L(t)U(t) \quad , \tag{3.7.52}$$

$$Z(t) = H(t)X(t) + \nu(t) \quad . \tag{3.7.53}$$

Filter

$$\dot{\hat{X}}(t) = F(t)\hat{X}(t) + L(t)U(t) + K(t)[Z(t) - H(t)\hat{X}(t)] \quad . \tag{3.7.54}$$

6) The Ricatti equation can only be solved analytically for simple problems, such as one with constant coefficients. However, various numerical techniques are available for more complicated problems.

7) It can be easily shown that

$$P_k^{-1}(+) = P_k^{-1}(-) + H_k^T R_k^{-1} H_k \quad . \tag{3.7.55}$$

From the above equation, we observe that the variance of the estimated error reduces by the term $H_k^T R_k^{-1} H_k$. If the measurement noise is large, i.e., R_k^{-1}

is small, then its contribution to $P_k(+)$ is negligible. However, if R_k^{-1} is large, it significantly reduces $P_k(+)$. If R_k is zero, then the problem becomes ill-conditioned and special care has to be taken to obtain a solution.

8) The Kalman filter and the Wiener filter are different solutions using the least mean squares estimate technique for the same problem. The Wiener approach leads to an integral equation whereas the Kalman solution is a differential equation. They are thus equivalent and one can be derived from the other. For the case of a Wiener filter, one performs the spectral factorization and then tries to synthesize the filter from the impulse response. For the case of a Kalman filter, one needs to solve a set of coupled nonlinear algebraic equations. As discussed, however, this can be conveniently performed recursively.

3.7.3 Solution of the Ricatti Equation with Constant Coefficients

The Ricatti equation is

$$\dot{P} = FP + PF^T - PH^TR^{-1}HP + GQG^T \tag{3.7.56}$$

with initial conditions $P(0) = P_0$. Substituting

$$P = XZ^{-1} \ , \tag{3.7.57}$$

$$Z(0) = 1 \ ,$$

$$PZ = X \quad \text{and} \tag{3.7.58}$$

$$\dot{P}Z + P\dot{Z} = \dot{X} \tag{3.7.59}$$

into (3.7.56), we obtain

$$\dot{X} = (PF + PF^T - PH^TR^{-1}HP + GQG^T)Z + P\dot{Z}$$

or

$$P[F^TZ - H^TR^{-1}HX + \dot{Z}] + [FX + GQG^TZ - \dot{X}] = 0 \ , \tag{3.7.60}$$

if

$$\dot{X} = FX + GQG^TZ \quad \text{and} \tag{3.7.61}$$

$$\dot{Z} = H^TR^{-1}HX - F^TZ \ . \tag{3.7.62}$$

We note that the substitution given by (3.7.57) has converted the nonlinear Ricatti equation into two linear differential equations.

3.7.4 Square Root Filtering

In the actual implentation of a Kalman filter, either by means of a digital computer or in any other way, the desired dynamic range for the variance P may pose problems. It is perfectly possible to start with a value of $P \sim 10^8$ and finally approach $P \sim 1$. If, in place of the variance P, one deals with the standard deviation (somewhat equivalent to the square root of P) which is then the quantity to be calculated and updated or propagated, the requirement on the dynamic range is reduced significantly. This technique for updating the square root of P is known as square root filtering. Note that the matrix square root A is defined as B where

$$A = BB^{\mathrm{T}} \quad .$$

This decomposition can be performed using Cholesky decomposition, as discussed in Appendix A. Note that the square root is not unique. However, it can be made unique by demanding that it be a triangular or symmetric matrix. To understand square root filtering, let us consider $Q_k = 1$ and that the measurement is a scalar quantity, i.e., R is scalar and H is a row vector $(1 \times n)$. With these simplifications, we obtain

$$P = P^- - P^- H^{\mathrm{T}}[HP^{-1}H^{\mathrm{T}} + R]^{-1} HP^{-1} \tag{3.7.63}$$

and

$$K = P^- H^{\mathrm{T}}(HP^- H^{\mathrm{T}} + R)^{-1} \quad . \tag{3.7.64}$$

Denoting square roots of P by $P^{1/2}$, we have

$$P = P^{1/2} P^{1/2\mathrm{T}} \quad \text{and} \quad P^- = P_-^{1/2} P_-^{1/2\mathrm{T}} \quad . \tag{3.7.65}$$

Substituting these in (3.7.63, 64) we obtain

$$K = \frac{P_-^{1/2}(HP_-^{1/2})^{\mathrm{T}}}{[HP_-^{1/2})(HP_-^{1/2})^{\mathrm{T}} + R]} \tag{3.7.66}$$

and

$$P^{1/2} P^{1/2\mathrm{T}} = P_-^{1/2}[I - (HP_-^{1/2})^{\mathrm{T}}(HP_-^{1/2})/r] P_-^{1/2\mathrm{T}} \tag{3.7.67}$$

where r is a scalar:

$$r = R + (HP_-^{1/2})(HP_-^{1/2})^{\mathrm{T}} = R + h_p \tag{3.7.68}$$

and

$$h_p = (HP_-^{1/2})(HP_-^{1/2})^{\mathrm{T}} \quad . \tag{3.7.69}$$

To make the right hand side of (3.7.67) into a square, we demand

$$I - (HP_-^{1/2})^{\mathrm{T}}(HP_-^{1/2})/r$$
$$= [I - a(HP_-^{1/2})^{\mathrm{T}}(HP_-^{1/2})][I - a(HP_-^{1/2})^{\mathrm{T}}(HP_-^{1/2})]^{\mathrm{T}}$$
$$= I - a(HP_-^{1/2})^{\mathrm{T}}(HP_-^{1/2}) - ah_p + a^2 h_p (HP_-^{1/2})^{\mathrm{T}}(HP_-^{1/2})$$

or

$$\left[a^2 h_p - 2a + \frac{1}{r}\right](HP_-^{1/2})^{\mathrm{T}}(HP_-^{1/2}) = 0$$

or

$$a = \frac{1 \pm \sqrt{1 - h_p/r}}{h_p} \tag{3.7.70}$$

because $(HP_-^{1/2})^{\mathrm{T}}(HP_-^{1/2})$ is not zero.

Choosing the + sign in (3.7.70), we obtain

$$a = \frac{1}{h_p}\left(1 - \sqrt{\frac{R}{r}}\right) \quad .$$

(3.7.71)

Then (3.7.67) yields

$$P^{1/2}(P^{1/2})^{\mathrm{T}} = P^{1/2}_{-}[I - a(HP^{1/2}_{-})^{\mathrm{T}}(HP^{1/2}_{-})]$$
$$\times [I - a(HP^{1/2}_{-})^{\mathrm{T}}(HP^{1/2}_{-})]^{\mathrm{T}}(P^{1/2}_{-})^{\mathrm{T}}$$

(3.7.72)

or

$$P^{1/2} = P^{1/2}_{-}[I - a(HP^{1/2}_{-})^{\mathrm{T}}(HP^{1/2}_{-})] \quad .$$

Note that $HP^{1/2}_{-}$ is a $1 \times n$ matrix. We thus obtain for our case

$$\begin{aligned} P^{-}_{k+1} &= \phi_k P_k \phi_k^{\mathrm{T}} + Q_k \\ &= \phi_k P_k \phi_k^{\mathrm{T}} \\ &= \phi_k P_k^{1/2}(P_k^{1/2})^{\mathrm{T}} \phi_k^{\mathrm{T}} \quad \text{or} \end{aligned}$$

$$P^{1/2}_{k+1}(-) = \phi_k P_k^{1/2} \quad .$$

(3.7.73)

The above equations are used to calculate and update the Kalman filter where the square root of P is used. As mentioned previously, this square root filtering puts less demand on the dynamic range of computations.

3.8 Two-Dimensional Signal Processing

We have so far considered signals that are functions of time only. For optics, however, two-dimensional signals which are functions of two space coordinates are also important. In this section we shall develop the fundamentals of two-dimensional signal processing with examples applicable to Fourier optics. Because time is not involved as an independent variable in most cases, the signals are not causal and no Hilbert transform relationships exist between the real and imaginary parts of a system response.

3.8.1 Analog Signals and Systems

We define the following two-dimensional functions:

Step function $\quad u(x,y) = \begin{cases} 1 & x \geq 0, \quad y \geq 0 \quad , \\ 0 & x < 0, \quad y < 0 \quad , \end{cases}$

(3.8.1)

Sign function $\quad \mathrm{sgn}(x,y) = \begin{cases} +1 & x > 0, \quad y > 0 \quad , \\ -1 & x < 0, \quad y < 0 \quad , \end{cases}$

(3.8.2)

$$\text{Rectangular pulse} \quad P(x, y) = \begin{cases} 1 & |x| \leq \frac{1}{2}, \quad |y \leq \frac{1}{2} \\ 0 & \text{otherwise} \end{cases}, \tag{3.8.3}$$

$$\text{Delta function} \quad \delta(x, y) = \begin{cases} 1 & x = 0, \quad y = 0 \\ 0 & \text{otherwise} \end{cases}. \tag{3.8.4}$$

Note that

$$u(x, y) = u(x)u(y) \quad, \tag{3.8.5}$$

$$\text{sgn}\,(x, y) = \text{sgn}\,(x)\,\text{sgn}\,(y) \quad, \tag{3.8.6}$$

$$P(x, y) = P(x)P(y) \quad, \tag{3.8.7}$$

$$\delta(x, y) = \delta(x)\delta(y) \quad. \tag{3.8.8}$$

The quantities on the right-hand sides of (3.8.5–8) are one-dimensional functions.

3.8.2 Linear Systems

For the system shown in Fig. 3.58, the input-output relationship is

$$g(x, y) = L[f(x, y)] \quad, \tag{3.8.9}$$

where L represents a functional with one-to-one mapping.

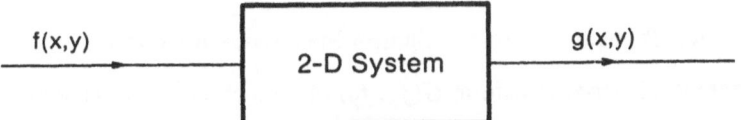

Fig. 3.58. Block diagram of a two-dimensional linear system

For a linear system

$$L[A_1 f_1(x, y) + A_2 f_2(x, y)] = A_1 L[f_1(x, y)] + A_2 L[f_2(x, y)] \tag{3.8.10}$$

where A_1 and A_2 are complex numbers and f_1 and f_2 are any two functions satisfying (3.8.9).

For a space-invariant system,

$$L[f(x - x_0, y - y_0)] = g(x - x_0, y - y_0) \quad, \tag{3.8.11}$$

where x_0 and y_0 are constants. The impulse response[5] $h(x, y; x', y')$ is defined as

$$h(x, y, x', y') = L[\delta(x - x', y - y')] \quad. \tag{3.8.12}$$

---

[5] Also called the point spread function or Green's function.

Thus we have the relationship

$$g(x, y) = \int_{-\infty}^{+\infty} \int_{-\infty}^{+\infty} h(x, y; x', y') f(x', y') dx' dy' \quad . \tag{3.8.13}$$

The above equation can be easily derived by noting that

$$f(x, y) = \int_{-\infty}^{+\infty} \int_{-\infty}^{+\infty} f(x', y') \delta(x - x', y - y') dx' dy' \quad . \tag{3.8.14}$$

For a space-invariant system

$$h(x, y; x', y') = h(x - x', y - y') \tag{3.8.15}$$

and (3.8.13) can be rewritten as

$$g(x, y) = \int_{-\infty}^{+\infty} \int_{-\infty}^{+\infty} h(x - x', y - y') f(x', y') dx' dy' \quad . \tag{3.8.16}$$

The last result is the two-dimensional convolution equation denoted by

$$g(x, y) = h(x, y) * f(x, y) \quad . \tag{3.8.17}$$

Note that, by definition,

$$\begin{aligned} g(x, y) &= h(x, y) * f(x, y) \\ &= f(x, y) * h(x, y) \quad . \end{aligned}$$

3.8.3 The Fourier Transform and the Spatial Frequency Response

The two-dimensional Fourier transform $G(f_x, f_y)$ of a function $g(x, y)$ is defined as

$$G(f_x, f_y) = \int_{-\infty}^{+\infty} \int_{-\infty}^{+\infty} g(x, y) e^{-j2\pi(f_x x + f_y y)} dx \, dy \quad . \tag{3.8.18}$$

It is assumed that $g(x, y)$ is bounded and goes to zero asymptotically as $x, y \to 0$. The inversion formula is given by

$$g(x, y) = \int_{-\infty}^{+\infty} \int_{-\infty}^{+\infty} G(f_x, f_y) e^{j2\pi(f_x x + f_y y)} df_x df_y \quad . \tag{3.8.19}$$

If the function $g(x, y)$ is separable, i.e.,

$$g(x, y) = g_1(x) g_2(y) \tag{3.8.20}$$

then

$$G(f_x, f_y) = G_1(f_x) G_2(f_y) \quad . \tag{3.8.21}$$

In many examples in optics, we can transform from rectangular coordinates (x, y)

into polar coordinates (r, θ). Then, if the functions are separable, we have

$$g(x, y) = g_3(r)g_4(\theta) \quad . \tag{3.8.22}$$

As θ is bounded in the region $0 \leq \theta \leq 2\pi$, we can write

$$g_4(\theta) = \sum_{k=-\infty}^{+\infty} c_k e^{jk\theta} \tag{3.8.23}$$

where

$$c_k = \frac{1}{2\pi} \int_0^{2\pi} g_4(\theta)e^{-jk\theta}\, d\theta \tag{3.8.24}$$

and k is a positive or negative integer, including zero. For this case

$$\mathcal{F}[g_3(r)g_4(\theta)] = \sum_{k=-\infty}^{+\infty} c_k(-j)^k e^{jk\phi} H_k[g_3(r)] \quad , \tag{3.8.25}$$

where

$$H_k[g_3(r)] = 2\pi \int_0^\infty r g_3(r) J_k(2\pi r \phi)\, dr \quad , \tag{3.8.26}$$

$$\phi = \sqrt{f_x^2 + f_y^2} \quad ,$$

$$\tan \phi = f_x/f_y \quad ,$$

and J_k is the kth order Bessel function. If the function is independent of θ or $g_4(\theta) = 1$, then we have

$$g(x, y) = g(r) \quad \text{and}$$

$$G(f_x, f_y) = G_r(\phi) = 2\pi \int_0^\infty r f(r) J_0(\phi r)\, dr \quad . \tag{3.8.27}$$

Here $J_0(r)$ is the zeroth-order Bessel function, given by

$$J_0(r) = \frac{1}{2\pi} \int_{-\pi}^{+\pi} e^{r \cos(\theta - \alpha)}\, d\theta \quad . \tag{3.8.28}$$

Note that $J_0(r)$ defined as in (3.8.28) is independent of α as $\cos \theta$ is periodic. Equation (3.8.27) defines what is known as the Hankel transform, except for the factor 2π. Denoting the Hankel transform by

$$G_H(\omega) = \int_0^\infty r g(r) J_0(\omega r)\, dr \tag{3.8.29}$$

we have

$$G_r(\phi_0) = 2\pi G_H(\omega) \quad \text{with} \tag{3.8.30}$$

307

$$\omega = \phi = \sqrt{f_x^2 + f_y^2} \quad . \tag{3.8.31}$$

The inversion formula is given by

$$g(r) = \int_0^\infty \omega G_H(\omega) J_0(r\omega) d\omega \quad . \tag{3.8.32}$$

For optical systems with circular symmetry in the plane perpendicular to the optical axis, the Hankel transform is very useful. Table 3.5 lists some useful functions and their Hankel transforms.

Table 3.5. Some useful functions, $f(r)$, and their Hankel transforms, $F_H(f_r)$

$f(r)$	$F_H(f_r)$
$1/r$	$1/f_r$
$\delta(r-a)$	$aJ_0(af_r)$
$\exp(-ar^2)$	$(1/2a)\exp(-f_r^2/4a)$
$\exp(jar^2)$	$(j/2a)\exp(-jf_r^2/4a)$
$\frac{\sin ar}{r}$	$\begin{cases} \dfrac{1}{\sqrt{f_r^2 - a^2}}, & f_r > a \\ 0, & f_r < a \end{cases}$
$f''(r) + \frac{1}{r}f(r)$	$-f_r^2 F_H(f_r)$
$f(\alpha r)$	$\frac{1}{\alpha^2} F_H(f_r)$
$\frac{J_n(r)}{r^n}$	$\begin{cases} \dfrac{(1 - f_r^2)^{n-1}}{2^{n-1}(n-1)!}, & f_r < 1 \\ 0, & f_r > 1 \end{cases}$
$P(a)$	$(a/f_r)J_1(f_r)$

The Hankel transform obeys the following relationships:

Symmetry property $G_H(r) \leftrightarrow g(\omega)$ (3.8.33)

Scaling $g(ar) \leftrightarrow \frac{1}{a^2} G_H(\frac{\omega}{a})$ (3.8.34)

Convolution $g_1(r) * g_2(r) \leftrightarrow 2\pi G_{H1}(\omega) G_{H2}(\omega)$ (3.8.35)

Perseval's Theorem $\int_0^\infty r g_1(r) g_2^*(r) dr = \int_{-\infty}^{+\infty} \omega G_{H1}(\omega) G_{H2}(\omega) d\omega$

(3.8.36)

The input-output relationship in the frequency domain for the space-invariant system shown in Fig. 3.59 can be written as

$$G(f_x, f_y) = H(f_x, f_y) F(f_x, f_y) \quad . \tag{3.8.37}$$

Here $H(f_x, f_y)$ is the Fourier transform of the impulse response $h(x,y)$. For an input given by

$$f(x,y) = e^{j2\pi(f_x x + f_y y)} \tag{3.8.38}$$

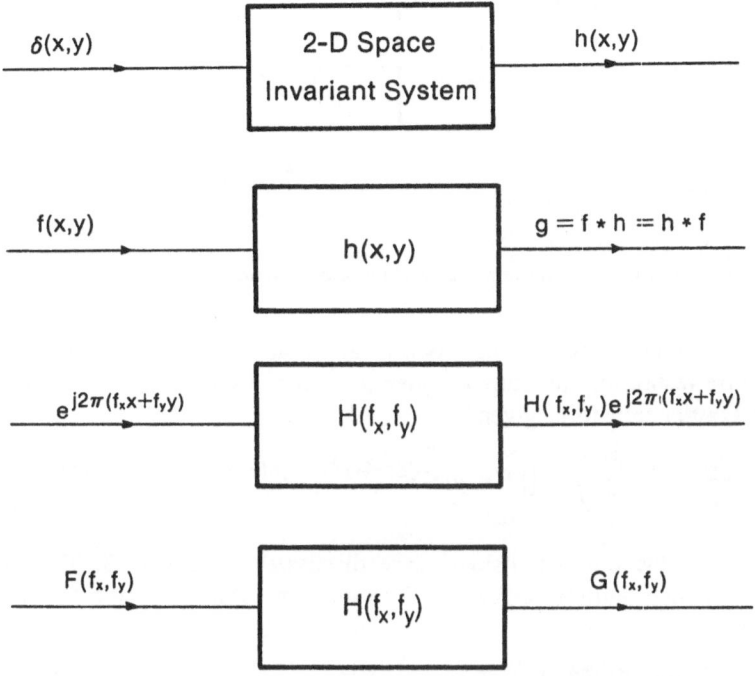

Fig. 3.59. Input-output relationship for a two-dimensional space-invariant system

the output is given by

$$g(x) = H(f_x, f_y)e^{j2\pi(f_x x + f_y y)} \quad , \tag{3.8.39}$$

where $H(f_x, f_y)$ is the frequency response of the two-dimensional system.

3.8.4 Examples of Fourier Transformation, Imaging, etc.

i) Chirp impulse response

$$h(x, y) = e^{j\beta(x^2 + y^2)} = e^{j\beta r^2} \quad , \tag{3.8.40}$$

$$H(f_x, f_y) = \left[\left(\frac{\pi}{\beta} \right)^{1/2} e^{-j\pi/4} \right] e^{j(\pi/\beta)(f_x^2 + f_y^2)} \tag{3.8.41}$$

$$H(f_r) = -j\left(\frac{\pi}{\beta} \right) e^{j(\pi/\beta)f_r^2} \quad , \tag{3.8.42}$$

$$f_r = \sqrt{f_x^2 + f_y^2} \quad .$$

ii) Two-dimensional Fourier transformation: multiply-convolve-multiply. Consider the system shown in Fig. 3.60. The signal $f(x, y)$ to be Fourier transformed is

309

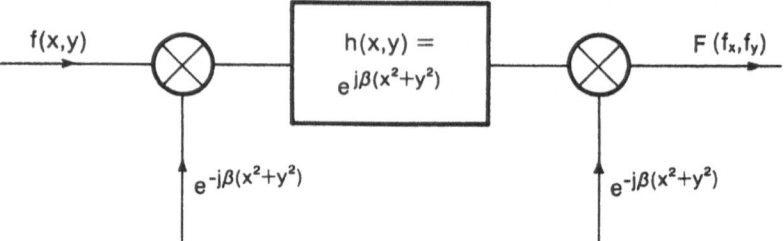

Fig. 3.60. Two-dimensional Fourier transformation: multiply-convolve-multiply

multiplied with a down-chirp and then used as an input to a system that has an up-chirp impulse response. The output is then multiplied by another down-chirp to obtain the Fourier transform given by

$$F\left(f_x = \frac{\beta x}{\pi}, f_y = \frac{\beta y}{\pi}\right) = \{[f(x,y)e^{-j\beta(x^2+y^2)}]*e^{j\beta(x^2+y^2)}\}e^{-j\beta(x^2+y^2)}$$

$$(3.8.43)$$

The proof follows the one-dimensional case discussed in Sect. 3.1.3. This is the multiply-convolve-multiply algorithm. The impulse response of the Fourier transformer is

$$h(x,y; x',y') = e^{-j2\beta(xx'+yy')} \quad . \tag{3.8.44}$$

Note that the system is not space-invariant.

iii) Convolve-multiply-convolve. The convolve-multiply-convolve algorithm is shown in Fig. 3.61. For this case too, the output is the Fourier transform given by

$$F\left(f_x = \frac{\beta x}{\pi}; f_y = \frac{\beta y}{\pi}\right) = \{[f(x,y)*e^{j\beta(x^2+y^2)}]e^{-j\beta(x^2+y^2)}\}*e^{j\beta(x^2+y^2)} \quad .$$

$$(3.8.45)$$

iv) Limited down-chirp. If the down-chirp in (3.8.45) is limited to $|x| < L_x/2$ and $|y| < L_y/2$, the output is given by

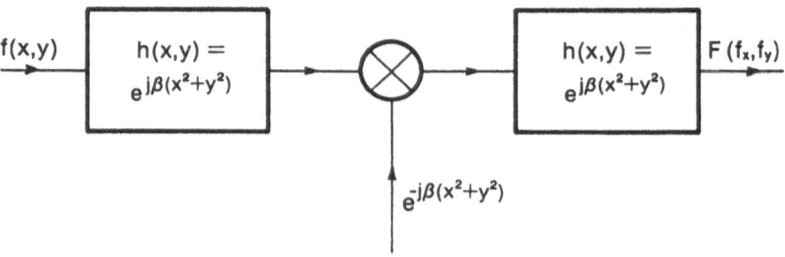

Fig. 3.61. Two-dimensional Fourier transformation: convolve-multiply-convolve

310

$$g(x,y) = \left\{ [f(x,y)*e^{j\beta(x^2+y^2)}]P\left(\frac{x}{L_x},\frac{y}{L_y}\right)e^{-j\beta(x^2+y^2)} \right\}*e^{j\beta(x^2+y^2)}$$

$$= \int_{-\infty}^{+\infty}\int_{-\infty}^{+\infty} f(x_1,y_1)I(x,x_1;y,y_1)e^{j\beta(x_1^2+y_1^2)}dx_1 dy_1 \qquad (3.8.46)$$

where

$$I(x,x_1;y,y_1) = e^{j\beta(x^2+y^2)}\iint e^{j\beta\mu(x^2+y^2)}P\left(\frac{x_2}{L_x},\frac{y_2}{L_y}\right)dx_2 dy_2 \qquad (3.8.47)$$

and

$$\mu(x_2,y_2) = x_2^2 + y_2^2 - 2x_2(x_1+x) - 2y_2(y_1+y) \quad . \qquad (3.8.48)$$

Using the principle of stationary phase for the case $\beta \gg 1$, as discussed in Appendix C, we obtain

$$I(x,x_1;y,y_1) \approx \sqrt{\frac{\pi}{\beta}}P\left(\frac{x_1+x}{L_x},\frac{y_1+y}{L_y}\right)e^{-j\beta(x_1^2+y_1^2+2xx_1+2yy_1)} + \cdots \quad .$$

$$(3.8.49)$$

Thus we obtain

$$g(xy) \approx \int_{-\infty}^{+\infty}\int_{-\infty}^{+\infty} f(x,y)e^{-j2\pi(f_x x_1+f_y y_1)}P\left(\frac{x+x}{L_x},\frac{y+y}{L_y}\right)dx_1 dy_1 \quad (3.8.50)$$

where

$$f_x = \frac{\beta x}{\pi} \quad \text{and} \quad f_y = \frac{\beta y}{\pi} \quad .$$

For $L_x, L_y \to \infty$ we note that

$$g(x,y) \to \mathcal{F}[f(x,y)]_{f_x=\beta x/\pi \,, f_y=\beta y/\pi} \quad . \qquad (3.8.51)$$

The impulse response of the system for this case is given by

$$h(x,y;x',y') = e^{-j2\beta(xx'+yy')}P\left(\frac{x+x_1}{L_x},\frac{y+y_1}{L_y}\right) \quad . \qquad (3.8.52)$$

If the multiplier has a general pupil function $W(x,y)$ rather than the rectangular function $P(x,y)$, the equations are modified as follows:

$$g(x,y) \approx \int_{-\infty}^{+\infty}\int_{-\infty}^{+\infty} f(x_1,y_1)e^{-j2\pi(f_x x_1+f_y y_1)}$$
$$\times W(x+x_1,y+y_1)dx_1 dy_1 \qquad (3.8.53)$$

and

$$h(x,y;x',y') \approx e^{-j2\beta(xx'+yy')}W(x+x',y+y') \quad . \qquad (3.8.54)$$

v) *Imaging.* Consider the system shown in Fig. 3.62. The impulse response of the system can be written as

$$h(x,y;y',y') = \{[\delta(x-x',y-y')*e^{j\beta(x^2+y^2)}]e^{-j\beta_2(x^2+y^2)}\}*e^{j\beta_3(x^2+y^2)}$$

$$= e^{j\beta_1(x^2+y^2)}e^{j\beta_3(x'^2+y'^2)}I(x,x';y,y') \qquad (3.8.55)$$

311

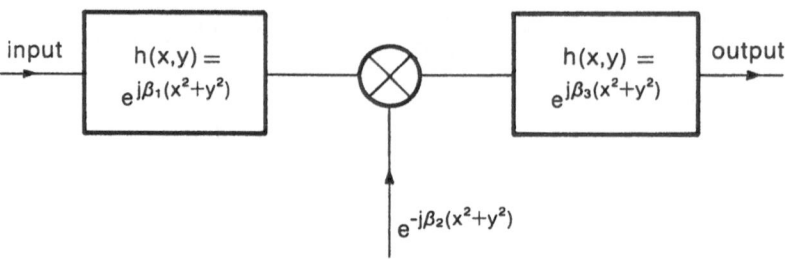

Fig. 3.62. Block diagram of an imaging system

where

$$I(x, y; x', y') = \iint e^{j(\bar{x}^2 + \bar{y}^2)(\beta_1 + \beta_3 - \beta_2)}$$
$$\times e^{-j2\beta_1(x\bar{x} + y\bar{y})} e^{-j2\beta_3(\bar{x}x' + \bar{y}y')} d\bar{x} \, d\bar{y} \quad . \tag{3.8.56}$$

If $\beta_1 + \beta_3 = \beta_2$, then

$$I(x, y; x', y') = \int_{-\infty}^{+\infty} \int_{-\infty}^{+\infty} \exp\left\{ -j2\beta_3 \left[\bar{x}\left(x + \frac{\beta_1}{\beta_3} x' \right) + \bar{y}\left(y + \frac{\beta_1}{\beta_3} y' \right) \right] \right\} d\bar{x} \, d\bar{y}$$

$$= M \int_{-\infty}^{+\infty} \int_{-\infty}^{+\infty} \exp\left\{ -j2\pi [\bar{f}_x(x + Mx') + \bar{f}_y(y + My')] \right\} d\bar{f}_x d\bar{f}_y$$

$$= M\delta(x + Mx', y + My') \tag{3.8.57}$$

where we have substituted

$$M = \frac{\beta_1}{\beta_3} \quad , \tag{3.8.58}$$

$$\bar{f}_x = \frac{\bar{x}}{\pi}\beta_3 \quad \text{and} \tag{3.8.59}$$

$$\bar{f}_y = \frac{\bar{y}}{\pi}\beta_3 \quad . \tag{3.8.60}$$

Finally we obtain

$$I(x, y; x', y') = \alpha M\delta(x + Mx', y + Mx')$$

$$= \alpha \frac{1}{M} \delta\left(\frac{x}{M} + x'; \frac{y}{M} + y' \right) \quad . \tag{3.8.61}$$

In the following we consider α to be constant, equal to 1. This is justified in many practical cases. Substituting (3.8.57) into (3.8.55), we have

$$h(x, y; x', y') = \frac{1}{M} \delta\left(\frac{x}{M} + x', \frac{y}{M} + y' \right) \quad . \tag{3.8.62}$$

Thus a point source of amplitude 1 at $x = x'$ and $y = y'$ is imaged at $x = -Mx'$

312

and $y = -My'$ and its value is $1/M$. If the input is $f(x,y)$, the output is given by

$$g(x,y) = \iint h(x,y; x,y')f(x',y')dx'dy'$$

$$= \frac{1}{M}\iint f(x',y')\delta\left(x' + \frac{x}{M}; y' + \frac{y}{M}\right)dx'dy'$$

$$= \frac{1}{M}f\left(-\frac{x}{M}, -\frac{y}{M}\right) \quad . \tag{3.8.63}$$

Hence the output is an exact replica of the input but with inverted axes and magnification M. The system is space-invariant because if the input is $f(x - x_0, y - y_0)$ the output is

$$g(x,y) = \frac{1}{M}f\left(-\frac{x - x_0}{M}, -\frac{y - y_0}{M}\right) \quad . \tag{3.8.64}$$

If we scale the input coordinates as

$$\tilde{x}' = -Mx' \quad \text{and} \quad \tilde{y}' = -My' \tag{3.8.65}$$

then (3.8.62) becomes

$$h(x,y; \tilde{x}', \tilde{y}') = M\iint e^{-j2\pi[(x-\tilde{x}')\bar{f}_x + (y-\tilde{y}')\bar{f}_y]}d\bar{f}_x d\bar{f}_y$$

$$= M\delta(x - \tilde{x}', y - \tilde{y}') \quad . \tag{3.8.66}$$

Thus we can write

$$g(x,y) = f(\tilde{x}', \tilde{y}') * h(\tilde{x}', \tilde{y}') \tag{3.8.67}$$

where

$$h(\tilde{x}', \tilde{y}') = M\delta(\tilde{x}', \tilde{y}') \quad . \tag{3.8.68}$$

vi) Imaging with a finite multiplier having pupil function $P(x,y)$. For this case, shown in Fig. 3.63, we have the multiplier

$$e^{-j\beta_2(x^2+y^2)}P\left(\frac{x}{L_x}, \frac{y}{L_y}\right) \quad .$$

Following the last example, one obtains for this case

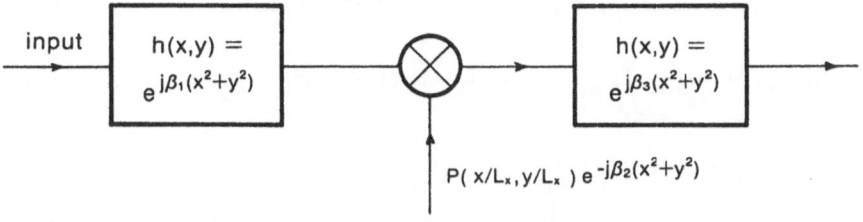

Fig. 3.63. Imaging with a finite multiplier having pupil function $P(x,y)$

$$h(x, y; x', y') = M \iint P\left(\frac{\pi \overline{f}_x/\beta_3}{L_x}, \frac{\pi \overline{f}_y/\beta_3}{L_y}\right)$$

$$\times \exp\left\{-j2\pi[\overline{f}_x(x + Mx') + \overline{f}_y(y + My')]\right\}d\overline{f}_x d\overline{f}_y \qquad (3.8.69)$$

or

$$h(x, y; \tilde{x}', \tilde{y}') = M \iint P\left(\frac{\pi \overline{f}_x/\beta_3}{L_x}, \frac{\pi \overline{f}_y/\beta_3}{L_y}\right)$$

$$\times \exp\left\{-j2\pi[\overline{f}_x(x - \tilde{x}') + \overline{f}_y(y - \tilde{y}')]\right\}d\overline{f}_x d\overline{f}_y$$

$$= h(x - \tilde{x}', y - \tilde{y}') \quad . \qquad (3.8.70)$$

As this is a space-invariant case, we can write

$$h(x, y) = M \iint P\left(\frac{\pi f_x/\beta_3}{L_x}, \frac{\pi f_y/\beta_3}{L_y}\right)$$

$$\times e^{-j2\pi(f_x x + f_y y)} df_x df_y$$

$$= M\mathcal{F}\left[P\left(\frac{\pi f_x/\beta_3}{L_x}, \frac{\pi f_y/\beta_3}{L_y}\right)\right] \quad . \qquad (3.8.71)$$

The frequency response $H(f_x, f_y)$ is given by

$$H(f_x, f_y) = \mathcal{F}[h(x, y)] = \mathcal{F}\left[M\mathcal{F}\left[P\left(\frac{\pi f_x}{\beta_3 L_x}, \frac{\pi f_y}{\beta_3 L_y}\right)\right]\right]$$

$$= MP\left(-\frac{\pi f_x}{\beta_3 L_x}, -\frac{\pi f_y}{\beta_3 L_y}\right) \quad . \qquad (3.8.72)$$

Thus the spatial frequency response of the imaging system is the pupil function itself with proper scaling of the coordinate axes.

3.8.5 Space-Variant Systems

For a space-variant system, the convolution integral relationship between the output, input and the impulse response does not apply. An example of a space-variant system is the Fourier transformer discussed in the previous section. For a space-variant impulse response we define the following Fourier transforms:

$$\hat{H}(f_x, f_y; x', y') = \iint h(x, y; x', y')e^{-j2\pi(f_x x + f_y y)}dx\, dy \quad , \qquad (3.8.73)$$

$$\hat{H}(x, y; f_x', f_y') = \iint h(x, y; x', y')e^{-j2\pi(f_x' x' + f_y' y')}dx'\, dy' \quad , \qquad (3.8.74)$$

$$\hat{\hat{H}}(f_x, f_y; f_x', f_y') = \iint \iint h(x, y; x', y')$$

$$\times e^{-j2\pi(f_x x + f_y y + f_x' x' + f_y' y')}dx\, dy\, dx'\, dy' \quad . \qquad (3.8.75)$$

314

We note that

$$\hat{H}(f_x, f_y; f'_x, f'_y) = \mathcal{F}[\hat{H}(f_x, f_y; x', y')]_{f'_x, f'_y}$$
$$= \mathcal{F}[\hat{H}(x, y; f'_x, f'_y)]_{f_x, f_y} \quad . \tag{3.8.76}$$

Using the above Fourier transform relationships, one obtains

$$G(f_x, f_y) = \iint \hat{H}(f_x, f_y; -f'_x, -f'_y)F(f'_x, f'_y)df'_x df'_y \quad , \tag{3.8.77}$$

$$g(x, y) = \iint \hat{H}(x, y; -f'_x, -f'_y)F(f_x, f_y)df'_x df'_y \tag{3.8.78}$$

and

$$g(x, y) = \iint \iint \hat{\hat{H}}(f_x, f_y; -f'_x, -f'_y)F(f'_x, f'_y)df_x df_y df'_x df'_y \quad . \tag{3.8.79}$$

For a space-invariant system

$$h(x, y; x', y') = h(x - x'; y - y') = h(x - x', y - y'; 0, 0) \quad . \tag{3.8.80}$$

Thus

$$\hat{\hat{H}}(f_x, f_y f'_x, f'_y) = H(f_x, f_y)\delta(f_x - f'_x, f_y - f'_y) \quad , \tag{3.8.81}$$

where

$$H(f_x, f_y) = \mathcal{F}[h(x, y)] \quad . \tag{3.8.82}$$

For this case, (3.8.77) becomes

$$G(f_x, f_y) = H(f_x, f_y)F(f_x, f_y) \quad . \tag{3.8.83}$$

It is of interest to consider the four functions h, $\hat{H}(f_x, f_y, x', y')$, $\hat{H}(x, y, f'_x, f'_y)$ and $\hat{\hat{H}}$ for the following four cases:

i) *Modulator, or multiplication by a function* $m(x, y)$. We have (Fig. 3.64)

$$g(x, y) = m(x, y)f(x, y) \quad , \tag{3.8.84}$$

$$h(x, y; x', y') = m(x', y')\delta(x - x', y - y') \quad , \tag{3.8.85}$$

$$\hat{H}(f_x, f_y, x', y') = m(x', y')e^{-j2\pi(f_x x' + f_y y')} \quad , \tag{3.8.86}$$

$$\hat{H}(x, y; f'_x, f'_y) = m(x, y)e^{j2\pi(f_x x' + f_y y')} \quad , \quad \text{and} \tag{3.8.87}$$

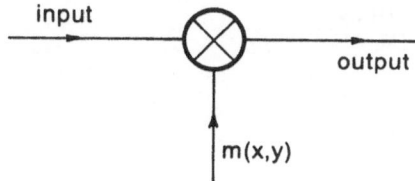

Fig. 3.64. Multiplication by a function $m(x, y)$

315

$$\hat{H}(f_x, f_y; f'_x, f'_y) = M(f_x - f'_x, f_y - f'_y) \quad . \tag{3.8.88}$$

Note that

$$M(f_x, f_y) = \mathcal{F}[m(x, y)] \quad . \tag{3.8.89}$$

ii) Invariant system. We have (Fig. 3.65)

$$g = h*f \quad , \tag{3.8.90}$$

$$h(x, y; x', y') = h(x - x', y - y') \quad , \tag{3.8.91}$$

$$\hat{H}(f_x, f_y; x', y') = H(f_x, f_y)e^{-j2\pi(f_x x' + f_y y')} \quad . \tag{3.8.92}$$

$$\hat{H}(x, y; f'_x, f'_y) = H(f_x, f_y)e^{j2\pi(f'_x x + f'_y y)} \quad , \tag{3.8.93}$$

$$\hat{\hat{H}}(f_x, f_y; f'_x, f'_y) = H(f_x, f_y)\delta(f_x - f'_x, f_y - f'_y) \quad . \tag{3.8.94}$$

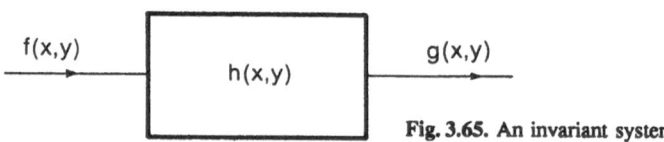

Fig. 3.65. An invariant system

iii) Fourier transformer. We have (Fig. 3.66)

$$g(x, y) = \iint f(x', y')e^{-j\beta(xx' + yy')}dx'dy' \quad , \tag{3.8.95}$$

$$h(x, y; x', y') = -j\beta e^{-j\beta(xx' + yy')} \quad , \tag{3.8.96}$$

$$\hat{H}(f_x, f_y; x', y') = -\frac{j}{\beta}\delta\left(\frac{f_x}{\beta} + x', \frac{f_y}{\beta} + y'\right) \quad , \tag{3.8.97}$$

$$\hat{H}(x, y; f'_x, f'_y) = -j\beta\delta(\beta x - f'_x, \beta y - f'_y) \quad , \tag{3.8.98}$$

$$\hat{\hat{H}}(f_x, f_y; f'_x, f'_y) = -\frac{j}{\beta}e^{-(j/\beta)(f_x f'_x + f_y f'_y)} \quad . \tag{3.8.99}$$

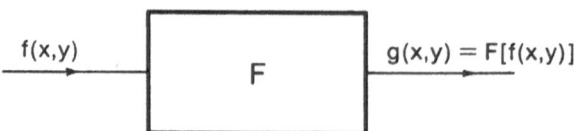

Fig. 3.66. A spatial two-dimensional Fourier transformer

iv) Magnifier. We have (Fig. 3.67)

$$g(x, y) = tf(tx, ty) \quad , \quad t = \text{const} \quad , \tag{3.8.100}$$

$$h(x, y; x', y') = t\delta(tx - x'; ty - y') \quad , \tag{3.8.101}$$

$$\hat{H}(f_x, f_y; x', y') = \frac{1}{t} e^{-(j/t)(f_x x' + f_y y')} \quad , \tag{3.8.102}$$

$$\hat{H}(x, y; f_x', f_y') = t e^{jt(f_x x' + f_y y')} \quad , \tag{3.8.103}$$

$$\hat{\hat{H}}(f_x, f_y; f_x', f_y') = \frac{1}{t}\delta\left(\frac{f_x}{t} - f_x', \frac{f_y}{t} - f_y'\right) \quad . \tag{3.8.104}$$

Fig. 3.67. Magnifier

3.8.6 Discrete Signals and Matrix Representation

We define the following two-dimensional discrete functions:

Step sequence $\quad u(n, m) = \begin{cases} 1, & n \geq 0, \quad m \geq 0 \quad , \\ 0, & n < 0, \quad m < 0 \quad , \end{cases}$ $\qquad$ (3.8.105)

$$u(n, m) = u(n)u(m) \quad .$$

Delta sequence $\quad \delta(n, m) = \begin{cases} 1, & n = 0, \quad m = 0 \quad , \\ 0, & \text{otherwise} \quad , \end{cases}$ $\qquad$ (3.8.106)

$$\delta(n, m) = \delta(n)\delta(m) \quad .$$

For any linear system, we have

$$g(n, m) = \sum_{k=-\infty}^{+\infty} \sum_{l=-\infty}^{+\infty} f(k, l)h(k, l; n, m) \quad , \tag{3.8.107}$$

where $h(k, l; n, m)$ is the space-variant impulse response. If the system is space-invariant, we have

$$h(k, l; n, m) = h(n - k, m - l) \tag{3.8.108}$$

and (3.8.107) becomes

$$g(n, m) = \sum_{k=-\infty}^{+\infty} \sum_{l=-\infty}^{+\infty} f(k, l)h(n - k, m - l)$$

$$= \sum_{k=-\infty}^{+\infty} \sum_{l=-\infty}^{+\infty} h(k,l) f(n-k, m-l) \quad . \tag{3.8.109}$$

Here $h(n,m)$ is the impulse response and (3.8.109) represents discrete convolution. The sampling theorem for the two-dimensional signals is given by

$$g(x,y) = \sum_{n=-\infty}^{+\infty} \sum_{m=-\infty}^{+\infty} g\left(\frac{n}{2B}, \frac{m}{2B}\right)$$
$$\times \left(2\pi B^2 \frac{J_1[2\pi B\sqrt{(x-n/2B)^2 + (y-m/2B)^2}]}{2\pi B\sqrt{(x-n/2B)^2 + (y-m/2B)^2}}\right) \tag{3.8.110}$$

and

$$g(x,y) = \sum_{n=-\infty}^{+\infty} \sum_{m=-\infty}^{+\infty} g\left(\frac{n}{2B_x}, \frac{m}{2B_y}\right)$$
$$\times \mathrm{sinc}\left[2B_x\left(x - \frac{n}{2B_x}\right)\right] \mathrm{sinc}\left[2B_y\left(y - \frac{m}{2B_y}\right)\right] \quad . \tag{3.8.111}$$

It is assumed that the signal is band-limited, i.e.,

$$G(f_x, f_y) = 0 \quad \text{for} \quad f_x \geq B_x \quad , \quad f_y \geq B_y \quad , \tag{3.8.112}$$

i.e.

$$G(f_x, f_y) = 0 \quad \text{for} \quad \sqrt{f_x^2 + f_y^2} \geq B \quad .$$

As for the case of one-dimensional signals, the sampling rates must be higher than the appropriate bandwidth $2B_x$, $2B_y$ or $2B$.

Similar to the matrix representation in the one-dimensional case, we can write $f(m,n)$ and $g(m,n)$ as two-dimensional matrices and $h(m,n,p,q)$ as a four-dimensional tensor. However, $g(m,n)$ can be scanned row by row to form a new matrix $g(m)$ of size $N^2 \times 1$, where $g(m,n)$ is $N \times N$. If $f(m,n)$ is $M \times M$, one has $f(m)$ as a $M^2 \times 1$ matrix. For this case, $h(m,n,p,q)$ can be written as a two-dimensional matrix of dimensions $N^2 \times M^2$. In matrix representation, the input-output relationship becomes

$$g(n) = \sum_m h(n,m) f(m) \quad . \tag{3.8.113}$$

Note that, even for the space-invariant case, $h(n,m)$ is not Toeplitz as in the one-dimensional case.

A particularly convenient way of obtaining $f(m)$ from $f(m,n)$ is to scan column by column. this can be performed by defning a vector $v_k(n \times 1)$ and a matrix $N_k(mn \times n)$ given by

$$v_k = \begin{bmatrix} 0 \\ 0 \\ \vdots \\ 1 \\ 0 \\ 0 \\ \vdots \end{bmatrix} \begin{matrix} 1 \\ 2 \\ \vdots \\ k \\ \vdots \\ \\ \end{matrix} \quad \text{and} \quad N_k = \begin{bmatrix} [0] \\ [0] \\ \vdots \\ [1] \\ \vdots \\ [0] \end{bmatrix} \begin{matrix} 1 \\ 2 \\ \vdots \\ k \\ \\ \vdots \end{matrix} \quad . \tag{3.8.114}$$

Here [0] represents a null $m \times n$ matrix and [1] represents a $m \times n$ identity matrix. For this case one obtains

$$f(m) = \sum_{k=1}^{n} N_k f(m,n) v_k \quad . \tag{3.8.115}$$

For example, consider

$$f(m,n) = \begin{bmatrix} a_1 & a_4 & a_1 \\ a_2 & a_5 & a_8 \\ a_3 & a_6 & a_9 \end{bmatrix} \quad . \tag{3.8.116}$$

Then

$$f(m) = \begin{bmatrix} a_1 \\ a_2 \\ a_3 \\ a_4 \\ a_5 \\ a_6 \\ a_7 \\ a_8 \\ a_9 \end{bmatrix} = \begin{bmatrix} a_1 \\ a_2 \\ a_3 \\ 0 \\ 0 \\ 0 \\ 0 \\ 0 \\ 0 \end{bmatrix} + \begin{bmatrix} 0 \\ 0 \\ 0 \\ a_4 \\ a_5 \\ a_6 \\ 0 \\ 0 \\ 0 \end{bmatrix} + \begin{bmatrix} 0 \\ 0 \\ 0 \\ 0 \\ 0 \\ 0 \\ a_7 \\ a_8 \\ a_9 \end{bmatrix} , \tag{3.9.117}$$

$$v_1 = \begin{bmatrix} 1 \\ 0 \\ 0 \end{bmatrix} , \quad v_2 = \begin{bmatrix} 0 \\ 1 \\ 0 \end{bmatrix} , \quad v_3 = \begin{bmatrix} 0 \\ 0 \\ 1 \end{bmatrix} , \tag{3.8.118}$$

$$N_1 = \begin{bmatrix} 1 & 0 & 0 \\ 0 & 1 & 0 \\ 0 & 0 & 1 \\ 0 & 0 & 0 \\ 0 & 0 & 0 \\ 0 & 0 & 0 \\ 0 & 0 & 0 \\ 0 & 0 & 0 \\ 0 & 0 & 0 \end{bmatrix} , \quad N_2 = \begin{bmatrix} 0 & 0 & 0 \\ 0 & 0 & 0 \\ 0 & 0 & 0 \\ 1 & 0 & 0 \\ 0 & 1 & 0 \\ 0 & 0 & 1 \\ 0 & 0 & 0 \\ 0 & 0 & 0 \\ 0 & 0 & 0 \end{bmatrix} , \quad N_3 = \begin{bmatrix} 0 & 0 & 0 \\ 0 & 0 & 0 \\ 0 & 0 & 0 \\ 0 & 0 & 0 \\ 0 & 0 & 0 \\ 0 & 0 & 0 \\ 1 & 0 & 0 \\ 0 & 1 & 0 \\ 0 & 0 & 1 \end{bmatrix} \quad . \tag{3.8.119}$$

We can easily check that

$$f(m) = N_1 f(m,n)v_1 + N_2 f(m,n)v_2 + N_3 f(m,n)v_3 \quad . \tag{3.8.120}$$

From (3.8.115) we also note that

$$f(m,n) = \sum_{k=1}^{n} N_k^T f(m)v_k^T \quad . \tag{3.8.121}$$

We note that v_k extracts the kth column from $f(m,n)$ and N_k places this column in the proper place of the $f(n)$ vector.

For the linear system given by (3.8.114) we can write

$$g(m) = H(m,n)f(n) \quad . \tag{3.8.122}$$

Here $g(m)$ is $M^2 \times 1$, $f(n)$ is $N^2 \times 1$ and $H(m,n)$ is an $M^2 \times N^2$ matrix. Similarly $g(m,n)$ is an $M \times M$ matrix and $f(m,n)$ is an $N \times N$ matrix. Using (3.8.115) one can write

$$g(m) = \sum_{k=1}^{N} H(m,n)N_k f(m,n)v_k \quad . \tag{3.8.123}$$

We also note that

$$g(m,n) = \sum_{l=1}^{M} M_l^T g(m)u_l^T \quad , \tag{3.8.124}$$

where M_l is similar to N_l except that its dimensions are $M^2 \times M$ rather than $N^2 \times N$. Hence u_l is similar to v_l but of dimension $M \times 1$.

Combining (3.9.123 and 124) we obtain

$$\begin{aligned}
g(m,n) &= \sum_{l=1}^{M} \sum_{k=1}^{N} (M_l^T H N_k) f(m,n)(v_k u_l^T) \\
&= \sum_{l=1}^{M} \sum_{k=1}^{N} H_{lk} f(m,n) u_k u_l^T \quad ,
\end{aligned} \tag{3.8.125}$$

where H_{lk} is an $M \times N$ matrix and the H matrix can be written as

$$H = \begin{bmatrix} [H_{11}] & [H_{12}] & [H_{13}] & \cdots \\ [H_{21}] & \cdots & \cdots & \cdots \\ \cdots & \cdots & \cdots & \cdots \\ \cdots & \cdots & \cdots & \cdots \end{bmatrix} \quad . \tag{3.8.126}$$

3.9 Stochastic Processes: Multidimensional

Sources in the optical region are more stochastic than deterministic. Coherent light sources have been available only since 1960, since the invention of lasers. General multidimensional stochastic processes related to optics are considered in this section, which also includes a discussion of the coherency matrix.

The electric field due to a light wave is a function of four variables: three space coordinates and time. Thus, in general,

$$E(x, y, z, t) = E(\boldsymbol{r}, t) = E(\boldsymbol{x}) \quad . \tag{3.9.1}$$

Following Sect. 3.3, for stochastic processes we define the autocorrelation function

$$G(\boldsymbol{x}_1, \boldsymbol{x}_2) = \langle E(\boldsymbol{x}_1) E^*(\boldsymbol{x}_2) \rangle \quad . \tag{3.9.2}$$

In optics literature, this is generally known as the coherence function. If the process is stationary, then

$$G(\boldsymbol{x}_1, \boldsymbol{x}_2) = G(\boldsymbol{r}_1, \boldsymbol{r}_2, t_2 - t_1) \quad . \tag{3.9.3}$$

If the process is homogeneous, then

$$G(\boldsymbol{x}_1, \boldsymbol{x}_2) = G(t_1, t_2, \boldsymbol{r}_2 - \boldsymbol{r}_1) \quad . \tag{3.9.4}$$

If we denote the Fourier transform of $E(r, t)$ by $E(r, \omega)$, i.e.,

$$E(\boldsymbol{r}, \omega) = \mathcal{F}[E(\boldsymbol{r}, t)] = \int E(\boldsymbol{r}, t) e^{-j2\pi f t} dt \quad , \tag{3.9.5}$$

we then have

$$G(\boldsymbol{r}_1, \boldsymbol{r}_2, \omega_1, \omega_2) = \langle E(\boldsymbol{r}_1, \omega_1) E(\boldsymbol{r}_2, \omega_2) \rangle \quad . \tag{3.9.6}$$

For a stationary process we have

$$G(\boldsymbol{r}_1, \boldsymbol{r}_2, \omega_1, \omega_2) = G(\boldsymbol{r}_1, \boldsymbol{r}_2, \omega_1) \delta(\omega_1 - \omega_2) \quad , \tag{3.9.7}$$

where

$$G(\boldsymbol{r}_1, \boldsymbol{r}_2, \omega) = \mathcal{F}[G(\boldsymbol{r}_1, \boldsymbol{r}_2, \tau)] \tag{3.9.8}$$

and is related to the power spectrum of the process.

The general coherence function is defined as

$$g(\boldsymbol{x}_1, \boldsymbol{x}_2) = \frac{G(\boldsymbol{x}_1, \boldsymbol{x}_2)}{[G(\boldsymbol{x}_1, \boldsymbol{x}_1) G(\boldsymbol{x}_2, \boldsymbol{x}_2)]^{1/2}} \quad . \tag{3.9.9}$$

If $E(\boldsymbol{x}_1)$ and $E(\boldsymbol{x}_2)$ are not correlated at all then

$$|G(\boldsymbol{x}_1, \boldsymbol{x}_2)| = |g(\boldsymbol{x}_1, \boldsymbol{x}_2)| = 0 \quad . \tag{3.9.10}$$

If $E(x_1)$ and $E(x_2)$ are deterministic,

$$|g(\boldsymbol{x}_1, \boldsymbol{x}_2)| = 1 \quad . \tag{3.9.11}$$

Note that according to the definition of the coherence function, if $E(\boldsymbol{x}) = cf(\boldsymbol{x})$ where c is a complex random variable, then

$$|g(\boldsymbol{x}_1, \boldsymbol{x}_2)| = 1 \quad . \tag{3.9.12}$$

Also, if $E(\boldsymbol{x}_2 = aE(\boldsymbol{x}_1))$, where a is deterministic,

$$|g(\boldsymbol{x}_1, \boldsymbol{x}_2)| = 1 \quad . \tag{3.9.13}$$

Thus if the coherence function has a magnitude 1, it does not necessarily mean that the process is completely deterministic. As

$$G(\boldsymbol{x}_1, \boldsymbol{x}_2) \leq [G(\boldsymbol{x}_1, \boldsymbol{x}_1)G(\boldsymbol{x}_2, \boldsymbol{x}_2)]^{1/2}$$

we have

$$0 \leq |g(\boldsymbol{x}_1, \boldsymbol{x}_2)| \leq 1 \quad .$$

For a stationary process, we have

$$g(\boldsymbol{r}_1, \boldsymbol{r}_2, \omega) = \frac{G(\boldsymbol{r}_1, \boldsymbol{r}_2, \omega)}{[G(\boldsymbol{x}_1, \boldsymbol{x}_1)]^{1/2}[G(\boldsymbol{x}_2, \boldsymbol{x}_2)]^{1/2}} \tag{3.9.14}$$

and

$$g(\boldsymbol{r}_1, \boldsymbol{r}_2, \tau) = \int g(\boldsymbol{r}_1, \boldsymbol{r}_2, \omega)e^{j\omega\tau}\,d\omega \quad . \tag{3.9.15}$$

Note that $g(\boldsymbol{r}_1, \boldsymbol{r}_2, \omega)$ is not bounded between 0 and 1.

At a particular frequency, ω, we define the spectral coherence

$$\begin{aligned}
g_{\mathrm{s}}(\boldsymbol{r}_1, \boldsymbol{r}_2, \omega) &= \frac{G(\boldsymbol{r}_1, \boldsymbol{r}_2, \omega)}{[G(\boldsymbol{r}_1, \boldsymbol{r}_2, \omega)G(\boldsymbol{r}_1, \boldsymbol{r}_2, \omega)]^{1/2}} \\
&= \frac{g(\boldsymbol{r}_1, \boldsymbol{r}_2, \omega)}{[g(\boldsymbol{r}_1, \boldsymbol{r}_1, \omega)g(\boldsymbol{r}_2, \boldsymbol{r}_2, \omega)]^{1/2}}
\end{aligned} \tag{3.9.16}$$

and $g_{\mathrm{s}}(\boldsymbol{r}_1, \boldsymbol{r}_2, \omega)$ is bounded, i.e.,

$$0 \leq g_{\mathrm{s}}(\boldsymbol{r}_1, \boldsymbol{r}_2, \omega) \leq 1 \quad . \tag{3.9.17}$$

If

$$G(\boldsymbol{x}_1, \boldsymbol{x}_2) = E(\boldsymbol{x}_1)E^*(\boldsymbol{x}_2) \quad , \tag{3.9.18}$$

i.e., $E(\boldsymbol{x})$ is deterministic, we have complete coherence and

$$|g(\boldsymbol{x}_1, \boldsymbol{x}_2)| = 1 \quad . \tag{3.9.19}$$

However, this deterministic field is not monochromatic. If this field is both deterministic and stationary, then it is also monochromatic, because

$$G(t_1, t_2) = \langle E(t_1)E(t_2) \rangle = E(t_1)E(t_2) = f(t_2 - t_1) \quad . \tag{3.9.20}$$

Thus

$$E(t) \propto e^{j\omega t} \quad . \tag{3.9.21}$$

A completely incoherent stationary field is not possible, because if

$$|G(\pmb{r}_1, \pmb{r}_2, \tau)| = 0 \quad \text{then}$$

$$|G(\pmb{r}_1, \pmb{r}_2, 0)| = 0 \quad.$$

Hence the field is identically zero. Excluding the point $\tau = 0$, a completely incoherent field has

$$|g(\pmb{r}_1, \pmb{r}_2, \tau)| = |g(\pmb{r}_1, \pmb{r}_2, \tau)|\delta(\pmb{r}_2 - \pmb{r}_1) \quad. \tag{3.9.22}$$

Let us consider partially coherent light which is stationary and thus monochromatic. As $E(\pmb{r}_1, t)$ obeys the wave equation, we have

$$\nabla_1^2 E(\pmb{r}_1, t) = \frac{1}{v^2} \frac{\partial^2}{\partial t^2} E(\pmb{r}_1, t) \quad, \quad \text{where} \tag{3.9.23}$$

$$\nabla_1^2 = \left(\frac{\partial^2}{\partial x_1^2} + \frac{\partial^2}{\partial y_1^2} + \frac{\partial^2}{\partial z_1^2} \right) \quad. \tag{3.9.24}$$

Multiplying both sides of the equation by $E^*(r_2, t + \tau)$ we obtain

$$\nabla_1^2 G(\pmb{r}_1, \pmb{r}_2, t) = \frac{1}{v^2} \frac{\partial^2}{\partial t^2} G(\pmb{r}_1, \pmb{r}_2, t) \quad. \tag{3.9.25}$$

Similarly one can show that

$$\nabla_2^2 G(\pmb{r}_1, \pmb{r}_2, t) = \frac{1}{v^2} \frac{\partial^2}{\partial t^2} G(\pmb{r}_1, \pmb{r}_2, t) \quad. \tag{3.9.26}$$

Thus the coherence function satisfies the wave equations given by (3.9.23).

We can also consider the case where the electric field $E(\pmb{r}, t)$ is generated by the source density $s(\pmb{r}, t)$, which is a stochastic process with coherence function $G_{ss}(\pmb{r}_1, \pmb{r}_2, t_1, t_2)$. For this case, we have

$$\nabla^2 E(\pmb{r}, t) - \frac{1}{v^2} \frac{\partial^2}{\partial t^2} E(\pmb{r}, t) = -s(\pmb{r}, t) \quad. \tag{3.9.27}$$

It can easily be shown that the following equations are satisfied:

$$\nabla_2^2 G_{sv}(\pmb{r}_1, \pmb{r}_2, t_1, t_2) - \frac{1}{v^2} \frac{\partial^2}{\partial t_2^2} G_{sv} = -G_{ss} \quad, \tag{3.9.28}$$

$$\nabla_1^2 G_{vv} - \frac{1}{v^2} \frac{\partial^2 G_{vv}}{\partial t^2} = -G_{sv} \quad, \quad \text{where} \tag{3.9.29}$$

$$G_{sv}(\pmb{r}_1, \pmb{r}_2, t_1, t_2) = \langle s(\pmb{r}_1, t_1) E^*(\pmb{r}_2, t_2) \rangle \tag{3.9.30}$$

and

$$G_{vv}(\pmb{r}_1, \pmb{r}_2, t_1, t_2) = \langle E(\pmb{r}_1, t_1) E^*(\pmb{r}_2, t_2) \rangle \quad. \tag{3.9.31}$$

If the source is stationary,

$$G_{ss}(\pmb{r}_1, \pmb{r}_2, t_1, t_2) = G_{ss}(\pmb{r}_1, \pmb{r}_2, \tau) \quad, \tag{3.9.32}$$

where $\tau = t_1 - t_2$. For this case we have

$$G_{sv}(\mathbf{r}_1, \mathbf{r}_2, t_1, t_2) = G_{sv}(\mathbf{r}_1, \mathbf{r}_2, \tau) \quad \text{and} \tag{3.9.33}$$

$$G_{vv}(\mathbf{r}_1, \mathbf{r}_2, t_1, t_2) = G_{vv}(\mathbf{r}_1, \mathbf{r}_2, \tau) \quad . \tag{3.9.34}$$

Equations (3.9.28, 29) for the stationary source case become

$$\nabla_2^2 G_{sv} - \frac{1}{v^2} \frac{\partial^2 G_{sv}}{\partial \tau^2} = -G_{ss} \quad \text{and} \tag{3.9.35}$$

$$\nabla_1^2 G_{vv} - \frac{1}{v^2} \frac{\partial^2 G_{vv}}{\partial \tau^2} = -G_{sv} \quad . \tag{3.9.36}$$

Although we started with a stochastic source, we note from (3.9.35, 36) that the evaluation of G_{sv} and G_{uv} involves the solution of the wave equation with deterministic sources. This is a very important result because it proves that all the diffraction results derived in Sect. 2.16 are directly applicable. In the following, we consider some important examples.

3.9.1 Point Source

Let us consider the source s to be given by

$$s(\mathbf{r}, t) = 4\pi p(t)\delta(\mathbf{r}) \quad , \tag{3.9.37}$$

where $p(t)$ is a stationary process with autocorrelation $R(\tau)$. Then we have

$$E(\mathbf{r}, t) = \frac{p(t - r/c)}{r} \quad \text{and} \tag{3.9.38}$$

$$E(\mathbf{r}, t + \tau)E^*(\mathbf{r}_2, t) = \frac{1}{r_1 r_2} p\left(t + \tau - \frac{r_1}{c}\right) p^*\left(t - \frac{r_2}{c}\right) \quad . \tag{3.9.39}$$

Taking the expected value of (3.9.39) we obtain

$$G_{vv}(\mathbf{r}_1, \mathbf{r}_2, \tau) = \frac{1}{r_1 r_2} R\left(\tau - \frac{r_1 - r_2}{c}\right) \quad . \tag{3.9.40}$$

Also

$$\langle |E(\mathbf{r}, t)|^2 \rangle = \frac{R(0)}{r^2} \quad . \tag{3.9.41}$$

Thus

$$g_{vv}(\mathbf{r}_1, \mathbf{r}_2, \tau) = \frac{1}{R(0)} R\left(\tau - \frac{r_1 - r_2}{c}\right) \quad . \tag{3.9.42}$$

For the two points defined by

$$r_1 - r_2 = \tau c \tag{3.9.43}$$

the electric fields are therefore completely coherent.

324

3.9.2 Partially Coherent Source Distribution

Let us consider a region of volume V with sources having a source density function s with source coherence function $G_{ss}(r_1, r_2, t_1, t_2)$. We are interested in finding $G_{vv}(r_{10}, r_{20}, t_1, t_2)$ and $G_{sv}(r_1, r_{10}, t_1, t_2)$, where the observation points r_{10} and r_{20} are outside the region of sources, as shown in Fig. 3.68.

Using (3.9.38) and (3.9.35) and the results of Sect. 2.16.2, we have

$$G_{sv}(r_1, r_{20}, t_1, t_2) = \frac{1}{4\pi} \int_V \frac{1}{r_3} G_{ss}\left(r_1, r_2, t_1 - t_2 - \frac{r_3}{c}\right) dV \quad , \qquad (3.9.44)$$

where

$$r_3 = |r_2 - r_{20}| \quad . \qquad (3.9.45)$$

Similarly, (3.9.36) yields

$$G_{vv}(r_{10}, r_{20}, t_1, t_2) = \frac{1}{4\pi} \int \frac{1}{r_4} G_{sv}\left(r_1, r_{20}, t_1 - \frac{r_4}{c}, t_2\right) dV \quad , \qquad (3.9.46)$$

where

$$r_4 = |r_1 - r_{10}| \quad . \qquad (3.9.47)$$

Substituting (3.9.44) into (3.9.46), we also obtain

$$G_{vv}(r_{10}, r_{20}, t_1, t_2)$$
$$= \frac{1}{16\pi^2} \int_V \int_{V_1} \frac{1}{r_3 r_4} G_{ss}\left(r_1, r_2, t_1 - \frac{r_4}{c}, t_2 - \frac{r_3}{c}\right) dV_1 dV \quad . \qquad (3.9.48)$$

If the sources are monochromatic but incoherent we then have

$$G_{ss}(r_1, r_2, t_1, t_2) = G_{ss}(r_1, r_2, \tau)$$
$$= q(r_1)\delta(r_1 - r_2)R(\tau) \quad . \qquad (3.9.49)$$

We thus obtain from (3.9.44 and 48) that

$$G_{sv}(r_1, r_{20}, \tau) = \frac{1}{4\pi r_3} q(r_1)R(\tau + r_3/c) \qquad (3.9.50)$$

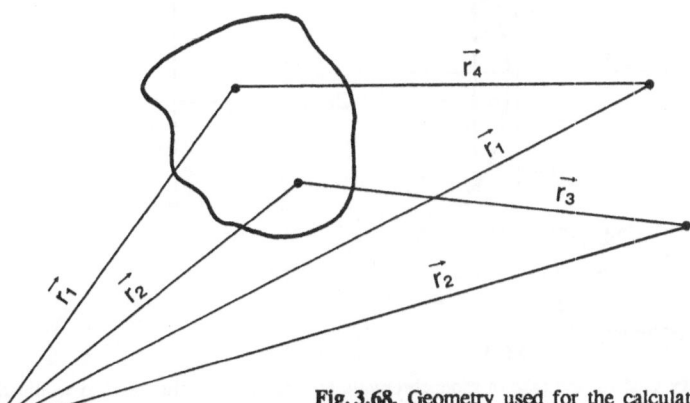

Fig. 3.68. Geometry used for the calculations with a partially coherent source distribution

and

$$G_{vv}(\mathbf{r}_{10}, \mathbf{r}_{20}, \tau) = \frac{1}{16\pi^2} \int \frac{1}{r_3 r_4} q(\mathbf{r}_1) R\left(\tau + \frac{r_3 - r_4}{c}\right) dV_1 \quad . \tag{3.9.51}$$

If $\mathbf{r}_{10} = \mathbf{r}_{20}$, (3.9.51) becomes

$$G_{vv}(\mathbf{r}_{10}, \mathbf{r}_{10}, 0) = \frac{R(0)}{16\pi^2} \int_V \frac{q(\mathbf{r}_1)}{r_4^2} dV \quad . \tag{3.9.52}$$

Note that $G_{vv}(\mathbf{r}_{10}, \mathbf{r}_{10}, 0)$ is the average intensity of the field due to incoherent sources and is obtained by summing the intensities due to each source.

Let us consider a special case where the light is monochromatic but incoherent. For this case

$$R(\tau) = e^{-j\omega\tau} \quad . \tag{3.9.53}$$

As shown in Fig. 3.69a, the source is at $z = 0$ and the observation points are in

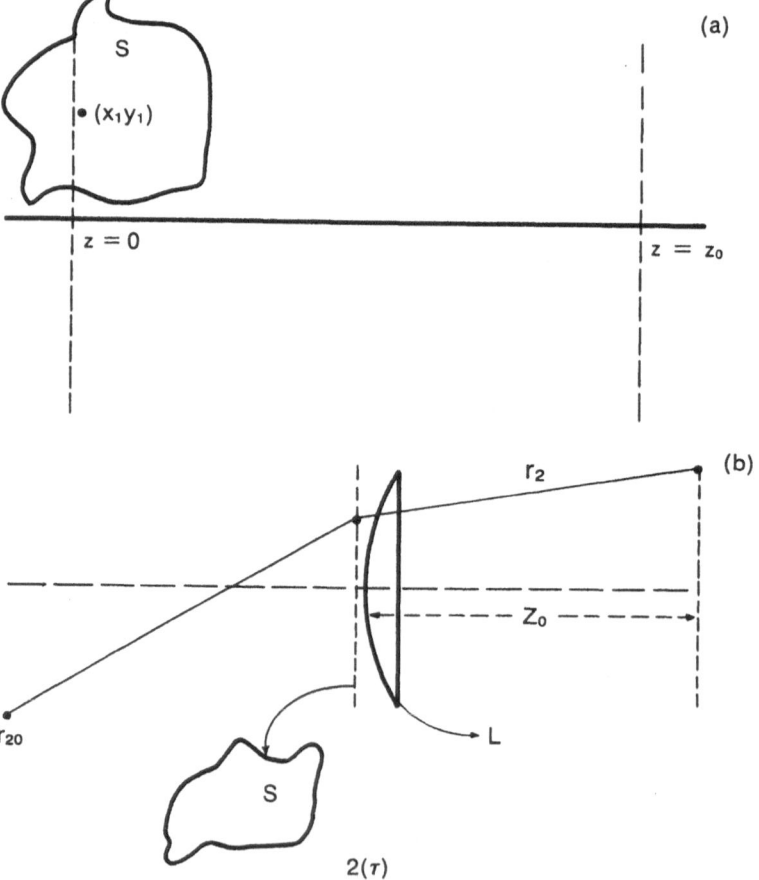

Fig. 3.69. (a) Geometry used for the case of monochromatic but incoherent light confined within S. (b) Configuration for the Van Cittert–Zerinke theorem

the $z = z_0$ plane. For this case, we have

$$G_{ss}(r_1, r_2, \tau) = q(x_1, y_1)\delta(x_1 - x_2)\delta(y_1 - y_2)e^{-j\omega\tau} \quad . \tag{3.9.54}$$

Inserting this expression for G_{ss} into (3.9.48), we obtain

$$G_{vv}(r_{10}, r_{20}, \tau) = \frac{e^{-j\omega\tau}}{16\pi^2 z_0^2} \iint q(x_1 y_1)e^{jk(r_3 - r_4)}dx_1 dy_1 \quad , \tag{3.9.55}$$

where we have assumed that x_1, y_1, x_2, y_2 are much smaller than z_0: the far field approximation. Equation (3.9.55) can also be written as

$$G_{vv}(r_{10}, r_{20}, \tau) = \frac{1}{16\pi^2 z_0^2} \iint q'(x_1 y_1)e^{-jkr_4}dx_1 dy_1 \quad , \tag{3.9.56}$$

where

$$q'(x_1, y_1) = q(x_1, y_1)e^{-jkr_3}e^{-j\omega\tau} \quad . \tag{3.9.57}$$

Comparing (3.9.57) and (2.17.2), we note that the coherence function G_{vv} follows the far field diffraction pattern due to a source with strength $q'(x_1 y_1)$.

The above result can also be interpreted quite differently as follows. This interpretation is generally known as the Van Cittert–Zerinke theorem. Let the source be placed at r'_{20} as shown in Fig. 3.69b. A lens with focal length f and a mask with transmission function $q(x, y)$ are placed at $z = 0$. Then the diffracted field at $z = z_0$ is given by

$$q(x, y) = e^{jkr_{20}}e^{-jk(x^2+y^2)/2f} \quad .$$

If we choose r'_{20} such that the source plane images in the image plane at $z = z_0$, then we have

$$r_2 + r'_{20} - \frac{x^2 + y^2}{2f} = \text{const} \quad .$$

Thus the diffracted field in the equivalent system is proportional to that given by (3.9.56).

3.9.3 Coherent Source

For this case, we have

$$G_{ss}(r_1, r_2, \tau) = p(r_1)p^*(r_2)R(\tau) \quad . \tag{3.9.58}$$

Using (3.9.44 and 48) we obtain

$$G_{sv}(r_1, r_{20}, \tau) = R(\tau)p(r_1)\frac{1}{4\pi} \int_V \frac{p^*(r_2)}{r_3}e^{-jkr_3}dV \tag{3.9.59}$$

and

$$G_{vv}(r_{10}, r_{20}, \tau) = R(\tau)\frac{1}{4\pi} \int_V \frac{p(r_1)}{r_4}e^{-jkr_4}dV$$

$$= \frac{1}{4\pi} \int_V \frac{p^*(r_2)}{r_3}e^{-jkr_3}dV \quad . \tag{3.9.60}$$

We note that

$$G_{vv}(r_{10}, r_{20}, 0) = R(0) \left| \frac{1}{4\pi} \int_V \frac{p(r_1)}{r_3} e^{jkr_3} dV \right|^2 . \tag{3.9.61}$$

Thus the average intensity of the light due to coherent sources is obtained by first summing the fields and then squaring the result.

3.9.4 Effect of a Mask

Let us consider the case shown in Fig. 3.70. Light of coherence function $G_{ii}(r_1, r_2, t_1, t_2)$ is incident on the transparency. If the mask has transmission function

$$T(x, y) = a(x, y)e^{-j\omega\tau(x,y)} , \tag{3.9.62}$$

where $a(x, y)$ and $\tau(x, y)$ are the attenuation and delay, respectively, then we have

$$G_{00}(r_1, r_2, t_1, t_2) = a(x_1 y_1) a(x_2 y_2)$$
$$\times G_{ii}(x_1, x_2, t_1 - \tau(x_1, y_2), t_2 - \tau(x_2 y_2)) . \tag{3.9.63}$$

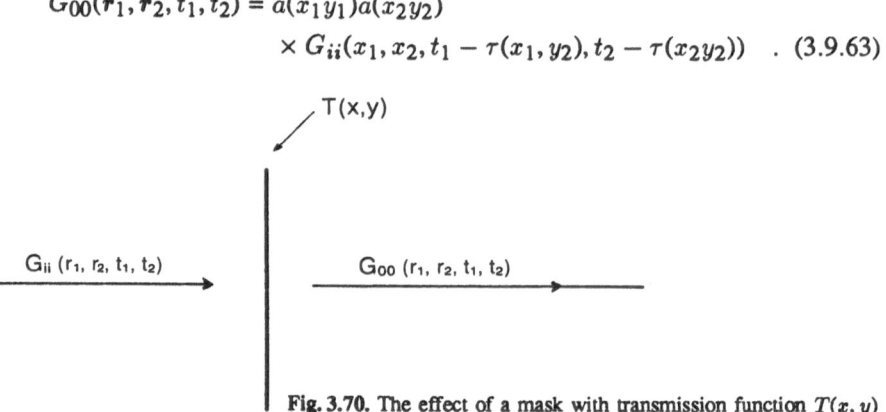

Fig. 3.70. The effect of a mask with transmission function $T(x, y)$

3.9.5 General Case

A general two-dimensional optical system is shown in Fig. 3.71, where the impulse response is given by $h(x, y)$. For an input $f(x, y)$ the output $g(x, y)$ is given by

$$g(x, y) = f(x, y) * h(x, y) . \tag{3.9.64}$$

Let us next consider the case where the input is partially coherent but monochromatic, i.e., its coherence function is given by

$$G_{ff}(x_1, y_1; x_2, y_2)e^{-j\omega\tau}.$$

Following the discussion in Sect. 3.3.1, it can easily be shown that

$$G_{fg}(x_1, y_1; x_2, y_2) = G_{ff}(x_1, y_1; x_2, y_2) * h(x_2, y_2) , \tag{3.9.65}$$

where $x_1 y_1$ is in the input and $x_2 y_2$ is in the output. Similarly

328

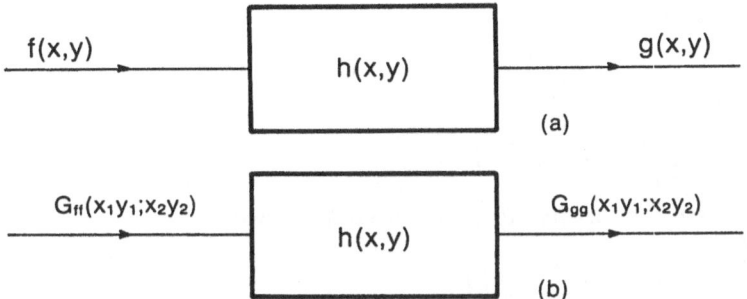

Fig. 3.71a,b. General two-dimensional optical system. (a) Block diagram for spatial input-output. (b) Block diagram for coherence function calculation

$$G_{gg}(x_1, y_1; x_2, y_2) = G_{fg}(x_1, y_1; x_2, y_2)*h(x_1, y_1)$$
$$= [G_{ff}(x_1, y_1; x_2, y_2)*h(x_2, y_2)]*h(x_1, y_1) \quad . \quad (3.9.66)$$

If the input is completely coherent, then

$$G_{ff}(x_1, y_1; x_2, y_2) = p(x_1, y_1)p^*(x_2, y_2) \tag{3.9.67}$$

and we obtain

$$G_{fg}(x_1, y_1; x_2, y_2) = p(x_1, y_1)[p^*(x_2, y_2)*h(x_2, y_2)] \quad , \tag{3.9.68}$$

$$G_{gg}(x_1, y_1; x_2, y_2) = [p(x_1, y_1)*h(x_1, y_1)][p^*(x_2, y_2)*h(x_2, y_2)] \tag{3.9.69}$$

and

$$G_{gg}(x_1, y_1; x_1, y_1) = |p(x_1, y_1)*h(x_1, y_1)|^2 \quad . \tag{3.9.70}$$

For the completely incoherent case

$$G_{ff}(x_1, y_1; x_2, y_2) = q(x_1, y_1)\delta(x_1 - x_2, y_1 - y_2) \quad , \tag{3.9.71}$$

$$G_{fg}(x_1, y_1; x_2, y_2) = q(x_1, y_1)h(x_2 - x_1, y_2 - y_1) \quad \text{and} \tag{3.9.72}$$

$$G_{gg}(x_1, y_1; x_2, y_2) = [q(x_1, y_1)h(x_2 - x_1, y_2 - y_1)]*h(x_1, y_1) \quad . \tag{3.9.73}$$

These results will be discussed further in the second volume, where they will also be analyzed in the spatial frequency domain.

For the general case where the light is not monochromatic, one should use $G(x_1, x_2, \omega)$ and a Fourier transformation to obtain the coherence function.

3.9.6 Coherency Matrix

For light propagating in the z direction, the electric field can be decomposed into the x and y directions. We have so far neglected this vectorial nature of the electric field in the definition of $G(x_1, x_2, t)$. One can rectify this situation by requiring G to be a vector. However, the coherence matrix formulation of this polarization problem is more convenient.

The coherency matrix C is defined by

$$C = \begin{bmatrix} \langle E_x E_x^* \rangle & \langle E_x E_y^* \rangle \\ \langle E_y E_x^* \rangle & \langle E_y E_y^* \rangle \end{bmatrix}$$
$$= \begin{bmatrix} C_{xx} & C_{xy} \\ C_{yx} & C_{yy} \end{bmatrix} = \langle [E][E]^\dagger \rangle \quad , \tag{3.9.74}$$

where we have assumed the light to be monochromatic and

$$[E] = \begin{bmatrix} E_x \\ E_y \end{bmatrix} \quad \text{and} \quad [E]^\dagger = [E_x^* E_y^*] \quad .$$

For a completely incoherent light, we have

$$\langle E_x \rangle = 0 = \langle E_y \rangle \quad , \tag{3.9.75}$$

$$\langle E_x E_y^* \rangle = 0 = \langle E_y E_x^* \rangle \quad \text{and} \tag{3.9.76}$$

$$\langle E_x E_x^* \rangle = \frac{I}{2} = \langle E_y E_y^* \rangle \quad , \tag{3.9.77}$$

where I is the total intensity and is given by

$$I = \langle E_x E_x^* \rangle + \langle E_y E_y^* \rangle \quad . \tag{3.9.78}$$

Thus we have

$$C = \frac{I}{2} \begin{bmatrix} 1 & 0 \\ 0 & 1 \end{bmatrix} \quad . \tag{3.9.79}$$

For a completely coherent but unpolarized light

$$c = |E_0|^2 \begin{bmatrix} 1 & 1 \\ 1 & 1 \end{bmatrix} \quad , \tag{3.9.80}$$

where E_0 is the magnitude of the electric field. The total intensity, I, is given by

$$I = \text{Tr}\{C\} = C_{xx} + C_{yy} \quad . \tag{3.9.81}$$

Using the Jones matrix formulation discussed in Sect. 2.15, it is easy to show that for the case of Fig. 3.72 with input light having coherency matrix C, the output light coherence matrix C' is given by

$$C' = M_j C M_j^\dagger \quad , \tag{3.9.82}$$

where $M_j^\dagger$ is the Hermitian conjugate, i.e.,

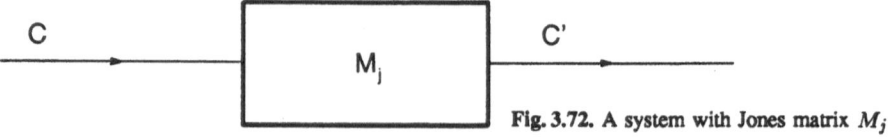

Fig. 3.72. A system with Jones matrix M_j

$$M_j = \begin{bmatrix} a & b \\ c & d \end{bmatrix} \quad \text{and} \tag{3.9.83}$$

$$M_j^\dagger = \begin{bmatrix} a^* & c^* \\ b^* & d^* \end{bmatrix} . \tag{3.9.84}$$

The intensity I' is then given by

$$I' = \mathrm{Tr}\{C'\} = \mathrm{Tr}\{M_j C M_j^\dagger\} = \mathrm{Tr}\{(M_j M_j^\dagger)C\} . \tag{3.9.85}$$

For example, if the input light is coherent but unpolarized and the system is a quarter-wave plate, we have an output coherency matrix given by

$$C' = \begin{bmatrix} e^{j\pi/4} & 0 \\ 0 & e^{-j\pi/4} \end{bmatrix} |E_0|^2 \begin{bmatrix} 1 & 1 \\ 1 & 1 \end{bmatrix} \begin{bmatrix} e^{-j\pi/4} & 0 \\ 0 & e^{j\pi/4} \end{bmatrix}$$

$$= |E_0|^2 \begin{bmatrix} 1 & j \\ -j & 1 \end{bmatrix} . \tag{3.9.86}$$

The total intensity of the output is given by

$$I = 2|E_0|^2 . \tag{3.9.87}$$

One generally defines a degree of polarization

$$P = \frac{I_{\mathrm{pol}}}{I} , \tag{3.9.88}$$

where I_{pol} is the polarized part of the intensity. One also defines a quantity called the normalized cross-correlation function μ_{xy}:

$$\mu_{xy} = \frac{X_{xy}}{\sqrt{C_{xx}}\sqrt{C_{xy}}} . \tag{3.9.89}$$

To obtain P one rotates x and y until μ_{xy} is maximized. For this situation, it can be shown that

$$P = \sqrt{1 - 4 \det C/(\mathrm{Tr}\{C\})^2} . \tag{3.9.90}$$

For (3.9.79),

$$P = 0 . \tag{3.9.91}$$

For (3.9.80),

$$P = 1 . \tag{3.9.92}$$

In general, however,

$$0 \leq P \leq 1 . $$

The coherency matrix defined above is complex. It is sometimes convenient to define real components as in the case of the Stokes vector for the electric field and the Mueller matrix for the optical system. Let us define the coherency matrix elements as a four-element column vector given by

$$C = \begin{bmatrix} C_{xx} \\ C_{xy} \\ C_{yx} \\ C_{yy} \end{bmatrix} \quad . \tag{3.9.93}$$

If we define a unitary matrix T given by

$$T = \begin{bmatrix} 1 & 0 & 0 & 1 \\ 1 & 0 & 0 & -1 \\ 0 & 1 & 1 & 0 \\ 0 & -j & j & 0 \end{bmatrix} \tag{3.9.94}$$

then we have a Stokes vector given by

$$[S] = TC \quad \text{or}$$

$$\begin{bmatrix} S_0 \\ S_1 \\ S_2 \\ S_3 \end{bmatrix} = \begin{bmatrix} 1 & 0 & 0 & 1 \\ 1 & 0 & 0 & -1 \\ 0 & 1 & 1 & 0 \\ 0 & -j & j & 0 \end{bmatrix} \begin{bmatrix} C_{xx} \\ C_{xy} \\ C_{yx} \\ C_{yy} \end{bmatrix} \quad . \tag{3.9.95}$$

Note that all the elements $S_0, \ldots, S_3$ of the Stokes vector are real. For the system in Fig. 3.73, the Mueller matrix is defined to be

$$S' = MS \quad \text{where} \tag{3.9.96}$$

$$M = T(M_J \times M_J^*)T^{-1} \quad . \tag{3.9.97}$$

The Kronecker product $M_J \times M_J^*$ is given by

$$M_j \times M_j^* = [M_j][M_j]^\dagger = \begin{bmatrix} a_1 \\ a_2 \\ a_3 \\ a_4 \end{bmatrix} [a_1^* a_2^* a_3^* a_4^*] \quad , \tag{3.9.98}$$

where

$$[M_j] = \begin{bmatrix} a_1 \\ a_2 \\ a_3 \\ a_4 \end{bmatrix} \longrightarrow \begin{bmatrix} a_1 & a_2 \\ a_3 & a_4 \end{bmatrix} \quad \text{(Jones matrix notation)} \quad . \tag{3.9.99}$$

By multiplying out the right-hand side of (3.9.99) one obtains

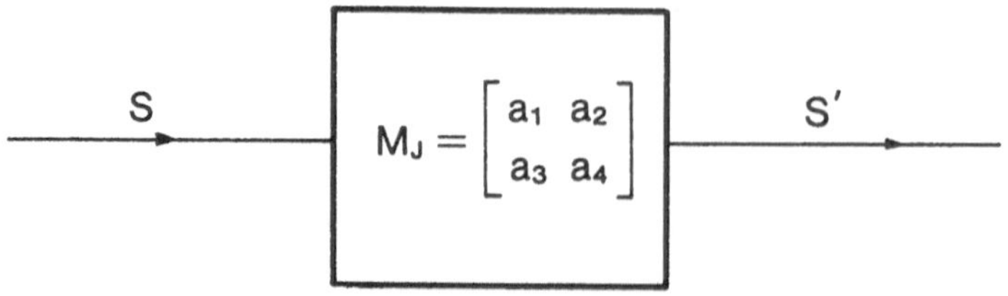

Fig. 3.73. Block diagram of a system for the determination of the Mueller matrix. S and S' are the input and output Stokes vectors

$$M = \begin{bmatrix} \frac{1}{2}(E_1 + E_2 + E_3 + E_4) & \frac{1}{2}(E_1 - E_2 - E_3 + E_4) & F_{13} + F_{42} & -G_{13} - G_{42} \\ \frac{1}{2}(E_1 - E_2 + E_3 - E_4) & \frac{1}{2}(E_1 + E_2 - E_3 - E_4) & F_{13} - F_{42} & -G_{13} + G_{42} \\ F_{14} + F_{32} & F_{14} - F_{32} & F_{12} + F_{34} & -G_{12} + G_{34} \\ G_{14} + G_{32} & G_{14} - G_{32} & G_{12} + G_{34} & F_{12} - F_{34} \end{bmatrix} ,$$

$$(3.9.100)$$

where

$$E_i = a_i a_i^* = |a_i|^2 \quad , \quad i = 1, 2, 3, 4 \quad ,$$

$$F_{ij} = F_{ji} = \text{Re}\{a_i a_j^*\} = \text{Re}\{a_j a_i^*\} \quad , \quad i, j = 1, 2, 3, 4 \quad , \quad \text{and}$$

$$G_{ij} = -G_{ji} = \text{Im}\{a_i^* a_j\} = -\text{Im}\{a_j^* a_i\} \quad .$$

For example, if

$$M_J = \begin{bmatrix} e^{j\delta} & 0 \\ 0 & e^{-j\delta} \end{bmatrix}$$

$$(3.9.101)$$

then

$$M = \begin{bmatrix} 1 & 0 & 0 & 0 \\ 0 & 1 & 0 & 0 \\ 0 & 0 & \cos 2\delta & -\sin 2\delta \\ 0 & 0 & \sin 2\delta & \cos 2\delta \end{bmatrix} .$$

$$(3.9.102)$$

Table 3.6 shows other cases discussed earlier in connection with the Jones matrix (Sect. 2.15.1). The coherency matrix and the Stokes vector for different light polarizations are shown in Table 3.7. The total intensity is given by

$$I = S_0 = C_{xx} + C_{yy} \quad .$$

$$(3.9.103)$$

The degree of polarization is given by

$$P = \frac{\sqrt{S_1^2 + S_2^2 + S_3^2}}{S_0} \quad .$$

$$(3.9.104)$$

If the light is completely polarized then $P = 1$ and we have

$$S_0^2 = S_1^2 + S_2^2 + S_3^2 \quad .$$

$$(3.9.105)$$

The last equation represents a sphere of radius S_0 in the subspace of S_1, S_2 and S_3, known as the Poincaré sphere. Each point on the sphere represents a definite state of polarization. For partially polarized light, the radius is smaller than S_0 and it goes to zero for unpolarized light.

A typical Poincaré sphere is shown in Fig. 3.74. Note that the longitude 2θ and latitude 2ε determine a point P_s that represents an ellipse of polarization with azimuth θ and ellipticity angle ε. The lines of longitude and latitude represent the equi-azimuth and equi-ellipticity contours, respectively. Note also that for Fig. 3.74 we have

Table 3.6. The Jones and Mueller matrix representations of certain optical instruments. (From [3.8])

Instrument	Jones representation	Mueller matrix
Compensator: Introduces a relative phase difference of 2δ	$\begin{bmatrix} e^{\delta} & 0 \\ 0 & e^{-\delta} \end{bmatrix}$	$\begin{bmatrix} 1 & 0 & 0 & 0 \\ 0 & 1 & 0 & 0 \\ 0 & 0 & \cos 2\delta & -\sin 2\delta \\ 0 & 0 & \sin 2\delta & \cos 2\delta \end{bmatrix}$
Rotator: Rotates the plane of polarization counterclockwise through angle θ about the z-axis	$\begin{bmatrix} \cos\theta & -\sin\theta \\ \sin\theta & \cos\theta \end{bmatrix}$	$\begin{bmatrix} 1 & 0 & 0 & 0 \\ 0 & \cos 2\theta & -\sin 2\theta & 0 \\ 0 & \sin 2\theta & \cos 2\theta & 0 \\ 0 & 0 & 0 & 1 \end{bmatrix}$
Polarizer: Takes the projection of the δ field in the direction making an angle α with the z-axis	$\begin{bmatrix} \cos^2\alpha & \sin\alpha\cos\alpha \\ \sin\alpha\cos\alpha & \sin^2\alpha \end{bmatrix}$	$\frac{1}{2}\begin{bmatrix} 1 & \cos 2\alpha & \sin 2\alpha & 0 \\ \cos 2\alpha & \cos^2 2\alpha & \sin 2\alpha\cos 2\alpha & 0 \\ \sin 2\alpha & \sin 2\alpha\cos 2\alpha & \sin^2 2\alpha & 0 \\ 0 & 0 & 0 & 0 \end{bmatrix}$
Absorber: η_z, η_y are the absorption coefficients in the z- and y-directions	$= e^{-\bar{\eta}}\begin{bmatrix} e^{-z} & 0 \\ 0 & e^{+z} \end{bmatrix}$ $\begin{bmatrix} e^{-\eta_z} & 0 \\ 0 & e^{-\eta_y} \end{bmatrix}$	$e^{-2\bar{\eta}}\begin{bmatrix} \cosh 2\epsilon & -\sinh 2\epsilon & 0 & 0 \\ -\sinh 2\epsilon & \cosh 2\epsilon & 0 & 0 \\ 0 & 0 & 1 & 0 \\ 0 & 0 & 0 & 1 \end{bmatrix}$
Define the mean absorption coefficients by $\bar{\eta} = \frac{1}{2}(\eta_z - \eta_y)$ and mean difference by $\epsilon = \frac{1}{2}(\eta_z - \eta_y)$		

Table 3.7. The coherency matrix and the Stokes vectors S for various light polarizations. (From [3.8])

State of polarization	J	C	S
Plane of polarization in the z direction	$\begin{bmatrix} 1 \\ 0 \end{bmatrix}$	$\begin{bmatrix} 1 & 0 \\ 0 & 0 \end{bmatrix}$	$\begin{bmatrix} 1 \\ 1 \\ 0 \\ 0 \end{bmatrix}$
Plane of polarization in the y direction	$\begin{bmatrix} 0 \\ 1 \end{bmatrix}$	$\begin{bmatrix} 0 & 0 \\ 0 & 1 \end{bmatrix}$	$\begin{bmatrix} 1 \\ -1 \\ 0 \\ 0 \end{bmatrix}$
Plane of polarization at 45° to the z-axis	$\begin{bmatrix} 1 \\ 1 \end{bmatrix}$	$\begin{bmatrix} 1 & 1 \\ 1 & 1 \end{bmatrix}$	$\begin{bmatrix} 1 \\ 0 \\ 1 \\ 0 \end{bmatrix}$
Plane of polarization at 135° to the z-axis	$\begin{bmatrix} 1 \\ -1 \end{bmatrix}$	$\begin{bmatrix} 1 & -1 \\ -1 & 1 \end{bmatrix}$	$\begin{bmatrix} 1 \\ 0 \\ -1 \\ 0 \end{bmatrix}$
Right circular polarization	$\begin{bmatrix} 1 \\ -j \end{bmatrix}$	$\begin{bmatrix} 1 & j \\ -j & 1 \end{bmatrix}$	$\begin{bmatrix} 1 \\ 0 \\ 0 \\ 1 \end{bmatrix}$
Left circular polarization	$\begin{bmatrix} 1 \\ j \end{bmatrix}$	$\begin{bmatrix} 1 & -j \\ j & -1 \end{bmatrix}$	$\begin{bmatrix} 1 \\ 0 \\ 0 \\ -1 \end{bmatrix}$

$$S_0 = \tfrac{1}{2} \quad,$$
$$S_1 = \tfrac{1}{2} \cos 2\varepsilon \, \cos 2\theta \quad,$$
$$S_2 = \tfrac{1}{2} \cos 2\varepsilon \, \sin 2\theta \quad \text{and}$$
$$S_3 = \tfrac{1}{2} \sin 2\varepsilon \quad.$$

Figure 3.75 shows N devices with individual Mueller matrix M_i. It can be shown that the equivalent total Mueller matrix M_{tot} is given by

$$M_{\text{tot}} = M_N M_{N-1} \ldots M_1 \quad. \tag{3.9.106}$$

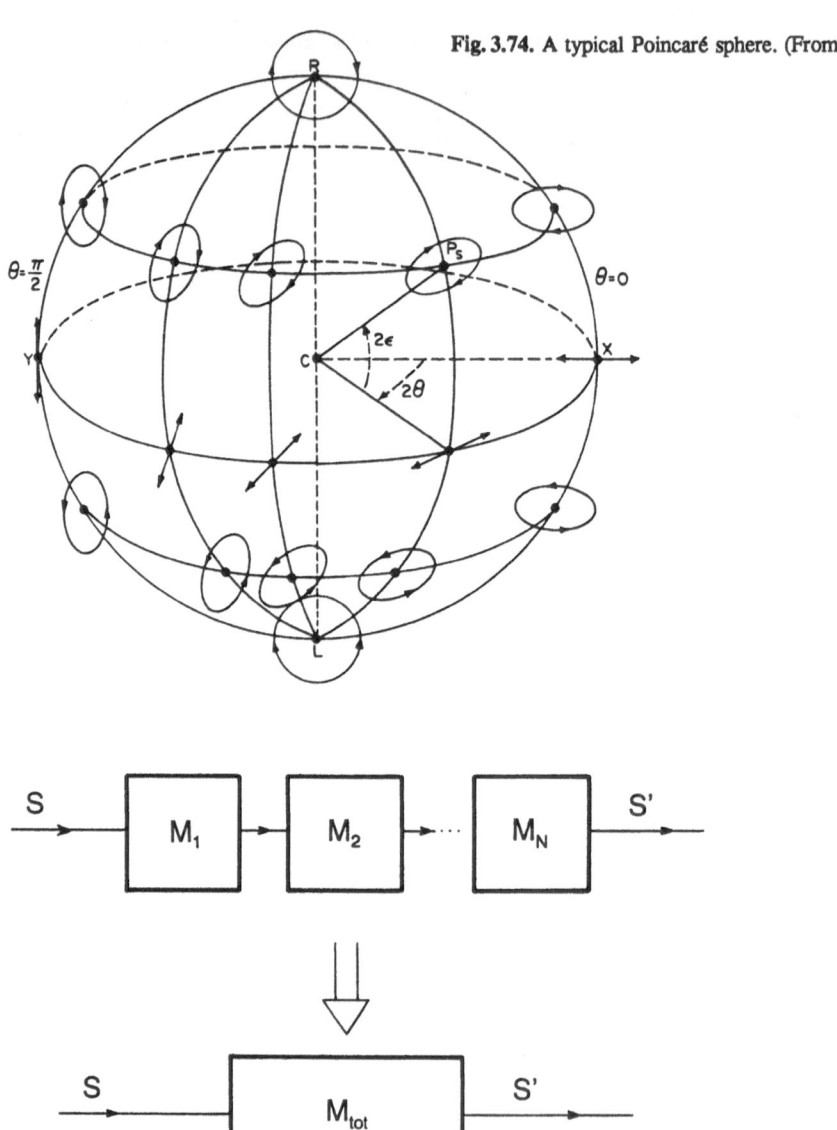

Fig. 3.74. A typical Poincaré sphere. (From [3.9])

Fig. 3.75. Geometry for defining the total Mueller matrix M_{tot}

3.10 The Ambiguity Function, Wigner Distribution Function and Triple Correlation

The ambiguity function of a signal $f(t)$ is defined as

$$\psi(\tau, \Delta f) = \int\limits_{-\infty}^{+\infty} f(t) f^*(t - \tau) e^{-j2\pi\Delta ft} dt \quad .$$ (3.10.1)

It is a function of τ and $\Delta\omega = 2\pi\Delta f$. For $\Delta\omega = 0$, we note that

$$\psi(\tau, 0) = f(t) \boxtimes f(t) \quad .$$ (3.10.2)

Thus the ambiguity function is the autocorrelation for $\Delta\omega = 0$. The ambiguity function arises in many applications of matched filtering where the signal to be match-filtered also has a frequency offset in its carrier due to the Doppler effect. In radar applications, for example, if the target is moving and $f(t)$ is the modulating envelope of the transmitted signal, then the received signal will be

$$f(t - \tau) e^{-j2\pi\Delta f(t-\tau)} \quad ,$$ (3.10.3)

where τ is the delay and $\Delta\omega$ is the frequency offset due to the Doppler shift in the carrier caused by the moving target. If this signal is passed through a system with an impulse response given by $f(-t)$ or correlated with the signal $f(t)$, then one obtains the ambiguity function except for a constant factor.

The Wigner distribution function $\phi(\tau, \Delta f)$ is defined by using the convolution of the frequency shifted signal:

$$\phi(\tau, \Delta f) = \int\limits_{-\infty}^{+\infty} f(t) f(\tau - t) e^{-j2\pi\Delta ft} dt \quad .$$ (3.10.4)

For $\Delta\omega = 0$, we have

$$\phi(\tau, 0) = f(t) * f(t) \quad .$$ (3.10.5)

Thus, the ambiguity function is the autoconvolution for $\Delta\omega = 0$. In the following we will discuss different properties of these functions and their interrelationships.

3.10.1 The Ambiguity Function

In the Fourier domain, the ambiguity function is given by

$$\psi(\tau, \Delta f) = \int\limits_{-\infty}^{+\infty} F^*(f - \Delta f) F(f) df \quad .$$ (3.10.6)

In general $\psi(\tau, \Delta f)$ is a complex quantity. Its squared magnitude $|\psi(\tau, \Delta\omega)|^2$ is a more relevant quantity for radar applications. The following two properties are easy to prove:

i) $|\psi(\tau, \Delta f)|^2 \leq |\psi(0,0)|^2$, $\qquad\qquad$ (3.10.7)

ii) $\displaystyle\int_{-\infty}^{+\infty} \int_{-\infty}^{+\infty} |\psi(\tau, \Delta f)|^2 d\tau\, df = |\psi(0,0)|^2$. $\qquad$ (3.10.8)

The second property can be interpreted as the conservation of total ambiguity in the $\tau - \Delta\omega$ plane and determining a target position (related to τ) and its velocity (related to $\Delta\omega$). Since the total volume under the $|\psi(\tau, \Delta f)|^2$ surface is a constant, if the volume is small in one region then it must peak up in some other region. Ideally one would like to design a signal $f(t)$ such that $|\psi(\tau, \Delta f)|^2$ is highly peaked at the origin and goes to zero rapidly elsewhere. In general this is not possible because, if it has a narrow peak near the origin, there will invariably be other peaks of smaller but nonzero amplitudes located elsewhere in the $\tau - \Delta f$ plane. Note that, the larger the size of the peak, the less ambiguous or more certain is the measurement. One can thus eliminate the undesired peaks by making the main peak broader. However, this results in a less accurate determination due to slower roll-off of the skirts of the peak.

Let us consider the example of a pulse of duration T, i.e.,

$$f(t) = \begin{cases} 1, & 0 \leq t \leq T , \\ 0, & \text{otherwise} . \end{cases} \qquad (3.10.9)$$

The ambiguity function for this signal is given by

$$\psi(\tau, \Delta f) = \frac{(1 - |\tau|/T)\sin(\frac{\pi \Delta f T}{2})}{\pi \Delta f / 2} . \qquad (3.10.10)$$

Depending upon the value of T, we have the long or short pulse case as shown in Fig. 3.76. The short pulse results in accurate range resolution while the long pulse results in accurate Doppler resolution. Consequently, this type of waveform is effective when one parameter is known and it is only necessary to estimate the other. To measure both range and Doppler, more sophisticated waveforms than a single pulse are necessary.

One such signal is the so-called burst waveform, which consists of a sequence of N ($N > 2$) short pulses and is shown in Fig. 3.77, along with the ambiguity function along the τ and $\Delta\omega$ axes. It can be seen from the figure that the resolution in time is determined by the width of a single pulse, while the resolution in frequency is determined by the length of the entire sequence. However, there are now multiple peaks in both time and frequency. The target can therefore be tracked with high resolution in both range and velocity provided the position and velocity of the target are known to be within the ambiguity introduced by the multiple peaks. Figure 3.78 shows the case for eight uniform pulse trains.

Another important example is the signal with a Gaussian envelope and linear frequency (quadratic phase) called a chirp or swept frequency pulse. For this case

338

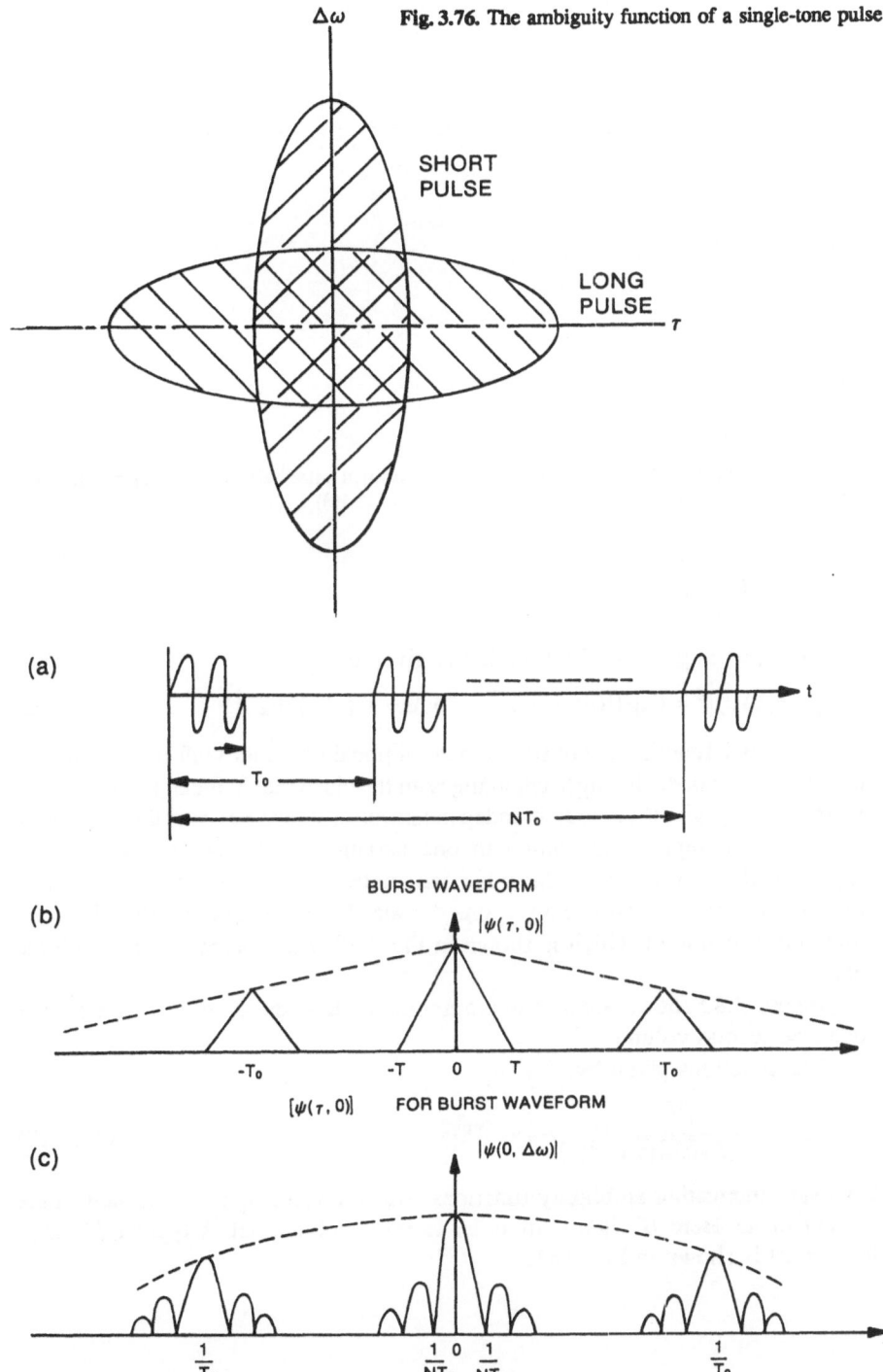

Fig. 3.76. The ambiguity function of a single-tone pulse

SHORT PULSE

LONG PULSE

(a)

T_0

NT_0

BURST WAVEFORM

(b) $|\psi(\tau, 0)|$

$-T_0$ $-T$ 0 T T_0

$[\psi(\tau, 0)]$ FOR BURST WAVEFORM

(c) $|\psi(0, \Delta\omega)|$

$\dfrac{1}{T_0}$ $\dfrac{1}{NT_0}$ 0 $\dfrac{1}{NT_0}$ $\dfrac{1}{T_0}$

Fig. 3.77. (a) The burst waveform. (b) $|\psi(\tau, 0)|$. (c) $|\psi(0, \Delta\omega)|$

339

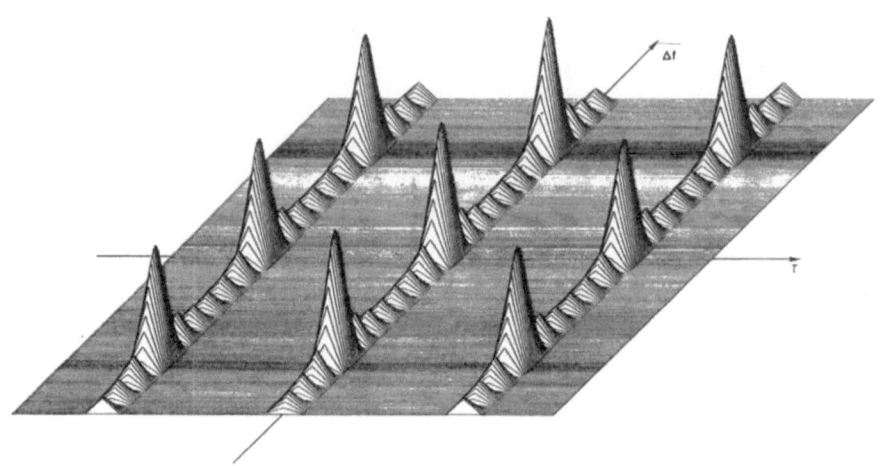

Fig. 3.78. Central part of $|\psi(\tau, A\omega)|$ for the uniform coherent pulse train $N = 8$, $T/T_0 = 0.1$, where T is the pulse duration and T_0 the repetition time. (From [3.10])

$$f(t) = \left(\frac{a^2}{\pi}\right)^{1/4} e^{-(a^2+jb)t^2/2} \quad . \tag{3.10.11}$$

The corresponding ambiguity function is given by

$$|\psi(\tau, \Delta\omega)|^2 = \exp\left([(a^2 + b^2)\tau^2 + 2b\Delta\omega\tau + (\Delta\omega)^2]/2a^2\right) \quad . \tag{3.10.12}$$

The ambiguity function is constant along elliptical contours with the orientation about the origin of each ellipse changing with the chirp rate b. It can be shown that the area of any given contour is independent of b; hence varying the frequency modulation to improve resolution in one parameter will necessarily result in worsening the resolution of the other parameter. As for the case of a single uniform pulse and a burst of pulses, one can also have a burst of chirp pulses, the ambiguity function of which is shown in Fig. 3.79. As expected, it has multiple peaks.

Figure 3.80 shows another important example – the case of 13-bit Barker code, see second volume.

The functions given by

$$f_n(t) = \frac{\gamma^{1/2}}{\sqrt{\pi^{1/2}n!2^n}} H_n(\sqrt{\gamma}t)e^{-\tau t^2/2} \tag{3.10.13}$$

have very interesting ambiguity functions which are rotationally invariant in the $\tau - \Delta f$ plane. Here H_n is the nth order Hermite polynomial. A typical $\psi(t, \Delta f)$ for $n = 10$ is shown in Fig. 3.81.

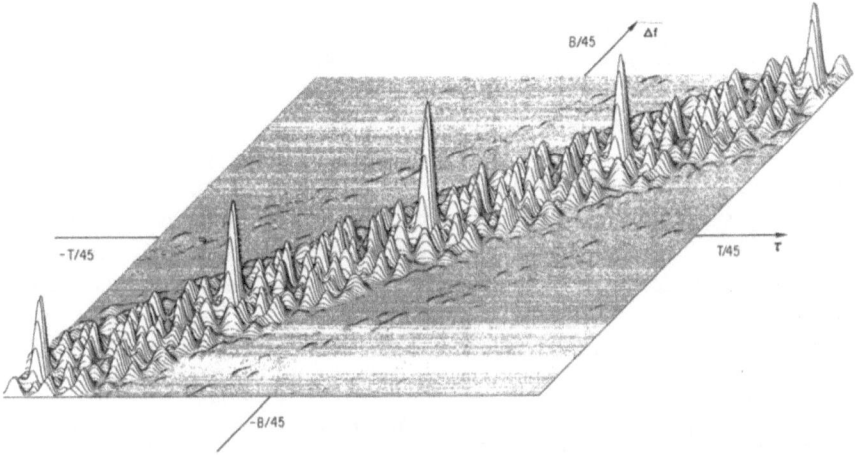

Fig. 3.79. The magnitude squared of the ambiguity function of a contiguous burst of chirp pulses. Then linear chirp pulses each having TB=10 and overall TB=1000. Frequency shifting is 10 times the pulse bandwidth. (From [3.10])

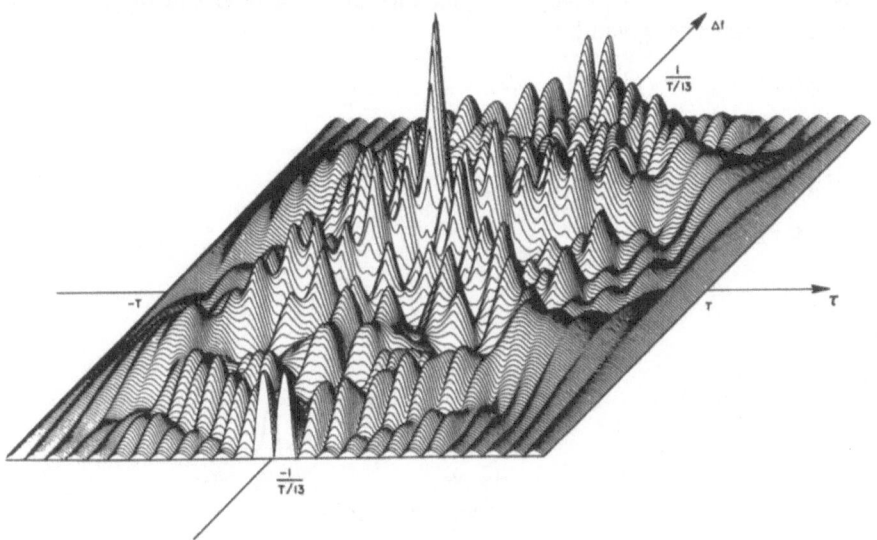

Fig. 3.80. The magnitude of the ambiguity function of 13-bit Barker code. T : bit time duration. (From [3.10])

Fig. 3.81. Cut-away of three-dimensional projected view of the magnitude of the ambiguity function $|\psi(t, \Delta f)|$ for $n = 10$ in (3.10.15). (From [3.11])

3.10.2 Wigner Distribution Function

Both the ambiguity function and the Wigner distribution function are two-dimensional representations of one-dimensional signals. Mathematically, it is true that $f(t)$ or $F(f)$ describes the signal completely. However, consider the example of a musical score, shown in Fig. 3.82, which describes the frequency to be played by the musicians at a particular time. For a purist, this musical score is mathematically unjustifiable because it violates the uncertainty principle (a short pulse will have a large spread in frequency). However, the musical score is undeniably very useful for musicians. The Wigner distribution function avoids this conflict between musicians and mathematical purists!

In place of (3.10.4) it is convenient to define the Wigner distribution function in a symmetrical form as

$$\phi(t, \Delta f) = \int f(t + \tau/2) f^*(t - \tau/2) e^{-j2\pi \Delta f \tau} d\tau$$
$$= \int F(\Delta f + f'/2) F^*(\Delta f - f'/2) e^{j2\pi f't} df' \quad . \tag{3.10.14}$$

This definition preserves complete symmetry between time and frequency. Let us define a function given by

$$g(t, \tau) = f(t + \tau/2) f^*(t - \tau/2)$$
$$= \int \phi(t, f) e^{j2\pi \tau f} df \quad . \tag{3.10.15}$$

By definition, $g(t, \tau)$ is Hermitian, i.e.,

$$g(t, \tau) = g^*(t, -\tau) \quad . \tag{3.10.16}$$

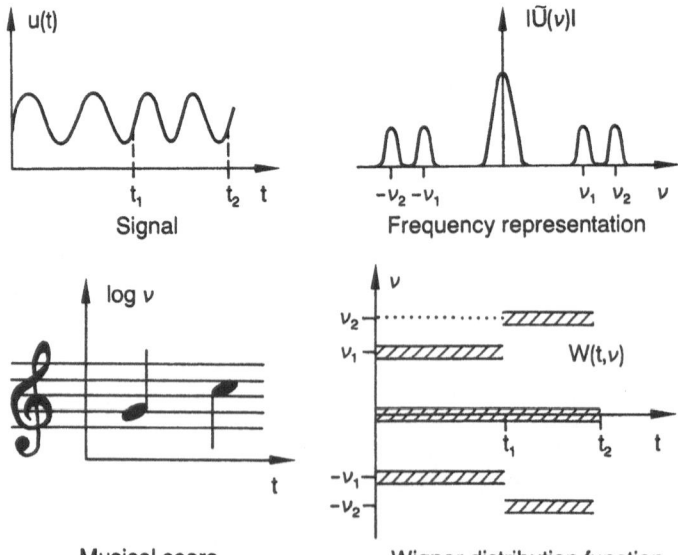

Fig. 3.82. Different descriptions of the same signal $f(t) = u(t)$ and $F(f) = U(\nu)$. (From [3.11])

342

From (3.10.17), we have

$$\phi(t, f) = \int g(t, \tau)e^{-j2\pi\tau f}\,d\tau \quad \text{or} \tag{3.10.17}$$

$$\phi^*(t, f) = \int g^*(t, -\tau)e^{j2\pi\tau f}\,d\tau$$
$$= \phi(t, f) \quad . \tag{3.10.18}$$

Thus $\phi(t, f)$ is real. If we symmetrize the ambiguity function as

$$\psi(t, \Delta f) = \int f(t + \tau/2)f^*(t - \tau/2)e^{-j2\pi\Delta f}\,d\tau, \tag{3.10.19}$$

we can then write down the following relationship between the ambiguity function and the Wigner distribution function:

$$\phi(t, \alpha) = \iint \psi(\tau, \beta)e^{j2\pi(t\tau - \beta\alpha)}\,d\tau\,d\beta \quad . \tag{3.10.20}$$

The relationship between the above functions is symbolically represented in Fig. 3.83. In the figure, F.T. ($\alpha \rightarrow \tau$) denotes the Fourier transformation mapping of α to τ coordinates.

The total energy, the power spectrum and the intensity of the signal, respectively, can be obtained from $\phi(t, f)$ as follows:

$$E_{\text{tot}} = \text{Total energy} = \iint \phi(t, f)dt\,df \quad . \tag{3.10.21}$$

$$|f(t)|^2 = \text{Intensity} = \int \phi(t, f)df \quad . \tag{3.10.22}$$

$$|F(f)|^2 = \text{Power spectrum} = \int \phi(t, f)dt \quad . \tag{3.10.23}$$

The signal itself can also be written as

$$f(t) = e^{j\gamma_0}g(t/2, \tau)/\sqrt{g(0, 0)} \quad \text{where} \tag{3.10.24}$$

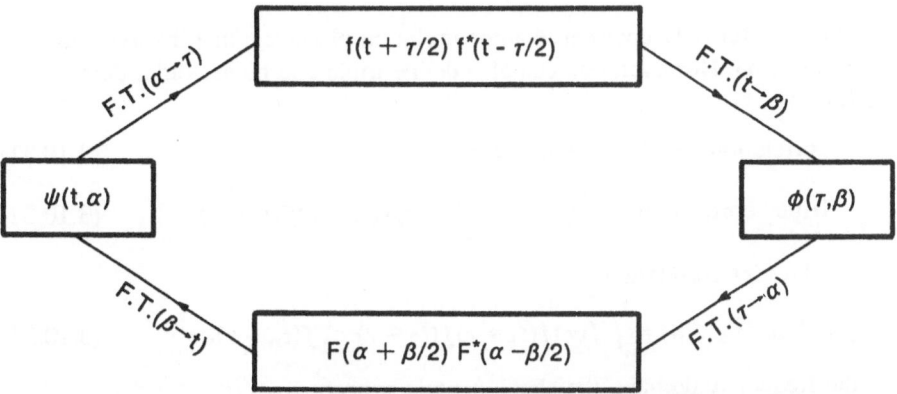

Fig. 3.83. Relationship between the Wigner distribution function $\phi(\tau, \beta)$ and the ambiguity function $\psi(t, \alpha)$

$$e^{\gamma_0} = f(0)/|f(0)| \quad .$$

(3.10.25)

The Wigner distribution function for the following three examples can easily be derived:

$$f(t) = e^{j2\pi f_0 t} \rightarrow \phi(t, f) = \delta(f - f_0) \quad ,$$

(3.10.26)

$$f(t) = \delta(t - t_0) \rightarrow \phi(t, f) = \delta(t - t_0) \quad ,$$

(3.10.27)

$$f(t) = e^{j2\pi(at^2/2 + bt + c)} \rightarrow \phi(t, f) = \delta(at + b - f) \quad .$$

(3.10.28)

The physical interpretation of the Wigner distribution function is somewhat similar to that of the ambiguity function and will be discussed further when we consider its implementation.

3.10.3 Two-Dimensional Ambiguity and Wigner Distribution Functions

The one-dimensional case can easily be extended to the two-dimensional or multidimensional case. In terms of the two-dimensional function and its Fourier transform discussed in Sect. 3.8, we have

$$\phi(x, y; f_x, f_y) = \iint f(x + x'/2, y + y'/2) f^*(x - x'/2, y - y'/2)$$
$$\times e^{-j2\pi(f_x x' + f_y y')} dx' dy'$$

(3.10.29)

and

$$\psi(x', y', f_x, f_y) = \iint f(x + x'/2, y + y'/2) f^*(x - x'/2, y - y'/2)$$
$$\times e^{-j2\pi(f_x x' + f_y y')} dx \, dy \quad .$$

(3.10.30)

It is to be noted that Fraunhofer and Fresnel diffraction results can be obtained from the Wigner distribution function.

3.10.4 Triple and Higher-Order Correlations

We have so far only considered autocorrelation of signals that involves just one multiplication. However, one can also define triple and higher-order correlations as follows:

Correlation $\quad C^2(t) = f \boxtimes g = \int f(\tau) f(t + \tau) d\tau \quad .$

(3.10.31)

Triple correlation $\quad C^3(t_1, t_2) = \int f(\tau) f(t_1 + \tau) f(t_2 + \tau) d\tau \quad .$

(3.10.32)

nth-order correlation

$$C^n(t_1, t_2 \ldots t_n) = \int f(\tau) f(t_1 + \tau) f(t_2 + \tau) \ldots f(t_n + \tau) d\tau \quad .$$

(3.10.33)

In the frequency domain, the above equations become

$$C^1(f) = F(f) F^*(f) \quad ,$$

(3.10.34)

$$C^2(f_1, f_2) = \mathcal{F}[C^2(t_1, t_2)]_{f_1 f_2} \quad \text{and} \tag{3.10.35}$$

$$C^n(f_1 \ldots f_n) = \mathcal{F}[C^n(t_1, t_2 \ldots t_n)]_{f_1 \ldots f_n} \tag{3.10.36}$$

Note that the right-hand side of (3.10.36) represents an n-dimensional Fourier transformation. $C^2(f_1, f_2)$ is generally known as the bispectrum. The above definitions can easily be extended to multidimensional signals. If $f(t)$ is a stochastic process and Gaussian, it can be shown that all odd-order correlations defined above are zero. However, for non-Gaussian signals, the triple correlation is not always zero; it has many useful and practical properties of interest. In the remainder of this section we shall restrict ourselves to the discussion of triple correlation.

The triple correlation and bispectrum of the following four elementary functions are shown in Figs. 3.84–87:

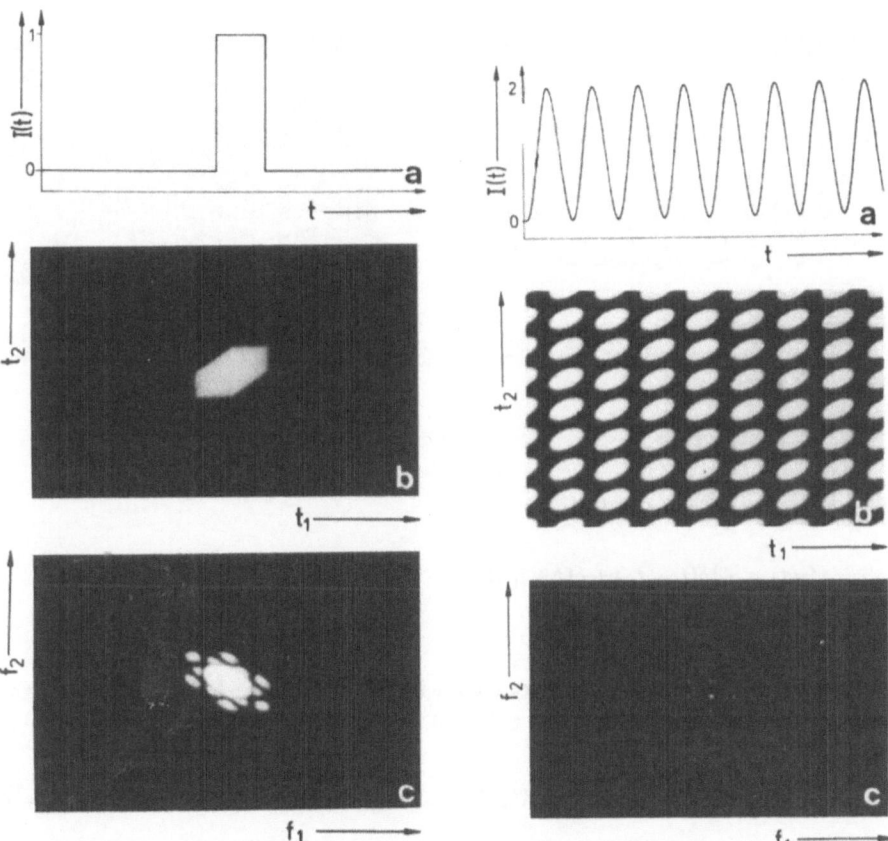

Fig. 3.84. Triple correlation (b) and bispectrum (c) for a square-box signal (a). (From [3.12])

Fig. 3.85. Triple correlation (b) and bispectrum (c) for a cosine signal on a bias (a). (From [3.12])

345

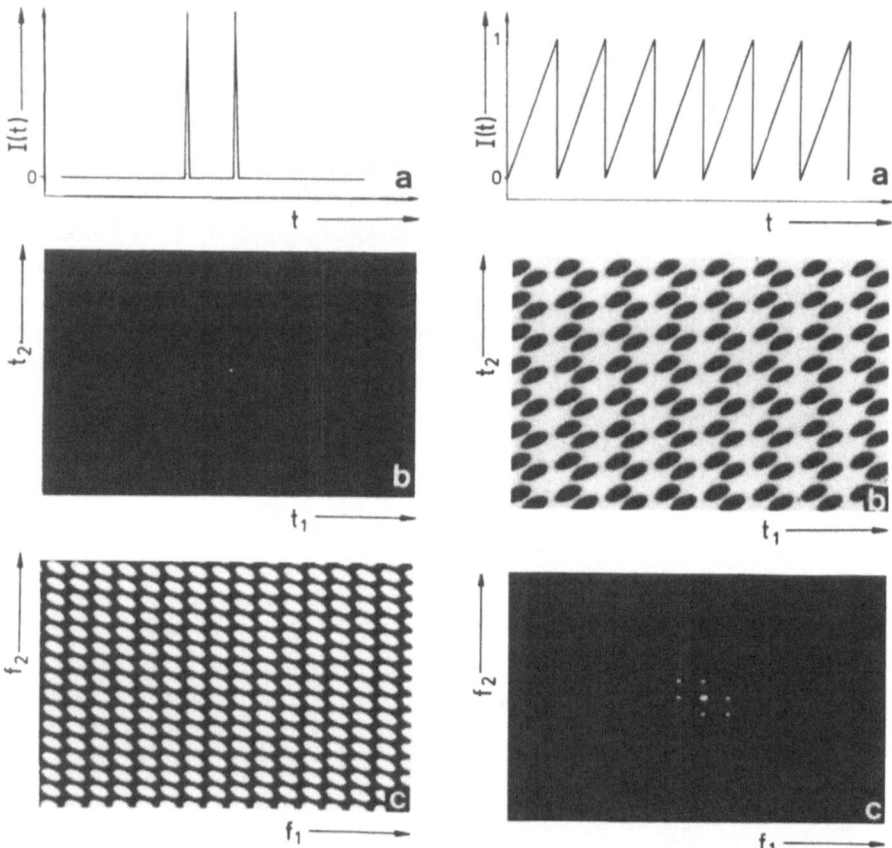

Fig. 3.86. Triple correlation (b) and bispectrum (c) for a signal consisting of two spikes (a). (From [3.12])

Fig. 3.87. Triple correlation (b) and bispectrum (c) for a periodical sawtooth signal (a). (From [3.12])

$$f_1(t) = P(T) \quad ,$$

$$f_2(t) = 1 + \cos (2\pi f_0 t) \quad ,$$

$$f_3(t) = A[\delta(t - t_0) + \delta(t + t_0)] \quad ,$$

$$f_4(t) : \text{sawtooth waveform} \quad .$$

For a linear time-invariant system with impulse response $h(t)$, it can easily be shown that the triple correlation $G^3(t)$ of an output $g(t)$ for an input $f(t)$ is given by

$$G^3(t_1, t_2) = F^3(t_1, t_2) * H^3(t_1, t_2) \quad \text{and} \tag{3.10.37}$$

$$G^3(f_1, f_2) = F^3(f_1, f_2) H^3(f_1, f_2) \quad , \tag{3.10.38}$$

where the terms with superscript 3 and variables t_1 and t_2 represent the triple correlation and those with superscript 3 and variables f_1 and f_2 represent the bispectrum. It can easily be shown that the bispectrum or triple correlation of a pure sinusoidal signal is zero. However, if one adds bias to it, then it becomes nonzero. If there is any nonlinearity in a system, then the output due to a sinusoidal input will contain a bias and the triple correlation, by being a sensitive test for detecting bias, might also be useful for detecting a weak nonlinearity in systems.

For an input $s(t)$ with additive signal-independent noise $n(t)$, the triple correlation is given by

$$S^3(t_1, t_2) = F^3(t_1, t_2) + \langle N^3(t_1, t_2) \rangle$$
$$+ \langle N \rangle [F^2(t_1) + F^2(t_2) + F^2(t_2 - t_1)]$$
$$+ \int f(t)dt [\langle N^2(t_1) \rangle + \langle N^2(t_2) \rangle$$
$$+ \langle N^2(t_2 - t_1) \rangle \Bigg] \quad , \quad \text{where} \tag{3.10.39}$$

$$s(t) = f(t) + n(t) \quad .$$

If

$$\langle N \rangle = 0 \quad , \quad \int f(t)dt = 0 \quad \text{and} \quad \langle N^3 \rangle = 0$$

then $S^3(t_1, t_2)$ might prove to be a useful detector for the signal $f(t)$. These conditions apply in many practical cases, and triple correlation turns out to be a more powerful technique than simple matched filtering.

Problems

3.1 Show that the system shown in Fig. 3.88, which uses the CMC (convolve-multiply-convolve) algorithm, can also be used to perform real-time Fourier transformation.

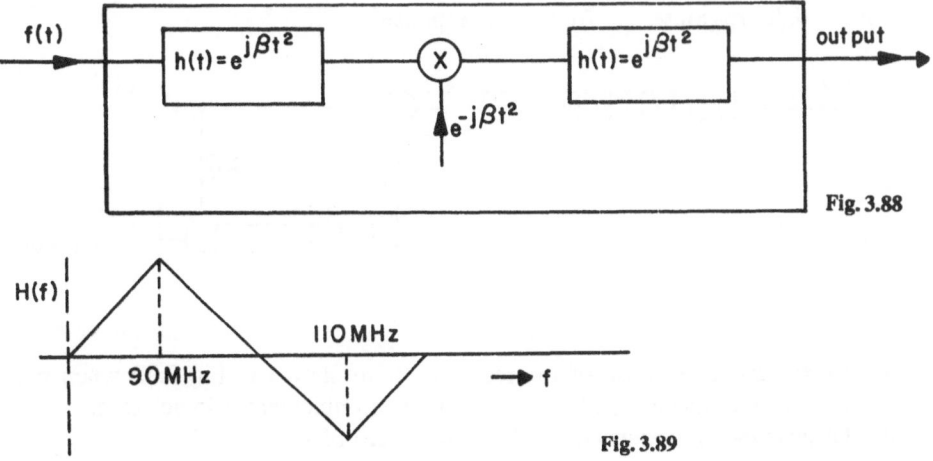

Fig. 3.88

Fig. 3.89

3.2 Design a SAW FM discriminator usable in the frequency range 90–110 MHz. Hint: Design a system with the frequency response shown in Fig. 3.89.

3.3

a) Find the autoconvolution and correlation of the signal shown in Fig. 3.90 (13 bit Barker code).

b) Design the mask for a SAW matched filter for the Barker code of (a). Center frequency = 100 MHz, $T = 0.1\,\mu s$, $v_{SAW} = 3000$ m/s.

c) Plot the impulse response of the device designed in (b).

Fig. 3.90

3.4 A recursive filter is built using two 8-stage CCD delay lines as shown in Fig. 3.91. Show that the z-response of the filter is given by

$$H(x) = \frac{1 - \lambda}{z^8 - \lambda} \quad .$$

Show that the locations of the poles for $\lambda = 0.252$ are given by

$$-0.322 + j0.778$$
$$-0.322 + j0.778$$
$$0.778 + j0.322$$
$$0.778 - j0.322$$
$$0.322 + j0.778$$
$$0.322 - j0.778$$
$$-0.778 + j0.322$$
$$-0.778 - j0.322 \quad .$$

Plot the delta response and frequency response.

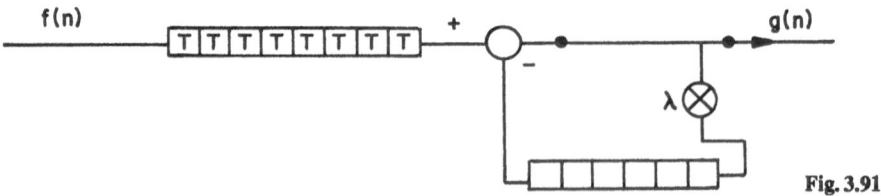

Fig. 3.91

3.5

a) Derive the z-response of the recursive filter shown in Fig. 3.92 when the transfer inefficiency ε of the CCDs used to implement it is not zero.

b) Discuss the consequences of different values of ε.

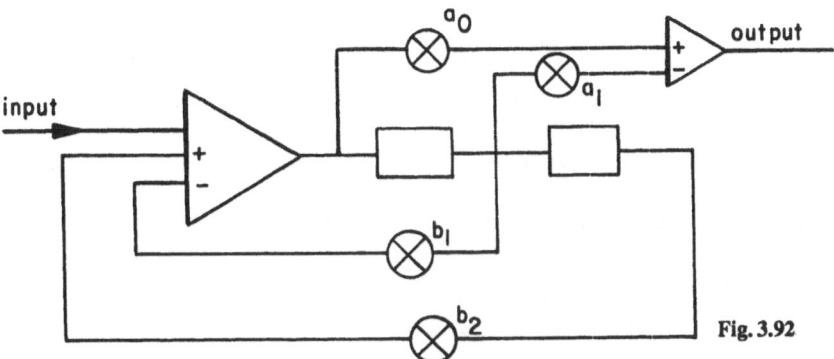

input

output

Fig. 3.92

3.6 A SAW delay is to be designed using a piezoelectric substrate having the SAW velocity $v = 3600$ m/s. It is decided that the delay time is to be $4\,\mu s$, the center frequency 180 MHz and the input and output interdigital transducers (IDTs) will have 4 and 8 finger pairs, respectively.

a) Make a drawing showing the exact dimensions of the mask to be made.
b) Calculate and plot the impulse response of the device neglecting any reflection.
c) Calculate the frequency response around 180 MHz only.
d) Calculate the size of triple-transit echo if the reflection coefficients of the transducers are $\frac{1}{4}$ and $\frac{1}{6}$ respectively.
e) What is the change in the answer to the question in (c) if triple-transit echo is included? Give an analytical expression for the change, if any.

3.7 A 3-phase CCD transversal filter is designed using a 10 tap structure. The CCD uses a 10 MHz clock and the dimensions of each cell are $10\,\mu m \times 100\,\mu m$.

a) Plot the impulse and frequency responses if all the weights have equal value. Give numerical values in the plot.
b) What is the highest frequency for which the CCD filter can be used? Discuss.
c) If the storage time of each cell is 1 s, what is the lowest frequency for which the CCD filter can be used? Discuss.
d) Repeat (a) and (b) for a transfer inefficiency of 10%.

3.8 For a spread spectrum communication system a 13 bit Barker code generator and a corresponding matched filter are to be designed. The center frequency of operation is to be 390 MHz with a bit rate of 78 MHz.

a) What technology (SAW, CCD or digital) would you choose? Explain.
b) SAW is chosen. Design the mask.
c) Plot the correlated output.
d) The system is to be used in a colored noise environment with the noise spectrum given by $N(\omega)$. Draw a block diagram outlining a particular design for the spread spectrum communication system.

349

3.9

a) The impulse response $h(t)$ of the device depicted in Fig. 3.93 is given by $\exp(j\beta t^2)$. For an input given by $f(t)\exp(-j\beta t^2)$, give an exact expression for the output.

b) Show in block diagrams how you would implement a real-time Hilbert transformer.

c) Calculate the output of the system shown in Fig. 3.94.

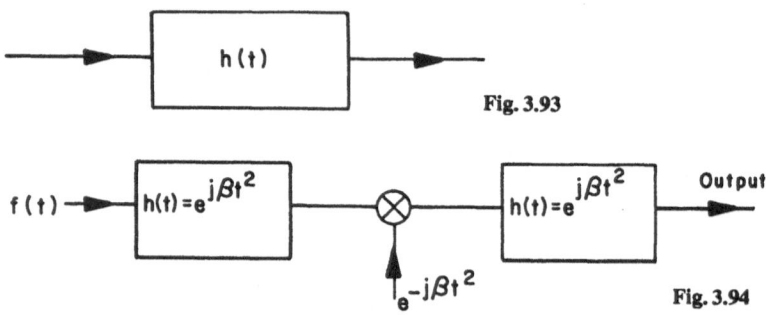

Fig. 3.93

Fig. 3.94

3.10 Consider the simple RC filter shown in Fig. 3.95. Show that the real and imaginary parts of the frequency response of this filter obey the Hilbert transform relationship.

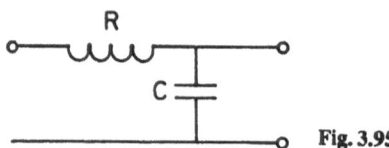

Fig. 3.95

3.11 A real-time Fourier transformer is implemented using the block diagram shown in Fig. 3.96. Find the impulse response of the Fourier transformer.

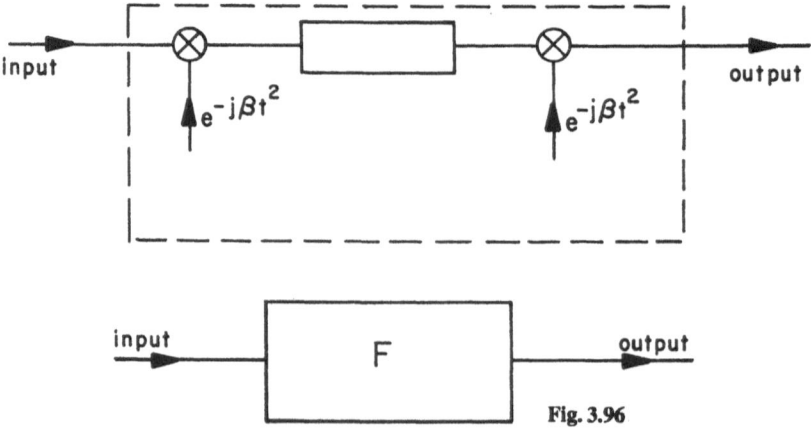

Fig. 3.96

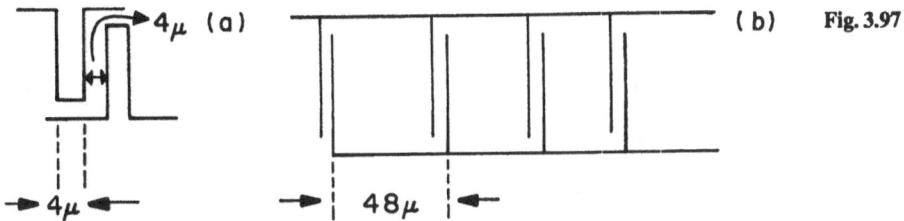

Fig. 3.97

4μ (a)

4μ

(b)

48μ

3.12 A SAW filter is designed using a piezoelectric substrate having the Rayleigh wave velocity $v = 3200$ m/s. The input transducer consists of 3 finger pairs with the finger pair dimensions shown in Fig. 3.97a. The output transducer consists of 4 one-finger pair transducers separated by 48 μm from each other (Fig. 3.97b).

a) Calculate and plot the impulse response.
b) Calculate and plot the frequency response (around the fundamental only).
c) Plot the output of the filter if the applied input is given by an rf pulse of duration 600 ns and center frequency 200 MHz.

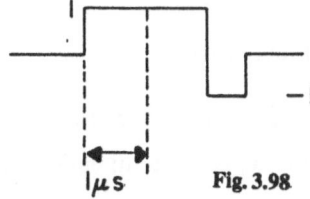

1μs   Fig. 3.98

3.13 A matched filter is to be designed for the waveform of Fig. 3.98 using a 4-phase CCD with three taps.

a) Design the matched filter, showing the tap weights in a figure.
b) What is the minimum clock frequency you need for the CCD?
c) Calculate the z-response of the designed filter for a transfer inefficiency of 10%.
d) Plot the impulse response for the case (c) using expansion of the z-response.

3.14

a) The impulse response $h(t)$ of the device in Fig. 3.99 is given by $\exp(j10^8 t^2)$. Give an exact expression for the output if the input is given by $\exp(+j10^8 t^2)$.
b) What is the impulse response of the matched filter for a signal identical to the output in (a) and limited to a duration of 10 μs?
c) What is the matched filtered output? Calculate and plot.

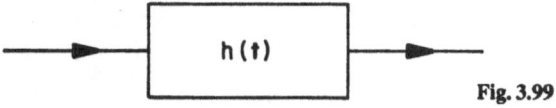

$h(t)$

Fig. 3.99

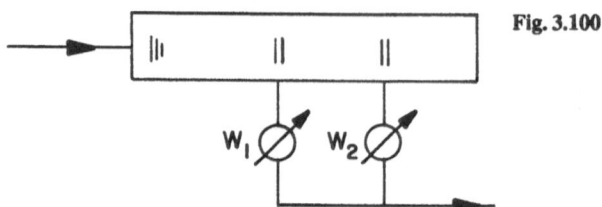

Fig. 3.100

3.15 A two-stage adaptive filter is to be designed using a SAW delay line as shown in Fig. 3.100. The correlation matrix for the input to the filter is given by

$$[R] = \begin{bmatrix} R_1 & R_2 \\ R_3 & R_4 \end{bmatrix} \quad .$$

a) What is the relationship between R_1, R_2, R_3, and R_4, if any?
b) Calculate the optimum weights, if the steering vector P is given by

$$[P] = \begin{bmatrix} P_1 \\ P_2 \end{bmatrix} \quad .$$

c) Derive and solve the time dependence of $W_1(t)$ and $W_2(t)$.

3.16 Consider a SAW resonator with separate input and output transducers.

a) Find the impulse response in terms of SAW velocity, physical dimensions, etc., with no loss.
b) Repeat (a) with loss taken into account.
c) Find the Q of the resonator from the impulse response.

3.17

a) Consider a real time function $h(t)$ which can be split up into $h_e(t)$ (even) and $h_o(t)$ (odd) parts:

$$h(t) = h_e(t) + h_o(t) \quad ,$$
$$h_e(t) = \tfrac{1}{2}[h(t) + h(-t)] = h_e(-t) \quad ,$$
$$h_o(t) = \tfrac{1}{2}[h(t) - h(-t)] = -h_o(-t) \quad .$$

Show that if $H(f) = R(f) + jX(f)$, then

$$R(f) = 2 \int_0^\infty h_e(t) \cos 2\pi f t \, dt \quad ,$$

$$X(f) = -2 \int_0^\infty h_o(t) \sin 2\pi f t \, dt \quad .$$

Also show that

$$H_e(f) = \mathcal{F}\{h_e(t)\} = R(f) = R(-f) \quad,$$
$$H_o(f) = \mathcal{F}\{h_o(t)\} = jX(f) = -jX(-f) \quad.$$

b) For this part of the problem assume $h(t)$ is causal. Then we have

$$h(t) = 2h_e(t) + 2h_o(t) \quad \text{for} \quad t>0 \quad.$$

Show that

$$X(f) = -\frac{2}{\pi} \int_0^\infty \int_0^\infty R(f') \cos f't \sin 2\pi f df' dt \quad,$$

$$R(f) = -\frac{2}{\pi} \int_0^\infty \int_0^\infty X(f') \sin f't \cos 2\pi ft df' dt \quad.$$

Assuming that $H(f)$ contains no impulses at the origin [i.e. $H(f) \to 0$ as $f \to \infty$] show that

$$X(f) = -4f \int_0^\infty \frac{R(f')}{4\pi^2 f^2 - f'^2} df' \quad,$$

$$R(f) = \frac{2}{\pi} \int_0^\infty \frac{f'X(f')}{4\pi^2 f^2 - f'^2} df' \quad.$$

If $H(f) \to R(\infty)$ for $f \to \infty$ then derive

$$R(f) = R(\infty) + \frac{2}{\pi} \int_0^\infty \frac{f'X(f')}{4\pi f^2 - f'^2} df' \quad.$$

3.18 A SAW filter is designed using a piezoelectric substrate having Rayleigh wave velocity $v = 3000$ m/s. The input transducer consists of five finger pairs with the finger pair dimensions shown in Fig. 3.101. The output transducer consists of 100 finger pairs with sinc function weighting centered symmetrically and containing one sidelobe on each side.

a) Calculate the operating center frequency of the device.
b) Calculate and plot the frequency response.

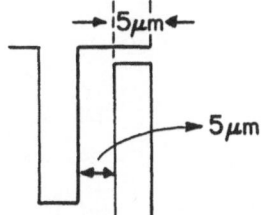

Fig. 3.101

3.19 A CCD recursive filter is to be designed. Its z-response is

$$H(z) = \frac{z(z^2 + z + 2)}{z^3 - z^2 - 5z - 1} \quad .$$

a) Show with a diagram how you would implement it considering ideal CCDs.
b) Derive an expression for the frequency response of the filter if the transfer inefficiency is 10^{-5} per phase and 4-phase CCDs are used in the design.

3.20 A matched filter for a 13-bit Barker code with bit rate 5 MHz is to be designed. The autocorrelation of the noise is given by Fig. 3.102. Explain the design procedure first in block diagrams and then how you would implement it.

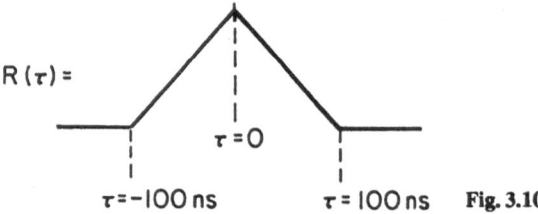

$R(\tau) =$

$\tau = 0$

$\tau = -100$ ns $\tau = 100$ ns   **Fig. 3.102**

3.21 A chirp filter with the frequency vs delay characteristics shown in Fig. 3.103 is used to perform real time Fourier transformation without dechirping (Fig. 3.104).

a) Calculate β.
b) Find the impulse response including the actual chirp filter characteristics.
c) Calculate the output for an input given by

$$\cos(2\pi \times 10^7 t)\, e^{j2\pi \times 10^8 t} \quad .$$

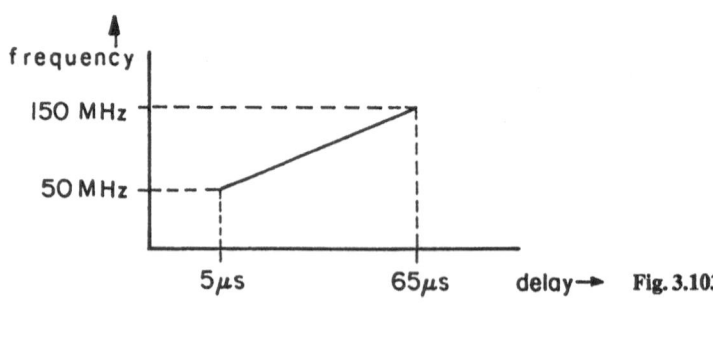

frequency

150 MHz

50 MHz

5 µs   65 µs   delay→   **Fig. 3.103**

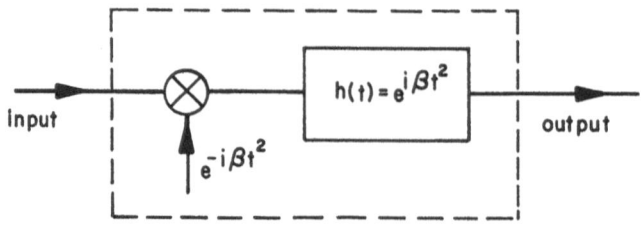

Input

$h(t) = e^{i\beta t^2}$

$e^{-i\beta t^2}$

output

Fig. 3.104

3.22 The input X to an adaptive processor has the following correlation matrix:

$$X(t) = f(t) + i(t) + n(t)$$

$$[R] = [F] + [I] + [N]$$

$$F = \begin{bmatrix} F_0 & F_1 & F_2 & 0 & 0 & 0 \\ F_1 & F_0 & F_1 & F_2 & 0 & 0 \\ F_2 & F_1 & F_0 & F_1 & F_2 & 0 \\ 0 & F_2 & F_1 & F_0 & F_1 & F_2 \\ 0 & 0 & F_2 & F_1 & F_0 & F_1 \\ 0 & 0 & 0 & F_2 & F_1 & F_0 \end{bmatrix}$$

$I_{ij} \neq 0$ for all i and j and N is a diagonal matrix .

We would like to use the Widrow-Hoff algorithm with a 6-tap filter as shown in Fig. 3.105 to obtain $f(t)$ and $i(t)$. Only $X(t)$ is available, not its components $f(t)$ or $i(t)$.

a) What d should we use?
b) Where should we take the output for $f(t)$?
c) Where should we take the output for $i(t)$?
d) Estimate $\varepsilon(t)$ for ideal conditions.
e) What are the optimum weights in the steady-state situation?

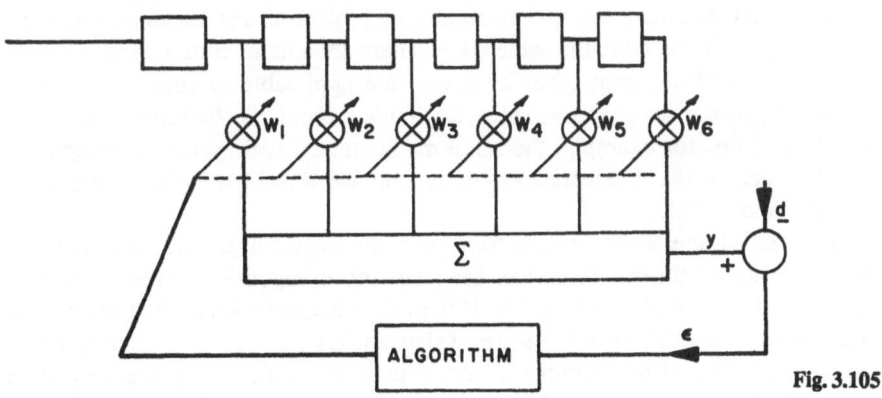

Fig. 3.105

4. Introduction to SAW and CCD Technology

The need for and advantages of carrying out real-time signal processing with faster speed and higher bandwidth culminated in the use of SAW and the invention of CCD devices in the 1960s. A major objective of this chapter is to introduce fundamentals of these devices to the reader. Another objective is to discuss applications of this technology in areas such as radar and sonar engineering, image processing, spread spectrum communications systems and filtering in general. Section 4.6 includes a discussion of the relative advantages and disadvantages of SAW and CCD devices and compares them with other signal processing methods, such as digital or optical signal processing. There are fundamental limitations in both of these types of device as well as some suitability to particular applications purpose. A thorough analysis is presented which provides enough information to enable the reader to choose a particular device for a particular purpose. Of course, as the technology advances, these analyses will need to be modified accordingly.

One might wonder why the diverse subjects of surface acoustic waves (i.e., a mechanical deformation wave or an elastic wave) and charge coupled devices (i.e., elements requiring the understanding of basic semiconductor physics and electronics) are included in the same book. There are a number of reasons for such a choice. First of all, both types of device are applicable to similar situations. For signal processing purposes, both are built around the building block of a delay line. Thus, for example, the same mathematical and systems concepts are needed to design the transversal or Finite Impulse Response (FIR) filters using either SAW or CCD.

Both CCD and SAW devices have been developed in the last two decades. Since their development, there has been healthy competition between the two because of their similar applications. However, it has been found that the combination of the two devices in the same system makes them complementary rather than competitive and in a number of areas, such as radar signal processing, their combination looks very attractive. Even at the device level, it appears that the concepts of SAW and CCD can be combined to obtain new hybrid devices that may offer superior performance. These are mostly still at the conceptual and development stage at present and will possibly be devices of the future.

4.1 History of CCD and SAW Devices

4.1.1 Charge Coupled Devices

CCDs were invented by *Boyle* and *Smith* of Bell Telephone Laboratories; they published their results in two classic papers [4.1, 2]. At the same time magnetic bubble development was also in progress and this helped to catalyze the invention of CCDs. In fact, CCDs are the electrical analog to the magnetic bubble concept.

However, before these first two CCD papers, there was similar activity in the Phillips Research Laboratory at Eindhoven, being conducted by *Sangster* and *Teer* [4.3]. They called their device the "Bucket Brigade Device" (BBD). It appears that similar developments were also going on in the General Electric Research Laboratory at Schenectady. *Tiemann* and his colleagues called their device a surface charge transfer device [4.4]. Thus in general, CCDs include BBDs and CTDs (Charge Transfer Devices). Probably the name CTD is more appropriate but we shall use CCD to include all these.

Although CCDs came into being around 1970, the particular concept of representing a bit of information, either 0 or 1, by the absence or presence of stored charge in a capacitor and moving this information sequentially like a shift register using clock voltages has been known to scientists for a long time. First of all, the use of stored charge on capacitors as a memory device was known to Zworykin and Wiener. An attempt to implement it using vacuum tube amplifiers and capacitors was successful but impractical due to bulkiness. It was realized that the same implementation serves as an analog variable delay line, which has many applications, one of which is in a stereo amplifier to simulate orchestral music with different delays for different notes from different instruments. Actually this particular application in the stereo amplifier was the motivation for the bucket brigade device. Conceptually there is not much difference between BBDs and CCDs in terms of the manufacturing operations needed to fabricate them.

It is of interest to include the following paragraph, taken from [4.1], which describes nearly all the possible applications of CCDs over and above their use as a simple delay line.

The basic shift register concept may be used to construct a recirculating memory or used as a delay line for times up to the storage time. Clearly, charge transfer in two dimensions is possible as well as the ability to perform logic. An imaging device may be made by having a light image incident on the substrate side of the device creating electron-hole pairs. The holes will diffuse to the electrode side where they can be stored in the potential wells created by the electrodes. After an appropriate integration time, the information may be read out via shift register action.

This expectation, dating from 1970, has not yet been fully realized. However, in certain areas such as imaging, the CCD is probably the most successful solid state device.

4.1.2 Surface Acoustic Waves

Unlike CCDs, which only came into existence in the 1960s, surface acoustic waves have been known to scientists since Lord Rayleigh presented his paper to the Royal Society in 1885. We quote from the last paragraph of that paper: "It is not improbable that the surface waves here investigated play an important part in earthquakes, and in the collision of elastic solids. Diverging in two dimensions only, they must acquire at a great distance from the source a continually increasing preponderance."

It is obvious that the reason for this study was not electronic devices, which were nonexistent at that time, but the understanding of earthquake phenomena. It transpires that probably the most significant part of an earthquake appears as a surface wave which probably does the most damage. In the 100 years since Lord Rayleigh's paper, seismologists have studied surface and other forms of waves extensively. For example, there now exist the Love wave, the Stoneley wave, the Lamb wave and others that explain different aspects of seismology. For example, the Stoneley wave is related to wave propagation at the interface of a solid and a liquid; it was developed primarily to understand the effects of an earthquake propagating at the bottom of the ocean.

Of course, our interest here is mainly in the propagation of high frequency ultrasound at a solid interface and its use in electronic devices. This includes not only the so-called Rayleigh wave, but also all the other waves mentioned and we generally use the term "SAW technology" rather loosely to include all of them. Although most devices built at present use Rayleigh waves, it is quite possible that other waves may be of importance in the future.

Rayleigh's original paper is a classic, in which he both derived and illustrated the following important properties of the Rayleigh wave:

i)   It is non-dispersive and the expression for the velocity can be related to the fundamental elastic constants.
ii) It is complex in the sense that the deformation includes both longitudinal (compressional) and transverse (shear) components.
iii) Both components decay away from the surface, but at different rates.
iv) The individual particles in the solid move in an elliptical orbit.

Lord Rayleigh's work included only isotropic solids. Since then the results have been extended to anisotropic solids and, more importantly, to piezoelectric solids. More recently it has been found that a simple surface wave, called the Bluestein-Gulyaev wave, can exist on a piezoelectric surface. This wave has only transverse components.

4.2 Why SAWs Became Popular and Useful in the 1960s

Lord Rayleigh discovered the surface wave in 1885 but it was only in the 1960s that scientists thought about using these waves for electronic applications. Why then did it take 80 years to realize the importance of the Rayleigh wave? One obvious answer is that electronics did not exist at the time of Lord Rayleigh and the consumer and military electronics thrust only started in the 1930s and 1940s. Also, the generation of ultrasound from electrical signals (or vice versa) requires transducers using piezoelectric materials, which were also unknown at the time of Lord Rayleigh. However, to answer the question in detail, we shall make a small digression into the use of bulk ultrasound in the electronics industry.

4.2.1 Bulk Ultrasound Devices

There are two forms of bulk ultrasound which can exist in a solid, namely a longitudinal and a transverse variety. In the longitudinal (or compressional) wave, the displacement of the particle motion is in the direction of wave propagation. As shown in Fig. 4.1, if an original solid is divided into minute grid points, the longitudinal wave propagation causes some grid points to approach each other (i.e., they are compressed together) while other points separate (i.e., are rarified or somewhat elongated). The reason for calling it a bulk wave is the lack of variation along the depth or the z axis. Similarly, Fig. 4.2 shows the transverse (or shear) wave. For this case, the particle motion is perpendicular to the direction of wave propagation and the grid squares do not become compressed and elongated but are rather twisted up or down. Theoretically, both the shear and compressional bulk nondispersive waves can exist only in a solid of infinite dimensions. However, for practical purposes, as long as the size of the medium or substrate is considerably greater than the wavelength, e.g., 100 times greater, the deviation is negligible.

The velocity of these bulk waves in a solid is around 3×10^3 m/s, which is about 10^5 times smaller than the velocity of electromagnetic waves in a solid.

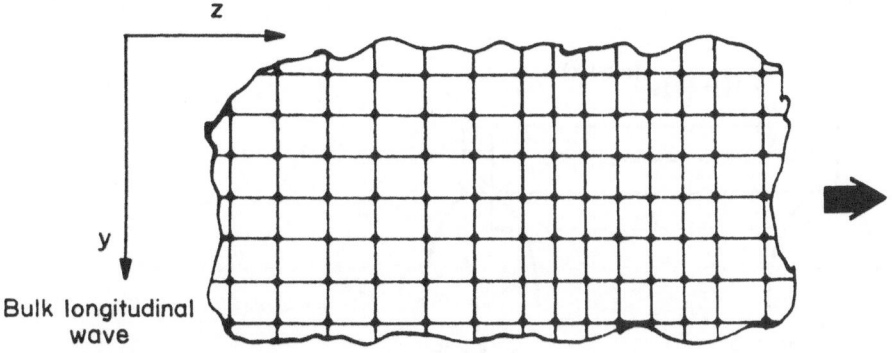

z

y

Bulk longitudinal
wave

Fig. 4.1. Bulk longitudinal ultrasound wave in a solid

359

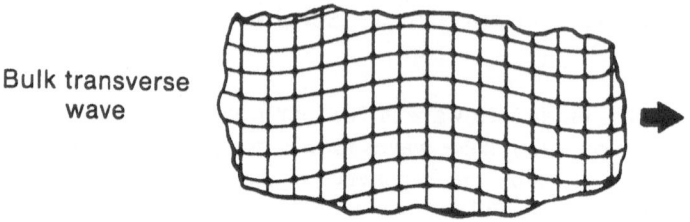

Bulk transverse
wave

Fig. 4.2. Bulk transverse ultrasound wave in a solid

This fact has been utilized to make bulk ultrasound delay lines. For example, to obtain a delay of 3 μs using electromagnetic waves one needs approximately 900 m of cable, whereas one needs only a solid of size 9 mm for an ultrasound delay line. However, the price paid for this compactness is the need for two transducers, one to convert electrical energy to ultrasound and the other one to convert the ultrasound back to electrical energy. This scheme is shown in Fig. 4.3. The transducers generally use a parallel slab of piezoelectric material such as lithium niobate ($LiNbO_3$) or quartz. The parallel sides are metallized, so that voltages can be applied. If a signal $f(t)$ modulates the carrier frequency or center frequency $\omega = 2\pi f$, and the combined voltage $f(t)e^{j\omega t}$ is applied to the input transducer 1, ultrasound is generated inside the solid which can be represented as

$$f\left(t - \frac{x}{v}\right)e^{j\omega(t - x/v)} = f\left(t - \frac{x}{v}\right)e^{j(\omega t - kx)} \tag{4.2.1}$$

where $k = 2\pi/\lambda = \omega/v$ and v is the velocity of wave propagation. The output of the second transducer for a delay line of length l will be given by

$$f\left(t - \frac{l}{v}\right)e^{j\omega\tau} \quad , \tag{4.2.2}$$

which excludes a proportionality factor incorporating any propagation or transduction losses. Here, the delay time τ is given by

$$\tau = \frac{l}{v} \quad . \tag{4.2.3}$$

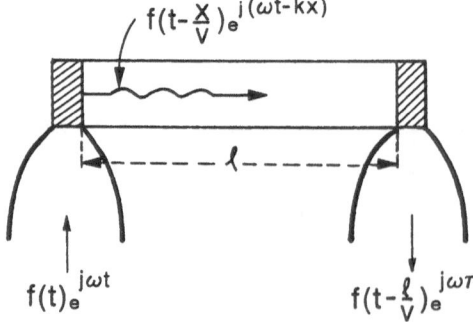

$$f\left(t - \frac{x}{v}\right)e^{j(\omega t - kx)}$$

$$f(t)e^{j\omega t} \qquad f\left(t - \frac{l}{v}\right)e^{j\omega\tau}$$

Fig. 4.3. Schematic ultrasound delay line

360

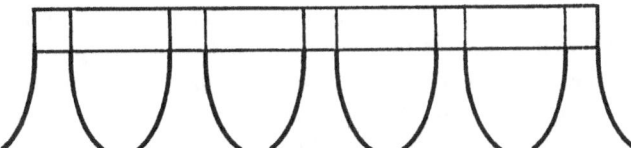

Fig. 4.4. Tapped ultrasound delay line

As discussed before, a small device can have long delays and these ultrasonic delay lines have been in use in electronic systems since the 1940s.

However, for many electronic applications, a number of these delay lines or a so-called tapped delay line is needed. To obtain a tapped delay line with N taps using this approach, one needs $N + 1$ transducers as shown in Fig. 4.4. At high frequency, each transducer needs careful gluing to the delay line substrate and it becomes rather cumbersome to fabricate.

A bulk wave in an unbounded medium is nondispersive. However, other waves such as plate or Lamb waves are dispersive. Some of these dispersive waves were also used in the 1950s for radar signal processing, especially for pulse compression of chirp signals. This is illustrated in Fig. 4.5. The dispersive property is shown in Fig. 4.6 as frequency vs delay time, which is related to

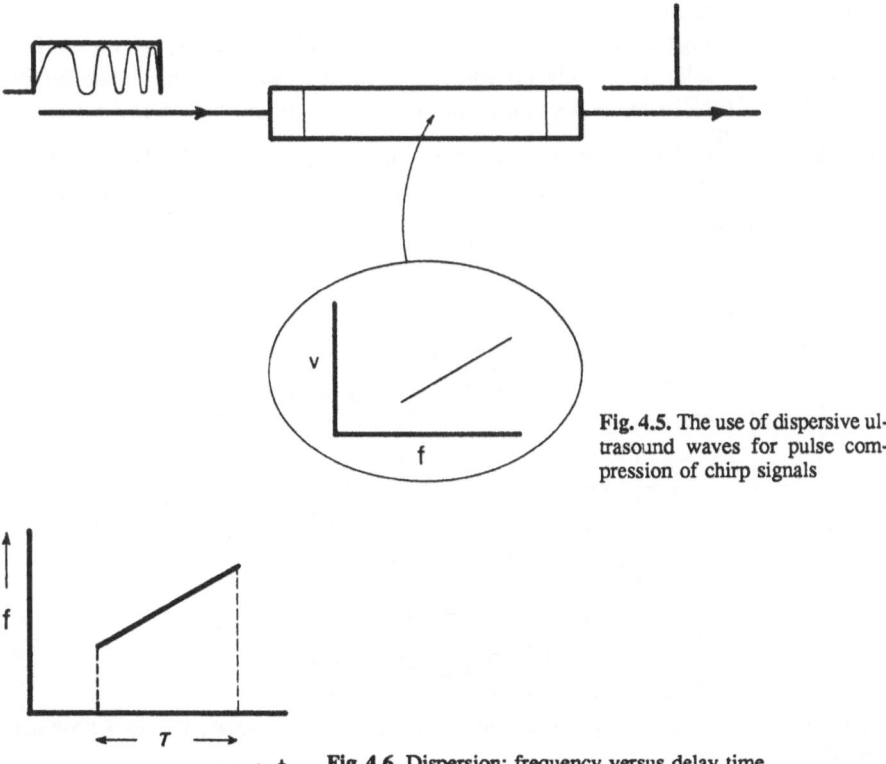

Fig. 4.5. The use of dispersive ultrasound waves for pulse compression of chirp signals

Fig. 4.6. Dispersion: frequency versus delay time

361

frequency vs velocity. If a signal pulse is applied to the input of the delay line whose instantaneous frequency is low at the beginning and increases progressively as a function of time, then, at the output, the beginning of the pulse is delayed more than the end of the pulse. If everything is matched, the whole pulse energy will appear at nearly the same instant in the output, thus achieving what is known as pulse compression (as discussed in Sect. 3.1.3). However, we observe that to obtain the best pulse compression, the dispersion and the signal must be precisely matched.

4.2.2 Advantages of SAWs

Compared to a bulk wave, a SAW is rather complex. This is illustrated in Fig. 4.7. The most important point to note is that the wave amplitude decays perpendicular to the interface. This decay constant is approximately one wavelength long. For example, the wavelength λ at $f = 100\,\text{MHz}$ is approximately $36\,\mu\text{m}$ for SAW propagating in the z direction on Y-cut LiNbO$_3$. "Y-cut" means that the normal to the surface is in the y direction. Thus, for a $100\,\text{MHz}$ SAW delay line, the SAW substrate thickness need not be more than a few hundred micrometers. The SAW velocity is approximately $0.9v_{\text{shear}}$ where v_{shear} is the shear wave velocity. As shown in Fig. 4.8, the grid points go through both shear and compressional motions. Actually, the Rayleigh wave has both longitudinal and transverse components of deformation.

One advantage of the SAW is that its amplitude is highest on the surface and thus it is easily accessible compared to the bulk wave. Three more factors account for SAWs' rise in popularity in electronic applications in the 1960s:

i) Availability of piezoelectric materials such as LiNbO$_3$ and ST-cut quartz.
ii) Invention of the interdigital transducer, including its tapped delay line implementation.

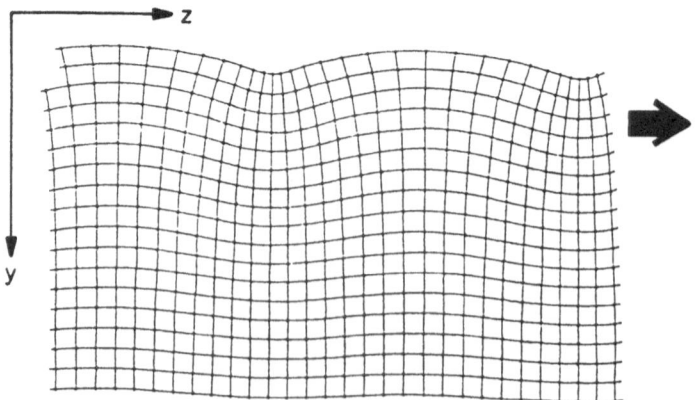

Fig. 4.7. Cross section of the displacements of a rectangular grid of material points for a SAW in an isotropic material

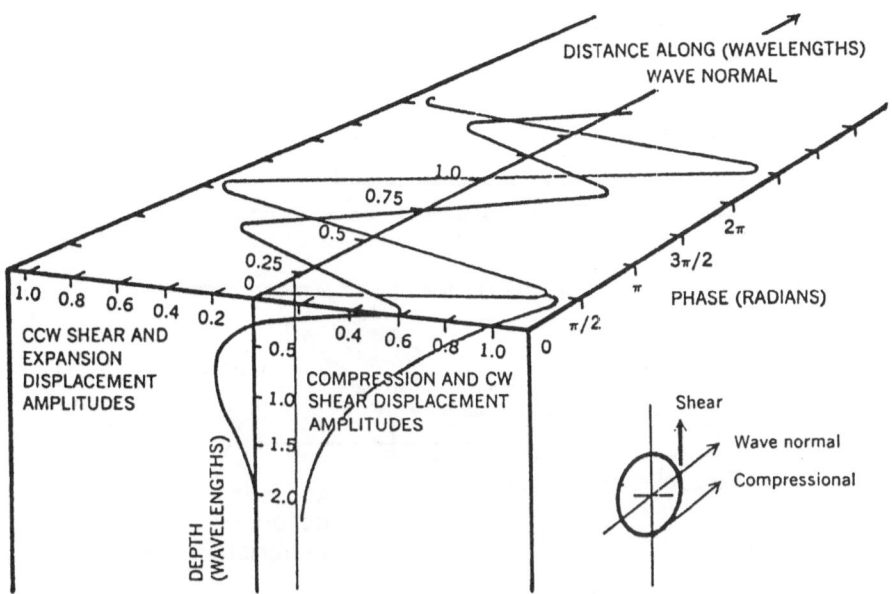

Fig. 4.8. In a SAW the grid points go through both shear and compressional motions. (From [4.9])

iii) Availability of photolithographic techniques, which were developed mostly for integrated circuit technology.

A good piezoelectric material is a necessary ingredient for any SAW device. Quartz crystal is the only piezoelectric material of any use that can be found naturally. However, most of the crystals needed for manufacturing devices are grown in the laboratory. ST-cut quartz, which is a compensated cut of quartz crystal, was found by means of a computer search and is the second most important SAW substrate. Lithium niobate ($LiNbO_3$) is probably the most important man-made crystal for SAW devices. It was discovered in the late 1950s. In the 1960s, other piezoelectric materials became available, such as PZT (lead zirconium titanate), zinc oxide, cadmium sulphide, gallium arsenide and lithium tantalate. This led to the possibility of immediate testing and manufacture of various prototypes of SAW devices.

Before the interdigital transducer (IDT) was invented[1], SAW were generated by using ordinary bulk-wave transducers oriented at a suitable angle and by other somewhat cumbersome methods. As the SAW velocity is slower than, for example, that of compressional waves, one can use the compressional bulk transducer at an angle

$$\phi = \tan^{-1} \frac{v_l}{v_R} \qquad (4.2.4)$$

---

[1] The first published paper on the IDT is [4.5]. However, in the ultrasonics community there is controversy about who really invented the IDT, because some scientists in Bell Telephone Laboratory apparently knew of it and had used it in classified research. The interesting point is that no patent claim has been filed on this invention, to the best of the author's knowledge.

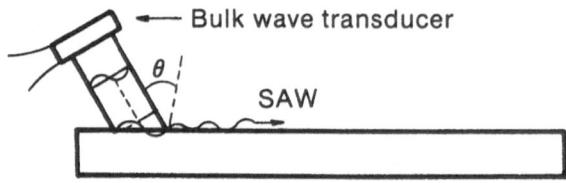

Fig. 4.9. Generation of a SAW using a compressional bulk wave transducer

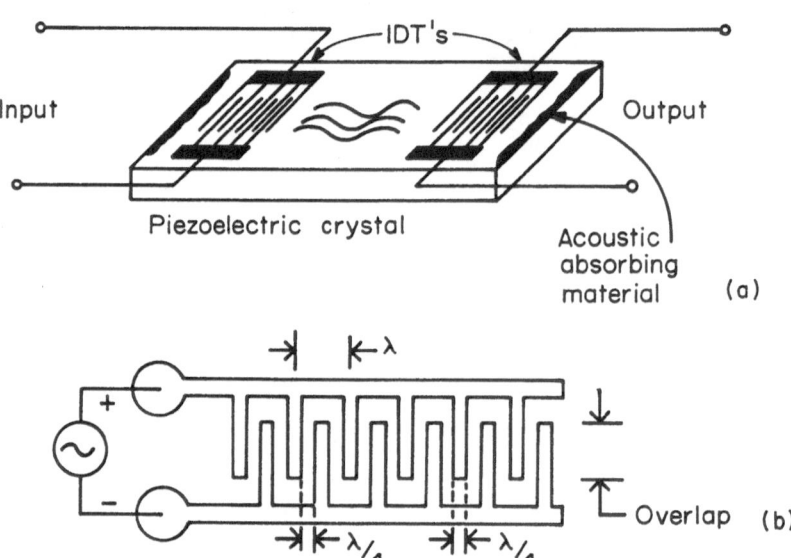

Fig. 4.10. (a) SAW delay line with two IDTs. (b) An IDT is simply a set of metal strips placed on the piezoelectric substrate. (From [4.14])

where v_l and v_R are longitudinal and Rayleigh wave velocities respectively, to generate a SAW as shown in Fig. 4.9. Compared to this method, the simple interdigital transducer is shown in Fig. 4.10 where a SAW delay line having two IDTs is also depicted.

A typical IDT is simply a set of metal strips placed on a piezoelectric substrate on which a SAW is to be generated. Alternate strips (commonly called fingers) are connected to each other by two other metal strips which form electrical inputs. The usual width of each metal finger is one quarter of a wavelength ($\lambda/4$), with fingers also separated by $\lambda/4$. When an rf voltage is applied to the two input terminals of this IDT, an electric field is set up between all the adjacent fingers simultaneously. The piezoelectric material is thus alternately compressed and elongated between the fingers and generates a SAW which starts travelling with a velocity v. Since the SAW is generated simultaneously by each pair of adjacent fingers, the sources are distributed. To obtain efficient surface wave generation, all these excitations must generate waves which add coherently. To examine this process, let the period of the rf voltage be T seconds. To propagate a distance of $\lambda/2$ it therefore takes $T/2$ seconds. Since this is also the time it takes for the electric field between two adjacent fingers to change polarity, all the generated waves will add constructively. For detection, the reverse is true; i.e., as the elastic deformation passes under the fingers, it induces in-phase voltages between each finger-pair. This type of operation makes the IDT a simple and truly remarkable SAW generator.

To construct a SAW delay line, one uses two IDTs on a piezoelectric substrate. If lithium niobate (LiNbO$_3$) is used as the piezoelectric material, the surface velocity is $v \sim 3.6 \times 10^3$ m/s so that, if $f = 100$ MHz, then $\lambda = 36\,\mu$m. Thus the width of each metal finger is $9\,\mu$m and 1 cm of separation between the IDTs will produce a delay of about 2.3 μs.

Why use many finger-pairs instead of only one pair? Actually, an IDT can be (and sometimes is) made with only a single finger-pair. However, with N finger-pairs, greater efficiency is achieved for the same applied voltage, since, in effect, N sources are adding up. However, as explained later, there is a bandwidth reduction as one uses more and more finger-pairs.

If we pause here for a moment and compare a SAW delay line with a bulk ultrasound delay line, we immediately see the advantages of the former. First, the SAW delay line can be made using a one-step photolithographic process, whereas bulk delay lines need the transducer to be glued to the material through which the wave propagates, a rather cumbersome and expensive process. Second, for a tapped delay line, it is both simpler and less expensive to implement the SAW delay line. For example, to produce a LiNbO$_3$ tapped delay line with $N = 10$ taps and a delay time between taps of $T_d = 1\,\mu$s, one needs an input transducer and nine output transducers situated 0.36 cm apart. The fabrication cost for a SAW delay line does not change for this case. However, for a bulk delay line, nine more transducers have to be built and glued, thus making this bulk tapped delay line considerably more expensive. In fact, this latter process is not even feasible if the distance between the transducers becomes too small.

In addition to fabricating a SAW tapped delay line using a single mask and one photolithographic operation, one can use this process to build a complete transversal filter. The concept of transversal filters has been developed fully in Sect. 3.4. However, we note here that to build a transversal filter with transfer function given by

$$H(\omega) = \sum_{n=0}^{N-1} C_n e^{-j\omega n T_d} \quad , \tag{4.2.5}$$

the delay line output at each of the N taps is multiplied electronically by the corresponding coefficient C_n, and the product is applied to a summing circuit. In general, the multipliers can be switched to different coefficient values so that one can implement a programmable filter. If, however, one is merely interested in a fixed filter response, then the coefficients can be incorporated in the transducer itself. This can be accomplished in many different ways, but we shall discuss the one called *apodization*.

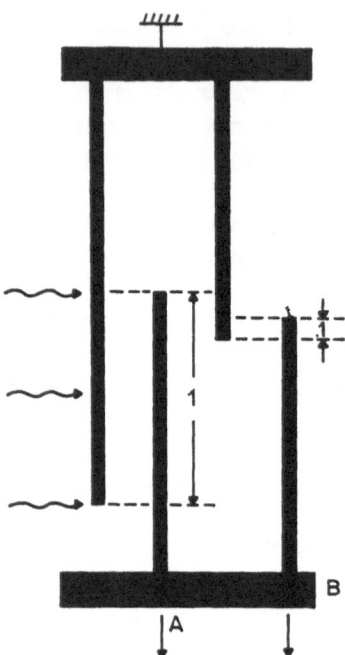

Fig. 4.11. Finger-pair overlap in an IDT determines the transversal filter coefficients

Consider Fig. 4.11 and assume that the same SAW is incident on both finger-pairs A and B. Note that the finger-pair A overlaps the entire width of the SAW wavefront whereas the finger-pair B overlaps only a small fraction (10%) of the SAW width. It is therefore expected that the voltage detected at A and B will correspond to coefficients $C_A = 1$ and $C_B = 0.1$. By using a single IDT with such variable overlaps, the overall voltage produced will be the sum of the contributions due to each finger-pair. This is a very convenient way of making inexpensive filters. The calculated coefficient values are incorporated directly into the mask used to fabricate the IDT pattern. To synthesize some specific transfer function, e.g., $H_1(\omega)$ [assuming $H_1(\omega)$ is suitably band limited], one can simply expand $H_1(\omega)$ in a Fourier series and end up with an expression similar to (4.2.5). In particular, if the impulse response of the filter is $h_1(t)$, the coefficient C_n in (4.2.5) can be shown to be given by

$$C_n = h_1(nT_d) \quad , \tag{4.2.6}$$

where T_d is now chosen to correspond to a correct sampling interval for reconstruction of the specific band-limited function $h_1(t)$. Thus, when we look at a SAW tapped delay line through a microscope, we actually see a sampled version of the impulse response of the device by observing the pattern of overlapping fingers. One can use the well-known techniques of digital filter design for the mask design of a specified filter. As a result, the production of transversal or finite impulse response filters with specific characteristics (within certain limitations) is readily achievable.

366

We now apply the idea of tapped delay lines to examine the frequency response of the basic IDT with N fingers discussed earlier. In the previous notation, taking all the coefficients $C_n = 1$, we have

$$H(\omega) = \sum_{n=0}^{N-1} e^{-j\omega n T_d} \quad \text{where} \quad T_d = \frac{1}{f_0} \quad , \tag{4.2.7}$$

where f_0 is the center frequency of the design. In the vicinity of f_0, the magnitude of the transfer function, from (4.2.7), is approximately

$$|H(\omega)| = N\frac{\sin(N\pi\Delta f/f_0)}{N\pi\Delta f/f_0} \tag{4.2.8}$$

where $\Delta f = f - f_0$ and $f = w/2\pi$. This "sinc-function" response centered around f_0 has a bandwidth inversely proportional to N, and corresponds to the rectangular impulse response of the tapped delay line whose tap coefficients are one. To design a sharp band-pass performance, the impulse response of the tapped delay line should look like the sinc function shown in Fig. 4.12. From the above discussion, it can be seen that there is indeed a bandwidth price that must be paid if more and more figures are used. Figure 4.13 shows the fractional

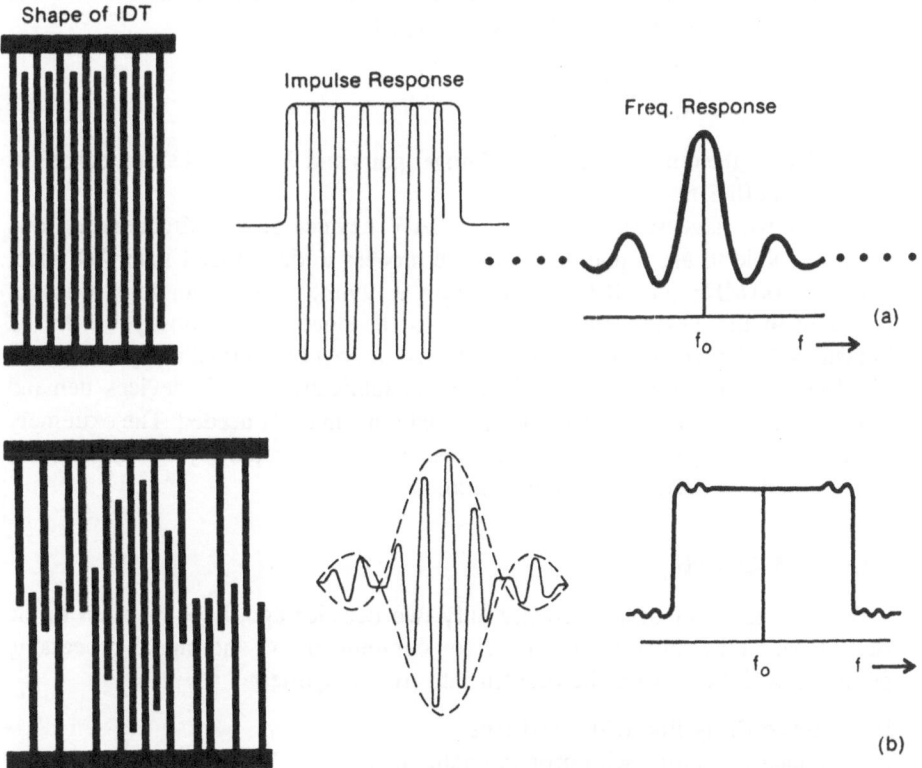

Fig. 4.12. Two IDT configurations and the corresponding filter characteristics. (From [4.14])

367

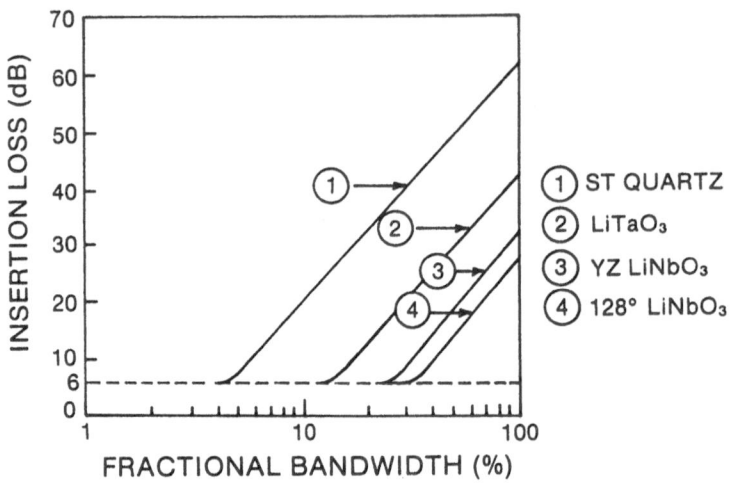

Fig. 4.13. Minimum theoretical insertion loss vs fractional bandwidth for different SAW substrates

bandwidth versus minimum theoretical insertion loss of a SAW delay line using different materials. It has been assumed that the IDTs are bidirectional. Using unidirectional transducers, to be discussed shortly, the minimum loss is 0 dB. The optimum fractional bandwidth $(\Delta f/f)_{opt}$ is

$$\left(\frac{\Delta f}{f}\right)_{opt} = \frac{1}{N_{opt}} = \frac{2K}{\sqrt{\pi}}$$

where N_{opt} is the optimum number of finger pairs and K is the electromechanical coupling coefficient.

Thus we have seen that to make SAW devices at high frequencies, one should be able to use a photolithographic process to draw metal lines $\lambda/4$ wide, (e.g., at 100 MHz on $LiNbO_3$, the width is $9\,\mu$m). This technology was not available in the 1940s and 1950s but was developed later, mostly by the IC industries. Thus, when the IDT was invented, it could be immediately tested and developed. Compared to semiconductor IC fabrication, SAW devices demand much less sophistication because usually only one mask is needed. The extremely difficult task of registration for a multi-mask photolithographic system is therefore not required for SAW device fabrication.

4.2.3 SAW Devices

SAW devices, a scientific curiosity only two decades ago, have matured to the stage where they are routinely used for communication and signal processing purposes. SAW filters can be divided into five categories:

i)   Tapped delay line with fixed taps
ii) Tapped delay line with programmable taps
iii) Resonators

368

iv) Convolvers

v) Transform-domain processors

A tapped delay line with fixed tap weights produces filters with a perscribed pass band frequency response and a rejection in the stop band which is unattainable by any other means. Resonators can be used in oscillator circuits or they can be cascaded to build narrow pass band filters. A convolver is a three terminal device which uses a nonlinear interaction to perform programmable convolution. Tapped delay lines with programmable taps can be used as programmable matched filters or, with added complexity incorporating a Widrow-Hoff LMS algorithm, as effective adaptive processors. Transform domain processors are subsystems which can perform real-time transformation (i.e., Fourier, Fresnel) using SAW chirp filters. In the following, brief descriptions of some of these devices are given.

a) Delay Line

The two important characteristics of the SAW delay line that limit its performance are (i) delay line losses and (ii) triple transit echo.

Delay Line Losses. There are four different loss mechanisms in a SAW delay line that one must reckon with. These are:

i) Inherent loss of 6 dB due to IDT

ii) Propagation loss

iii) Misalignment or steering loss

iv) Diffraction loss

The SAW delay line has an inherent loss of 6 dB because the ordinary IDT generates waves in both the forward and backward directions, thus causing a 3 dB loss. By reciprocity, another 3 dB loss is caused at the receiving IDT. There are different ways to reduce or eliminate this 6 dB loss by making so-called unidirectional transducers. Four important techniques are two IDTs separated by $\lambda/4$, three-phase excitation, multistrip couplers, and the SPUDT (Single Phase Unidirectional Transducer).

In Fig. 4.14, the unidirectional transducer actually consists of two IDTs separated by a gap $\lambda/4$ to introduce a phase difference of 90°. The second IDT is also terminated with an output load such that it reflects the SAW completely. This reflection property of IDT is discussed later in this section in connection with the triple transit echo. For this transducer, the excited backward wave is reflected back by the second transducer with proper phase to reinforce the forward travelling one. Thus the inherent 3 dB loss is ideally reduced to zero by 100% reflection. However, in this arrangement, the frequency response of the combination IDT becomes more restrictive.

The three phase excitation IDT is more complex and is shown in Fig. 4.15. Firstly, one needs an insulating layer on the substrate for the third phase, thus

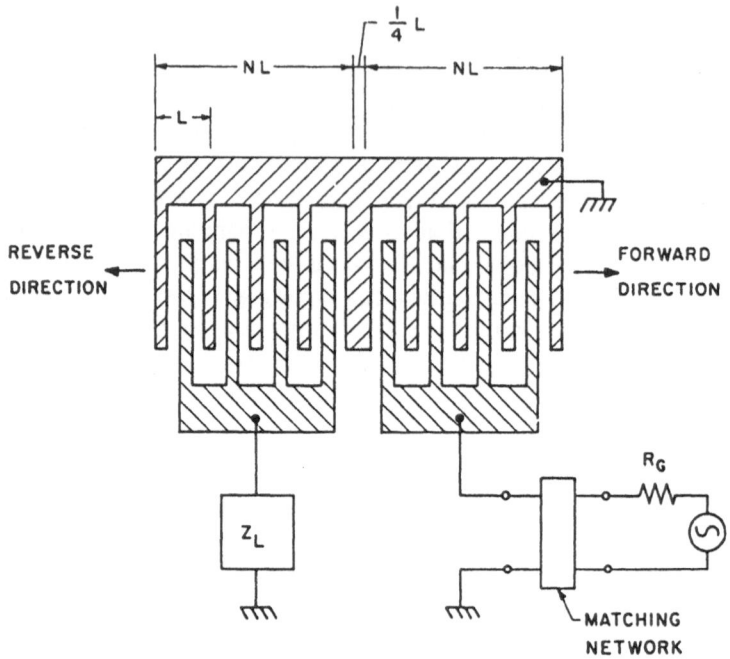

Fig. 4.14. Schematic of a unidectional transducer using a reflector IDT spaced λ/4 from the driven IDT. (From [4.15])

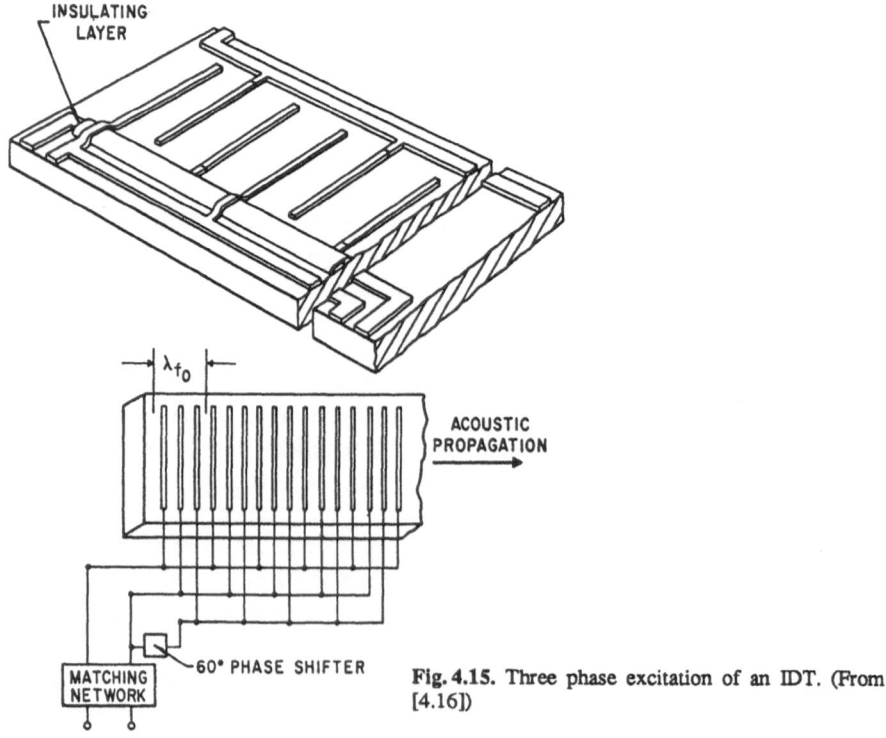

Fig. 4.15. Three phase excitation of an IDT. (From [4.16])

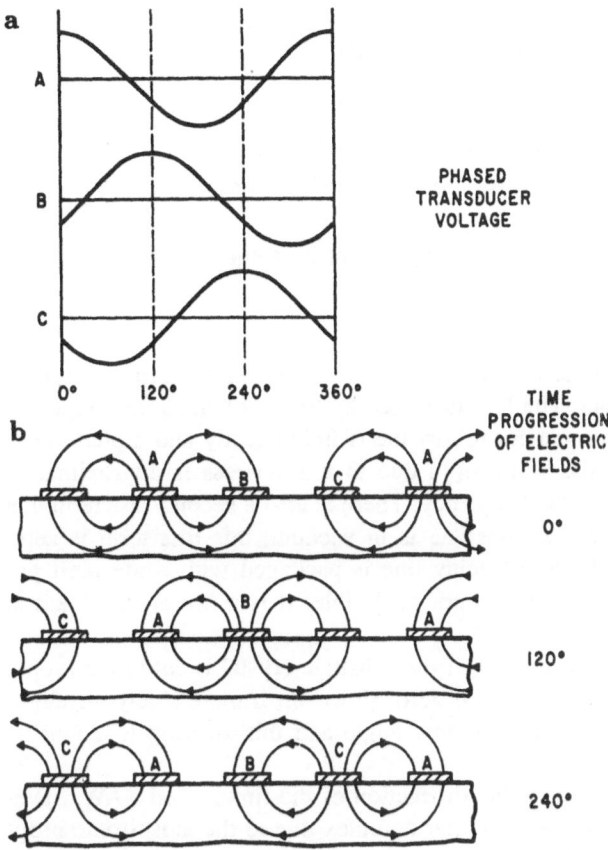

Fig. 4.16a,b. Three phase IDT. (a) Phased transducer voltage. (b) Time progression of electric fields between the fingers. (From [4.16])

necessitating a complex fabrication process. Secondly, one needs a matching network with a 60° phase shifter to convert the regular rf to a three-phase signal, each phase being separated by 120° and denoted by A, B and C, respectively, in Fig. 4.16. Each of the phases is connected to the set of fingers, causing electric fields between them as shown in Fig. 4.16. In this case, only the forward travelling wave propagates and the backward wave cancels out. Of course, the forward wave can be cancelled if the phases of the waves are changed by 120°.

The multistrip coupled unidirectional IDT discussed after clarifying the working concepts of the multistrip coupler.

The SPUDT is the most recent development. As shown in Fig. 4.17, it uses dummy floating electrodes which are not electrically connected. The reflections caused by these dummy electrodes, if properly designed, reinforce the SAW in the forward direction and completely cancel the backward propagation.

After the *inherent* loss of an IDT, the next important loss mechanism is the propagation loss. This loss is higher for a longer delay line and is generally

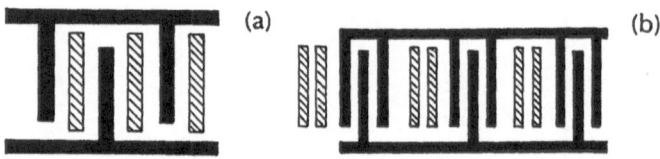

Fig. 4.17. SPUDT. (a) Single floating electrode. (b) Double floating electrode

quantified by a constant $\alpha(f)$, which is dependent on frequency and is defined as

$$A(x) = A_0 e^{-\alpha(f)x} \tag{4.2.9}$$

where $A(x)$ is the amplitude of the SAW at a distance x from the origin, with the amplitude at $x = 0$ being A_0. The loss is caused by the fact that as the frequency is increased the deformation cannot follow the rf field exactly and a phase term is introduced. In general, the value $\alpha(f)$ due to this process is approximately proportional to the square of the frequency. There is also a second loss term due to air loading. If the delay line substrate is in vacuum, this loss term is zero. However, in most cases, the SAW delay line is packaged with some inert gas such as nitrogen, so that this term is not negligible. It is more predominant at low frequencies.

Beam steering loss is ideally zero if the substrate crystal is cut in the proper direction so that the power flow angle is zero. However, if there is any misalignment, there is a finite nonzero power flow angle and this steering loss will be significant for a long delay line.

Diffraction loss is caused by the diffraction of the finite sized SAW wavefront. The calculation of this loss is rather complex due to the anisotropic propagation of SAW in crystals. However, accurate numerical models are available [4. 6] which predict the delay line loss to within a fraction of a decibel.

Triple Transit Echo. When an rf signal is applied to a SAW delay line input, we expect the same signal to appear in the output but delayed by the delay time τ and with the amplitude attenuated by the losses. However, other undesirable outputs appear, the most important and troublesome of which is the triple transit echo. This appears in the output at a time $t = 3\tau$ and is caused by the reflections of SAW from the IDTs. For the delay line with two matched IDTs, this echo is only 12 dB lower than the main signal. It is of interest to understand physically why so much of the SAW energy can be reflected by the IDT. Each metal finger causes a small amount of reflection due to the different characteristic impedance of the equivalent transmission line, caused by the $\Delta v/v$ change due to the piezoelectric shortening effect and the mass loading of the metal film. However, these small reflections add up due to phase matching (i.e., for the same reason that the IDT is such a good phase-matched generator) and a large reflection coefficient can result. To avoid this phase-matching condition one sometimes uses a so-called split finger for the receiving IDTs (Fig. 4.18), where each finger is split into two sections.

372

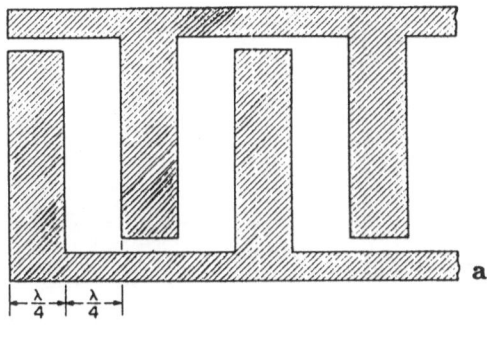

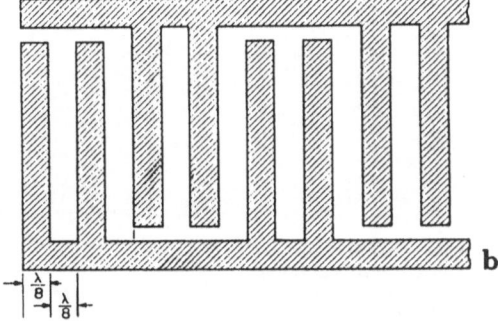

Fig. 4.18. Transducers with (a) a single electrode per half-wavelength and (b) double (split) electrodes

Multistrip couplers can also be used to trap the triple transit echo and they will be discussed in Sect. 4.2.3c.

b) SAW Filters

In Sect. 4.2.2, we discussed how a tapped delay line can be designed to perform as a SAW filter, but that description gave a simplified picture of SAW filters. For a good design other details must be considered, such as the bandwidth limitations of the input transducer, the resulting frequency response, and the individual finger reflection, all of which lead to a complex initial design. In a proper design, the IDT is precision modelled by computer and the design changed iteratively until the desired performance is obtained. The art of computer-aided SAW filter design has reached the stage where insertion losses as low as a few decibels, delays of tens of microseconds, and a center frequency that ranges from 10 MHz to a few gigahertz have been achieved. However, at times these specifications conflict with one another, so that, for example, a device with a high center frequency will invariably also have a high insertion loss. For routine implementation, the center frequency is now limited to 1000 MHz due to fabrication complexity, and the insertion loss tends to be in the 20 dB region. The insertion loss depends upon the substrate material used, the electrical matching networks employed, the center frequency of the devices, and the fractional bandwidth. The dynamic range of these devices tends to be in the vicinity of 70 dB. Figure 4.19 shows a typical filter response. At present, this filter is nearly impossible to duplicate by any other technology.

373

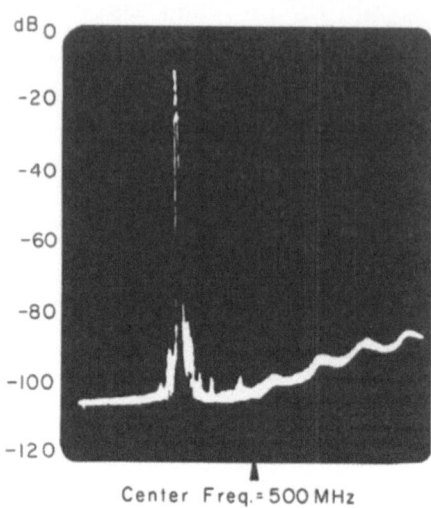

dB 0
-20
-40
-60
-80
-100
-120

Center Freq. = 500 MHz
100 MHz / Div

Fig. 4.19. Typical response of a SAW filter. (From [4.17])

Table 4.1. SAW filter parameters

Parameter definition	Symbol	Current practical limits
Center frequency	f_0	10–1000 MHz
Rejection bandwidth at given dB	$(\Delta f)_R$	
Passband width at given dB	$(\Delta f)_P$	
Fractional bandwidth	$\frac{(\Delta f)_P}{f_0}$	0.1 %–65 %
Shape factor	$\frac{(\Delta f)_R}{(\Delta f)_P}$: 1	1.1 : 1 to 4.0 : 1
Rejection	REJ	40–70 dB
Insertion loss	IL	Bidirectional: 15–35 dB
		Unidirectional: 3–10 dB
Amplitude ripple	ε_P	0.1–1 dB
Phase ripple	$\Delta\theta$	1°–10°
Group delay ripple	$\Delta\tau$	± 5 ns minimum
Transition bandwidth	$(\Delta f)_T$	200 kHz minimum

For filter specifications in a sophisticated electronic system one has to worry not only about insertion loss and bandwidth but also about many other parameters. These are shown in Fig. 4.20. The symbols in the figure are described in Table 4.1. The table also shows typical values that have been achieved for SAW filters with proper design and fabrication. Chirp filters are special filters that play an important role in transform domain processing and are discussed in Sect. 4.6.

c) Multistrip Couplers

We have mentioned multistrip couplers (MSCs) before in connection with the unidirectional IDT and with their use in suppressing undesirable bulk modes in SAW filters. In this section, we shall develop a simple understanding of a MSCs and discuss its various applications. We shall see that a MSC is just a set of

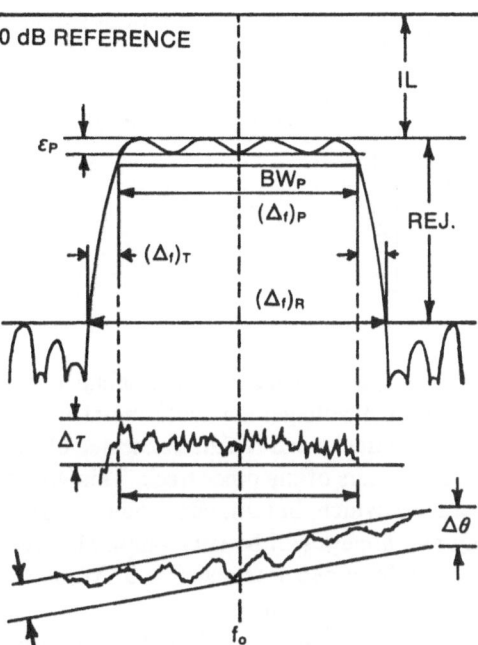

Fig. 4.20. SAW frequency response parameters

0 dB REFERENCE

IL

ε_P

BW_P

$(\Delta_f)_P$

REJ.

$(\Delta_f)_T$

$(\Delta_f)_R$

Δ_T

$\Delta\theta$

f_o

parallel metal fingers which are not electrically connected to each other. To fully appreciate the way MSCs work, we need now to consider the two-dimensional wavefront of SAW propagation.

Simple Theory. A typical MSC is shown in Fig. 4.21. If the transducer A_1 is excited and if the MSC is properly designed, then all the SAW energy will be transferred to track B and appear as output in the transducer B_2. The output of transducer A_2 will be zero in this case. If only the transducer B_1 is excited, the SAW again transfers to track A after propagating under the MSC. The MSC is just a bunch of metal fingers which are not connected electrically, but each finger extends into both tracks. At first thought, it might seem impossible for mechanical SAW energy to be transferred from one track to another by means of simple metal fingers. However, it is appropriate to recognize that, because the

track A

A_1

A_2

B_1

B_2

track B

Fig. 4.21. Multistrip directional coupler with interdigital transducers at input and output ports. (From [4.7])

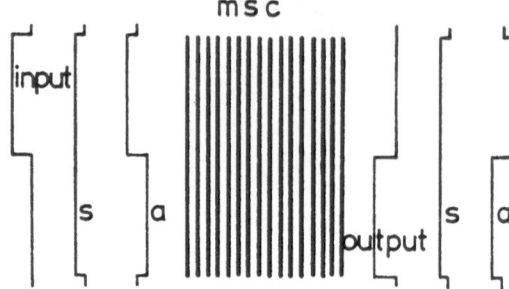

m s c

Fig. 4.22. Input and output field distributions resolved into symmetric (s) and antisymmetric (a) modes for a MSC in which 100% transfer from track *A* to track *B* occurs. (From [4.7])

substrate is piezoelectric, as the SAW propagates under the same metal fingers in one track, it produces an electric field and voltages on these metal fingers. Because the same metal finger is on both the tracks and is thus connected electrically, the voltage is also developed on the fingers of the other track. This voltage can induce an electric field in the other track which, in turn, can produce a SAW. Of course, the SAW generated by only two fingers will be very small. However, using a multitude of them, it is quite possible to generate a strong SAW, so that transfer of energy from one track to the other is feasible.

In other words, due to the presence of metal fingers in both the tracks and due to the piezoelectric effect, there is coupling between the SAW in one track and the other. Because of this it is possible to have transfer of energy for a long MSC. The operation of a MSC and the coupling effect can be better understood if we consider Fig. 4.22, where the SAW wavefront generated by the transducer in only track *A* has been separated into symmetric and antisymmetric parts. The wavefront of the symmetric part extends uniformly over both tracks and has a magnitude *A*/2, where *A* is the magnitude of the SAW in track *A*. The antisymmetric part has the same magnitude *A*/2 and extends over both tracks; however, in track *B* it is 180° out of phase with respect to its component in track *A*, which has the same phase as the symmetric one. If we then combine the symmetric and antisymmetric parts, cancellation occurs in track *B* and the full amplitude of the SAW appears in track *A*.

While both the symmetric and the antisymmetric parts propagate under the MSC, we note that there is an important difference between the two. The symmetric part induces voltages on the metal fingers which are of the same phase in both tracks, whereas the antisymmetric part induces voltages which are of opposite phase on those portions of the same finger that occur on different tracks. The electric current flowing in the fingers for the antisymmetric wave will be nearest to the unstiffened velocity, because it will behave as if it cannot support any current parallel to the fingers. Another way to look at it is as if the substrate were coated with a material which has a very low conductivity transverse to the direction of propagation. For the symmetric part, however, the velocity will be more like a stiffened velocity. In actual fact the metal fingers have gaps in between them, so the symmetric and antisymmetric wave velocities will not be exactly equal to the stiffened and unstiffened velocities which correspond to completely metallized or bare substrates. Using these arguments it can be shown that complete transfer of power from track 1 to track 2 can take place for a length L_T of a MSC given by

$$L_T = \frac{\lambda}{K^2} \quad , \tag{4.2.10}$$

where the coupling coefficient $K^2 \approx 2\Delta v/v$.

If d is the MSC repeat distance, then the number of fingers N_T needed for a complete transfer is

$$N_T = \frac{\lambda}{K^2 d} \quad . \tag{4.2.11}$$

For a length x of the MSC, the power in the two tracks is given by

$$P_1 \propto \cos^2 \frac{\pi x}{L_T} \quad ,$$
$$P_2 \propto \sin^2 \frac{\pi x}{L_T} \quad . \tag{4.2.12}$$

There are many applications of MSC devices, some of which are listed below:

1. 100% coupler
2. Bulk wave suppressor
3. Coupling between different substrates
4. Aperture transformation
5. Precision attenuator – phase correction by offsetting
6. Delay line tap
7. Beamwidth compressor
8. Magic tee
9. Beam redirection
10. Reflector
11. Unidirectional transducer
12. Reflecting track changer
13. Echo trap
14. Better filter design
15. Multiplexing
16. Strip coupled amplifier and convolver
17. Compressed convolver

The details of these can be found in [4.7, 8].

d) SAW Oscillators

SAW devices can be used as the high-Q resonating circuit in an oscillator in two distinct ways. In the first case a SAW delay line is used as shown in Fig. 4.23. In the second case a SAW resonator (i.e., a planar cavity formed by two SAW grating reflectors) is used. SAW oscillators have many advantages over the bulk wave quartz crystal oscillators in the frequency range 100 MHz up to possibly

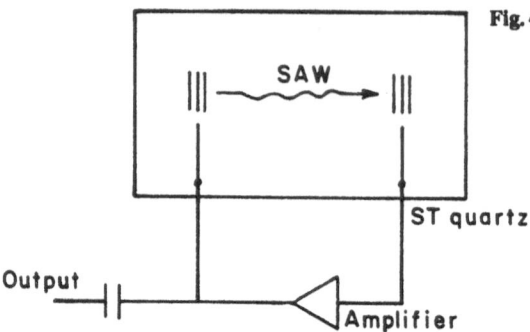

Fig. 4.23. SAW oscillator using a delay line

5 GHz. To operate bulk crystals in this frequency range at fundamental resonance requires crystals too thin for commercial fabrication. Thus the bulk resonators must be used at harmonics, which in turn requires multipliers and associated filters for operation. Elimination of these multipliers and filters is a big advantage for SAW oscillators. This elimination, in turn, produces reductions in the overall size, weight, power and cost.

The delay line oscillator is a feedback oscillator and will oscillate at frequencies

$$f_n \sim n \frac{v}{L} \quad , \tag{4.2.13}$$

where n is an integer, provided the gain of the amplifier offsets the delay line loss and the overall phase condition of multiples of 2π around the feedback loop is satisfied. To choose a particular frequency, some filtering functions are directly incorporated into the IDTs of the delay line.

In SAW resonators, the SAW reflects back and forth between two grating reflectors forming a planar cavity. In a conventional bulk wave resonator, waves are confined in a piece of the crystal with two parallel surfaces. These two surfaces define the cavity since the equivalent impedance mismatch between the crystal and vacuum forms a perfect reflector. Because of the complex particle motion, the SAW largely decomposes into reflected longitudinal and shear waves when incident upon an abrupt surface discontinuity. Therefore, to reflect the SAW back as a SAW, it is necessary to use an array having a large number of small periodic surface perturbations. These perturbations can be produced by shorting the piezoelectric field through metal strips, mass loading by insulating strips, topological discontinuities such as etched grooves and periodic crystal inhomogeneities such as ion implanted regions.

Typical one-port and two-port SAW resonators are shown in Fig. 4.24. In a two-port resonator, separate IDTs are used for generation and detection of the SAW. The number of strips in the grating reflector is in the hundreds.

The performance of a resonator is determined almost entirely by the characteristics of the reflectors. The Q factor of the resonant cavity with no propagation loss can be written as

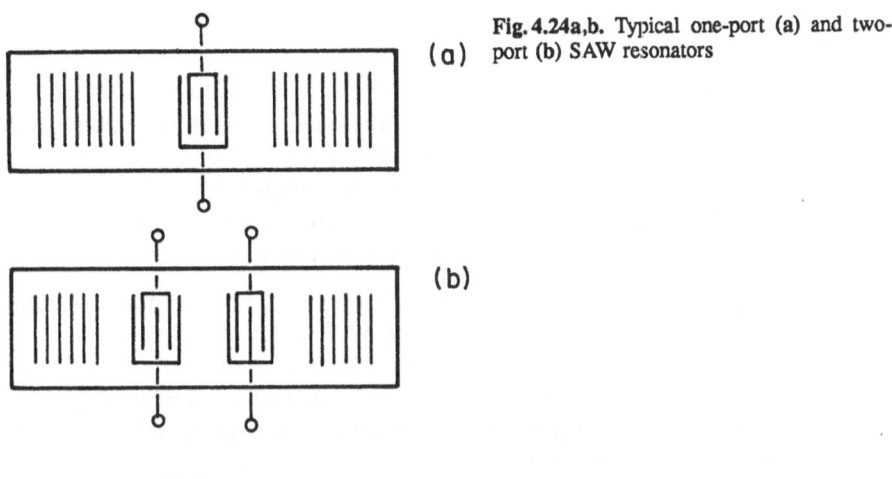

Fig. 4.24a,b. Typical one-port (a) and two-port (b) SAW resonators

$$Q = \frac{2\pi l}{\lambda(1 - |r_f|^2)} \quad , \tag{4.2.14}$$

where l is the effective cavity length and $|r_f|$ is the amplitude reflection factor. Any energy lost in the reflector due to mode conversion or absorption reduces r_f and therefore Q. The cavity length l is the sum of the separation between the reflectors and the penetration distance into the reflectors, the latter being dependent on the reflector property. Compared to Q's of a few thousand for delay line oscillators, oscillator Q's of the order of 30 000 have been reported using resonators. With resonators, oscillators can be made with stability and noise characteristics as good as oscillators using quartz overtone crystals, but at higher frequencies. SAW resonators are components finding increasing use as the frequency determining element in UHF and VHF oscillators.

e) Convolvers

The nonlinear interaction of SAWs in a propagating medium under proper circumstances can be configured to make convolvers[2]. The nonlinearity employed can be a large-amplitude nonlinearity or that due to acousto-electric interaction. There are three different configurations of SAW acousto-electric convolvers: separate medium, combined medium and strip-coupled. In the following we discuss each kind briefly.

The large amplitude nonlinearity is due to the breakdown of Hooke's law at high power levels. One generally uses piezoelectric substrates for the nonlinear interaction media. This has three advantages: the nonlinearity due to piezoelectricity is larger than the elastic nonlinearity. More importantly, the second-harmonic strain wave has an associated electric field which is easy to detect and integrate by measuring the total voltage or the total current flowing through the interacting region. Finally, for either bulk or surface waves a separate transducer

---

[2] See [4.9] for a general discussion of convolvers.

may not be needed, which is a great advantage for surface waves, for example on LiNbO$_3$.

These types of convolvers have been demonstrated at different frequencies, but they have large inherent losses due to the small size of the nonlinearity constant K. Also, the dynamic range is limited by the failure of the elastic media. To improve the performance of these convolvers, a higher power level must be achieved. For this purpose multistrip couplers, horns, or curved transducers are used. The following observations are of interest in this context.

i) *Multistrip coupler.* To increase the input power level one can use two multi-strip couplers as shown in Fig. 4.25. The SAW generated at the transducers is divided into two parts by the first multistrip coupler denoted by MSC$_1$. The second multistrip coupler, one for each channel, marked BWC$_1$ in the figure, compresses the beamwidth to concentrate the energy of the wave into a region near the metal plate.

ii) *Horn.* One can use a parabolic or differently shaped horn to concentrate the SAW power in the interaction region, which also acts as a waveguide for the SAW. This is shown in Fig. 4.26.

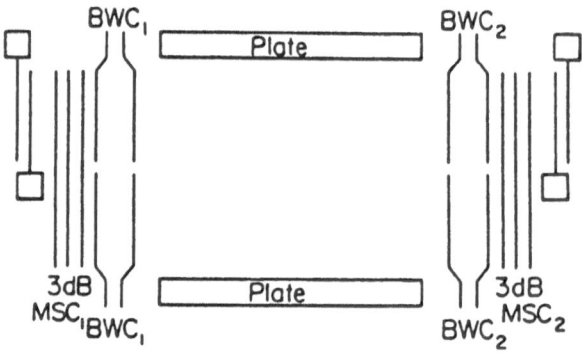

MSC – Multistrip coupler
BWC – Beamwidth compressor using MSC

Fig. 4.25. Convolver using multistrip coupler compression

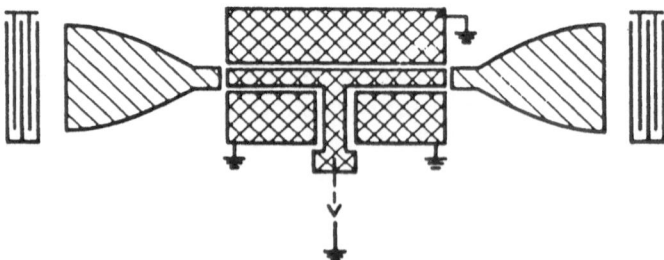

Fig. 4.26. Convolver using horn compression

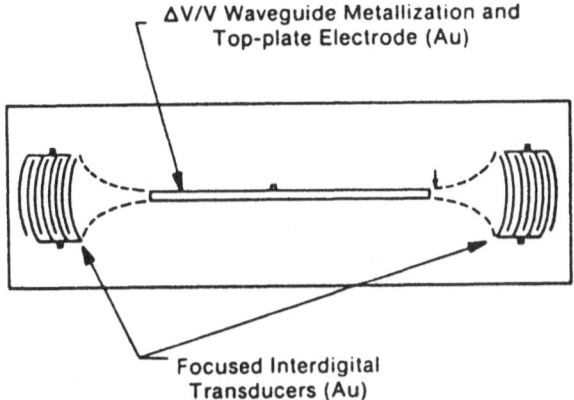

ΔV/V Waveguide Metallization and
Top-plate Electrode (Au)

Focused Interdigital
Transducers (Au)

Fig. 4.27. Convolver using curved transducers

iii) *Curved transducer.* The curved transducer basically focuses the surface acoustic wave onto the waveguide, thus achieving a high power level, as illustrated in Fig. 4.27.

The phenomenon of acoustoelectric nonlinearity is of a different nature. This nonlinearity is produced by the interaction of the electric field associated with a propagating SAW and the resultant space charge produced by the displacement of free carriers in a semiconductor. The current density in a semiconductor is proportional to the product of the carrier density and the electric field. The carrier density itself is a function of the electric field, since the carriers are rearranged to create a space charge. Thus, a normal component of the electric field at the semiconductor surface creates a potential which is proportional to the square of the field. In implementing an acoustoelectric convolver, two possibilities exist. One can use a piezoelectric semiconductor itself to generate the SAW – the so-called combined medium structure. Alternatively, one can have a separate medium structure as shown in Fig. 4.28, where the semiconductor is in close proximity to a SAW LiNbO$_3$ substrate. One can also directly evaporate semiconductor materials such as InSb on LiNbO$_3$ to obtain a convolver. A variation

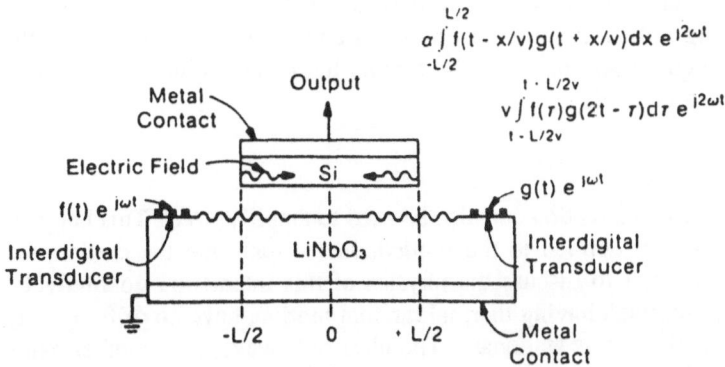

$$\alpha \int_{-L/2}^{L/2} f(t - x/v)g(t + x/v)dx \, e^{j2\omega t}$$

$$v \int_{t - L/2v}^{t \cdot L/2v} f(\tau)g(2t - \tau)d\tau \, e^{j2\omega t}$$

Fig. 4.28. Separate medium structure convolver. (From [4.19])

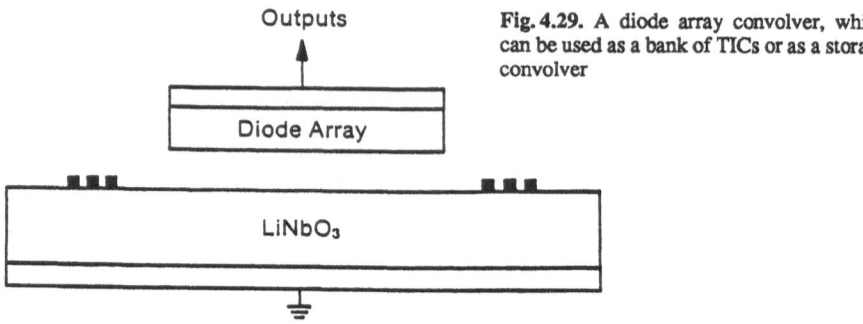

Outputs

Diode Array

LiNbO₃

Fig. 4.29. A diode array convolver, which can be used as a bank of TICs or as a storage convolver

of the separate medium structure is to evaporate ZnO film directly on silicon or GaAs wafers. The ZnO film is piezoelectric so that by using interdigital transducers one can have a somewhat integrated convolver which is compatible with silicon or GaAs integrated circuit technology.

If the semiconductor wafer is replaced by a vidicon-like wafer which contains a periodic array of either p-n diodes or Schottky diodes, as shown in Fig. 4.29, one can obtain storage convolvers. These have enormous potential in many different applications. They can be used to convolve, correlate and even perform adaptive signal processing. Basically, one function is stored as space charge in the diodes. This can be done in one of two ways. In the first method, a SAW modulated by the function of interest is launched on the LiNbO₃ to convert the temporal function to a spatial function. The electric field associated with the SAW is weak so that the diodes will be forward biased only with the aid of an additional dc voltage. Since this bias is applied across the semiconductor-LiNbO₃ structure, it has to be of the order of 100 V to bring the diodes close to threshold. When the bias is switched on for a short time, the spatial function is imaged onto the diodes in a manner resembling flash photography, from which it derives its name – the flash method.

The second method makes use of the fact that, when two counterpropagating waves interact nonlinearly, a dc spatial pattern at twice the spatial frequency is set up whose amplitude is proportional to the convolution of the two envelope functions. This dc pattern can be used to bias the diodes. Thus the function can be stored by using a delta function at one input and the desired function at the other. This technique, also known as the second harmonic method, is the more popular of the two methods.

f) Amplifiers

SAW acousto-electric interaction can also be used to amplify SAW. This happens when a dc bias field is applied to the semiconductor such that the carriers are moving. It is of interest to discuss the physics of this interaction qualitatively. The free carriers, although having thermal random motion, have no drift velocity and the carrier distribution is stationary. The ultrasonic wave, on the other hand, is propagating with the wave velocity. Because of the acoustoelectric coupling, ultrasound tries to drag the stationary distribution of free carriers. In other words, some energy is transferred from the ultrasound wave to the free carriers, which manifests itself as a drift velocity in the direction of wave propagation. The ultrasound energy loss results in an acoustoelectric attenuation and the gain in energy by the free carriers results in an acoustoelectric current if the terminals are short circuited, or in an acoustoelectric voltage if the terminals are open circuited. This generation of a dc voltage due to ultrasound is somewhat analogous to the Hall effect, in the sense that the dc voltage is zero when the magnetic field, for the Hall effect (or the ultrasonic wave amplitude, for the acoustoelectric effect), goes to zero. Also, the ultrasound wave velocity slows down due to free-carrier drag.

If we can also apply a dc electric field across the semiconductor in which ultrasound is propagating, amplification is possible in place of attenuation of waves, because the drift velocity of carrier distribution causes the ultrasound to gain energy if its velocity is lower than the drift velocity. Consequently amplification will only occur when the drift velocity v_d is larger than the wave velocity v_0.

The most interesting application of SAW amplifiers is to make a lossless SAW delay line using semiconductor (i.e., InSb, GaAs) films deposited directly on a SAW delay line substrate. A very compact, video bandwidth ($\sim 200\,\text{MHz}$), long-duration ($\sim 1\,\text{ms}$) delay line can be fabricated using this approach. The long delay is obtained by reflecting the SAW back and forth on the delay line substrate.

4.3 Charge Coupled Devices

The invention of charge coupled devices, bucket brigade devices and devices falling under the general heading of charge transfer devices was stimulated by the desire to effectively implement an analog shift register. A typical block diagram of an analog shift register is shown in Fig. 4.30. Here the capacitors are equal and they store charges which are proportional to the signal. By closing the switch S_1 to the on position, one transfers the charge from C_1 to C_2 thus transferring or shifting the signal by one stage. Note that, if all the switches are to be operated by the same clock, we must have two clock cycles to turn on two sets of switches denoted by S_1 and S_2, respectively. This provides isolation so that not all the capacitors are connected simultaneously.

Figure 4.31 shows the actual implementation of an analog shift register such as that shown in Fig. 4.30. Here the switches are junction transistors (Fig. 4.31a) and MOSFETs (Fig. 4.31b), respectively. Note that the storage capacitors need not be separate. For the case of the junction transistor switch, the capacitor can be the collector-to-emitter capacitance itself. This facilitates fabrication, as shown in Fig. 4.32, and gives what is called a bucket brigade device. The reason for this name is obvious. The signal charge is the fluid which is contained in the bucket

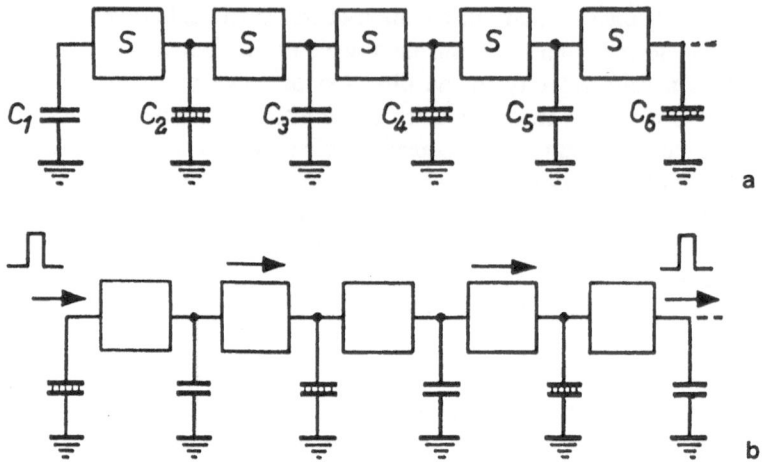

Fig. 4.30. Block diagram of an analog shift register

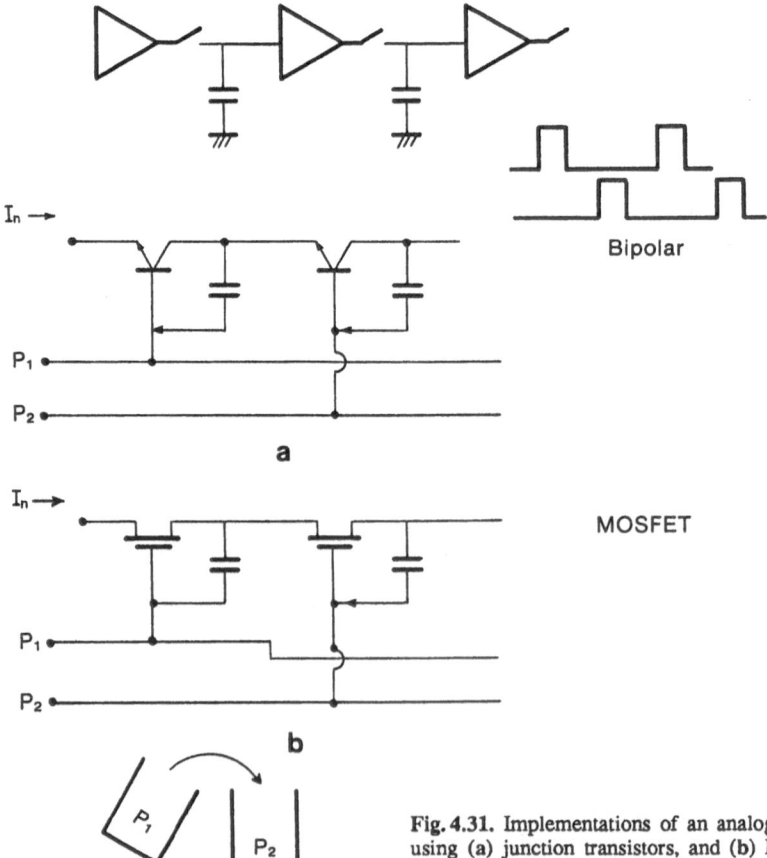

Fig. 4.31. Implementations of an analog shift register using (a) junction transistors, and (b) MOSFETs. (c) Illustration of the bucket brigade analogy

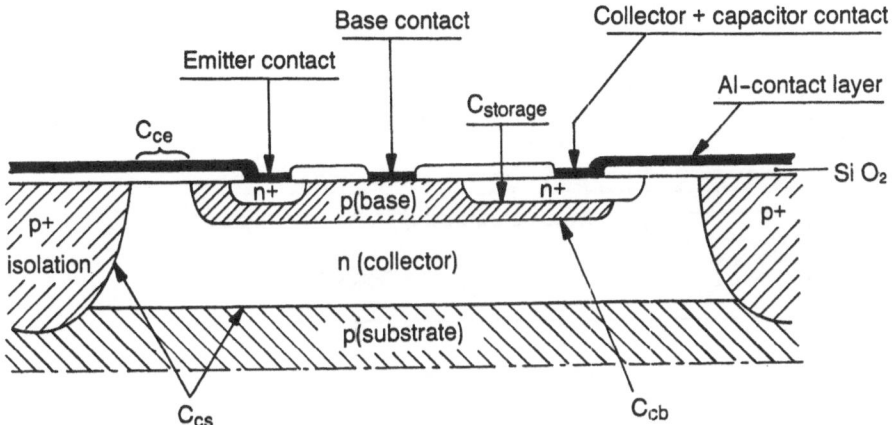

Fig. 4.32. Cross section of an integrated "bucket"

(the capacitor). By connecting the switch, one pours the fluid from one bucket to the other, thus transferring the fluid on one stage (Fig. 4.31c). This is repeated again to complete the bucket brigade analogy. Note the clock voltages applied to the emitters of the junction transistors to switch alternate junction transistors on and off.

This integration of the analog shift register using solid state devices was developed by *Sangster* and *Teer* [4.3] for application as analog delay lines. The objective of this development was to use these delay lines in stereo amplifiers to simulate the effect of distance between different instruments in an actual orchestra. The idea of *Boyle* and *Smith* [4.1] was somewhat different. They also represented the signal by the stored charge in a capacitor, however, since they were influenced by the magnetic bubble development occurring at that time, they preferred to implement the switch not by a transistor but merely by coupling between the capacitors. They showed that just as in the operation of a magnetic bubble, by applying electric voltage to the metal plates on an oxide layer, charge in a CCD can be moved around in such a way that the CCD acts as an analog shift register. This concept of simple charge coupled devices is shown in Fig. 4.33, where the metal plates are the capacitor plates which store the charge. The charge transfer takes place when the proper gate voltage is applied to the adjacent metal plate. However, let us now make a digression to clarify the operation of the MOS capacitor, which is essential to a proper understanding of CCDs.

Let us consider the MOS capacitor formed on a *p*-type substrate, as shown in Fig. 4.34a. Application of a negative voltage to the metal (also called the gate electrode) attracts the holes to the surface, forming an accumulation layer as shown in Fig. 4.34b. Application of a positive voltage to the gate repels the holes and a depletion layer is formed where only fixed ionized acceptor ions are present (Fig. 4.34c). As this positive voltage is increased, the depletion layer width increases to satisfy the need for more negative charges on the semiconductor side of the capacitor plate. For a large value of the applied positive voltage,

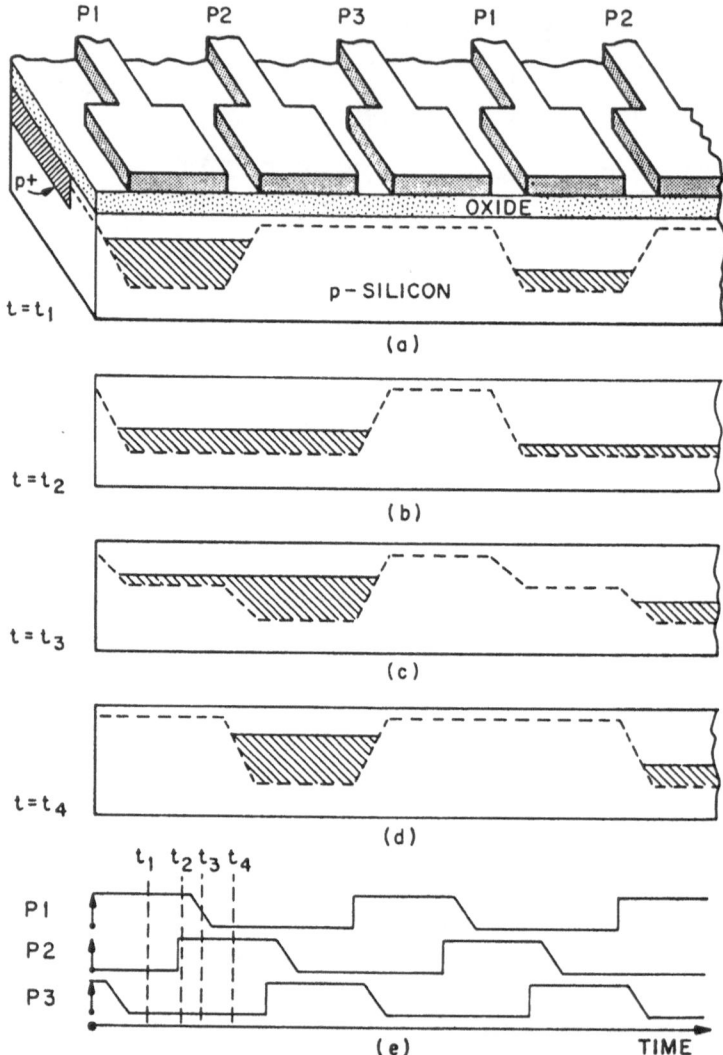

Fig. 4.33. (a) A three-phase n-channel CCD with the charge-carrying potential wells shown in the cross section of the silicon substrate. (b–d) Potential wells at subsequent time intervals illustrating the transfer of charge. (e) The corresponding time slots marked in the waveform diagram. (From [4.13])

we achieve what is called deep depletion. Note that this depletion condition is a transient state because electrons are generated near and inside the depletion region due to thermal generation and for other reasons. These electrons are attracted to the surface and they tend to destroy the deep depletion state. Eventually, in the equilibrium condition, enough electrons appear near the surface to collapse the depletion width down to a very small value. This condition is generally called inversion of the surface and is shown in Fig. 4.34d. This inversion is used for the operation of the MOSFETs but inversion is to be avoided at all costs for

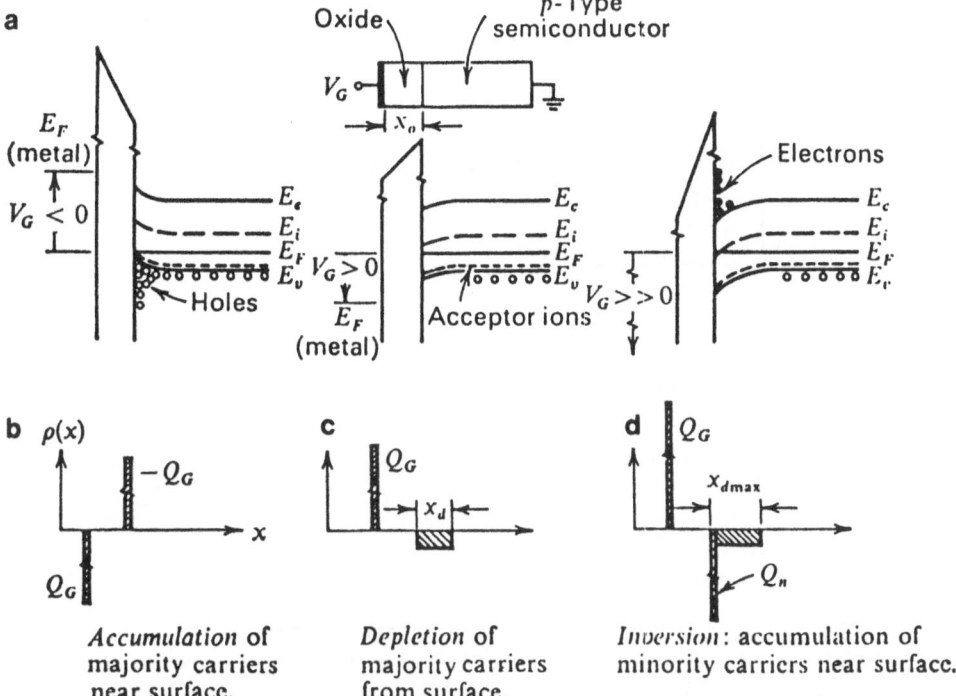

Fig. 4.34a–d. Energy bands and charge distribution in an MOS structure under various bias conditions, in the absence of surface states and work function difference. (From [4.20])

the operation of CCDs. The time for which the deep depletion state exists is called the storage time and the CCD can be operated only up to this time. This is because the minority carriers are used as the signal charges and signal charges can never be mixed with inversion charges, which represent a noise source for a CCD. In general, the signal charge level is represented as the level of the fluid in a bucket (as shown in Fig. 4.35a). In reality, the signal charge is at the surface and the dotted line (bucket outline) represents the no-signal condition of the deep depletion state. As the signal charges are stored in the capacitor, the depletion collapses to the top of the fluid level. However, the fluid and the bucket offer a very good visual representation of the signal charge and we shall use it often.

The transfer of charges takes place through charge coupling. To understand it, consider Fig. 4.35b where we have represented two capacitors far apart. One of the capacitors has the signal charge, and we have applied an appropriate voltage to the other one to put it into deep depletion as shown by the empty bucket. No transfer of charges can take place because there is a larger barrier in between. However, if we now bring these two capacitors close enough so that the fringing electric fields couple, causing the barrier to disappear (Fig. 4.35c), then the charges can transfer from one bucket to the adjacent one. This is the crucial point in the process and the one which justifies the name "charge coupled devices". Note that we have been able to transfer charges between two capacitors

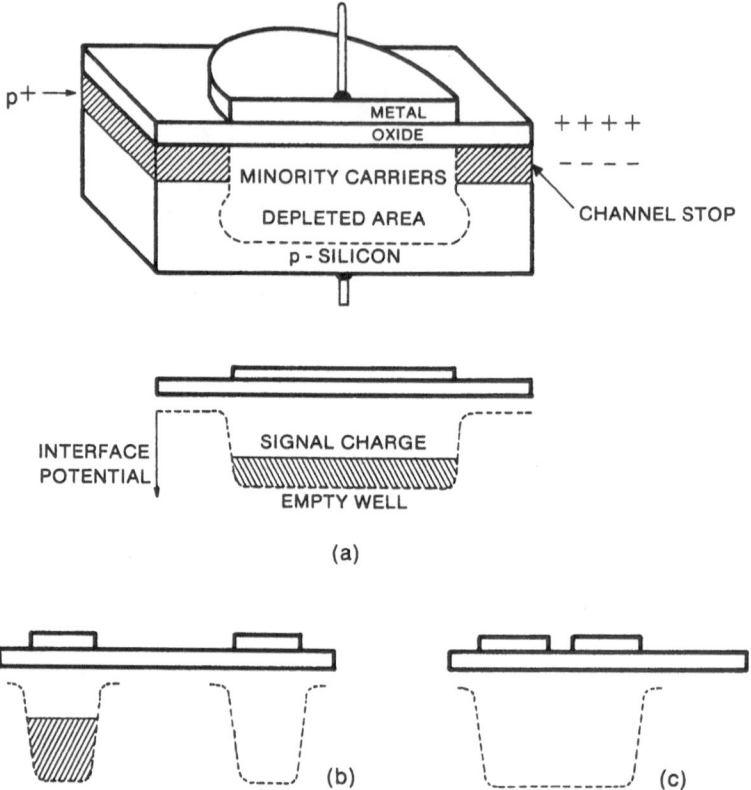

Fig. 4.35. (a) Potential well with the signal charge represented schematically as a liquid sitting at the bottom of the well. (b) The potential wells of two capacitors far apart. (c) When two capacitors are brought close together the potential barrier disappears

just by applying electric voltages to the gate electrodes. Neither a switch nor a transistor is needed. This ability to move charges around at will is a fundamentally new concept and is discussed later.

Coming back to Fig. 4.33, we can understand how the CCD operates. Three capacitors are needed for each stage because otherwise the charge can flow from C_3 to both C_2 and C_4. To make the flow unidirectional, no voltage is applied to C_2 when the transfer is taking place between C_3 and C_4. Thus we need a three-phase clocking scheme, as shown in Fig. 4.33. Note that, by reversing the clock cycles, it is possible to make the charges move in the backward direction. Also, the two phases of the clock must overlap for the transfer to take place. This is an inconvenience for the design of clock circuitry and therefore four-phase clocking is often used, as shown in Fig. 4.36. However, for the four phases, one needs to have four transfers before a single delay element can be simulated. One can also use two-phase clocking, as shown in Fig. 4.37, where the device elements have built-in charge transfer directionality using stepped oxide thickness. Thicker oxide causes a smaller depletion width for the same applied voltage. However,

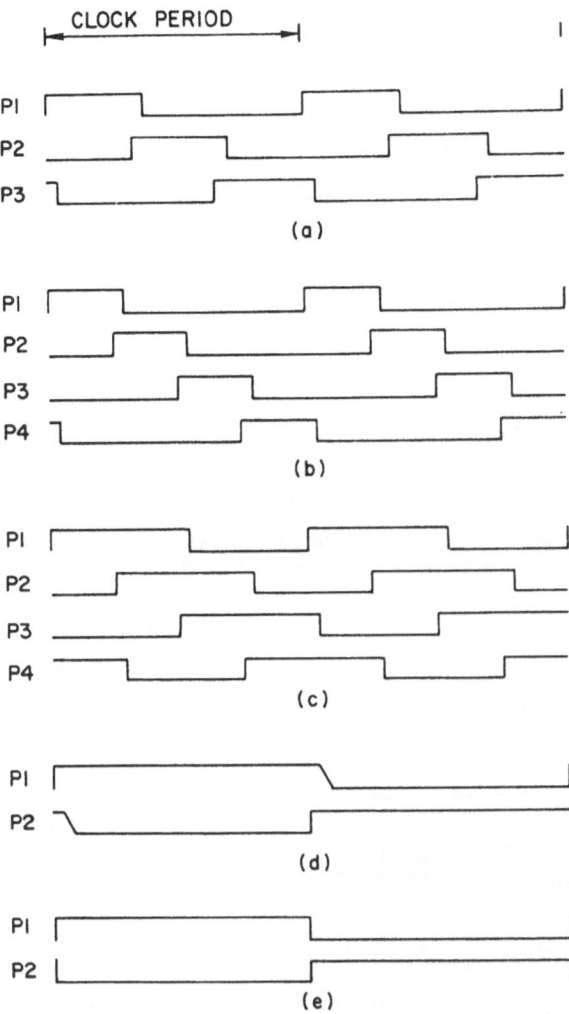

Fig. 4.36a–e. Operating pulse waveforms for various CCDs. (a) Three-phase CCDs. (b) Four-phase CCDs operated in a normal four-phase manner, and (c) operated in a mode that yields twice the signal handling. (d) CCDs operated with pulse overlap (push clocks), and (e) operated without pulse overlap (drop clocks). (From [4.13])

note that for the two-phase case we have lost the simplicity of the devices involving just metal plates on the oxide layer.

We have not yet discussed how the input and output of the CCD work. The input voltage to the device is sampled at the clock frequency and then, using an input diode, another CCD or a similar stage, this sampled voltage is converted into a proportional signal charge, which is transferred to the first storage capacitor of the first stage. An important input scheme called the fill and spill or potential equilibration technique is shown in Fig. 4.38. Similarly, the signal charge in the output stage is converted back to an equivalent voltage output using either

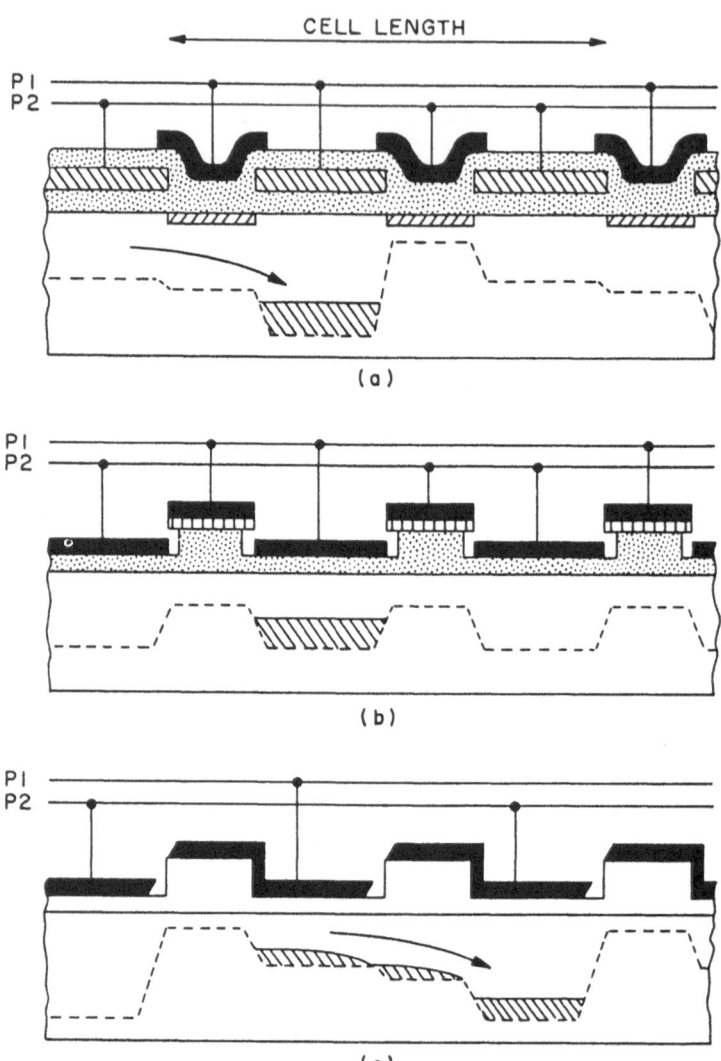

CELL LENGTH

(a)

(b)

(c)

Fig. 4.37. Two-phase CCD electrode structures employing (a) overlapping electrodes connected in pairs and an optional implant to enhance the potential step, (b) paired electrodes isolated by an undercut on the oxide step, and (c) truly stepped electrodes obtained by oblique evaporation. (From [4.13])

destructive or nondestructive sensing. In the destructive case, the charge packet is lost; a typical scheme is shown in Fig. 4.39.

For nondestructive sensing, the passage of the charge into a particular storage capacitor causes a proportional voltage to be induced on the gate plates. This induced voltage is used as the output. A particular case of nondestructive sensing is shown in Fig. 4.40, where one can also implement a positive or negative coefficient multiplied by the output. A CCD transversal filter is shown in Fig. 4.41a, where the coefficients follow a sinc function pattern to implement a

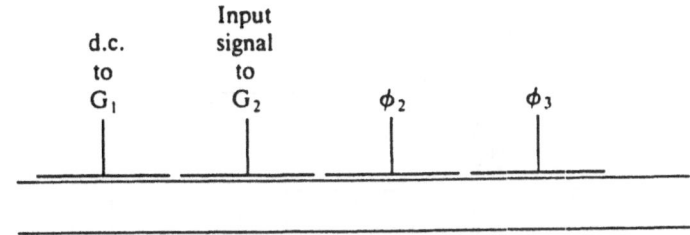

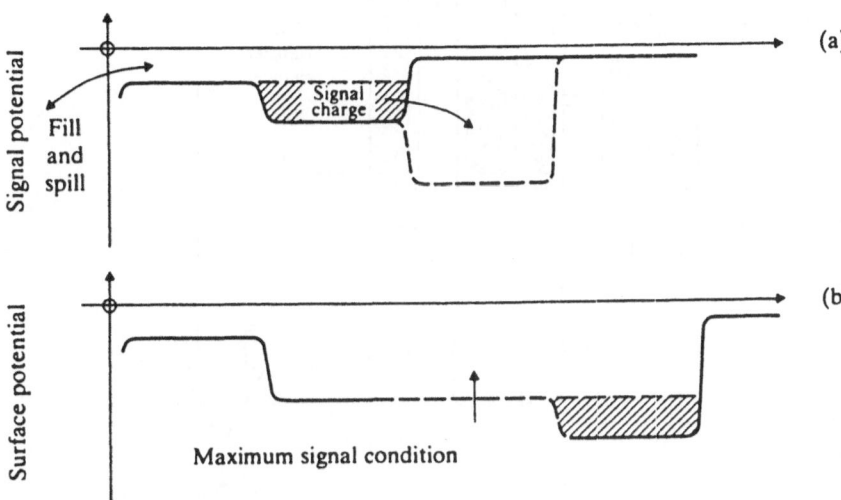

Fig. 4.38. A fill and spill technique which always equilibrates at the same surface potential. (From [4.21])

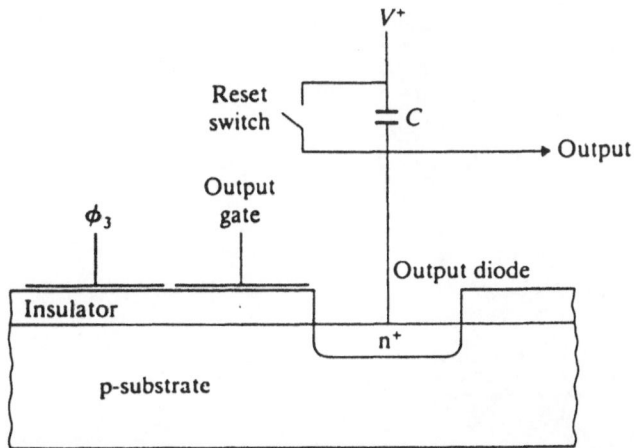

Fig. 4.39. A capacitor reset destructive sensing technique. (From [4.21])

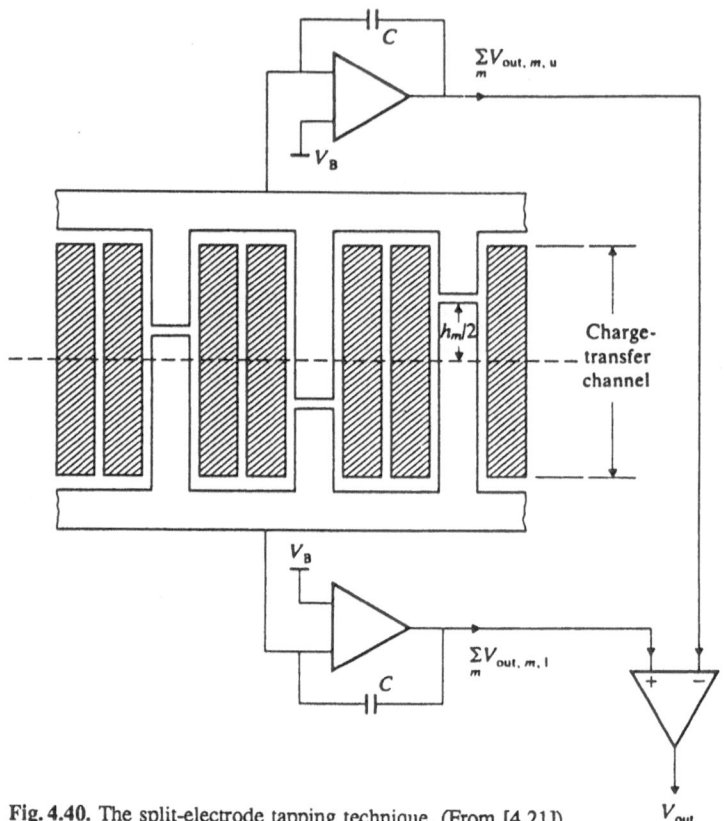

Fig. 4.40. The split-electrode tapping technique. (From [4.21])

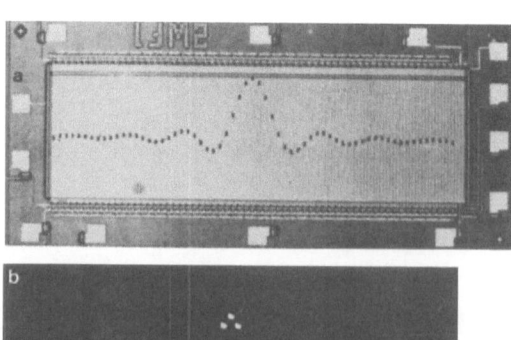

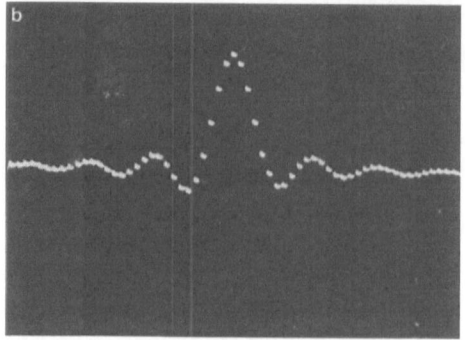

Fig. 4.41. (a) Photograph of a low-pass split-electrode transversal filter. (b) The observed impulse response, which corresponds to the profile of the splits in the electrodes. (From [4.10])

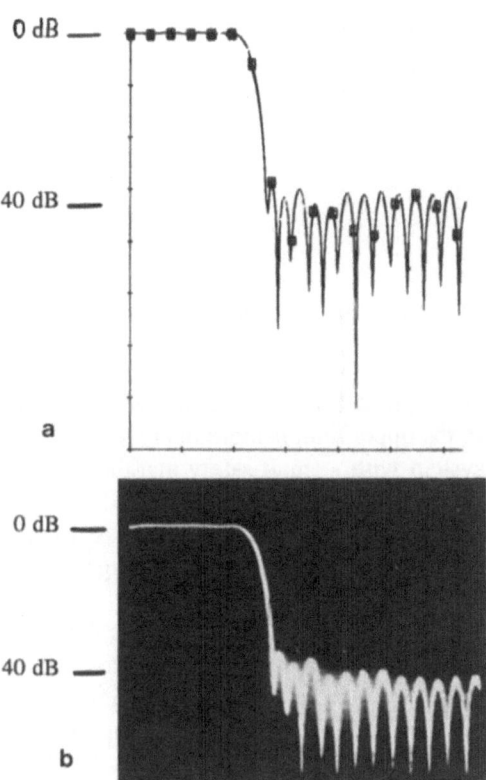

Fig. 4.42. (a) Calculated transfer function of the device shown in Fig. 4.41. (b) The observed transfer function. The device was operated at 32 kHz and the bandwidth was 3.2 kHz. (From [4.10])

low pass filter. The break in the third phase electrode is related to the value of the coefficient. As an example, if the break is at the middle, an equal voltage is induced on both halves and as these voltages are subtracted using a differential amplifier, an output with zero coefficient is produced. Note that the marks where the breaks occur represent the impulse response of the CCD device, the time scale being determined by the clock period. For the pattern of marks shown in Fig. 4.41a the impulse response is shown in Fig. 4.41b and, as expected, it is an exact replica of the gate electrode split pattern of the device. The corresponding theoretical and experimental frequency response are shown in Fig. 4.42. Comparing CCD and SAW transversal filters we note the following nine points.

i) The CCD is a base-band device whereas SAW devices always work around a center frequency in the tens of megahertz region. Thus the filter shown in Fig. 4.42 can never be implemented by a SAW device. Only band-pass filters are possible using SAW devices.

ii) The CCD is a discrete time signal processing device whereas a SAW device is a time analog device. To calculate the bandwidth of the signal a CCD can process, one must consider the Nyquist sampling rate, the clock frequency and the number of phases used in one delay stage. Thus, for a 4-phase CCD

393

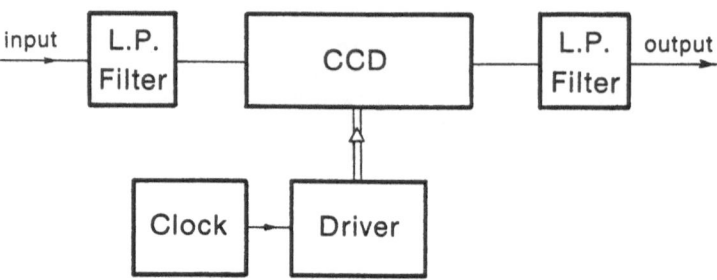

Fig. 4.43. Block diagram showing the input and output low pass filters to be used for a CCD filter

with a clock frequency of 20 MHz, the highest bandwidth signal which can be processed is 20/2 = 10 MHz. However, the upper limit is more likely to be 8 MHz for this case because one designs a system with a small safety margin from the exact Nyquist sampling rate. For a CCD, one must use a low-pass filter at the input to guarantee the Nyquist rate for a fixed sampling frequency or to stop "aliasing". To convert the discrete time signals back into the analog signals, one needs another low-pass filter in the output, as shown in Fig. 4.43.

iii) As discussed before, CCDs operate by the transfer of charges from one stage to another. However, this transfer is not instantaneous, but takes a finite amount of time. Thus the clock frequency cannot be increased indefinitely and an upper limit for that frequency exists for each device, depending on how it is fabricated and the substrate and dimensions that are used. Similarly, as CCDs are transient devices, there is a lower limit on the clock frequency, determined by the storage time. A CCD filter can be operated at any frequency between these two limits. As the clock frequency can be changed electronically at will, this makes CCD filters very attractive for some applications, such as expansion and compression of bandwidth. The reason for this is shown in block diagram form in Fig. 4.44: the center frequency of the band-pass filter can be controlled by a factor of 1000 or more by changing the clock frequency. For a similar case with SAW devices, the center frequency is fixed and related to the SAW velocity, which is also more or less fixed and can be altered by only about 0.1%, using an applied field. For the expansion or compression of a signal, one loads the signal in the CCD at a certain clock frequency and then changes the clock frequency before the output is taken, as shown in Fig. 4.45.

iv) For an ideal charge transfer device, all the signal charge must transfer from one cell to the next. However, some charges get trapped in the bulk traps or surface states. Also, it takes a finite amount of time to transfer the charges from one cell to the next. For high frequency clocking, there is not enough time for all the charges to transfer. The charge transfer efficiency, η, which is defined as the fraction of the charge transferred to the next cell is thus never exactly 1, but it can be as high as 0.99999 or even higher. It is customary to define a quantity ε, called the charge transfer inefficiency. This denotes the fraction of charge left behind for each transfer. Thus

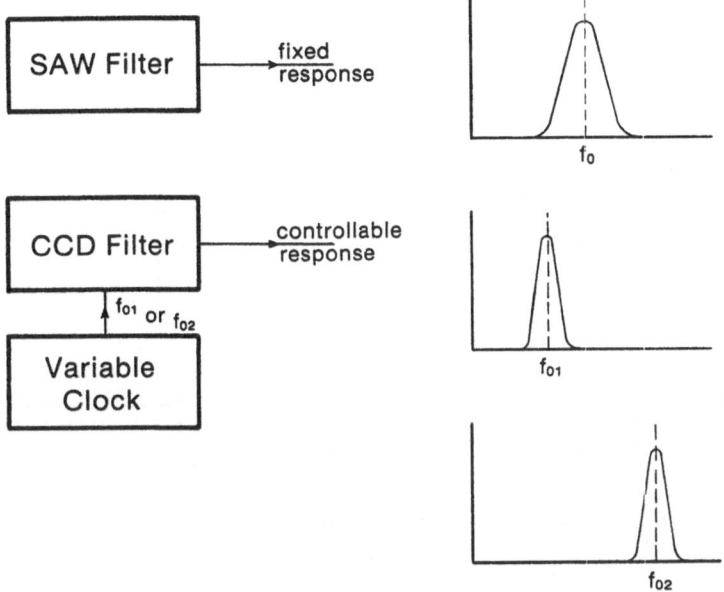

Fig. 4.44. The center frequency of a SAW filter is fixed, but the center frequency of a CCD filter can easily be changed by simply varying the clock frequency

Variable Speed Clock

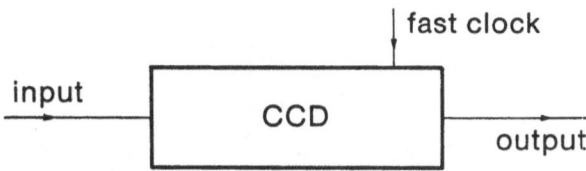

Fig. 4.45. To compress a signal the input is given to the CCD at one clock frequency then the clock frequency is increased before the output is taken

$$\varepsilon = 1 - \eta \quad ;$$

ε is a small quantity and for good devices it ranges from 10^{-3} to 10^{-6}. As mentioned before η (or ε) is a function of clock frequency. However, it turns out that even if we use a long time for transfer or use a very low clock frequency, some charges get trapped in the empty surface states and η reaches a steady-state value less than 1. The situation is very similar to the case of fluid poured into an empty bucket. If we now try to transfer the fluid from this bucket, some fluid will be left clinging to the bucket's surface. However, the probability of transferring all the fluid increases if we first wet the surface of the bucket. The same technique, called *fat zero*, is applied to charge transfer devices. For this case, we intentionally add a small amount of charge to fill up the surface states.

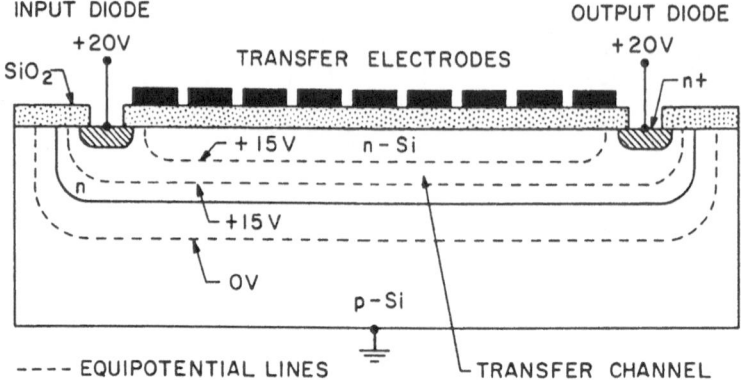

INPUT DIODE

+20V

SiO₂

TRANSFER ELECTRODES

OUTPUT DIODE

+20V

n+

+15V n−Si

n

+15V

0V

p−Si

---- EQUIPOTENTIAL LINES

TRANSFER CHANNEL

Fig. 4.46. Longitudinal cross section through a bulk channel CCD showing the channel layer in contact with the reverse biased input-output diodes and equipotential lines which indicate the actual position of the transfer channel. (From [4.11])

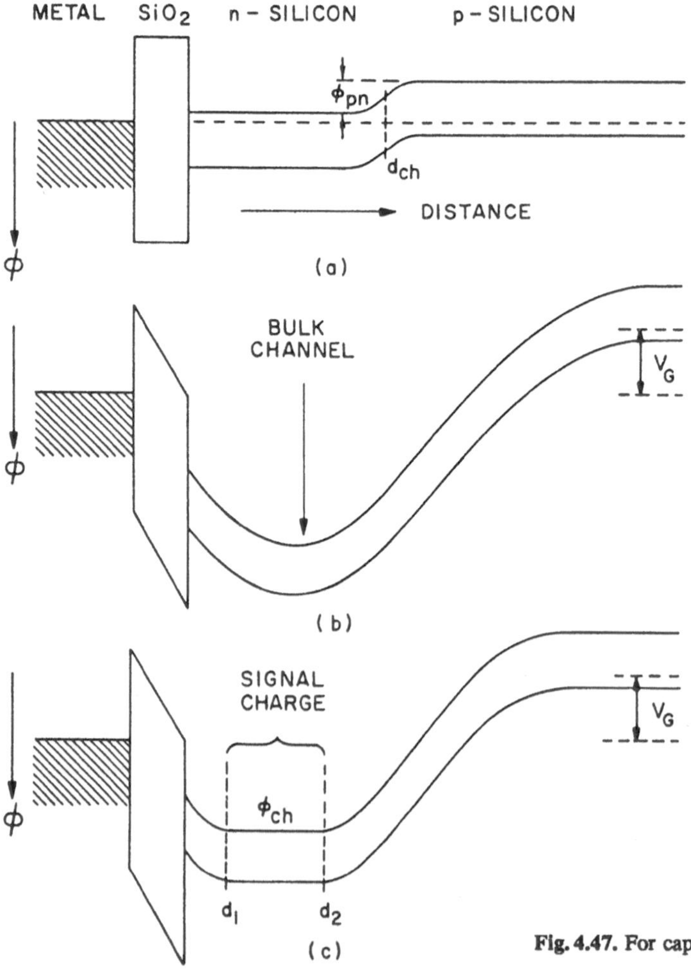

METAL   SiO₂   n−SILICON   p−SILICON

ϕ_{pn}

d_{ch}

DISTANCE

ϕ

(a)

BULK
CHANNEL

V_G

ϕ

(b)

SIGNAL
CHARGE

V_G

ϕ_{ch}

ϕ

d_1 d_2

(c)

Fig. 4.47. For caption see opposite page

Thus, when the signal level is actually zero, the actual charge in the bucket is not zero but a small, finite, known amount.

v) *Buried channel devices* or *peristaltic devices*. The detrimental effects of the surface states show up in the charge transfer efficiency not only because of the trapping in surface states, but also because of the reduced surface mobility of electrons. To avoid this, one uses buried (or bulk) channel devices as shown in Fig. 4.46. For this device, an extra n region layer is needed, thus making the fabrication of the device more complex. If this n region is completely depleted by the diodes, as shown in Fig. 4.46, the potential profile is given as shown in Fig. 4.47. The minima of the potential is not at the surface, but away from it. Hence the signal charge is stored at a distance from the surface, thus avoiding the detrimental surface effects. The electrons transfer in this bulk channel from one cell to the next in a fashion similar to how food is transferred through the intestines. *Esser* [4.12] used this analogy when he coined the name *peristaltic device* for the buried channel device. Another important advantage over surface channel devices is that the fringing electric field between the cells overlaps more because the channel is further away from the gate. Consequently, buried channel devices have many good attributes and most devices are now made using this technique.

vi) *Recursive filters*. Charge coupled devices can easily be configured to act as a recursive filter. (A general recursive filter is shown in Fig. 4.48.) This is an advantage over SAW filters which are mostly transversal or FIR (finite impulse response). The recursive filters are also called IIR (infinite impulse response) because the impulse response of an ideal recursive filter is infinitely long because of the feedback.

vii) We have discussed before that, due to problems associated with the charge transfer rate and efficiency, the bandwidth of the signal that can be easily processed with CCDs tends to be low. However, as sonar signals are in general of low bandwidth, CCDs are ideal for sonar applications, as shown in Fig. 4.49, where CCDs are used as the delay elements for the phased array sonar. The phased array sonar can be focused at any point within its view by changing the delays, which changes the effective focal length of the phased array lens and steers it sideways.

viii) CCD *imaging devices*. So far, we have discussed only an electrical signal as an input to charge transfer devices. However, if the gate electrode is transparent and light is incident through this gate, a number of electron-hole pairs proportional to the number of incident photons will be generated. The electrons will be attracted to the surface in a properly biased p-substrate device and will constitute the signal charge proportional to the light intensity. The incident light intensity can be integrated up to the storage time and then transferred to the output of the device. A typical arrangement is shown in Fig. 4.50, although many

---

Fig. 4.47. Energy band diagram perpendicular to the surface in a bulk channel device, (a) in the unbiased state, (b) after draining all mobile carriers from the channel layer, and (c) in the presence of some signal charge. (From [4.13])

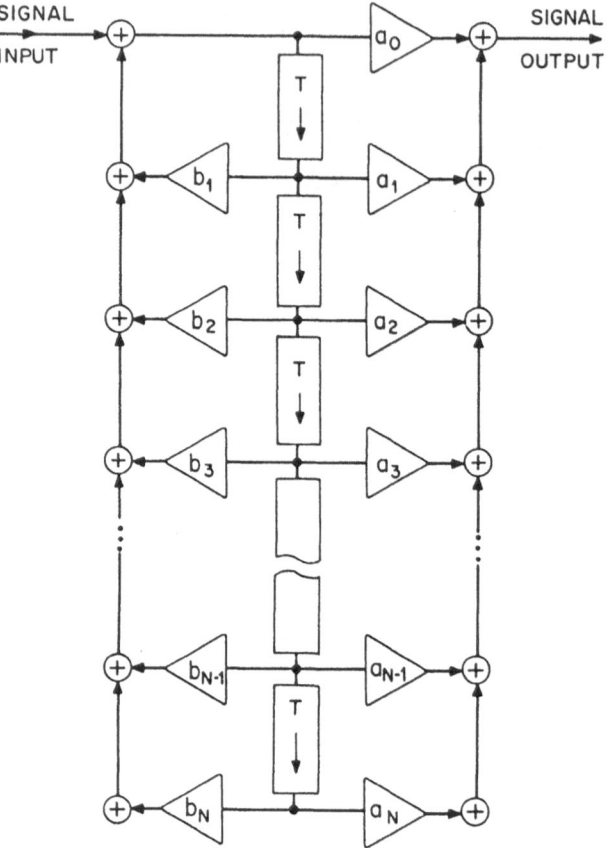

Fig. 4.48. General form of the recursive filter with both feedback and feedforward of the delayed signal

different structures are possible. In place of the front-side illumination discussed above, one can also have rear-side illumination where the light is incident through the substrate, which has been thinned down to reduce absorption.

Imaging devices are one of the best and most successful applications of charge transfer devices. As the incident-light photon energy must be greater than the energy bandgap of the semiconductor material, one must use semiconductors like mercury cadmium telluride for infrared radiation with wavelength greater than 1 μm. Also, a proper imaging device must be able to record two-dimensional images. A typical infrared CCD imager uses an array of 64×64 HgCdTe devices. Optical CCD imagers have been fabricated with more than 1000×1000 elements.

ix) CCD *memory and digital logic elements*. As the charge cannot be stored indefinitely in a CCD, a memory built using CCDs must be recirculating. A typical arrangement is shown in Fig. 4.51. Note that, if a block of data is needed at the same time rather than in a random access manner, CCD memories have a distinct advantage. However, because of the enormous improvements of both static and dynamic RAMs since the invention of CCDs, the CCD memory has

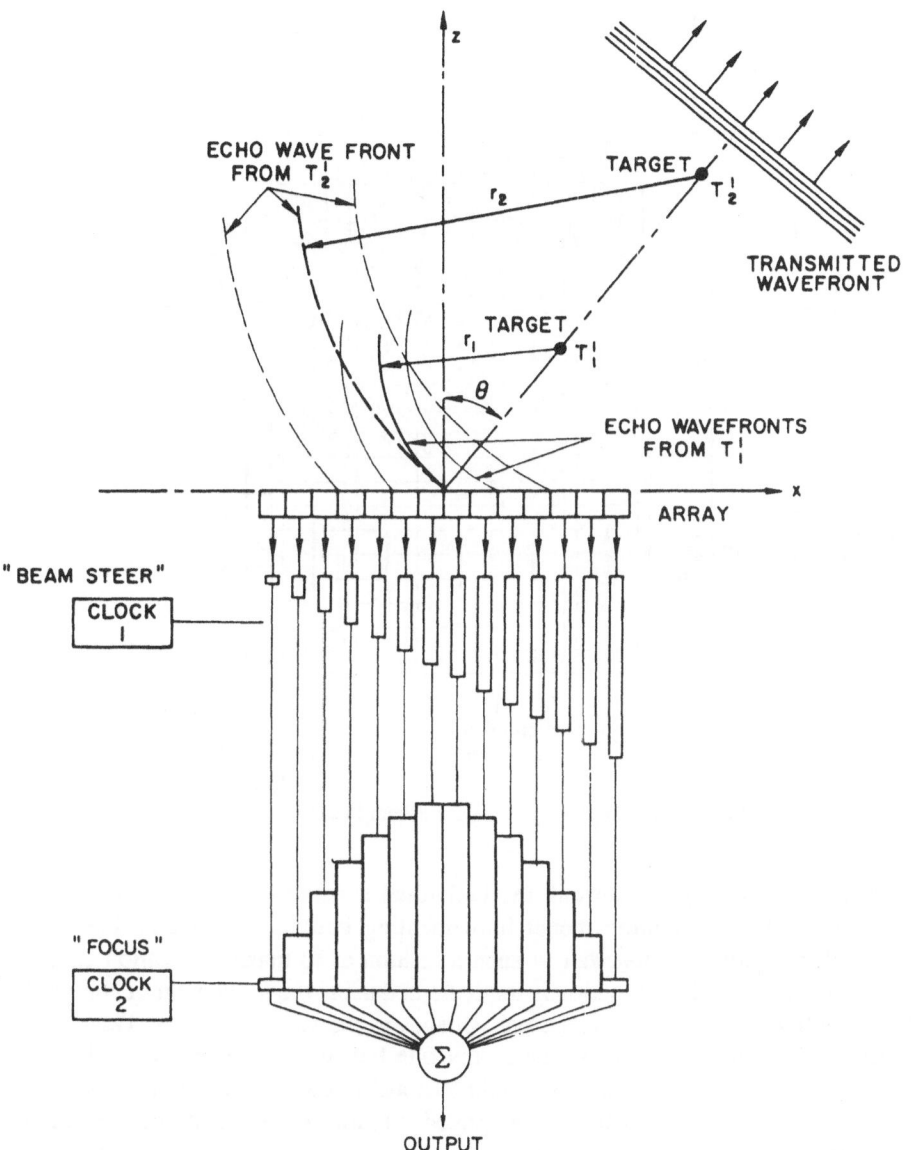

Fig. 4.49. Phased array sonar. Dynamic focusing with a quadratically tapped variable delay line. (From [4.22])

not yet been a commercial success. The same comments apply to CCD logic elements, such as AND/OR operations, adders, and multipliers, which have been fabricated and tested.

Before we conclude this section it is important to point out that the CCD is a true integrated circuit in the sense that a particular function (such as a delay line) is implemented. In this functional form, one does not, and sometimes cannot,

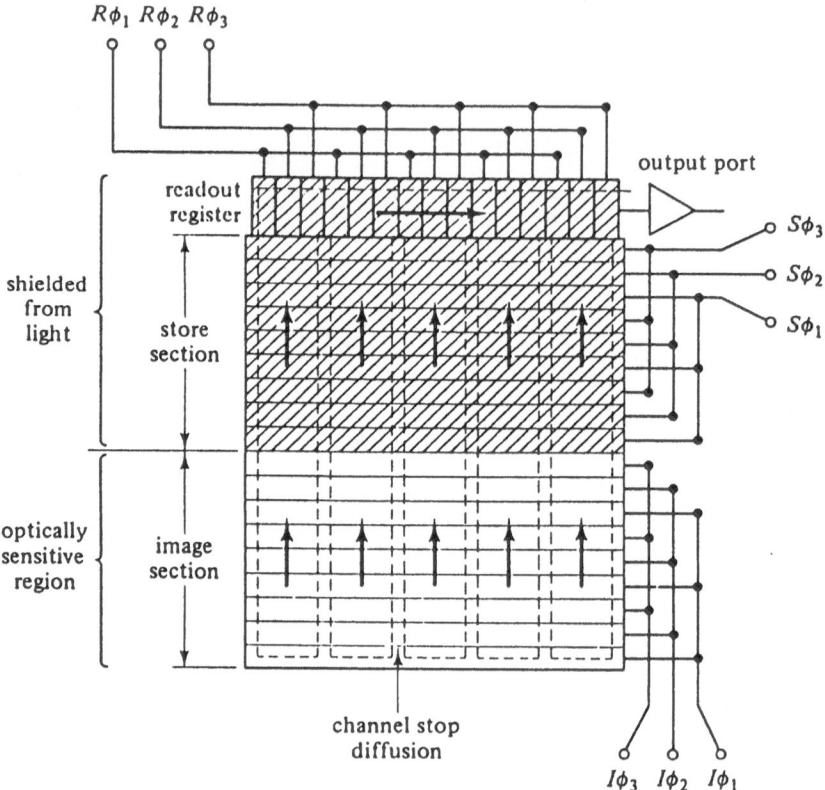

Fig. 4.50. Typical design of a CCD imaging device. (From [4.23])

consider the voltage or current, the individual resistors, inductors or capacitors. Before CCDs, integration meant implementing individual elements like R, L, C and the transistors together in such a fashion as to make fabrication cheaper. However, in a CCD, a function is implemented using the movement of charges by the application of a set of electrical voltages to the gate electrodes. The concept can be developed further to include not one but many channels through which charges can be moved. Such three-dimensional circuits, an example of which is shown in Fig. 4.52, will be the integrated circuits or solid-state devices of the future.

For proper storage and transfer of charges in a charge transfer device, one needs a depletion layer formed by the application of a voltage. We have so far discussed only the depletion layer formed by the field effect in a MOS. Other possibilities are reverse-biased p-n junction and Schottky diodes. The latter have been used for GaAs devices, as shown in Fig. 4.53. It is still very difficult to fabricate reliable reproducible MOS devices on GaAs and this drawback is avoided

Fig. 4.52. CCD as a three-dimensional integrated circuit. (From [4.24])

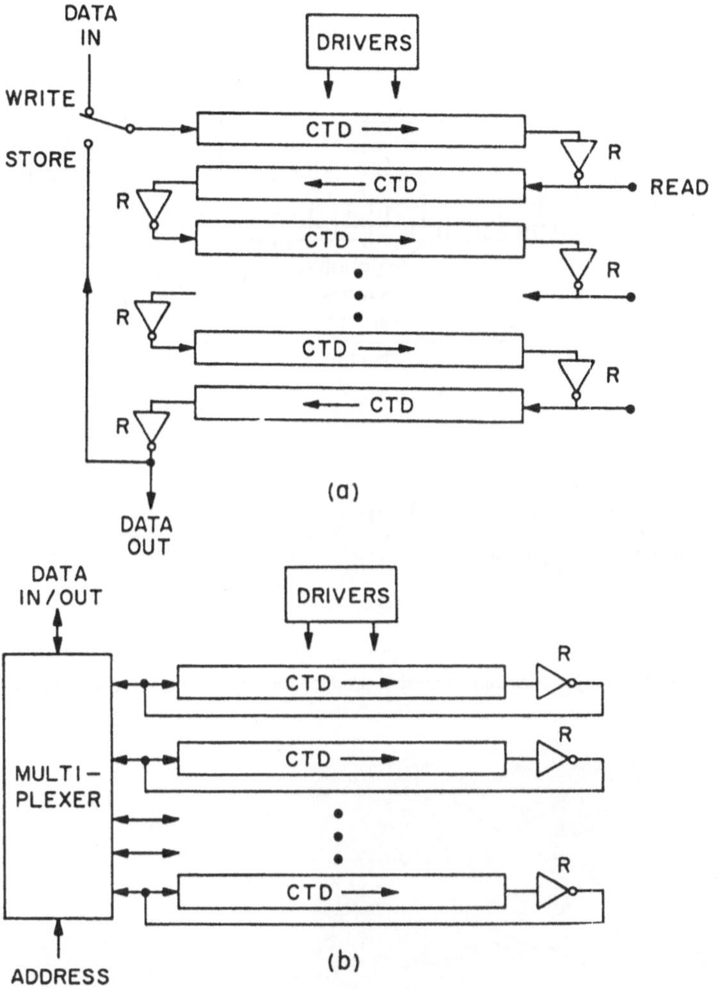

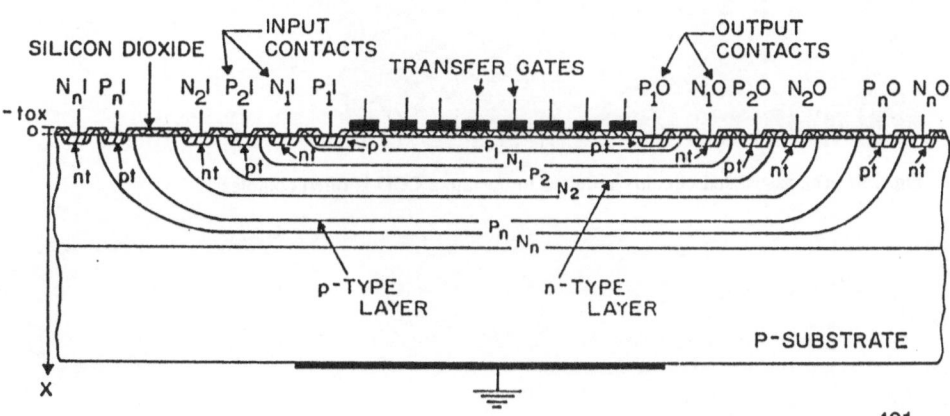

Fig. 4.51a,b. Arrangements of memories built from CCDs. (a) Serpentine organization and (b) loop organization, which lead to memories with fast access times but requiring decoding to address individual output ports. (From [4.13])

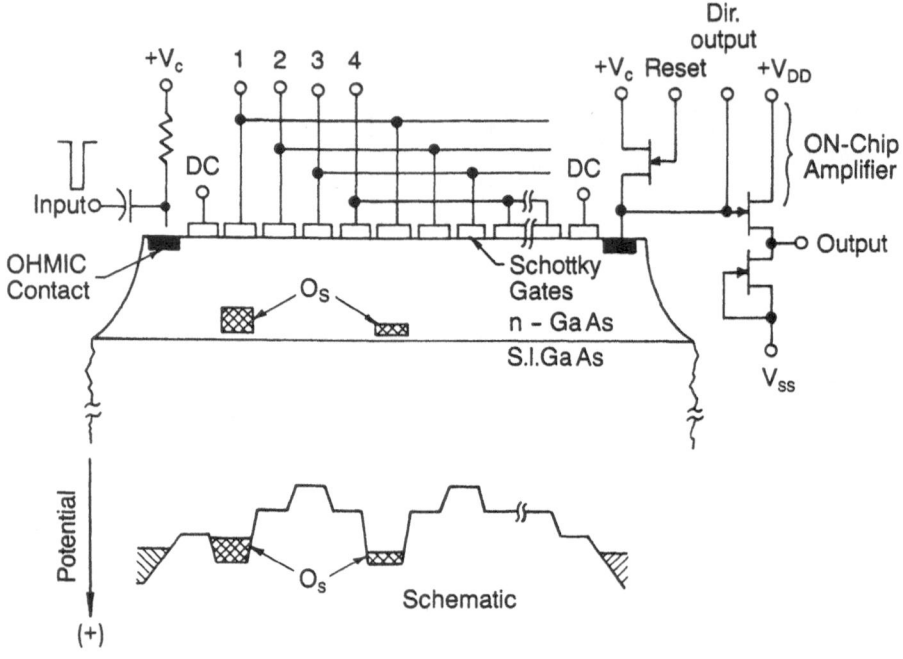

Fig. 4.53. In GaAs devices Schottky diodes can be used to produce the depletion layer. (From [4.25])

Fig. 4.54. The associated circuitry required to operate a CCD is often complex

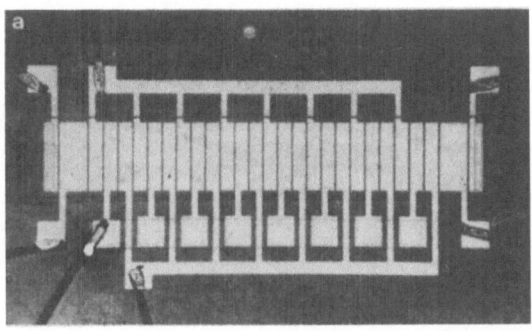

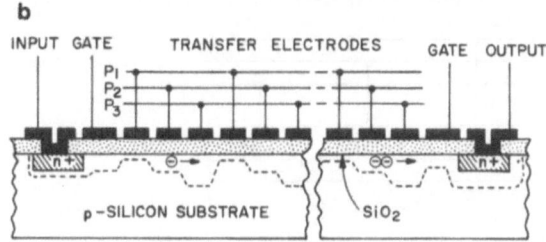

Fig. 4.55. First charge coupled device comprising eight three-phase elements and input-output gates and diodes shown (a) in plan view and (b) schematically in cross section. (From [4.2])

by using Schottky diodes. The device shown in Fig. 4.53 has been clocked in excess of 4 GHz.

One of the main disadvantages of CCDs is the requirement for high level clock voltages, on the order of 20 V. Thus, the associated circuitry is complex and sometimes overwhelms the applications. This is obvious from Fig. 4.54, which shows the small CCD and the large number of other ICs needed for its operation. The large clocking voltages also cause clocking noise, which, if not properly taken care of, might also be a problem.

The original proposal of the CCD by *Boyle* and *Smith* [4.1] was the simple structure of metal gates only on the oxide layer over the silicon substrate. However, as the metal gates need to be near each other for electric field coupling, this gap being about 1 μm, its fabrication becomes difficult. Also the open structure shown in Figs. 4.33 and 4.55 causes instability due to contamination such as sodium ions inside this gap. To avoid this, in general, the whole surface is covered with polysilicon, which is then ion-implanted to produce gate electrodes as shown in Fig. 4.56b. A better arrangement, which is also self aligning with respect to the gap, is shown in Fig. 4.56c. Here, as the gap is produced by the second oxide layer grown on the polysilicon, it can be very thin, on the order of hundreds of angstroms, without requiring any special mask of such close tolerance.

Finally, we must mention the video cameras which use CCD imagers and memories. Actually, electronic photography (for both still and moving pictures) is the most successful commercial application of CCDs.

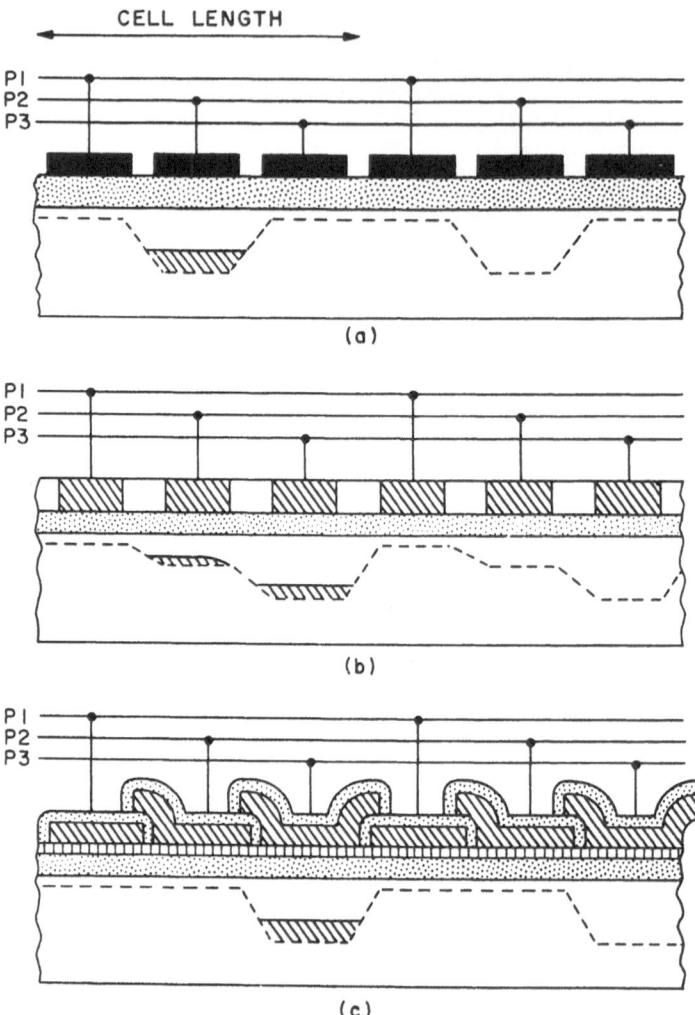

Fig. 4.56. Three-phase CCD electrode structures employing (a) a single level of metal with narrow gaps, (b) selectively doped regions in a layer of intrinsic polysilicon, and (c) three overlapping levels of oxidized polysilicon. (From [4.13])

4.4 Magneto-Static Waves

Guided magneto-static waves (MSWs) in ferrimagnetic films have some properties similar to SAWs and thus can be used to fabricate the usual SAW devices discussed earlier, i.e., tapped delay line, reflectors, chirp filters, and convolvers, etc. These waves are slow, dispersive, magnetically dominated, electromagnetic waves which propagate in magnetically biased ferrite materials at microwave frequencies. The key to the success of NSW devices has been the development of

404

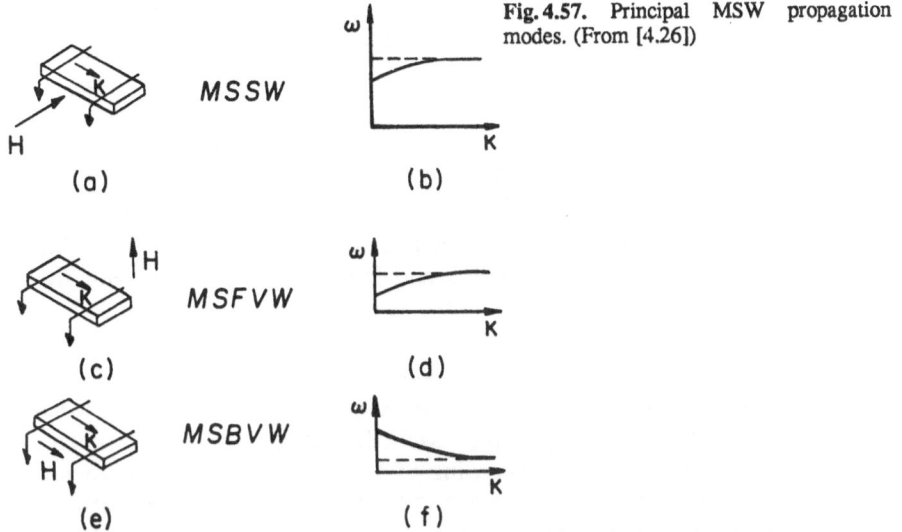

Fig. 4.57. Principal MSW propagation modes. (From [4.26])

MSSW

(a)

(b)

MSFVW

(c)

(d)

MSBVW

(e)

(f)

high quality low linewidth (which translates to low loss), large area YIG (yittrium iron garnet) grown epitaxially on gadolinium gallium garnet (GGG) substrates using liquid phase epitaxy technology. There are three major propagating modes that have been used in device applications. These are (i) magneto-static surface wave (MSSW), (ii) magneto-static forward volume wave (MFSFW), and (iii) magneto-static backward volume wave (MSBVW) modes. Which of the particular modes propagate is determined by the orientation of the bias field relative to the propagation direction and the finite slab. These situations are shown in Fig. 4.57.

The modes are dispersive and they have a limited passband with a center frequency that can be tuned by means of the magnetic bias field. This tunability permits propagation of a particular wavelength at any desired frequency within a frequency range. All modes have typical group velocities in the range 3×10^3 to 3×10^5 m/s and wavelengths $1\,\mu m$ to $1\,mm$ for a $10\,\mu m$ thick epitaxial film. The frequency range is 1–20 GHz, and for the same delay line with a fixed transducer, any frequency can be chosen provided a suitable bias field is applied. The transducers are typically single microstrip line and they lend themselves to easy fabrication by standard photolithographic processes.

For the MSSW mode, the bias field is oriented perpendicularly to the direction of the wave propagation in the plane of the film (Fig. 4.57a). This mode exhibits highly anisotropic propagation with energy conentrations that are bound to the top surface of the epitaxial film in the single forward-propagating wave mode and to the bottom surface of the film in the single reverse-propagating wave mode. The dispersion curve is shown in Fig. 4.57b.

The MSFVW mode is characterized by near isotropic propagation in the slab with multiple mode energy distributions concentrated in the interior of the film, similar to those found in a parallel plate waveguide. For this case, the

405

bias field is applied perpendicular to the film (Fig. 4.57c). The dispersion curve (Fig. 4.57d) exhibits a negative group delay slope, the same as that of MSSW.

For MSBVW, the bias field is oriented along the direction of propagation in the plane of the film (Fig. 4.57e). Similarly to MSFVW, it is characterized by multiple modes with energy concentration inside the film. This mode has a positive group delay slope (Fig. 4.57f).

In the next section we discuss the fundamentals of MSWs. This is followed by a discussion of devices. However, it is of interest to discuss the relative advantages and disadvantages of SAWs and MSWs in connection with signal processing applications. MSW technology has an edge over competing technologies in terms of propagation losses and operational bandwidths in the microwave region (1–20 GHz), the most common operating frequency range of MSW devices. SAW devices incur enormous losses beyond 1 GHz. Thus, with MSWs, one is able to perform analog signal processing directly at microwave frequencies with bandwidths approaching or exceeding 1 GHz. To use SAWs for microwaves, one must down-convert from microwave to intermediate frequencies (IF).

MSW devices are as compact and rigid as SAW devices. Typical propagation delays for MSW delay lines are between ten and several hundred ns/cm compared to about 1 μs/cm for SAW delay lines. MSW devices involve planar structures similar to SAW devices, and they are compatible with the current microwave and hybrid monolithic integrated circuits. As MSW transducers are typically single microstrip lines, much less stringent photolithographic processing techniques are required for their fabrication than for SAW devices. MSW devices need a biasing magnetic field and they are dispersive. However, this may be an advantage if tunability of the center frequency is a requirement. Furthermore, the nonreciprocal propagation characteristics of MSW may be exploited for the construction of nonreciprocal components such as directional couplers.

4.4.1 MSW Field Equations and Dispersion Relations

As the speed of MSW propagation in a medium is much slower than that of light, a quasi-static approximation is valid, which leads to a simplification of Maxwell's equations:

$$\nabla \times \boldsymbol{h} \approx 0 \quad , \quad \text{or} \tag{4.4.1}$$

$$\boldsymbol{h} = -\nabla \phi \quad , \tag{4.4.2}$$

where $\boldsymbol{h}$ is the small-signal time-varying magnetic field vector and ϕ is a scalar potential. The effective permeability tensor obtained from the simplified spin equation of motion under small-signal conditions and using linearization is given by

$$\overline{\overline{\mu}} = \begin{vmatrix} \mu & -j\alpha & 0 \\ j\alpha & \mu & 0 \\ 0 & 0 & \mu \end{vmatrix} \quad \text{where} \tag{4.4.3}$$

$$\mu = (\omega_3^2 - \omega^2)/(\omega_0^2 - \omega^2) \quad ,$$

$$\alpha = -\omega_m\omega/(\omega_0^2 - \omega^2) \quad ,$$

$$\omega_3 = [\omega_0(\omega_0 + \omega_m)]^{1/2} \quad ,$$

$\omega_0 = \gamma\omega_0 H_0$ is the gyro-frequency, $\omega_m = \gamma\mu_0 M_0$ is the magnetization frequency, H_0 is the dc bias magnetic field applied in the z direction, M_0 is the saturation magnetization produced in the z direction and $\gamma = 2.8\,\text{MHz/gauss}$ is the gyro-magnetic ratio. The relationship between the small-signal magnetic induction and the small-signal magnetic field vector is

$$b = \mu_0\overline{\overline{\mu}}h \quad . \tag{4.4.4}$$

The other two equations needed are

$$\nabla \cdot b = 0 \quad , \quad \text{and}$$

$$\nabla \times e = 0 \quad ,$$

where e is the small-signal electric field vector.

The dispersion relations obtained using the above equations are quite complex if the ground planes are included. If the ground planes are at infinity (Fig. 4.58), the dispersion relations are given by

MSSW	$k = \dfrac{1}{2d}\ln\dfrac{(\omega_m/2)^2}{[\omega_0+\omega_m/2]^2-\omega^2},$	(4.4.5)
MSFVW	$k = (\beta d)^{-1}\tan^{-1}[-2\beta/(1+\mu)],$	(4.4.6)
MSBVW	$k = (\beta/d)\tan^{-1}[2\beta/(1+\mu)],$	(4.4.7)

where d is the thickness of the YIG film and $\beta = (-\mu)^{1/2}$. Note that the disper-

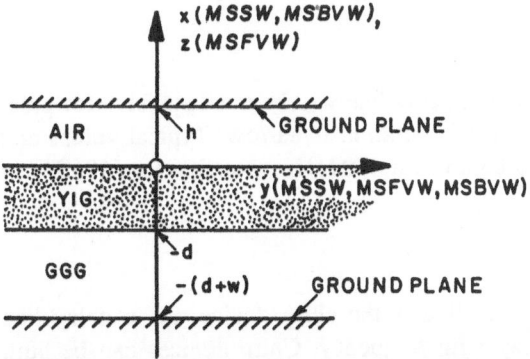

Fig. 4.58. The guided MSW single-YIG-film two-ground-plane geometry. The bias field $H_0 = H_0 z_0$ is applied along the z direction in a right-handed coordinate system for all three MSW-type configurations. The propagation is along x for MSFVW and in an arbitrary direction in the yz plane for MSSW and MSBVW. (From [4.27])

sion relation changes significantly if the ground planes are brought near the YIG film.

4.4.2 MSW Devices

Signal processing devices such as delay lines (dispersive and nondispersive), filters and resonators can be realized with MSW components. As discussed before, these devices require a dc bias magnetic field. The center frequency of operation is dependent on this bias field. The frequency of operation is also proportional to the YIG saturation magnetization (M_s). Thus, by doping YIG films with non-magnetic ions, the value of M_s is reduced, making the frequency of operation lower. For example, in a pure YIG film, a bias field from 250 to 5500 Oe tunes the frequency of operation from 2 to 18 GHz; if the YIG film is doped with Ga, the tuning range is from 0.3 to 2 GHz using a bias field of 10–500 Oe.

A typical practical device is shown in Fig. 4.59[3]. The transducers are fabricated on a dielectric (sapphire and alumina microstrip substrate) with a ground plane. The YIG films are integrated with the dielectric substrate into an electromagnet to supply the dc bias field such that MSSWs are generated. In the following, we discuss some further typical devices.

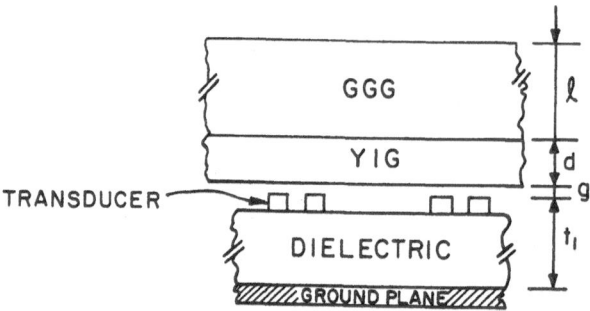

Fig. 4.59. MSSW delay-line schematic: the flip-over configuration. (From [4.28])

a) Nondispersive Delay Lines

Although the MSSW is dispersive, the delay line will be nondispersive for practical purposes if the frequency band of operation is narrow. Typical values are: center frequency 600 MHz, 10 dB bandwidth 470 MHz, insertion loss 12 dB and delay time 100 ns.

b) Dispersive Delay Lines

The most important dispersive delay line is the chirp device, i.e. a delay line exhibiting linear variation of delay with frequency. Chirp devices can be built using the intrinsic dispersion of the waves or using the technique of reflective

---

[3] This flip-over configuration is used by Hewlett-Packard Co.

array compressors (RACs) discussed in Sect. 4.6.1. The intrinsic dispersion can be controlled by placing a proper ground plane in proximity to the YIG film. The typical device time-bandwidth products are about 200–500 with a bandwidth up to 1 GHz in the 2–18 GHz center-frequency range.

For the reflective arrays, etched or ion-milled grooves, metal film patterns and ion-implanted zones have been used as reflecting elements.

c) Band Pass Filters

Microwave filters can be built using interdigital transducers, meander lines or multi-element gratings. Typical bandwidths are 25 MHz at a center frequency of 1 GHz with an insertion loss of about 20 dB using MSSW.

d) Resonators and Oscillators

High-Q oscillators can be built by using either a delay line or a resonator cavity as discussed in Sect. 4.2.3. A typical MSSW resonator is shown in Fig. 4.60. The gratings could be made in different ways to reflect energy, as discussed in connection with the RAC. Typical Qs are in the range of 400–800 with a center frequency of 3 GHz.

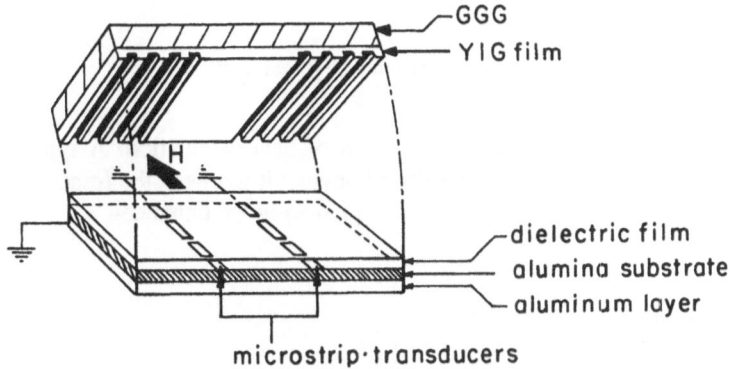

Fig. 4.60. Schematic representation of a MSSW resonator. (From [4.29])

e) Nonlinear Devices

Nonlinear devices include convolvers and frequency selective limiters. A typical convolver using MSFVWs is shown in Fig. 4.61 and its operation is very similar to the SAW convolver discussed earlier. Both MSSWs and MSFVWs have been used for this purpose and were found to have high nonlinear efficiency.

Frequency selective limiters work on the principle of spin wave excitation at a particular frequency and high power in a magnetic insulator immersed in a static bias field. This insulator is placed in the path of the propagating rf wave. At low power levels, the spin wave excitation in YIG is linear, as discussed previously. Above some critical threshold level, spin waves at approximately half the applied frequency are excited due to the spin precession amplitude saturation. These spin

409

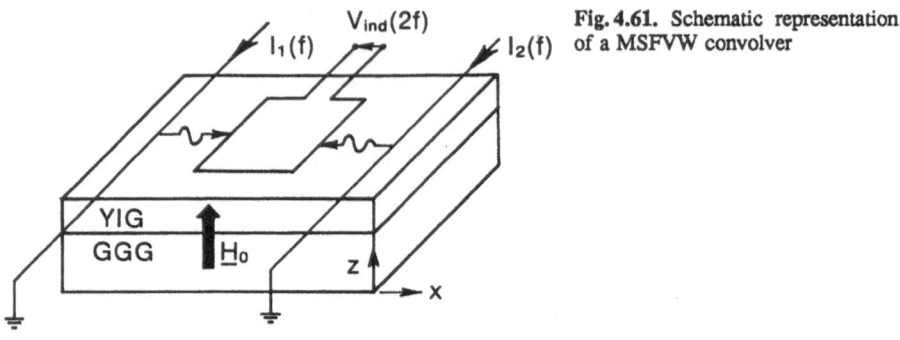

$V_{ind}(2f)$

$I_1(f)$ $I_2(f)$

Fig. 4.61. Schematic representation of a MSFVW convolver

YIG

GGG $\underline{H}_0$

z

x

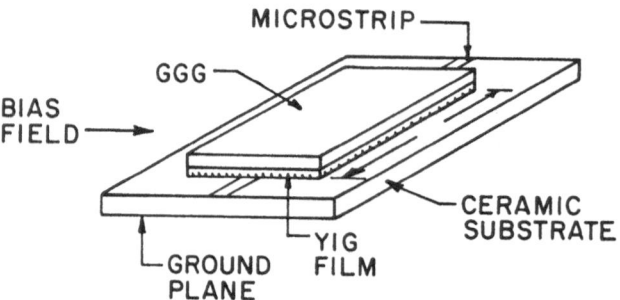

MICROSTRIP

GGG

BIAS FIELD

CERAMIC SUBSTRATE

GROUND PLANE

YIG FILM

Fig. 4.62. Schematic representation of a thin-film YIG limiter. (From [4.30])

waves couple energy to the substrate in the form of heat dissipation. A typical device is shown in Fig. 4.62. This is a broadband device; it works in the frequency region of 2–4 GHz and can also be used as a signal-to-noise enhancer.

4.5 ACT Devices

Acoustic Charge Transport (ACT) technology is a marriage between CCD and SAW technologies, where the transport of charges takes place as in a CCD but with the transporting electric potential induced by a SAW rather than supplied by clock voltages applied to the gates. Thus, for a simple ACT delay line, the device is simply a buried channel GaAs CCD without any gates, as shown in Fig. 4.63. However, a SAW interdigital transducer is provided to generate the SAW. This SAW, as it propagates on the GaAs substrate, induces an electric potential as shown schematically in Fig. 4.63. The substrate is biased such that the surface layer in which the SAW propagates is depleted. The SAW potential augments this depletion at every wavelength and thus forms so-called buckets of charge, which can be transported as the SAW propagates. The input and output of the device are very similar to that of a CCD.

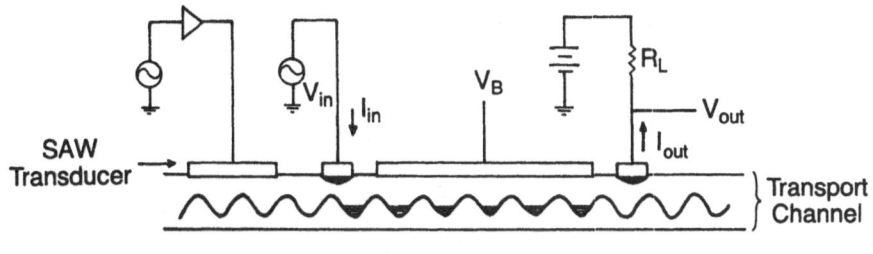

Fig. 4.63. An acoustic charge transport device. (From [4.31])

There is a fundamental difference between ACT and CCD operation, which in some applications may be a limiting factor. In CCDs the speed of charge transport can be controlled and altered at will just by changing the clocking frequency of the gate voltages. For ACT, it is fixed and is the same as the SAW velocity on the substrate. However, an ACT device, if used as a simple delay line, has the enormous fabrication advantage that it requires no gate electrodes, which demand a precise photolithographic process. However, if multiple taps are needed with each tap having some fixed weights for a band-pass or programmable filter, then this advantage is somewhat reduced.

To match the maximum of the electric potential produced by the SAW at the charge transport channel, GaAs epitaxial layers are usually used as shown in Fig. 4.64. The SAW potential profile in (100) cut, $\langle 110 \rangle$ propagating GaAs is shown in Fig. 4.65. For a SAW frequency of 300 MHz, the maximum is at a depth of $-4\,\mu$m. This is also the width of the n-type GaAs layer, which is followed by a thin p-type GaAs layer grown on a semi-insulating GaAs substrate.

Because of the SAW drive limitations, the maximum value of the induced potential is about 1 V, while that for conventional CCD is much higher. This sharply decreases the capacity of the wells to store charge. The transfer efficiency is mostly limited by the diffusion to the nearest neighboring wells. However, an experimental transfer efficiency of 0.99 at 300 MHz has been reported.

For the tapped delay line with fixed weighting, a split electrode structure similar to Fig. 4.40 is used. The applications of ACT devices are similar to those discussed earlier in connection with SAW and CCD devices. Some of these are:

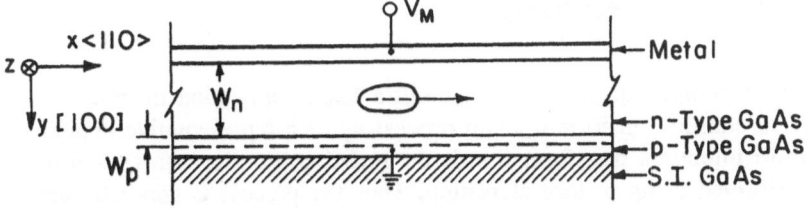

Fig. 4.64. ACT epitaxial layer structure. (From [4.31])

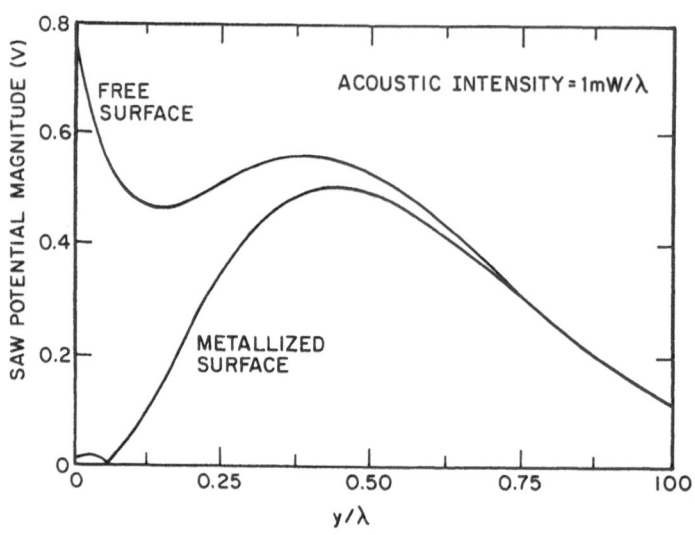

Fig. 4.65. SAW potential profile in (100) cut, ⟨100⟩ propagating GaAs. (From [4.31])

- delay lines

- multiplexers

- demultiplexers

- transversal filters

- recursive filters

- correlators

- convolvers

Currently, ACT devices in GaAs have been demonstrated at a 358 MHz SAW frequency, with a bandwidth of about 50 MHz and a delay of 0.7 μs, corresponding to a 1000 element CCD with charge transfer efficiency exceeding 0.996. A tapped version of the delay line has also been demonstrated using nondestructive floating sense electrodes.

4.6 Comparison of Technologies

It is difficult to compare technologies without a fixed application in mind. We have therefore chosen transform domain processing as a representative application. The most important transformation is the real-time Fourier transformation, which is performed using a chirp algorithm. Thus the process is generally also referred to as a chirp transformation. The heart of the chirp transformation is the

412

chirp device, which can be fabricated using the following major technologies: [4]

1. SAW devices
2. Bulk ultrasound
3. CCDs
4. Acoustically coupled CCDs
5. Magnetostatic surface waves
6. Superconducting tapped delay lines
7. Fiber optic tapped delay lines
8. Acousto-optics
9. Integrated circuits

The quality of a chirp device is classified by the bandwidth, time duration and the time-bandwidth product of the device. Figure 1.10 shows the capabilities of the different technologies. For immediate applications (within the next few years) superconducting, fiber-optic and magneto-surface wave delay lines will probably still not be available.

Superconducting tapped electromagnetic delay lines are like ordinary coaxial cables except that superconducting niobium striplines on a thin dielectric substrate are used for low loss. Compared to acoustic devices, these have very high bandwidth (10 GHz). However, as the velocity is also very high compared to elastic waves, time delays are on the order of 100 ns. A chirp device is shown in Fig. 4.66 where backward-wave couplers between two adjacent electromagnetic delay lines are used in cascade. The devices can be weighted with typical values as follows: center frequency 4 GHz, bandwidth 2.5 GHz, time duration 40 ns, TB product 100, insertion loss 10 dB. Note that the recent discovery of high-T_c superconductors will be a big factor for the practical applications of supeconducting signal processing devices.

Fiber-optic tapped delay lines also use electromagnetic waves in the optical region and have an enormous bandwidth capability (~ 100 GHz). Again, the time duration is quite limited ~ 100 ns) due to the high electromagnetic wave velocity. Different tapping mechanisms have been proposed, although no chirp has been demonstrated experimentally.

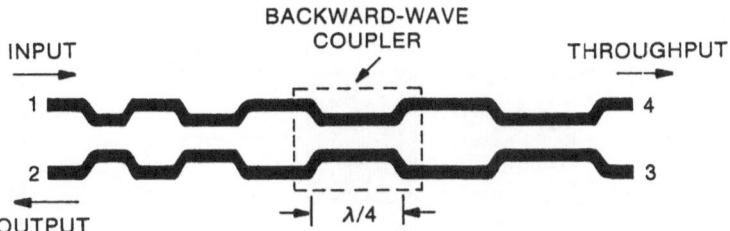

Fig. 4.66. Chirp filter formed by cascading backward-wave couplers between adjacent electromagnetic delay lines

---

[4] We have excluded in this discussion chirp devices in the optical frequency range. These will be discussed in detail in the second volume.

Magneto-surface wave devices (Sect. 4.4) also promise very large center frequencies, ranging from 350 MHz to 30 GHz. The devices are based on propagation of magneto-static waves in epitaxially grown ferrimagnetic thin films such as yttrium iron garnet (YIG). Chirp devices use either the dispersive property of the waves or grating reflectors. Of course one needs a magnetic field for the operation of these devices. Varying the magnetic field provides tunability. Generally, these devices have large losses which somewhat limit their dynamic range.

4.6.1 SAW Technology

For chirp transformation purposes, SAW is the most mature technology. Nearly all the experimental work in connection with chirp transformation has been done using SAW chirp devices. The bandwidth achievable is on the order of 500 MHz and a time duration can be obtained on the order of 80 μs. A TB product of more than 16 000 has been demonstrated experimentally. Chirp devices using SAW technology can be subdivided into three categories:

1. tapped delay lines
2. slanted IDT delay lines
3. reflective array compressors (RACs)

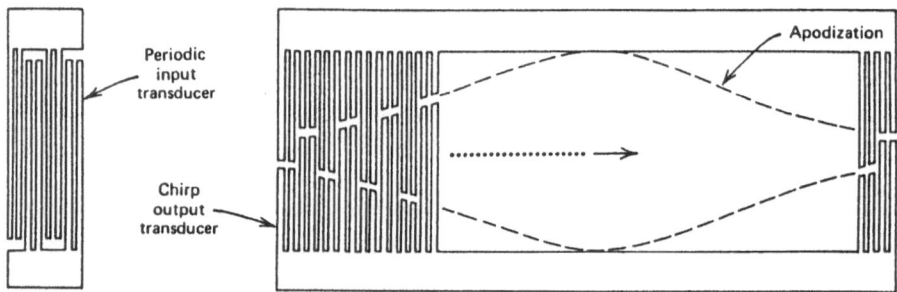

Fig. 4.67. Chirped tapped delay line. Periodic/chirp configuration shown with double electrodes. (From [4.32])

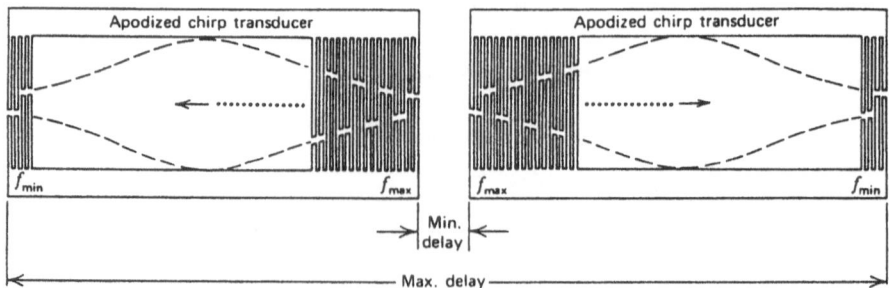

Fig. 4.68. Chirped tapped delay line. Chirp/chirp configuration (down-chirp) with "mirror image" apodization and double electrodes. (From [4.32])

414

A chirped tapped delay line is shown in Fig. 4.67. Here one interdigital transducer (IDT) is chirped as shown. The fingers have double electrodes to reduce reflection. A variation is given in Fig. 4.68, where both the input and output transducers are chirped. Although phase and amplitude linearity are somewhat of a problem for these IDT chirp devices due to Fresnel ripple, they are the easiest to fabricate once the mask is designed and thus the cheapest of the three implementations.

Slanted IDT avoids some of the problems of the tapped delay line discussed above because different frequency waves travel the same length of the IDT. This is shown in Fig. 4.69. The slanted devices need more piezoelectric substrate than the straight IDT ones, although once the mask is properly made, the fabrication cost should not be prohibitive.

A reflective array compressor (RAC) is shown in Fig. 4.70. From this structure, one obtains twice the time duration for the same substrate length compared

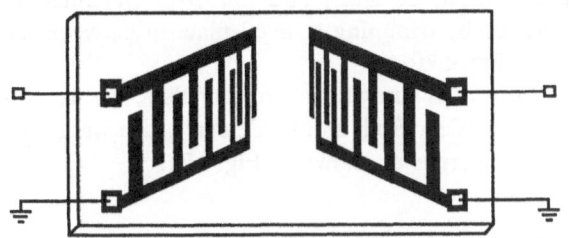

Fig. 4.69. Slanted interdigital transducer. (From [4.32])

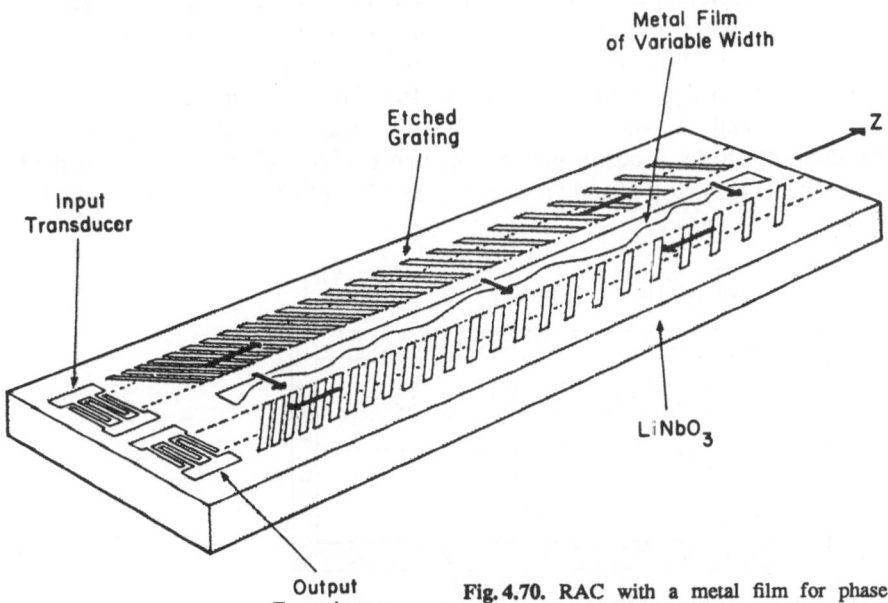

Metal Film of Variable Width

Etched Grating

Z

Input Transducer

LiNbO$_3$

Output Transducer

Fig. 4.70. RAC with a metal film for phase compensation. (From [4.33])

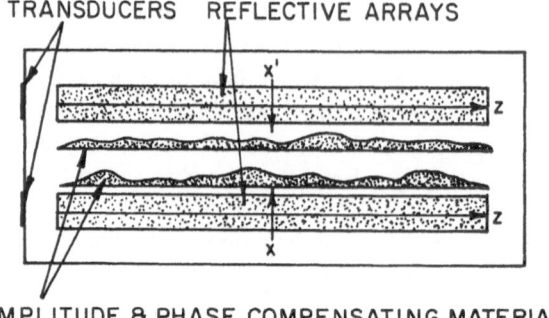

TRANSDUCERS   REFLECTIVE ARRAYS

Fig. 4.71. SAW amplitude and phase compensation

AMPLITUDE & PHASE COMPENSATING MATERIALS

to the other cases. A SAW is generated by the IDT and is selectively reflected by two gratings. These reflection gratings can be either metal fingers or etched or ion-milled grooves. Metal fingers are the easiest to fabricate, however, the ion-milled ones have the highest amplitude and phase linearity. The linearity is, in general, individually improved by trimming a metal plate in between the reflection gratings, as indicated in Fig. 4.70.

In all the SAW implementations, it is straightforward to include the weighting function in the device itself, if needed. Both amplitude and phase weighting can be included by compensating materials as shown in Fig. 4.71.

4.6.2 Bulk Ultrasound Devices

There are different ways one can fabricate chirp devices using bulk ultrasound. These are: (i) use of dispersive modes in a plate, such as the first symmetric Lamb mode in a thin strip of metal, (ii) use of a grating reflection device. The most successful device in this group is "IMCON", manufactured by Anderson Laboratory and shown in Fig. 4.72. IMCON stands for IMpedance CONtrol. By controlling the acoustic impedance, one can control reflection. The device consists of a thin strip of solid material (generally rolled spring steel) in which the

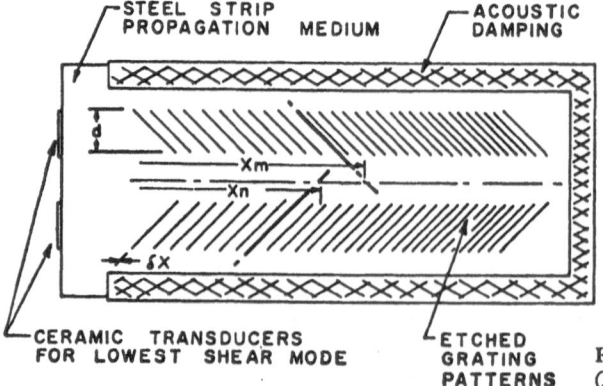

Fig. 4.72. IMCON configuration. (From [4.34])

nondispersive zero-order plate shear mode is excited and received by transducers bonded at the ends. The thickness of the strip is made less than a wavelength so that no other modes are excited. The grating pattern is photolithographically etched on the strip. The periodicity of the grating is varied, so that reflection for a particular frequency occurs only at a particular point. Because steel plates are used, typical time durations are 100–500 μs. However, the bandwidth is limited to about 15 MHz, and the time-bandwidth product is typically 1500–2500. However, by cascading 18 devices, a time duration of 10000 μs, bandwidth of 2.5 MHz and time-bandwidth product of 25 000 have been achieved.

4.6.3 Charge Coupled Devices

Charge coupled devices (Sect. 4.3) deal with discrete-time signals, although they are analog devices. Thus, to have a bandwidth of 10 MHz, the sampling must be at least 20 MHz but in practice 25 MHz is better. The devices are generally two phase, three phase or four-phase. CCDs have been fabricated operating at a clock frequency approaching 4 GHz using GaAs. However, the devices with silicon are usually limited to less than 100 MHz with a typical value of about 50 MHz. For a chirp device, one needs a tapped delay line. Weighting of the taps can be performed by splitting the electrodes for the third phase of each cell, as shown in Fig. 4.40, and taking the difference between them. A different approach is to use charge packet splitting, which has been shown to have some advantages with respect to linearity and dynamic range. A somewhat cumbersome, but effective approach is to use just the tapped delay line outputs from the CCD, while using the external circuits to provide for the weighting function.

For discrete signals, the discrete Fourier transform is used with the chirp algorithm. Implementation of the chirp transform algorithm using CCDs is shown in Fig. 4.73. For this purpose $2N - 1$ CCD stages are needed, where N is the total number of points transformed.

However, if only the power spectrum is important, one can use a sliding Chirp z Transform (CZT). The sliding CZT needs only N CCD stages. Assuming that the spectrum is effectively constant over the interval $2NT_c$, then the sliding CZT is a very good approximation to the discrete Fourier transform. Note also in Fig. 4.75 that, as CCDs operate in baseband, one must work with real and imaginary parts and process them separately using sine and cosine chirps. This is a disadvantage compared to SAW devices.

A 32 point Discrete Fourier Transform (DFT) chip using CCD chirp filters implemented with split finger techniques has been developed by Texas Instruments. This chip contains all of the necessary multipliers, clock, etc. in a single chip. A 500 stage CCD filter for spectral analysis has also been demonstrated by TI using the split finger weighting technique. These devices can be expected to operate up to a sample rate of 10 MHz.

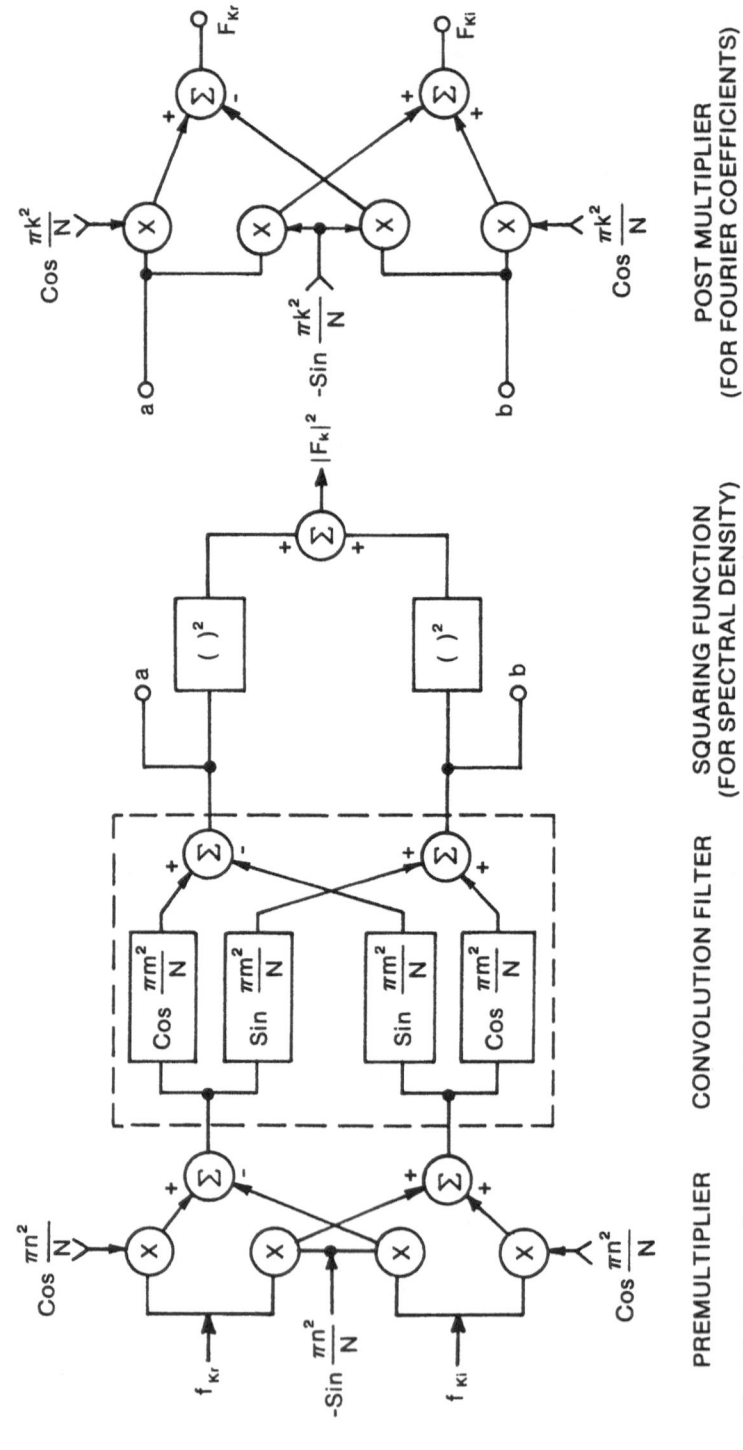

Fig. 4.73. Chirp transform by means of CCDs

PREMULTIPLIER   CONVOLUTION FILTER   SQUARING FUNCTION
(FOR SPECTRAL DENSITY)

POST MULTIPLIER
(FOR FOURIER COEFFICIENTS)

418

4.6.4 Acoustic Charge Transport

See also Sect. 4.5. For ordinary CCDs, charge transport is by the application of sequential periodic gate voltages to the device. For ACT, no gate structure is needed because the periodic potential arises from the acousto-electric voltage generated by the SAW-semiconductor interaction. As the SAW travels, the charge packets travel with the SAW. The fabrication becomes quite simple because only *input* and *output* are needed for a 1000 element delay. For a conventional 4-phase clock CCD, one would need an additional 4000 gates for applying the clock voltages. However, for the conventional CCD, one can change the clock frequency at will and thus vary the delay time, but for ACT/CCD the clock frequency is fixed and given by $1000T$, where T is the time period of the SAW.

ACT in GaAs has been demonstrated at a 358 MHz SAW frequency, a bandwidth of ~ 50 MHz, and a delay of 0.7 μs, corresponding to a 1000 element CCD with charge transfer efficiency exceeding 0.996. A tapped version of the delay line has also been demonstrated using nondestructive floating sense electrodes. Tapped delay lines can easily be fabricated and one can envision using external circuits as weights for building a chirp filter. However, the time duration of the device is a problem because the speed of charge transfer is constrained to be the SAW velocity.

4.6.5 Acousto-optics

Acousto-optic interaction can also be used to obtain real-time Fourier transformation. As shown in Fig. 4.74, the light from the laser source is incident on an acousto-optic modulator. This acousto-optic modulator can be either bulk or SAW and it can operate in either the Bragg or the Raman-Nath regime.

The light from the modulator is Fourier-transformed using a lens and spatially filtered before it is incident on a linear array of photodetectors. Part of the

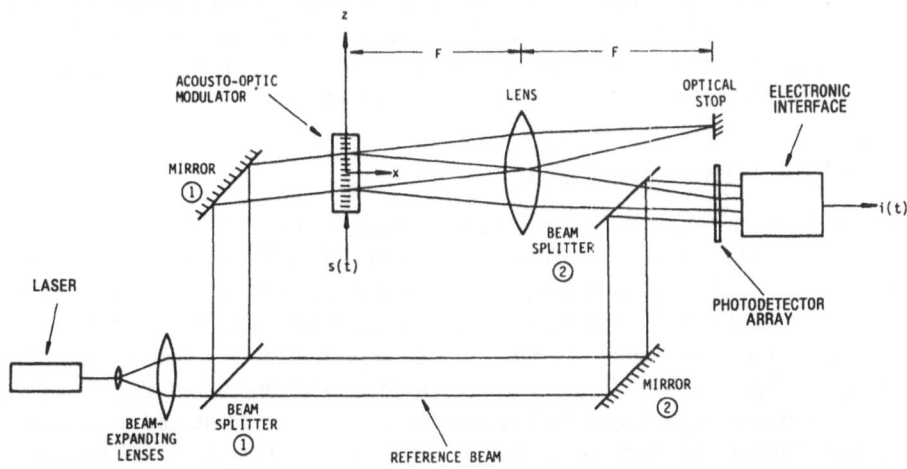

Fig. 4.74. Real-time Fourier transformation using acousto-optic interaction. (From [4.18])

419

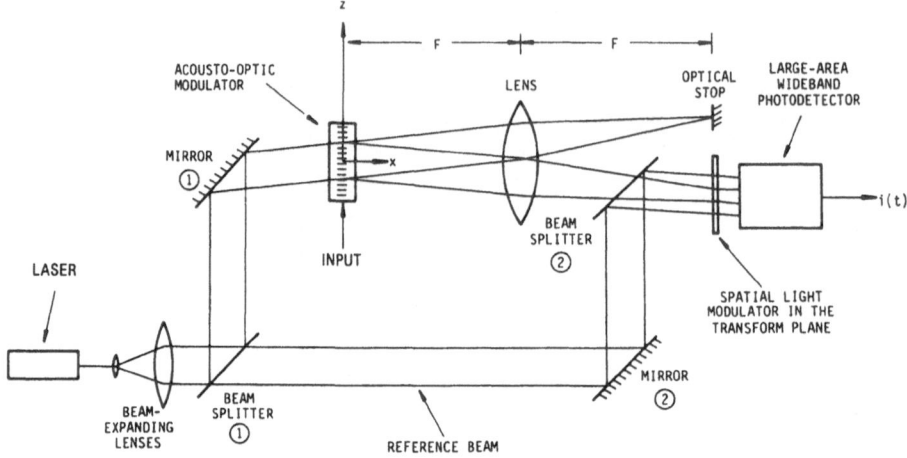

Fig. 4.75. Optical layout for narrowband jammer excision

laser light, split into a separate path, is also incident on the photodetector array so that heterodyne detection can be performed. The output of the photodetector, band-pass filtered around the center frequency, is the Fourier transform of the signal. Although the schematic diagram in the figure is somewhat cumbersome, it could potentially be on a GaAs chip using integrated optics. The whole function, in a single chip, is still to be demonstrated.

The layout shown in Fig. 4.74 can easily be modified to perform narrowband excision, as shown in Fig. 4.75, where the output is a single, wideband, large area photodetector. A mask is placed in front of the photodetector to block off the jamming frequencies. The mask can be programmable if a spatial light modulator is used. Note that the output, after filtering, is in the time domain.

Acousto-optic modulators have been built with bandwidths of a few gigahertz and nominal parameter values of time duration 20 μs and a time-bandwidth (TB) product of 20 000. Note that the TB product is equivalent to the number of points that would be transformed in a DFT or FFT. Thus, a device with a TB product of 2000 effectively performs a 2000-point DFT.

4.6.6 Digital Devices: IC/VHSIC

The IC implementation uses the fast Fourier transform (FFT) algorithm to compute the well-known discrete Fourier transform (DFT). With standard relations between odd and even coefficients, the original N-point DFT can be reduced to two $N/2$-point DFTs. This odd and even decomposition is the key to an effective FFT algorithm. The kernel operation can be represented as a flow diagram as shown in Fig. 4.76. This is the well-known "butterfly" operation, so called because of its shape. An entire FFT regardless of N can be computed by a series of repetitions of the butterfly operation, reuqiring a total of $N \log_2 N$ multiplications.

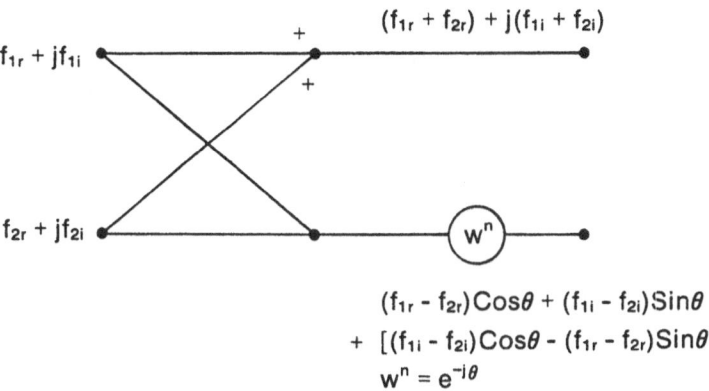

Fig. 4.76. Flow diagram showing the FFT kernel or "butterfly" operation

To implement this "butterfly" using ICs one has many choices. One possibility is silicon technology, which is available at present and is quite mature compared to GaAs technology (which has the advantage of higher speed and thus bandwidth). For silicon technology, IC chips are available which can perform a 1024-point FFT in 19 ms, or one can use Advanced Micro Devices' AM29500, which can do a 1024-point FFT in 2.0493 ms when properly configured. The other alternative is to design custom ICs or use gate array technology to obtain customized ICs where systolic, pipelined or parallel operations can be implemented. Theoretically, by using enough parallel operations (at a proportionally higher cost) one can perform the FFT in one clock cycle, which can be on the order of 10 ns, depending on whether CMOS, bipolar or ECL fabrication technique is used. Some of these special VLSI chips are being fabricated under a VHSIC program by different companies. A typical example is the FFT chip set of 6 large chips fabricated by TRW. It is capable of performing a 512-point complex FFT with better than 12 bit accuracy and a clock speed exceeding 10 MHz ($\sim 5 \times 10^6$ butterflies/s). ITT uses 1.25 μm CMOS technology.

For real valued signals, the Hartley transform is easier to compute and implement in real time. Note that a fast discrete Hartley transform analogous to a FFT exists and computes faster. If a DFT is needed, it can be computed from the Hartley transform quite simply. Recently, it has also been shown that a FFT can be implemented using a Hartley transform which does not require complex arithmetic or storage of complex values.

It is of interest to discuss the relative merits of digital IC versus other implementations. One of the main advantages of digital ICs is that, because one will have to use ICs for all other necessary functions, it is possible to integrate the whole transform domain chip processor in a very small volume using only silicon technology. CCDs and ACT devices will have the same advantage. However, for other technologies, such as SAW devices, one must use piezoelectric materials such as $LiNbO_3$ in addition to silicon ICs. The main disadvantage of digital ICs is the need for high speed A/D converters.

Transformations can also be performed using matrix processors, which can be implemented using digital optical matrix processors or systolic arrays using either silicon ICs or optical methods. However, these might be more suitable for two-dimensional transforms with applications in image processing.

Appendices

A. Matrices

Matrices originated as a notational convenience for writing a set of linear equations and solving them. Let us consider the following set of linear equations which map $Y = (y_1, y_2, \ldots y_n)$ to $X = (x_1, x_2, \ldots x_n)$:

$$y_1 = a_{11}x_1 + a_{12}x_2 + a_{13}x_3 + \ldots + a_nx_n$$
$$y_2 = a_{21}x_1 + a_{22}x_2 + a_{23}x_3 + \ldots + a_{2n}x_n$$
$$\vdots$$
$$y_i = a_{i1}x_1 + a_{i2}x_2 + a_{i3}x_3 + \ldots + a_{in}x_n \qquad\qquad (A.0.1)$$
$$\vdots$$
$$y_n = a_{n1}x_1 + a_{n2}x_2 + a_{n3}x_3 + \ldots + a_{nn}x_n$$

or

$$y_i = \sum_{j=1}^{N} a_{ij}x_j \quad , \qquad i = \text{row label} \quad ,$$
$$\qquad\qquad\qquad\qquad j = \text{column label} \quad , \qquad\qquad (A.0.2)$$

or

$$Y = AX \qquad\qquad (A.0.3)$$

$$
\begin{bmatrix} y_1 \\ y_2 \\ \vdots \\ y_n \end{bmatrix}
=
\begin{bmatrix}
a_{11} & a_{12} & a_{13} \cdots a_{1n} \\
a_{21} & a_{22} & a_{23} \cdots a_{2n} \\
\vdots & & \\
a_{n1} & a_{n2} & a_{n3} \cdots a_{nn}
\end{bmatrix}
\begin{bmatrix} x_1 \\ x_2 \\ x_3 \\ \vdots \\ x_n \end{bmatrix} . \qquad (A.0.4)
$$

The A matrix shown above is a square matrix with n columns and n rows. In general, however, a matrix is defined as a set of numbers consisting of M rows and N columns. If $M = N$, it is called a square matrix. In the example shown Y and X are $N \times 1$ matrices. Any one column matrix is thus a vector. In general, the elements of a matrix will be denoted by a_{ij}. The matrix A refers to the entire assembly of the numbers a_{ij} arranged in a very definite manner in rows and columns. Matrices can be operated algebraically. This means that the important operations of algebra (addition, subtraction, multiplication, division, raising to a power and taking the root) can be extended to matrices under certain conditions.

The two fundamental operations from which everything else can be derived are addition and multiplication. The addition of two matrices $A(m \times n)$ and $B(m \times n)$ forms a new matrix $C(m \times n)$ given by

$$C = A + B \ ,$$
$$C_{ij} = A_{ij} + B_{ij} \ .$$
(A.0.5)

The result of multiplication of a matrix $A(m \times n)$ with a matrix $B(n \times k)$ is a matrix $C(m \times k)$ given by

$$C = AB \quad \text{or}$$
$$C_{ij} = \sum_{k=1}^{n} A_{ik} B_{kj} \ .$$
(A.0.6)

Note that multiplication is defined for conformable matrices only, i.e., A has the same number of columns as B has rows. Only for a square matrix can one define both AB and BA. But note that in general

$$AB \neq BA \ .$$

Customary algebra is based on six fundamental postulates. These are

1. Commutative law of addition: $a + b = b + a$
2. Associative law of addition: $(a + b) + c = a + (b + c)$
3. Commutative law of multiplication: $ab = ba$
4. Associative law of multiplication: $(ab)c = a(bc)$
5. Distributive law of multiplication: $(a + b)c = ac + bc$, $c(a + b) = ca + cb$
6. The nonfactorability of zero: If $ab = 0$, then $a = 0$, or $b = 0$, or $a = b = 0$.

The associative law of multiplication does not apply for matrices. More important is the fact that the property of nonfactorability of zero does not hold good for a matrix. To understand the profound influence of this fact, we first discuss the eigenvalue analysis of a matrix. For a matrix A, consider the equation

$$AX = \lambda X$$
(A.0.7)

where λ is a diagonal matrix of constant diagonal elements. A diagonal matrix λ is defined as

$$\Lambda_{ij} = \lambda_i \quad \text{if } i = j \ ,$$
$$= 0 \quad \text{otherwise} \ .$$
(A.0.8)

One can define an identity matrix I, which is a diagonal matrix whose diagonal elements are equal to 1. Thus a 4×4 identity matrix is given by

$$I = \begin{bmatrix} 1 & 0 & 0 & 0 \\ 0 & 1 & 0 & 0 \\ 0 & 0 & 1 & 0 \\ 0 & 0 & 0 & 1 \end{bmatrix} \ .$$
(A.0.9)

The identity matrix has the unusual property that, for any matrix A,

$$AI = IA = A \quad .$$

(A.0.10)

Thus (A.0.8) can be written as

$$AX = \lambda I X$$

(A.0.11)

where in the above equation λ is a scalar constant.

Equation (A.0.11) written in component form is

$$a_{11}x_1 + a_{12}x_2 + a_{13}x_3 + \ldots + a_{1n}x_n = \lambda x_1$$
$$a_{21}x_1 + a_{22}x_2 + a_{23}x_3 + \ldots + a_{2n}x_n = \lambda x_2$$
$$\vdots$$

(A.0.12)

or

$$(a_{11} - \lambda)x_1 + a_{12}x_2 + a_{13}x_3 + \ldots + a_{1n}x_n = 0$$
$$a_{21}x_1 + (a_{22} - \lambda)x_2 + a_{23}x_3 + \ldots + a_{2n}x_n = 0$$
$$a_{31}x_1 + a_{32}x_2 + (a_{33} - \lambda)x_3 + \ldots + a_{3n}x_n = 0$$
$$\vdots$$

(A.0.13)

The above set of linear equations represent n equations with n unknowns. To have a nontrivial solution we must have the determinant of the system equal to zero:

$$\begin{vmatrix} (a_{11} - \lambda) & a_{12} & a_{13} & \cdots & a_{1n} \\ a_{21} & (a_{22} - \lambda) & a_{23} & \cdots & a_{2n} \\ a_{31} & a_{32} & (a_{33} - \lambda) & \cdots & a_{3n} \\ \vdots & \vdots & \vdots & \vdots & \\ a_{n1} & a_{n2} & a_{n3} & \cdots & (a_{nn} - \lambda) \end{vmatrix} = 0 \quad . \quad \text{(A.0.14)}$$

We could have obtained the same result by writing symbolically

$$(A - \lambda I)X = 0 \quad \text{or}$$

(A.0.15)

$$|A - \lambda I| = 0 \quad .$$

(A.0.16)

If we expand the determinant we obtain the characteristic polynomial in λ of A given by

$$(-1)^n[\lambda^n + C_{n-1}\lambda^{n-1} + \ldots + C_0] = 0 \quad .$$

(A.0.17)

If we ignore the factor $(-1)^n$ we obtain for λ

$$\lambda^n + C_{n-1}\lambda^{n-1} + \ldots + C_0 = 0 \quad ;$$

(A.0.18)

C_n, C_{n-1}, etc. are dependent on a_{ij} values. The solution of (A.0.18) gives us n and only n values of λ:

$$\lambda = \lambda_1, \ldots, \lambda_n \quad .$$

(A.0.19)

These λ_i are called the eigenvalues. For every value of λ_i, we can solve (A.0.7) to obtain $(X)_i$ which we denote by u_i. These u_i are called eigenvectors. The elements of a particular u_i will be given by

$$u_i = x_1^i, x_2^i, x_3^i, \ldots \quad .$$

We can form a new matrix $U(n \times n)$ whose columns are u_i and a diagonal matrix Λ whose diagonal elements are λ_i. It can be easily shown that

$$AU = \Lambda U \quad . \tag{A.0.20}$$

A.1 The Hamilton-Cayley Theorem

The general solution of the equation

$$(A - \lambda_1 I)x = 0 \tag{A.1.1}$$

is $X = \alpha_1 u_1$, where α_1 is an arbitrary constant. Thus the general solution for the equation

$$HX = 0 \quad \text{where}$$
$$H = (A - \lambda_1 I)\ldots(A - \lambda_n I) \tag{A.1.2}$$

is given by

$$X = \alpha_1 u_1 + \ldots + \alpha_n u_n \quad , \tag{A.1.3}$$

where α_i are arbitrary constants. However, for an n-dimensional system, X represented by (A.1.3) represents any arbitrary vector. Thus H must vanish or A must satisfy the Cayley-Hamilton equation given by

$$(A - \lambda_1 I)(A - \lambda_2 I)\ldots(A - \lambda_n I) = 0 \tag{A.1.4}$$

or

$$A^n + C_{n-1}A^{n-1} + \ldots + C_0 I = 0 \quad . \tag{A.1.5}$$

Thus the matrix satisfies its own characteristic equation identically given by (A.0.18).

The existence of the Cayley-Hamilton equation distinguishes matrix algebra from ordinary algebra even when operations with only a single matrix are involved. If x is an ordinary algebraic quantity then any general polynomial of order m cannot be represented by a polynomial of order k where $k < m$. But for a matrix, $A(n \times n)$, any polynomial of order $m > (n - 1)$ can be exactly reduced to a polynomial of not more than $(n - 1)$th degree.

Another important difference concerns the process of division, which is related to the problem of finding the inverse of a matrix. For any matrix A,

$$X^{-1} = (A - \lambda_i I)^{-1} \tag{A.1.6}$$

is not defined. Note that none of the elements of X need be zero. This is quite different from ordinary algebra where $1/x$ is not defined only when x is identically zero.

A.2 Some Definitions

If the rows and columns of a matrix A are interchanged we obtain the transpose of matrix A, denoted by $\tilde{A}$. Thus if

$$A = [a_{ij}] \quad \text{then}$$

$$\tilde{A} = [a_{ji}] \quad . \tag{A.2.1}$$

The matrix A is said to be symmetric if $A = \tilde{A}$. If $A = -\tilde{A}$ it is called skew-symmetric. Note that

$$(ABC)\tilde{} = \tilde{C}\tilde{B}\tilde{A} \quad . \tag{A.2.2}$$

The conjugate A^* of matrix A is given by

$$A^* = [a_{ij}^*] \quad . \tag{A.2.3}$$

Here a_{ij}^* means the complex conjugate of a_{ij}. If all a_{ij} are real then $A = A^*$. However, if all a_{ij} are complex then $A = -A^*$.

The conjugate transpose of A is called the associate matrix of A written as $\tilde{A}^*$. If $A = \tilde{A}^*$, it is called Hermitian. If $A = -\tilde{A}^*$, it is called skew-Hermitian.

The rank r of a matrix A is the size of the largest nonzero determinant that can be formed from A. Thus r is the smaller of m or n. For a square matrix $n \times n$, if $r = n$ then the matrix is nonsingular.

A.3 Matrix Inversion

Consider the linear set of equations

$$Y = AX \quad .$$

In many signal processing problems we are interested in solving the equations for X, or symbolically

$$X = A^{-1}Y \quad . \tag{A.3.1}$$

Thus if we can invert the matrix, we can solve the problem.

If any of the eigenvalues of $A(n \times n)$ are zero then inversion is not defined. In the strict mathematical sense the inversion problem cannot be solved only if A is exactly zero. From a practical standpoint, however, one encounters great difficulty not only if A has a zero eigenvalue but if A has one or more very small eigenvalues. We shall discuss these problems later, together with the problem of inversion when A is not an $n \times n$ matrix but an $m \times n$ matrix.

The matrix A is called a singular matrix if any of its eigenvalues are zero. If the matrix is nonsingular and it has n eigenvalues denoted by λ_i where $i = 1$ to n, then we can define the corresponding U matrix associated with the eigenvectors as discussed previously:

$$AU = \Lambda U \quad . \tag{A.3.2}$$

Similarly we can find for the transpose $\tilde{A}$ of matrix A, the eigenvalues and the corresponding eigenvectors written in the form of matrix V. Thus

$$\tilde{A}V = V\Lambda \quad . \tag{A.3.3}$$

By using the identity

$$\tilde{\tilde{A}} = A \tag{A.3.4}$$

we obtain from (A.3.3), taking the transpose of both sides,

$$\tilde{V}A = \Lambda\tilde{V} \quad . \tag{A.3.5}$$

Note that

$$\tilde{\Lambda} = \Lambda \quad . \tag{A.3.6}$$

Postmultiplying (A.3.5) by U we have

$$\tilde{V}AU = \Lambda\tilde{V}U \quad . \tag{A.3.7}$$

Similarly, from (A.3.2) by premultiplying with $\tilde{V}$, we also obtain

$$\tilde{V}AU = \tilde{V}U\Lambda \quad . \tag{A.3.8}$$

Thus

$$\Lambda\tilde{V}U = \tilde{V}U\Lambda \quad \text{or}$$

$$\Lambda\tilde{V}U - \tilde{V}U\Lambda = 0 \quad \text{or} \tag{A.3.9}$$

$$\Lambda W - W\Lambda = 0 \quad . \tag{A.3.10}$$

We note that

$$\Lambda W = \begin{bmatrix} \lambda_1 w_{11} & \lambda_2 w_{12} & \cdots & \lambda_1 w_{1n} \\ \lambda_2 w_{21} & \lambda_2 w_{22} & \cdots & \lambda_2 w_{2n} \\ \vdots & & & \\ \lambda_n w_{n1} & \lambda_n w_{n2} & \cdots & \lambda_n w_{nn} \end{bmatrix} \tag{A.3.11}$$

and

$$W\Lambda = \begin{bmatrix} \lambda_1 w_{11} & \lambda_2 w_{12} & \cdots & \lambda_n w_{1n} \\ \lambda_1 w_{21} & \lambda_2 w_{22} & \cdots & \lambda_n w_{2n} \\ \vdots & & & \\ \lambda_1 w_{n1} & \lambda_2 w_{n2} & \cdots & \lambda_n w_{nn} \end{bmatrix} \quad . \tag{A.3.12}$$

Equation (A.3.10) can be written as

$$\begin{bmatrix} 0 & (\lambda_1 - \lambda_2)w_{12} & (\lambda_1 - \lambda_3)w_{13} & \cdots & (\lambda_1 - \lambda_n)w_{1n} \\ (\lambda_2 - \lambda_1)w_{21} & 0 & (\lambda_2 - \lambda_3)w_{23} & \cdots & (\lambda_2 - \lambda_n)w_{2n} \\ \vdots & & & & \\ (\lambda_n - \lambda_1)w_{n1} & (\lambda_n - \lambda_2)w_{n2} & (\lambda_n - \lambda_3)w_{n3} & \cdots & 0 \end{bmatrix} = 0 \quad . \tag{A.3.13}$$

Since $\lambda_i \neq \lambda_k$ for $i \neq k$, $w_{ik} = 0$ for $i \neq k$. Thus $W = \tilde{V}U$ must be diagonal. Redefining the W_{ii} values such that they are all equal to 1 we obtain

$$\tilde{V}U = \tilde{U}V = I \quad \text{or} \tag{A.3.14}$$

$$\begin{aligned} \tilde{V} &= U^{-1}, \quad V = \tilde{U}^{-1}, \\ U &= \tilde{V}^{-1} \quad \text{and} \quad \tilde{U} = V^{-1}. \end{aligned} \tag{A.3.15}$$

Also

$$A = U\Lambda\tilde{V} \quad \text{and} \tag{A.3.16}$$

$$\tilde{A} = V\Lambda\tilde{U}. \tag{A.3.17}$$

Now we are in a position to complete the inversion of matrix A:

$$A^{-1} = (U\Lambda\tilde{V})^{-1} \quad \text{or} \tag{A.3.18}$$

$$A^{-1} = \tilde{V}^{-1}\Lambda^{-1}U^{-1} = U\Lambda\tilde{V}. \tag{A.3.19}$$

Note that (A.3.19) can also be obtained starting with

$$A^{-1}X = \lambda X. \tag{A.3.20}$$

A.4 Gaussian Elimination Method

Although the matrix inversion method discussed above is algebraically convenient, for numerical computation better methods exist. One in particular is the Gaussian elimination method, discussed below. In this method we write the matrix equation as

$$Ax - b = 0 \quad \text{or} \tag{A.4.1}$$

$$\begin{aligned} a_{11}x_1 + a_{12}x_2 + a_{13}x_3 \ldots a_{1n}x_n - b_1 &= 0 \\ a_{21}x_1 + a_{22}x_2 + a_{23}x_3 \ldots a_{2n}x_n - b_2 &= 0 \\ &\vdots \\ a_{n1}x_1 + a_{n2}x_2 + a_{n3}x_3 \ldots a_{nn}x_n - b_n &= 0. \end{aligned}$$

Any of the above equations can be multiplied by a factor and added to another equation. The above two properties form the two fundamental operations on which the Gaussian elimination method is based. If we choose the largest element of the matrix, a_{ik}, and divide the ith row by a_{ik} then in the new equation $a_{ik} = 1$. The aim of the method is to make all other elements of the kth column equal to zero by multiplying the new equation by the particular elements and subtracting.

Note that the inverse of the matrix is also obtained at the same time. The Gaussian elimination process gives the inverse of a matrix in the smallest number of operations. However, the rounding errors accumulate and, for a large matrix, the results might be meaningless because of this rounding error accumulation problem. The method of successive orthogonalization of a matrix keeps the rounding errors under control, no matter how large the matrix is.

A.5 Successive Orthogonalization of a Matrix

Let us consider the equation given by

$$Ax = b \quad .$$

(A.5.1)

By writing (A.5.1) in detail we have

$$a_{11}x_1 + a_{12}x_2 + \ldots a_{1n}x_n = b_1$$
$$a_{21}x_1 + a_{22}x_2 + \ldots a_{2n}x_n = b_2$$
$$\vdots$$
$$a_{n1}x_1 + a_{n2}x_2 + \ldots a_{nn}x_n = b_n \quad .$$

We define the u_n vectors having the following n components in an n-dimensional space:

$$u_1 = (a_{11}, a_{21}, a_{31}, \ldots a_{n1})$$
$$u_2 = (a_{12}, a_{22}, a_{32}, \ldots a_{n2})$$
$$u_n = (a_{1n}, a_{2n}, a_{3n}, \ldots a_{nn})$$
$$b = (b_1, b_2, b_3, \ldots b_n) \quad .$$

(A.5.2)

Equation (A.5.1) can now be interpreted as finding a linear combination of the u_n vectors such that the resultant vector is b:

$$x_1 u_1 + x_2 u_2 + \ldots + x_n u_n = b \quad .$$

(A.5.3)

To solve this problem, one constructs the adjoint set $v_1, v_2, \ldots, v_n$. The rows of the inverse matrix are given by this adjoint set of vectors. If the base vectors are orthogonal then

$$u_i \cdot u_k \begin{cases} = 0, & i \neq k \quad , \\ = 1, & i = i \end{cases} \quad .$$

(A.5.4)

For this self-adjoint system we have

$$v_i = u_i \quad .$$

(A.5.5)

The inverse matrix for this case is simply the transpose of the original matrix

$$A^{-1} = \tilde{A} \quad .$$

The method of successive orthogonalization uses an auxiliary orthogonal reference system. The problem is solved in this new orthogonal system and then returned to the original frame of reference.

The first vector of the orthogonal system is chosen as

$$w_1 = \frac{u_1}{\sqrt{u_1 \cdot u_1}} \quad .$$

(A.5.6)

w_2 is chosen in the plane of the first two vectors u_1 and u_2 and orthogonal to w_1. The process is repeated n times, to obtain w_n all of length 1.

In general, each new vector is constructed as follows:

a)   First construct w_i' given by

$$w_i' = u_i - p_{i1}w_1 - p_{i2}w_2 - \ldots - p_{ii-1}w_{i-1} \quad . \tag{A.5.7}$$

The coefficients p_{ik} are given by

$$p_{ik} = u_i \cdot w_k \quad .$$

b)   Normalize w_i', i.e.,

$$w_i = \frac{w_i'}{|w_i' \cdot w_i'|^{1/2}} = \frac{w_i'}{|p_{ii}|} \tag{A.5.8}$$

where $p_{ii} = |w_i'|$.

We note that

$$u_i = p_{i1}w_1 + p_{12}w_2 + \ldots + p_{ii}w_i \quad . \tag{A.5.9}$$

It is observed that the elements p_{ik} form a triangular matrix given by

$$P = \begin{vmatrix} p_{11} & 0 & 0 & 0 & 0 & \cdots \\ p_{21} & p_{22} & 0 & 0 & 0 & \cdots \\ p_{31} & p_{32} & p_{33} & 0 & 0 & \cdots \end{vmatrix} \quad . \tag{A.5.10}$$

We also note the following matrix equation:

$$A = W\tilde{P} \quad , \tag{A.5.11}$$

where W is given by

$$W = \begin{bmatrix} w_{11} & w_{12} & w_{13} & \cdots & w_{1n} \\ w_{21} & w_{22} & w_{23} & \cdots & w_{2n} \\ \vdots & & & & \\ w_{n1} & w_{n2} & w_{n3} & \cdots & w_{nn} \end{bmatrix} \quad . \tag{A.5.12}$$

By construction

$$W\tilde{W} = I \quad \text{or} \quad W^{-1} = \tilde{W} \quad . \tag{A.5.13}$$

If we define

$$Q = P^{-1} \tag{A.5.14}$$

then we have

$$A^{-1} = \tilde{Q}\tilde{W} \quad . \tag{A.5.15}$$

From (A.5.11) and (A.5.13) we note that

$$\tilde{A}A = (W\tilde{P})W\tilde{P} = P\tilde{W}W\tilde{P} = P\tilde{P} \quad . \tag{A.5.16}$$

As $\tilde{A}A$ is symmetric and positive-definite we see that any positive-definite matrix can be written as a multiplication of a triangular matrix P and its transpose $\tilde{P}$.

This is a very important property and is used often in connection with correlation matrices. Thus if R is a correlation matrix, using (A.5.14) we obtain

$$R = P\tilde{P} \quad \text{and}$$

$$(R)^{-1} = (P\tilde{P})^{-1} = (\tilde{P})^{-1}P^{-1} = \tilde{Q}Q \quad . \tag{A.5.17}$$

We shall discuss shortly how to obtain a Q matrix or how to invert a triangular matrix, but first we consider the problem of rounding errors. To ensure that w_i' is orthogonal to a previous w_k even if the previous vectors w_k are not orthogonal, the least squares principle must be used. The method consists of choosing p_{ik} of (A.5.8) by minimizing the length of w_i', i.e.,

$$|u_i - p_{i1}w_1 - p_{i2}w_2 - \ldots - p_{i\,i-1}w_{i-1}|^2 = \text{minimum} \quad . \tag{A.5.18}$$

The above condition gives the following equations:

$$p_{i1}|w_1|^2 + p_{i2}w_1 \cdot w_2 + \ldots + p_{i\,i-1}w_1 \cdot w_{i-1} = u_i w_1 \quad .$$
$$p_{i1}w_2 \cdot w_1 + p_{i2}|w_2|^2 + \ldots + p_{i\,i-1}w_2 \cdot w_{i-1} = u_i w_2 \quad .$$
$$p_{i1}w_{i-1} \cdot w_1 + p_{i3}w_{i-1} \cdot w_2 + \ldots + p_{i\,i-1}|w_{i-1}|^2 = u_i w_{i-1} \quad . \tag{A.5.19}$$

As the w_k vectors are nearly orthogonal, the above equations can be solved easily in an approximate fashion to obtain w_i. Once w_i is obtained, one checks the orthogonality using

$$w_i \cdot w_k = \varepsilon_{ik} \quad (k = 1, 2, \ldots, i-1)$$
$$= 1 + \varepsilon_{ii} \quad (k = i) \quad . \tag{A.5.20}$$

As these ε_{ik} are small, the p_{ik}'s are corrected as follows. First obtain the approximate value

$$p_{ik}' = u_i w_k \quad (k < i) \quad . \tag{A.5.21}$$

The corrected p_{ik} is then obtained by

$$p_{ik} = p_{ik}' - (\varepsilon_{ki}p_{i1}' + \ldots \varepsilon_{k\,i-1}p_{i\,i-1}') \quad . \tag{A.5.22}$$

The inversion of the triangular matrix can easily be accomplished by noting that

$$u_1 = p_{11}w_1 \quad ,$$
$$u_2 = p_{21}w_1 + p_{22}w_2 \quad ,$$
$$\vdots$$
$$u_n = p_{n1}w_1 + p_{n2}w_2 + \ldots + p_{nn}w_n \quad . \tag{A.5.23}$$

We note that the first equation is easily solvable for w_1. Once w_1 is known, the second equation can be used to obtain w_2, and so on. The inverse can be written as

$$w_1 = q_{11}u_1 \quad ,$$

$$w_2 = q_{21}u_1 + q_{22}u_2 \quad ,$$

$$\vdots$$

$$w_n = q_{n1}u_1 + q_{n2}u_2 + \ldots + q_{nn}u_n \quad . \tag{A.5.24}$$

Thus the inverse of a triangular matrix is also a triangular matrix. In place of the above scheme, the following numerical algorithm is more efficient. We note that $q_{ii} = p_{ii}^{-1}$. We thus write a square matrix as follows:

$$\begin{bmatrix} q_{11} & p_{21} & p_{31} & \cdots & p_{n1} \\ q_{21} & q_{22} & p_{32} & \cdots & p_{n2} \\ q_{31} & q_{32} & q_{33} & \cdots & p_{n3} \\ \vdots & & & & \\ q_{n1} & q_{n2} & q_{n3} & \cdots & q_{nn} \end{bmatrix} . \tag{A.5.25}$$

To obtain q_{ik} $(i > k)$ use the rows i and k. Start from the pivotal element q_{ii} and use the elements of the lower triangle for the ith row and the elements of upper triangle for the kth row:

$$q_{ik} = -[q_{ii}p_{ik} + q_{i\,i-1}p_{i-1\,k} + \ldots]q_{kk} \quad . \tag{A.5.26}$$

The multiplication continues until the pivotal element q_{kk} is reached.

A.6 Circulant Matrices and Fourier Matrices

A circulant matrix is one which is obtained by starting with an arbitrary first row and obtaining the rest of the rows by a simple circular shift of the previous row by a single element. For an example, the following matrix $[C]$ is circulant:

$$C = \begin{bmatrix} c(0) & c(1) & c(2) & \cdots & c(N-1) \\ c(N-1) & c(0) & c(1) & \cdots & c(N-2) \\ \vdots & & & & \\ c(1) & c(2) & c(3) & \cdots & c(0) \end{bmatrix} . \tag{A.6.1}$$

Note that by definition a circulant matrix is Toeplitz. However, not every Toeplitz matrix is a circulant matrix.

A Fourier matrix $[F]$ is defined in the following fashion. Consider N, a fixed integer which is greater than or equal to 1. We define

$$W = e^{j2\pi/N} = \cos\frac{2\pi}{N} + j\,\sin\frac{2\pi}{N} \quad . \tag{A.6.2}$$

Thus we have

$$W^N = 1 \quad ,$$

$$WW^* = e^0 = 1 \quad ,$$

$$W^* = e^{-j2\pi/N} = W^{-1} \quad ,$$

$$(W^*)^k = e^{-j2\pi k/N} = W^{-k} = W^{N-k} \quad ,$$

$$1 + W + W^2 + \ldots + W^{N-1} = \frac{1 - W^N}{1 - W} = 0 \quad . \tag{A.6.3}$$

The matrix F is given by

$$F = \frac{1}{\sqrt{N}} \begin{bmatrix} 1 & 1 & 1 & 1 & \cdots & 1 \\ 1 & W^{-1} & W^{-2} & W^{-3} & \cdots & W^{-(N-1)} \\ 1 & W^{-2} & W^{-4} & W^{-6} & \cdots & W^{-2(N-1)} \\ \vdots & & & & & \\ 1 & W^{-(N-1)} & W^{-2(N-1)} & W^{-3(N-1)} & \cdots & W^{-(N-1)(N-1)} \end{bmatrix}$$

or

$$F_{ij} = \frac{1}{\sqrt{N}} W^{-(j-1)(i-1)} \quad . \tag{A.6.4}$$

Similarly one defines F^*, the conjugate and transpose Fourier matrix, whose ijth element is given by

$$F_{ij}^* = \frac{1}{\sqrt{N}} W^{(i-1)(j-1)} \tag{A.6.5}$$

or

$$F^* = \frac{1}{\sqrt{N}} \begin{bmatrix} 1 & 1 & 1 & 1 & \cdots & 1 \\ 1 & W & W^2 & W^3 & \cdots & W^{N-1} \\ 1 & W^2 & W^4 & W^6 & \cdots & W^{2(N-1)} \\ \vdots & & & & & \\ 1 & W^{N-1} & W^{2(N-1)} & W^{3(N-1)} & \cdots & W^{(N-1)(N-1)} \end{bmatrix}$$

Note that

$$FF^* = F^*F = I \quad \text{or}$$
$$F^* = F^{-1} \quad . \tag{A.6.6}$$

If any matrix g which is a column matrix (i.e., $N \times 1$) is postmultiplied with a F matrix, we obtain the discrete Fourier transform G of g. This is obvious from the discussion in Sect. 3.2.3. Thus

$$G = Fg \quad . \tag{A.6.7}$$

Similarly by multiplying with F^{-1} or F^* one obtains the inverse discrete Fourier transform. Note

$$F^{-1}G = F^{-1}Fg = g \quad . \tag{A.6.8}$$

A very important property of the circulant matrix is that it can be diagonalized with the Fourier matrix. Thus

$$C = F^* \Lambda F \quad . \tag{A.6.9}$$

where Λ is a diagonal matrix with diagonal elements λ_i, $i = 0, \ldots, N-1$. It can be shown that, if λ_i is written as a column matrix $\lambda(N \times 1)$, then

$$\lambda = FC_r \quad , \tag{A.6.10}$$

where C_r is the column matrix whose elements are the first row of C. Thus (A.6.10) is written in full as

$$\begin{bmatrix} \lambda_0 \\ \lambda_1 \\ \lambda_2 \\ \vdots \\ \lambda_{n-1} \end{bmatrix} = \begin{bmatrix} 1 & 1 & 1 & 1 & \cdots \\ 1 & W^{-1} & W^{-2} & W^{-3} & \cdots \\ 1 & W^{-2} & W^{-4} & W^{-6} & \cdots \\ \vdots & & & & \\ 1 & W^{-(N-1)} & W^{-2(N-1)} & & \cdots \end{bmatrix} \begin{bmatrix} c_0 \\ c_1 \\ c_2 \\ \vdots \\ c_{N-1} \end{bmatrix} . \tag{A.6.11}$$

Note that to obtain the inverse C^{-1} of C one uses the following relationship:

$$\begin{aligned} C^{-1} &= (F^* \Lambda F)^{-1} \\ &= F^{-1} \Lambda^{-1} F^{*-1} = F^* \Lambda^{-1} F \quad . \end{aligned} \tag{A.6.12}$$

Let us consider a system with input $F_p(N \times 1)$, output $G_p(N \times 1)$ and the system matrix given by $H_p(N \times N)$. If H_p is circulant we can write

$$H_p = F^* \Lambda F \quad . \tag{A.6.13}$$

Thus

$$\begin{aligned} G_p &= H_p F_p \\ &= F^* \Lambda F F_p \\ &= F^{-1}(\Lambda(F F_p)) \quad \text{or} \end{aligned} \tag{A.6.14}$$

$$F F_p = \Lambda^{-1} F G_p \quad . \tag{A.6.15}$$

Note that $F F_p$ is the discrete Fourier transform of the input F_p. The matrix Λ is the eigenvalue matrix of the circulant matrix and thus can be obtained by the discrete Fourier transform of the first row of the circulant matrix. Thus (A.6.15) is similar to that obtained in Sect. 3.1.2, i.e.,

$$g(t) = \mathcal{F}^{-1}[F(f)h(f)] \quad .$$

However, note that in general H_p is not circulant, but for the time-invariant case it is Toeplitz. If H_p is Toeplitz, $N \times N$ and banded, then the following algorithm called "circulant decomposition" can be used to invert H_p. Let H_p be given as ($p, q \ll N$ and $h_q \neq 0$ and $h_{-p} \neq 0$)

$$H_p = \begin{bmatrix} h_0 & h_{-1} & \cdots & h_{-p} & 0 & \cdots & 0 \\ h_1 & & & & & & \vdots \\ \vdots & & & & & & 0 \\ h_q & & & & & & h_{-p} \\ 0 & & & & & & \vdots \\ \vdots & & & & & & h_{-1} \\ 0 & \cdots & 0 & h_q & \cdots & h_1 & h_0 \end{bmatrix} . \tag{A.6.16}$$

Let $k = \max(p, q)$, then one can use the circulant matrix H_c, extended from H_p:

$$H_c = \begin{bmatrix} \overset{\longleftarrow \quad N \quad \longrightarrow}{} & & & & & | & \overset{\longleftarrow \; k \; \longrightarrow}{} & \\ h_0 & h_{-1} \cdots h_{-p} & 0 \cdots 0 & & & | & h_q & \cdots & h_1 \\ h_1 & & & & & | & & & \vdots \\ \vdots & & h_p & & & | & & & h_q \\ h_q & & & & & | & & & 0 \\ 0 & & & & & | & & & \vdots \\ \vdots & & & & & | & & & 0 \\ 0 & & & & & | & & & h_{-p} \\ h_{-p} & & & & & | & & & \vdots \\ \vdots & & & & & | & & & h_{-1} \\ h_{-1} & \cdots h_{-p} & 0 \cdots 0 & & & | & h_q & \cdots & h_1 h_0 \end{bmatrix} . \quad (A.6.17)$$

The inverse of the H_c matrix can be easily obtained using the discrete Fourier transform and (A.6.12). It can also be shown that

$$H_c^{-1} = \begin{bmatrix} \overset{\longleftarrow N \longrightarrow}{} & & \overset{\longleftarrow k \longrightarrow}{} \\ B_{11} & | & B_{12} \\ - - - & & - - - \\ B_{21} & | & B_{22} \end{bmatrix} \begin{matrix} \uparrow \\ N \\ \downarrow \\ \uparrow \\ k \\ \downarrow \end{matrix} , \quad (A.6.18)$$

where B_{22} is a new Toeplitz matrix but of smaller size, as $k < N$. To obtain H_p^{-1}, we note

$$H_p^{-1} = B_{11} - B_{12} B_{22}^{-1} B_{21} . \quad (A.6.19)$$

Thus the advantage of using (A.6.19) to obtain H_p^{-1} is that one need only invert a much smaller matrix, $B_{22}(k \times k)$.

The last equation is obtained by using the method of matrix inversion by partitioning. For example, for a matrix A defined by submatrices A_1, A_2, B_1 and B_2

$$[A]_{(m_1+m_2),(m_1+m_2)} = \begin{bmatrix} \overset{\longleftarrow m_1 \longrightarrow}{} & & \overset{\longleftarrow m_2 \longrightarrow}{} \\ A_1 & | & B_1 \\ - - - & & - - - \\ A_2 & | & B_2 \end{bmatrix} \begin{matrix} \uparrow \\ m_1 \\ \downarrow \\ \uparrow \\ m_2 \\ \downarrow \end{matrix} , \quad (A.6.20)$$

the inverse is given by

$$[A^{-1}]_{(m_1+m_2),(m_1+m_2)} = \begin{bmatrix} \overset{\leftarrow m_1 \rightarrow}{A_1^{-1} + A_1^{-1}B_1C_2} & \vdots & \overset{\leftarrow m_2 \rightarrow}{-C_1} \\ -\,-\,-\,-\,-\,-\,-\,- & \vdots & -\,-\,- \\ -C_2 & \vdots & \overline{B}_2^{-1} \end{bmatrix} \begin{matrix} \uparrow \\ m_1 \\ \downarrow \\ \uparrow \\ m_2 \\ \downarrow \end{matrix} \qquad (A.6.21)$$

where

$$C_1 = A_1^{-1}B_1\overline{B}_2^{-1} \quad , \qquad (A.6.22)$$

$$C_2 = \overline{B}_2^{-1}A_2 \quad \text{and} \qquad (A.6.23)$$

$$\overline{B}_2 = B_2 - A_2 A_1^{-1}B_1 \quad . \qquad (A.6.24)$$

A.7 Pseudo-Inverse, Singular-Value Decomposition, Overdetermination and Principle of Least Squares: Kalman Filtering

We have already discussed the solution of the matrix equation

$$Ax = Y \quad , \qquad (A.7.1)$$

where A is an $n \times n$ matrix, x is $N \times 1$ and y is $n \times 1$. As long as A is not singular, i.e., if the eigenvalues are all nonzero, then a unique solution exists. However, in many practical applications, especially in signal processing, A is often singular. Sometimes A is also overdetermined, i.e., A is a $m \times n$ matrix with $m > n$. In this case, the usual matrix inversion method discussed before cannot be applied directly and actually an inverse does not exist. However, one can always define a useful pseudo-inverse. The pseudo-inverse and singular-value decomposition techniques are different names for the solution of the same problem. We shall use the principle of least mean square error to derive the results. However, before we do that we quote from *Lanezos* [A.1] to show the importance of this problem.

The difficulties inherent in many large-scale linear systems are comparable to the difficulties of an orator who in his speech tries to cover too large a variety of items. In the beginning, his speech goes on rather fluently. However, as he checks off more and more items on his list, he becomes more and more tired and occasionally loses track of his thoughts. He does not remember all the incidents he wanted to tell at the right moment and thus omits certain items and repeats instead with different words the things he has previously said. Since he does not stick very closely to the truth, he comes into contradictions by forgetting in what direction he slanted the story in his earlier remarks (violating the old oratorical principle *Mendacem oportet esse memorem* – The liar should have a good memory). In the last ten minutes his mind goes completely blank, he garbles everything, and finally sits down to thunderous applause. The liar of bad memory of our story is the "noise" which interferes with the accuracy of our measurements and distorts

437

the true course of events. Since noise is of a random nature, the distortion is not consistent but occurs once in one, once in the other direction. There is *one* danger encountered in large-scale recordings of physical events. The *other* danger is that the information we have at our disposal is *insufficient* for actual determination of all the unknowns of the problem. In our story the speaker omitted to comment on certain items of his journey and replaced these comments by retelling with different words certain episodes on which he commented before. In analogy to this situation, it can happen (and it frequently does happen) that the statements of our system of equations are insufficient for complete determination of all the unknowns of our problem. We count the number of equations and find that we have just as many equations as unknowns. Hence we think that our system is balanced and allows a unique solution. Yet it can happen that certain equations merely repeat in different words the statements made before, without adding anything essentially new to the previous statements. In this case our system is underdetermined and not in the position to yield a complete solution to our problem.

To obtain a solution to the problem

$$AX - Y = 0 \quad ,$$

where A is a $m \times n$ matrix with $m > n$, we form the mean square error given by

$$\tilde{\varepsilon}\varepsilon = (AX - \tilde{Y})(AX - Y) \quad . \tag{A.7.2}$$

The minimum of the above equation is obtained by differentiating with X and setting the result equal to zero:

$$|\varepsilon|^2 = \tilde{\varepsilon}\varepsilon = \tilde{X}\tilde{A}AX - \tilde{X}\tilde{A}Y - \tilde{Y}AX - \tilde{Y}Y \quad ,$$

$$\nabla_X |\varepsilon|^2 = \tilde{A}AX - \tilde{A}Y = 0 \tag{A.7.3}$$

or

$$X = (\tilde{A}A)^{-1}\tilde{A}Y \quad . \tag{A.7.4}$$

Note that $\tilde{A}A$ is a square matrix ($n \times n$). As

$$(\tilde{A}A) = \tilde{A}A \tag{A.7.5}$$

it is always symmetric and thus positive-definite and thus has nonzero positive real eigenvalues. The matrix $(\tilde{A}A)^{-1}\tilde{A}$ is a particular example of a pseudo-inverse.

Similarly, for the underdetermined case when $m < n$, one can obtain

$$X = \tilde{A}^* = (A\tilde{A}^*)^{-1}Y \quad . \tag{A.7.6}$$

The matrix $\tilde{A}^*(AA^*)^{-1}$ is also called the pseudo-inverse, and is applicable for the case $m < n$. Note that $A\tilde{A}^*$ is also symmetric and positive-definite.

In solving an actual problem, one first obtains the eigenvectors and eigenvalues of $\tilde{A}A$ and $A\tilde{A}$. Thus

$$(\tilde{A}A)P = \Lambda_n P \quad \text{and} \tag{A.7.7}$$

$$(A\tilde{A})Q = \Lambda_m Q \quad . \tag{A.7.8}$$

Note that P is an $m \times m$ matrix and Q is an $n \times n$ matrix, $\tilde{P}P = I = \tilde{Q}Q$, and Λ_n is an $n \times n$ diagonal matrix with diagonal elements λ_i^2, $i = 1, \ldots, k$, where k is the rank of A. It is also assumed that $\lambda_1 > \lambda_2 > \ldots > \lambda_k$. All other elements are zero. Also

$$\tilde{A}A = P\Lambda_n \tilde{P} \quad ,$$
$$A\tilde{A} = Q\Lambda_n \tilde{Q} \quad , \tag{A.7.9}$$

$$A = Q\Lambda^{1/2}\tilde{P} \quad \text{and}$$
$$\tilde{A} = P\Lambda^{1/2}\tilde{Q} \quad . \tag{A.7.10}$$

Here $\Lambda^{1/2}$ is an $m \times n$ matrix with only the first k diagonal elements nonzero. Thus the optimum solution in the least squares sense is given by

$$\begin{aligned} X &= (\tilde{A}A)^{-1}\tilde{A}Y \\ &= (P\Lambda\tilde{P})^{-1}P\Lambda^{1/2}\tilde{Q}Y \\ &= P\Lambda^{-1}P^{-1}P\Lambda^{1/2}\tilde{Q}Y \\ &= P\Lambda^{-1/2}\tilde{Q}Y \quad . \end{aligned} \tag{A.7.11}$$

The advantage of the singular value decomposition method is that if the eigenvalues are very nearly zero they can be discarded. This makes the solution dependent on $1/\sqrt{\lambda_i}$ stable even when $\lambda_i \to 0$.

In an actual problem, it is not necessary to determine both the P and Q matrices. For a particular λ_i, one can be determined from the other using the two equations

$$AP_i = \sqrt{\lambda_i}Q_i \quad \text{and}$$

$$\tilde{A}Q_i = \sqrt{\lambda_i}P_i \quad .$$

Instead of minimizing $\tilde{\varepsilon}\varepsilon$ it is sometimes convenient to minimize a weighted version of the error given by $Q\tilde{\varepsilon}Q\varepsilon$ where Q is a symmetric $m \times m$ nonsingular and often diagonal matrix. For this case, the solution is given by

$$X = (\tilde{A}QA)^{-1}\tilde{A}QY \quad . \tag{A.7.12}$$

Thus the results discussed before are directly applicable. The only change is to solve for the eigenvalues and eigenvectors for $(\tilde{A}QA)$ rather than $\tilde{A}A$.

A very important case of the weighted least squares solution is the one which is recursive. For this case, let us consider that at the kth iteration a solution has been obtained given by

$$x_k = (\tilde{A}QA)^{-1}\tilde{A}QY_k \tag{A.7.13}$$

for the equation

$$Y_k = AX + \varepsilon_k \quad . \tag{A.7.14}$$

Let us also assume that an additional set of equations given by

$$Y_{k+1} = H_{k+1}X + \varepsilon_{k+1} \tag{A.7.15}$$

has become available at the $(k+1)$th iteration. To obtain the new optimum solution at the $(k+1)$th iteration given by x_{k+1} we must minimize J, given by

$$J = [\tilde{\varepsilon} \vdots \tilde{\varepsilon}_{k+1}] \begin{bmatrix} Q & 0 & \vdots & \varepsilon \\ 0 & Q_{k+1} & \vdots & \varepsilon_{k+1} \end{bmatrix} \quad . \tag{A.7.16}$$

Here Q_{k+1} is the new weighting matrix. The new x_{k+1} is given by

$$x_{k+1} = \left\{ \tilde{B} \begin{bmatrix} Q & \vdots & 0 \\ 0 & \vdots & Q_{k+1} \end{bmatrix} B \right\}^{-1} [\tilde{B}] \begin{bmatrix} Q' & \vdots & 0 \\ 0 & \vdots & Q_{k+1} \end{bmatrix} \begin{bmatrix} Y_k \\ Y_{k+1} \end{bmatrix} \quad , \tag{A.7.17}$$

where

$$\begin{bmatrix} Y_k \\ Y_{k+1} \end{bmatrix} = \begin{bmatrix} A \\ H_{k+1} \end{bmatrix} [X] + \begin{bmatrix} \varepsilon \\ \varepsilon_{k+1} \end{bmatrix} = [B]X + \begin{bmatrix} \varepsilon \\ \varepsilon_{k+1} \end{bmatrix} \quad , \tag{A.7.18}$$

$$B = \begin{bmatrix} A \\ H_{k+1} \end{bmatrix} \quad \text{and} \tag{A.7.19}$$

$$B^{\mathrm{T}} = [A^{\mathrm{T}}; H_{k+1}^{\mathrm{T}}] \quad . \tag{A.7.20}$$

We note that

$$X_{k+1} = [\tilde{A}QA + \tilde{H}_{k+1}Q_{k+1}H_{k+1}]^{-1}[\tilde{A}QY_k + \tilde{H}_{k+1}Q_{k+1}Y_{k+1}] \quad . \tag{A.7.21}$$

Defining $P_k^{-1} = A^{\mathrm{T}}QA$ we obtain

$$X_{k+1} = [P_k^{-1} + \tilde{H}_{k+1}Q_{k+1}H_{k+1}][\tilde{A}QY_k + \tilde{H}_{k+1}Q_{k+1}Y_{k+1}] \quad . \tag{A.7.22}$$

Using the matrix identity given by

$$[P^{-1} + \tilde{H}QH]^{-1} = P - P\tilde{H}[HP\tilde{H} + Q^{-1}]^{-1}HP \quad , \tag{A.7.23}$$

we obtain

$$X_{k+1} = \{P_k - P_k\tilde{H}_{k+1}[H_{k+1}P_k\tilde{H}_{k+1} + Q_{k+1}^{-1}]^{-1}H_{k+1}P_k\} \\ \times \{\tilde{A}QY_k + \tilde{H}_{k+1}Q_{k+1}Y_{k+1}\} \quad . \tag{A.7.24}$$

As

$$P_k\tilde{A}QY_k = X_k \quad ,$$

we have

$$X_{k+1} = X_k - P_k\tilde{H}_{k+1}[H_{k+1}P_k\tilde{H}_{k+1} + Q_{k+1}^{-1}]^{-1}H_{k+1}X_k \\ + P_k\tilde{H}_{k+1}\{I - [H_{k+1}P_k\tilde{H}_{k+1} + Q_{k+1}^{-1}]^{-1}H_{k+1}P_k\tilde{H}_{k+1}\}Q_{k+1}Y_{k+1} \quad . \tag{A.7.25}$$

Writing

$$I = [H_{k+1} P_k \tilde{H}_{k+1} + Q_{k+1}^{-1}] - 1 [H_{k+1} P_k \tilde{H}_{k+1} + Q_{k+1}] \qquad \text{(A.7.26)}$$

we have

$$X_{k+1} = X_k + P_k \tilde{H}_{k+1} [H_{k+1} P_k \tilde{H}_{k+1} + Q_{k+1}^{-1}][Y_{k+1} - H_{k+1} X_k] \quad . \text{(A.7.27)}$$

For the next step of the recursion we need P_{k+1},

$$\begin{aligned}
P_{k+1} &= [\tilde{A} : \tilde{H}_{k+1}] \begin{bmatrix} Q & \vdots & 0 \\ 0 & \vdots & Q_{k+1} \end{bmatrix} \begin{bmatrix} A \\ \tilde{H}_{k+1} \end{bmatrix}^{-1} \\
&= [\tilde{A} Q A + \tilde{H}_{k+1} Q_{k+1} H_{k+1}]^{-1} \\
&= [P_k^{-1} + \tilde{H}_{k+1} Q_{k+1} H_{k+1}]^{-1} \quad .
\end{aligned} \qquad \text{(A.7.28)}$$

Finally we can write

$$X_{k+1} = X_k + K_k [Y_{k+1} - H_{k+1} Y_k] \quad , \qquad \text{(A.7.29)}$$

where

$$K_k = P_k \tilde{H}_{k+1} [H_{k+1} P_k \tilde{H}_{k+1} + Q_{k+1}^{-1}]^{-1} \qquad \text{(A.7.30)}$$

and

$$P_{k+1} = P_k - P_k \tilde{H}_{k+1} [H_{k+1} P_k \tilde{H}_{k+1} + Q_{k+1}^{-1}] H_{k+1} P_k \quad . \qquad \text{(A.7.31)}$$

The above set of equations are known as Kalman filtering equations for the recursive solution of the problem.

A.8 Coordinate Transformation

In matrix notation the equation of a general second order central surface becomes

$$\tilde{X} A X = 1 \quad , \qquad \text{(A.8.1)}$$

where A is an $n \times n$ symmetric matrix and X an n-dimensional vector.

In many problems, we need to find the principal axes of this quadratic surface. The principal axes are defined by the directions in which the normal to the surface is parallel to the radius vector. Thus the principal axes are defined by

$$A X = \lambda X \quad , \qquad \text{(A.8.2)}$$

as X is the radius vector, λ a constant and the direction cosines of the normal are proportional to AX. We note that (A.8.2) is the same as the eigenvalue problem for the matrix A. The complete solution is given by n values of λ and n associated vectors $U_1, U_2, \ldots, U_n$. The eigenvalue λ_n is associated with the reciprocal of the square of the distance from the origin to the point on the quadratic surface where the principal axis intersects it. This is easily seen by noting that the point of intersection is given by (A.8.1) and (A.8.2). Multiplying (A.8.2) by $\tilde{X}$ we obtain

$$\tilde{X}AX = \lambda \tilde{X}X = 1 \quad \text{or}$$

$$\sum x_i^2 = 1/\lambda \quad .$$

For a symmetric matrix A

$$AX = \lambda X \quad \text{and}$$

$$\tilde{A}X = \lambda X \quad .$$

Thus $U_1, U_2, \ldots, U_n$ vectors are an orthonormal set.

It can easily be shown that the eigenvalues of a symmetric matrix are always real. Actually if the matrix A is complex but Hermitian, i.e.,

$$\tilde{A}^* = A \quad ,$$

then the eigenvalues are also real. For Hermitian matrices

$$V = U^* \quad \text{and}$$

$$\tilde{U}U^* = I \quad .$$

If we define A to be symmetric

$$U = UX \quad ,$$

we obtain from (A.8.1)

$$\tilde{X}AX = 1$$

$$\tilde{X}\tilde{U}AUX = 1$$

$$(\widetilde{UX})A(UX) = 1$$

$$\tilde{Y}AY = 1 \quad \text{or}$$

$$\frac{y_1^2}{\lambda_1} + \frac{y_2^2}{\lambda_2} + \ldots = 1 \quad .$$

B. Orthogonal Functions and Polynomials

B.1 Sturm-Liouville Equation

Most of the eigenvalue problems in this book can be characterized by differential operators D of the form

$$D = \frac{d}{dx}\left(A(x)\frac{d}{dx}\right) + B(x) \quad . \tag{B.1.1}$$

Equation (B.1.1) defines the Sturm-Liouville operator. The physical problems we consider require that $A(x)$ be positive within the given interval (a, b). We note that

$$vDu - uDv = \frac{d}{dx}(A(x)(vu' - uv')) \quad ,$$ (B.1.2)

where u' denotes du/dx. The right-hand side of (B.1.2) is a total derivative and thus

$$\int_a^b (vDu - uDv)dx = [A(x)(vu' - uv')]_a^b \quad .$$ (B.1.3)

Any differential operator D that allows the transformation of integral on the left-hand side of (B.1.3) into a pure boundary term given by the right-hand side of (B.1.3) is called self-adjoint. If the boundary conditions are such that the right-hand side of (B.1.3) also vanishes then it is said to have self-adjoint boundary conditions. For a self-adjoint operator with self-adjoint boundary conditions we have

$$\int_a^b (vDu - uDv)dx = 0 \quad .$$ (B.1.4)

Equation (B.1.4) is also known as Green's identity. The eigenvalue problem associated with the self-adjoint operator D starts with the differential equation

$$D\phi = \lambda\phi \quad .$$ (B.1.5)

If a solution ϕ exists for (B.1.5) satisfying the boundary conditions, it is called an eigenfunction and λ the eigenvalue. However, only a certain selected value λ_i satisfies all the conditions. This infinite set of λ_i are all real and each one corresponds to an eigenfunction ϕ_i.

Let us consider two different eigenvalues λ_i, λ_k and their corresponding eigenfunctions ϕ_i, ϕ_k. Substituting $u = \phi_i$ and $v = \phi_k$ into (B.1.4) we obtain

$$\int_a^b (\lambda_i\phi_i\phi_k - \lambda_k\phi_k\phi_i)dx = 0 \quad .$$ (B.1.6)

This yields the orthogonality condition given by

$$\int_a^b \phi_i(x)\phi_k(x)dx = 0 \quad \text{for} \quad i \neq k \quad .$$ (B.1.7)

We can normalize the eigenfunction by demanding

$$\int_a^b \phi_j^2(x)dx = 1 \quad .$$ (B.1.8)

This forms an orthonormal set of eigenfunctions and the orthonormal property can be written as

$$\int_a^b \phi_j(x)\phi_k(x)dx = \delta_{jk} \quad , \tag{B.1.9}$$

where δ_{jk} is the Kronecker delta function. Any arbitrary function $f(x)$ can be expanded in an infinite expansion and can be represented in the form

$$f(x) = \sum_{k=1}^{\infty} c_k \phi_k(x) \quad , \tag{B.1.10}$$

where the coefficients c_k are given by

$$c_k = \int_a^b f(x)\phi_k(x)dx \quad . \tag{B.1.11}$$

Under certain general conditions, it can be shown that the set ϕ_k is complete and the expansion given by (B.1.10) actually converges to $f(x)$. If $f(x)$ is the solution to a differential equation given by

$$Df(x) = p(x) \quad , \tag{B.1.12}$$

then $p(x)$ represents the input to a linear system with $f(x)$ as the output. Substituting in (B.1.4) $u = f$ and $v = \phi_k$, we have

$$\int_a^b (\phi_k Df - f D\phi_k)dx = 0 \quad , \quad \text{or} \tag{B.1.13}$$

$$\int_a^b (\phi_k p - f\lambda_k \phi_k)dx = 0 \quad , \quad \text{or}$$

$$\int_a^b f\phi_k dx = \frac{1}{\lambda_k} \int_a^b \phi_k p\, dx \quad , \quad \text{or}$$

$$c_k = \frac{1}{\lambda_k} \int_a^b \lambda_k p\, dx \quad .$$

Thus we obtain

$$f(x) = \sum_{k=1}^{\infty} c_k k(x) = \int_a^b p(y)\left(\sum_{k=1}^{\infty} \frac{\phi_k(x)\phi_k(y)}{\lambda_k}\right) dy \quad . \tag{B.1.14}$$

Defining the Green's function or impulse response as

$$G(x, y) = \sum_{k=1}^{\infty} \frac{\phi_k(x)\phi_k(y)}{\lambda_k} \quad , \tag{B.1.15}$$

we have

$$f(x) = \int_a^b G(x,y)p(y)dy \quad . \tag{B.1.16}$$

If

$$p(x) = \delta(x - x_0) \quad , \tag{B.1.17}$$

then

$$f(x) = \int_a^b \delta(y - x_0)G(x,y)dy = G(x,x_0) \quad . \tag{B.1.18}$$

Note that $G(x,x_0)$ also represents the solution to the equation

$$D\,G(x,x_0) = \delta(x - x_0) \quad .$$

In the following we discuss a few examples of Sturm-Liouville differential equation.

B.2 Fourier Series

$$a = -\pi \quad , \quad b = \pi \quad , \quad A(x) = -1 \quad , \quad B(x) = 0 \quad .$$

The boundary conditions are

$$\phi(-\pi) = \phi(\pi) \quad \text{and}$$
$$\phi'(-\pi) = \phi'(\pi) \quad . \tag{B.2.1}$$

The eigenvalues and eigenfunctions are defined by

$$-\frac{d^2\phi(x)}{dx^2} = \lambda u \quad . \tag{B.2.2}$$

The solution of (B.2.2) is given by

$$\phi = A \cos \sqrt{\lambda}x + B \sin \sqrt{\lambda}x \quad , \tag{B.2.3}$$

where A and B are arbitrary constants.

Using the boundary conditions given by (B.2.1) we obtain

$$\lambda_k = k^2 \quad (k = 0, 1, 2, \ldots) \quad . \tag{B.2.4}$$

We note that for every eigenvalue there are two functions given by

$$\phi_k = \cos kx \quad \text{and} \quad \overline{\phi}_k = \sin kx \quad . \tag{B.2.5}$$

B.3 Hypergeometric Series

Before we consider the hypergeometric series we note that the most general second order differential operator is given by

$$p(x)u''(x) + q(x)u'(x) + [r(x) + \lambda]u(x) = 0 \quad . \tag{B.3.1}$$

To change (B.3.1) into the form of (B.1.1) we multiply it by $\varrho(x)$ and demand

$$\varrho(x)q(x) = [\varrho(x)p(x)]' \quad \text{or}$$

$$\frac{\varrho'}{\varrho} = \frac{q - p'}{p} \quad . \tag{B.3.2}$$

If we identify

$$A(x) = \varrho(x)p(x) \quad , \tag{B.3.3}$$

we obtain from (B.3.1)

$$\frac{d}{dx}[A(x)u'(x)] + \varrho(x)[r(x) + \lambda]u(x) = 0 \quad . \tag{B.3.4}$$

The orthogonality condition for this case becomes

$$\int_a^b \varrho(x)\phi_k(x)\phi_i(x)dx = 0 \quad . \tag{B.3.5}$$

The condition given by (B.3.5) is known as weighted orthogonality. For the hypergeometric function, one has

$$p(x) = x(1 - x) \quad ; \quad q(x) = r - (\alpha + \beta + 1)x$$

$$r(x) = 0 \quad \text{and} \quad \lambda = -\alpha\beta \quad .$$

The differential equation satisfied by the eigenfunctions is given by

$$x(1 - x)\phi'' + [\gamma - (\alpha + \beta + 1)x]\phi' = \lambda\phi = \alpha\beta\phi \quad . \tag{B.3.6}$$

For this case

$$\varrho(x) = x^{\gamma-1}(1 - x)^{\alpha+\beta-\gamma} \tag{B.3.7}$$

and

$$A(x) = x^{\gamma}(1 - x)^{\alpha+\beta+1-\gamma} \quad . \tag{B.3.8}$$

The solution of (B.3.6) is given by

$$\phi(\alpha, \beta, \gamma; x) = 1 + \frac{\alpha\beta}{\gamma \cdot 1}x + \frac{\alpha(\alpha + 1)\beta(\beta + 1)}{\gamma(\gamma + 1) \cdot 1 \cdot 2}x^2$$

$$+\frac{\alpha(\alpha + 1)(\alpha + 2)(\alpha + 1)(\alpha + 2)}{(\gamma + 1)(\gamma + 2)1 \cdot 2 \cdot 3}x^3 + \dots \quad . \tag{B.3.9}$$

The solution can be obtained by expanding ϕ in terms of a power series, substituting it in (B.3.5) and equating powers on both sides. ϕ given by (B.3.9) converges for all values of $|x| < 1$ and diverges for all values of $|x| > 1$. The parameters α, β, γ can assume arbitrary real or complex values, except that γ cannot be a negative integer. Let us assume

$$\gamma > 0 \quad \text{and} \quad \alpha + \beta + 1 - \gamma = \delta > 0 \quad .$$

Now if we choose $\alpha = -n$, and $\beta = n + \gamma + \gamma - 1$ the hypergeometric series terminates with the power of x^n and we obtain the solution of the eigenvalue problem given by the polynomials

$$P_n^{(\gamma,\delta)}(x) = F(-n, n + \gamma + \delta - 1, \gamma; x) \quad . \tag{B.3.10}$$

Equation (B.3.10) defines the Jacobi polynomials and has the following weighted orthogonal property:

$$\int_0^1 P_n(x)P_m(x)\varrho(x)dn = 1 \quad . \tag{B.3.11}$$

If we substitute $\gamma = \delta$ and $x' = (1 - x)/2$, then (B.3.9) and (B.3.10) can be rewritten as

$$P_n^{(\gamma)}(x) = F\left(-n, n + 2\gamma - 1, \gamma; \frac{1-x}{2}\right)$$

and

$$\int_{-1}^{+1} P_n^{(\gamma)}(x)P_m^{(\gamma)}(x)\varrho(x)dx = 0 \quad ,$$

where $\varrho(x) = (1 - x^2)^{\gamma-1}$ and $\beta = n + 2\gamma - 1$. Note that $P_n^{(\gamma)}(x)$ are even or odd polynomials according to the even or odd character of n.

B.4 Legendre Polynomials

If $\gamma = 1$ and $\varrho(x) = 1$ we obtain

$$P_n(x) = F\left(-n, n + 1; \frac{1-x}{2}\right) \quad .$$

Because of its importance we discuss further details below. The differential equation for this case is given by

$$(1 - x^2)\phi'' - 2x\phi' = \lambda\phi \quad .$$

The solutions are

$$\phi_n(x) = P_n(x) \quad \text{and} \quad \lambda_n = -k(n + 1) \quad .$$

The normalization condition is

$$\int_{-1}^{+1} [P_n(x)]^2 dx = \frac{2}{2n + 1} \quad .$$

The recurrence relation is given by

$$(n + 1)P_{n+1} = (2n + 1)xP_n(x) - nP_{n-1}(x) \quad \text{and}$$

$$nP_n(x) = xP_n'(x) - P_{n-1}'(x) \quad .$$

B.5 Hermite Polynomials

$$\varrho(x) = e^{-x^2} \quad ; \quad a = \infty \quad ; \quad b = -\infty \quad .$$

$$\phi'' - 2x\phi' = \lambda\phi = -2n\phi \quad .$$

$$\phi_n = H_n(x) \quad \text{and} \quad \lambda_n = -2n \quad .$$

$$\int\limits_{-\infty}^{+\infty} H_n(x)H_m(x)e^{-x^2}\,dx \begin{cases} = 0 \quad , & n \neq m \quad , \\ = \sqrt{\pi}2^n n! \quad , & n = m \quad . \end{cases}$$

$$H_{n+1} = 2xH_n(x) - 2nH_{n-1}(x)$$

$$H_n(x) = (-1)^n e^{x^2}\frac{\partial^n}{\partial x^n}(e^{-x^2})$$

$$H_0(x) = 1$$

$$H_1(x) = 2x$$

$$H_2(x) = 4x^2 - 2$$

$$H_3(x) = 8x^3 - 12x$$

$$H_4(x) = 16x^4 - 48x^2 + 12$$

$$H_5(x) = 32x^5 - 160x^3 + 120x$$

$$H_6(x) = 64x^6 - 480x^4 + 720x^2 - 120$$

B.6 Laguerre Polynomials

$$\varrho(x) = e^{-x} \quad ; \quad a = 0 \quad ; \quad b = \infty \quad .$$

$$x\phi'' + (1-x)\phi' = \lambda\phi$$

$$\phi_n = L_n(x) \quad ; \quad \lambda_n = -n$$

$$\int\limits_0^\infty e^{-x} L_n(x)L_m(x)\,dx = \delta_{nm}$$

$$(n+1)L_{n+1}(x) + (x - 1 - 2n)L_n(x) + nL_{n-1}(x) = 0$$

$$L_n(x) = \frac{e^x}{n!} \frac{d^n}{dx^n}(x^n e^{-x})$$

$$L_0(x) = 1$$

$$L_1(x) = 1 - x$$

$$L_2(x) = -2x + \frac{x^2}{2}$$

B.7 Generalized Laguerre Polynomials

$$\varrho(x) = e^{-x} x^\alpha \quad ; \quad a = 0 \quad ; \quad b = \infty$$

$$x\phi'' + (\alpha + 1 - x)\phi' = \lambda\phi$$

$$\phi_n = L_n^\alpha(x) \quad ; \quad \lambda_n = -n$$

$$\int_0^\infty e^{-x} L_n^\alpha(x) L_m^\alpha(x) dx \begin{cases} = 0 \quad , & n \neq m \quad , \\ = \dfrac{\Gamma(\alpha + n + 1)}{n!} \quad , & n = m \quad . \end{cases}$$

$$(n+1)L_{n+1}^\alpha(x) + (x - \alpha - 1 - 2n)L_n^\alpha(x) + (n+\alpha)L_{n-1}^\alpha(x) = 0$$

$$L_n^\alpha(x) = \frac{e^x x^{-\alpha}}{n!} \frac{d^n}{dx^n}(e^{-x} x^{n+\alpha})$$

$$L_0^\alpha(x) = 1$$

$$L_1^\alpha(x) = \alpha + 1 - x$$

$$L_2^\alpha(x) = \frac{1}{2}(\alpha + 1)(\alpha + 2) - (\alpha + 2)x + \frac{x^2}{2}$$

B.8 Chebyshev Polynomials

$$\cos(n+1)\theta + \cos(n-1)\theta = 2\cos\theta\cos n\theta$$

$$x = \cos\theta \quad ; \quad T_n(x) = \cos n\theta$$

$$T_{n+1}(x) = 2xT_n(x) - T_{n-1}(x)$$

$$\sin(n+1)\theta + \sin(n-1)\theta = 2\cos\theta\sin n\theta$$

$$U_n(x) = \sin n\theta$$

$$U_{n+1}(x) = 2xU_n(x) - U_{n-1}(x)$$

$$U_0(x) = 1 \quad ; \quad U_1(x) = 2x$$

$$T_0(x) = 1 \quad ; \quad T_1(x) = x$$

$$a = -1 \quad , \quad b = 1 \quad .$$

If

$$a = 0 \quad \text{and} \quad b = 1 \quad ,$$

$$x = \frac{1 - \cos\theta}{2} = \sin^2\frac{\theta}{2} \quad .$$

Shifted Chebyshev polynomial $T_n^*(x)$:

$$T_{n+1}^*(x) = 2(1 - 2x)T_n^*(x) - T_{n-1}^*(x)$$

$$U_{n+1}^*(x) = 2(1 - 2x)U_n^*(x) - U_{n-1}^*(x)$$

$$U_0^*(x) = 1 \quad ; \quad U_1^*(x) = 2(1 - 2x)$$

$$T_0^*(x) = 1 \quad ; \quad T_1^*(x) = 1 - 2x$$

$$\varrho(x) = \frac{1}{\sqrt{1 - x^2}} \quad ; \quad \gamma = \frac{1}{2}$$

$$T_n(x) = F\left(-n, n, \frac{1}{2}, \frac{1-x}{2}\right) \quad .$$

These polynomials have the greatest efficiency in approximating arbitrary functions.

$T \rightarrow$ Chebyshev polynomials of the first kind .

$U \rightarrow$ Chebyshev polynomials of the second kind .

$$\gamma = \frac{3}{2} \quad ; \quad U_n(x) = (n + 1)F\left(-n, n + 2, \frac{3}{2}; \frac{1-x}{2}\right) \quad .$$

U's are well suited for the polynomial representation of a function which assumes large values in the neighborhood of $x = \pm 1$ and remains small everywhere else.

$$T \rightarrow x(1 - x)\phi'' + [\tfrac{1}{2} - x]\phi' = -n^2\phi = \lambda\phi \quad .$$

$$U \rightarrow x(1 - x)\phi'' + [\tfrac{3}{2} - 2x]\phi' = -n(n + 2)\phi = \lambda\phi \quad .$$

If

$$x = \frac{1 + \cos\theta}{2} = \cos^2\theta/2 \quad ,$$

$$T_k^*(x) = \cos k\theta = T_k(2x - 1)$$

and

$$y = f(x) = \sum_i a_i x^i = \sum_i b_i T_i^*(x) \quad .$$

Note that b_i *converges very very rapidly.*

B.9 Bessel Functions

$$x^2 \phi'' + x\phi' + x^2\phi = \lambda\phi$$

$$\phi_n = J_n(x) \quad ; \quad \lambda = n^2$$

$$J_n = \sum_{m=0}^{\infty} \frac{(-1)^m x^{n+2m}}{m!(n+m)!}$$

$$Y_n = \frac{2}{\pi}\left(\gamma + \ln\frac{x}{2}\right) J_n(x)$$

$$- \frac{1}{\pi}\sum_{m=0}^{n-1} \frac{(n-m-1)!}{m!}\left(\frac{2}{x}\right)^{n-2m}$$

$$- \frac{1}{\pi}\sum_{m=0}^{\infty} \frac{(-1)^m (x/2)^{n+2m}}{m!(n+m)!}\left(1 + \frac{1}{2} + \frac{1}{3}\cdots\frac{1}{m} + 1 + \frac{1}{2} + \frac{1}{3}\cdots\frac{1}{n+m}\right)$$

$$\gamma = 0.5772 \quad \text{and} \quad Y_n(x \to 0) \to \infty$$

$$J_{n+1} = \frac{2n}{x}J_n(x) - J_{n-1}(x) \quad .$$

For large arguments $(x \gg 1, n)$

$$J_n(x) \to \sqrt{\frac{2}{\pi x}}\cos\left(x - \frac{n\pi}{2} - \frac{\pi}{4}\right) \quad ,$$

$$Y_n(x) \to \sqrt{\frac{2}{\pi x}}\sin\left(x - \frac{n\pi}{2} - \frac{\pi}{4}\right) \quad .$$

For this reason it is convenient to introduce particular linear combinations of J_n and Y_n through the definition

$$H_n^{(1)}(x) = J_n(x) + jY_n(x) \quad .$$

$$H_n^{(2)}(x) = J_n(x) - jY_n(x) \quad .$$

These combinations also form a set of solutions to Bessel equations and are known as Hankel functions or Bessel functions of the third kind. For large x

$$H_n^{(1)}(x) \to \exp\left[j\left(x - \frac{n\pi}{2} - \frac{\pi}{4}\right)\right] \quad ,$$

$$H_n^{(2)}(x) \to \exp\left[-j\left(x - \frac{n\pi}{2} - \frac{\pi}{4}\right)\right] \quad .$$

For

$$x \ll 1 \quad ,$$

$$J_n(x) \to \frac{(x/2)^n}{n!}$$

and

$$Y_n(x) \to \begin{cases} \dfrac{2}{\pi}\left(0.5772 + \ln\dfrac{x}{2}\right) \quad , & n = 0 \quad , \\[2ex] -\dfrac{(n-1)!}{\pi}\left(\dfrac{2}{x}\right)^n \quad , & n \neq 0 \quad . \end{cases}$$

If x is imaginary then, writing $x = jy$ where y is real, one defines the following modified Bessel functions:

$$I_n(y) = (j)^{-n} J_n(jy)$$

and

$$K_n(y) = \frac{\pi}{2}(j)^{n+1} H_n^{(1)}(jy) \quad .$$

The reason for defining K_n with a Hankel function rather than a Bessel function of the second kind, Y_n, is because an expression similar to I_n yields a complex function. Note that I_n and K_n as defined are real functions:

$$I_n(y) \to \frac{e^y}{\sqrt{2\pi y}} \quad \text{and}$$

$$K_n(y) \to \sqrt{\frac{\pi}{2y}}e^{-y}$$

where $y \gg 1$ and n.

Figure B.1 plots the first few orders of J_n, Y_n, $H_n^{(1)}$, $H_n^{(2)}$, I_n and K_n. We note that J_n and Y_n are oscillatory, very similar to sine and cosine functions. However Y_n possesses a singularity at $x = 0$ and must be excluded from solutions in physical problems unless a source exists at the origin. Hankel functions are like $e^{\pm jx}$ and I_n and K_n are similar to sinh and cosh functions. Some other relationships of interest are

$$J_{-n}(x) = (-1)^n J_n(x)$$

$$Y_{-n}(x) = (-1)^n Y_n(x)$$

$$I_{-n}(x) = I_n(x)$$

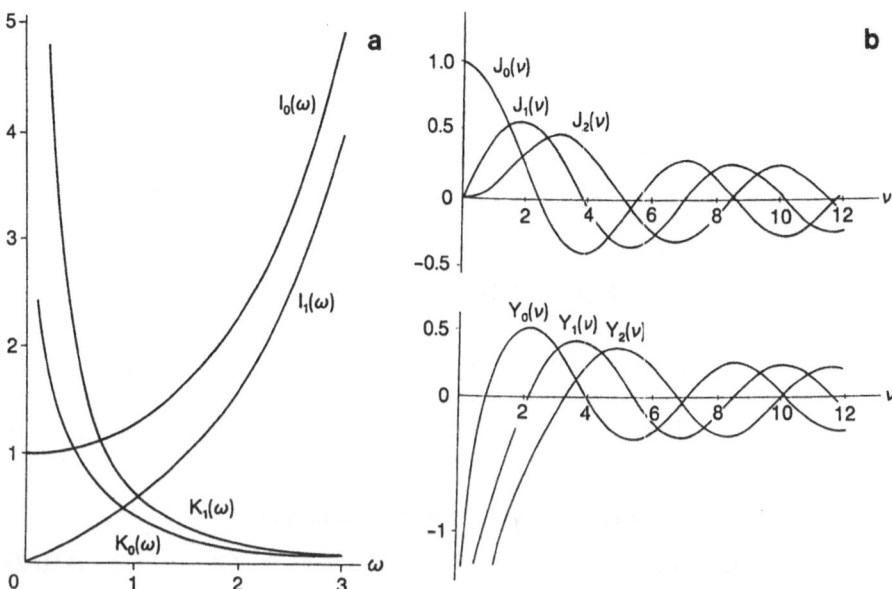

Fig. B.1. (a) Low order modified Bessel functions, and (b) low order Bessel functions of first and second kinds

Table B.1. The roots γ_{mn} of $J_n(x)$; $J_n(\gamma_{nm}) = 0$

m/n	0	1	2	3	4	5
1	2.4048	3.8317	5.1356	6.3802	7.5883	8.7715
2	5.5201	7.0156	8.4172	9.7610	11.0647	12.3386
3	8.6537	10.1735	11.6198	13.0152	14.3725	15.7002
4	11.7915	13.3237	14.7960	16.2235	17.6160	18.9801
5	14.9309	16.4706	16.9398	19.4094	20.8269	22.2178
6	18.0711	19.6159	21.1170	22.5827	24.0190	25.4303

$$K_{-n}(x) = K_n(x)$$

$$nJ_n(x) = \frac{n}{2}[J_{n-1}(x) + J_{n+1}(x)]$$

$$J_n'(x) = \tfrac{1}{2}[J_{n-1}(x) - J_{n+1}(x)]$$

$$\frac{d}{dx}(x^n J_n(x)) = x^n J_{n-1}(x)$$

$$J_0'(x) = -J_1(x) \quad .$$

$J_n(\gamma_{nm}x/x_0)$ forms a set of orthogonal functions with $\phi(x) = x$ and $a = 0$, $b = x_0$. Thus

$$\int\limits_0^{x_0} x J_n\left(\gamma_{nm}\frac{x}{x_0}\right) J_n\left(\gamma_{np}\frac{x}{x_0}\right) dx$$

$$= \frac{x_0^2}{2} J_{n+1}^2(\gamma_{nm})\delta_{mp} \quad,$$

where γ_{nm} are the roots of $J_n(x)$, i.e., $J_n(\gamma_{nm}) = 0$. See Table B.1.

C. Principle of Stationary Phase

In many cases we need to evaluate the following integral for $t \to$ large value:

$$I = \int\limits_{-\infty}^{+\infty} A(\omega)e^{jt\mu(\omega)} d\omega \quad, \tag{C.0.1}$$

where $\mu(\omega)$ is a general function of ω with the conditions

$$\mu'(\omega_0) = 0 \quad \text{and} \tag{C.0.2}$$

$$\mu''(\omega_0) \ne 0 \quad. \tag{C.0.3}$$

For t very large, $\exp[jt\mu(\omega)]$ oscillates rapidly and thus the contribution to the integral is very small except near $\omega = \omega_0$. Thus we can write

$$I \sim A(\omega_0)\sqrt{\frac{2\pi}{t|\mu''(\omega_0)|}}e^{j[t\mu\omega_0)+\pi/4]} \quad. \tag{C.0.4}$$

The above approximation is generally known as the principle of stationary phase.
For n stationary points located at $\omega_i = \omega_1, \omega_2, \ldots, \omega_n$, we have

$$\mu'(\omega_i) = 0 \quad, \tag{C.0.5}$$

$$\mu''(\omega_i) \ne 0 \quad, \tag{C.0.6}$$

$$I \sim \sum_{i=1}^n A(\omega_i)\sqrt{\frac{2\pi}{t|\mu''(\omega_i)|}}e^{j[t\mu(\omega_i)+\pi/4]} \quad. \tag{C.0.7}$$

If the stationary point $\omega = \omega_0$ is such that

$$\mu'(\omega_0) = \mu''(\omega_0)\ldots = \mu^{n-1}(\omega_0) = 0 \tag{C.0.8}$$

and

$$\mu^n(\omega_0) \ne 0 \quad, \tag{C.0.9}$$

we have

$$I \sim A(\omega_0)e^{jt\mu(\omega_0)}\frac{\Gamma(1/n)e^{j\pi/2n}}{n[t\mu^n(\omega_0)/n!]^{1/n}} \quad. \tag{C.0.10}$$

If the limits of integration are from a to b and $a < \omega_0 < b$, then for n odd $\exp(j\pi/2n)$ in (C.0.10) is to be replaced by $\cos \pi/2n$.

In the two-dimensional case we have the following results:

$$I = \iint\limits_{R} A(x, y)e^{jt\mu(x,y)}\, dx\, dy \qquad \text{(C.0.11)}$$

$$\left.\begin{array}{l} \dfrac{\partial \mu}{\partial x} = \mu_x(x_0, y_0) = 0 \quad, \\[2mm] \dfrac{\partial \mu}{\partial y} = \mu_y(x_0, y_0) = 0 \quad, \end{array}\right\} \quad x_0, y_0 \in R \quad, \qquad \text{(C.0.12)}$$

$$(\mu_{xx}\mu_{yy} - \mu_{xy}^2)_{x_0, y_0} \neq 0 \quad ; \quad \mu_{yy}|_{x_0 y_0} \neq 0 \quad, \qquad \text{(C.0.13)}$$

$$I \sim \frac{2\pi j A(x_0 y_0)e^{jt\mu(x_0 y_0)}}{t\sqrt{\mu_{xx}\mu_{xy} - \mu_{yy}^2}} \quad . \qquad \text{(C.0.14)}$$

D. Vectors

D.1 Important Results

$$\nabla(\phi\psi) = \psi\nabla\phi + \phi\nabla\psi \qquad \text{(D.1.1)}$$

$$\nabla \cdot (\phi A) = \phi\nabla \cdot A + \nabla\phi \cdot A \qquad \text{(D.1.2)}$$

$$\nabla \cdot (A \times B) = B \cdot \nabla \times A - A \cdot \nabla \times B \qquad \text{(D.1.3)}$$

$$\nabla \times (A \times B) = A\nabla \cdot B - B\nabla \cdot A + (B \cdot \nabla)A - (A \cdot \nabla)B \qquad \text{(D.1.4)}$$

$$\nabla \cdot (\nabla \times A) = 0 \qquad \text{(D.1.5)}$$

$$\nabla \times \nabla \times A = \nabla(\nabla \cdot A) - \nabla^2 A \qquad \text{(D.1.6)}$$

$$\int_V \nabla \cdot A\, dV = \int_S A \cdot dS \quad \text{Divergence theorem} \qquad \text{(D.1.7)}$$

$$\int_S (\nabla \times A) \cdot dS = \oint_C A \cdot dl \quad \text{Stokes theorem} \qquad \text{(D.1.8)}$$

D.2 Green's Theorem: Scalar

Using the divergence theorem, we have

$$\int_V \nabla \cdot (\phi \nabla \psi) dV = \int_S \phi \nabla \psi \cdot dS \quad . \tag{D.2.1}$$

Also, using (D.1.2), we have

$$\nabla \cdot (\phi \nabla \psi) = \phi \nabla^2 \psi + \nabla \phi \cdot \nabla \psi \quad .$$

Thus

$$\int_V (\phi \nabla^2 \psi + \nabla \phi \cdot \nabla \psi) dV = \int_S \phi \nabla \psi \cdot dS \quad . \tag{D.2.2}$$

Interchanging ϕ and ψ in (D.2.2) we have

$$\int_V (\psi \nabla^2 \phi + \nabla \psi \cdot \nabla \phi) dV = \int_S \psi \nabla \phi \cdot dS \quad . \tag{D.2.3}$$

Subtracting (D.2.3) from (D.2.2), we obtain the following identity, known as Green's theorem:

$$\int_V (\phi \nabla^2 \psi - \psi \nabla^2 \phi) dV = \int_S (\phi \nabla \psi - \psi \nabla \phi) \cdot dS \tag{D.2.4}$$

or

$$\int_V (\phi \nabla^2 \psi - \psi \nabla^2 \phi) dV = \int_S \left(\phi \frac{\partial \psi}{\partial n} - \psi \frac{\partial \phi}{\partial n} \right) dS \quad . \tag{D.2.5}$$

D.3 Green's Theorem: Vector

From (D.1.3) we have

$$\nabla \cdot (A \times B) = B \cdot \nabla \times A - A \cdot \nabla \times B \quad .$$

Choose

$$A = F \quad \text{and} \quad B = \nabla \times G \quad .$$

Then (D.1.3) becomes

$$\nabla \cdot (F \times \nabla \times G) = \nabla \times G \cdot \nabla \times F - F \cdot \nabla \times \nabla \times G \quad . \tag{D.3.1}$$

Choose

$$A = G \quad \text{and} \quad B = \nabla \times F \quad .$$

Then (D.1.3) becomes

$$\nabla \cdot (G \times \nabla \times F) = \nabla \times G \cdot \nabla \times F - G \cdot \nabla \times \nabla \times F \quad . \tag{D.3.2}$$

Subtracting (D.3.1) from (D.3.2) and integrating over volume we have

$$\int\limits_V (\boldsymbol{F} \cdot \nabla \times \nabla \times \boldsymbol{G} - \boldsymbol{G} \cdot \nabla \times \nabla \times \boldsymbol{F}) dV$$

$$= \int\limits_V \nabla \cdot [\boldsymbol{G} \times \nabla \times \boldsymbol{F} - \boldsymbol{F} \times \nabla \times \boldsymbol{G}] dV \quad . \tag{D.3.3}$$

Using the divergence theorem we obtain

$$\int\limits_V (\boldsymbol{F} \cdot \nabla \times \nabla \times \boldsymbol{G} - \boldsymbol{G} \cdot \nabla \times \nabla \times \boldsymbol{F}) dV$$

$$= + \int\limits_S (\boldsymbol{G} \times \nabla \times \boldsymbol{F} - \boldsymbol{F} \times \nabla \times \boldsymbol{G}) \cdot d\boldsymbol{S} \quad . \tag{D.3.4}$$

Denoting the inward normal by l_n, we have

$$\int\limits_V (\boldsymbol{F} \cdot \nabla \times \nabla \times \boldsymbol{G} - \boldsymbol{G} \cdot \nabla \times \nabla \times \boldsymbol{F}) dV$$

$$= - \int\limits_S (\boldsymbol{G} \times \nabla \times \boldsymbol{F} - \boldsymbol{F} \times \nabla \times \boldsymbol{G}) \cdot l_n dS \quad . \tag{D.3.5}$$

E. Symmetry Properties of Different Coefficients in Crystal Classes. (From [A.1])

Table E.1. Form of the elastic-coefficient (c_{ij}) matrices

Key to Notation

	•	zero component
	●	nonzero component
	●——●	equal components
	●——○	components numerically equal, but opposite in sign
For s	⊙	twice the numerical equal of the heavy dot component to which it is joined
For c	⊙	the numerical equal of the heavy dot component to which it is joined
For s	✕	$2(s_{11} - s_{12})$
For c	✕	$\frac{1}{2}(c_{11} - c_{12})$

All the matrices are symmetrical about the leading diagonal.

Triclinic

Both classes

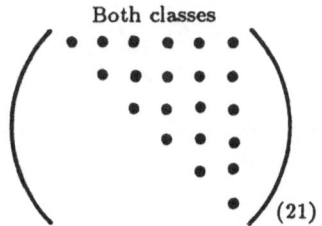

(21)

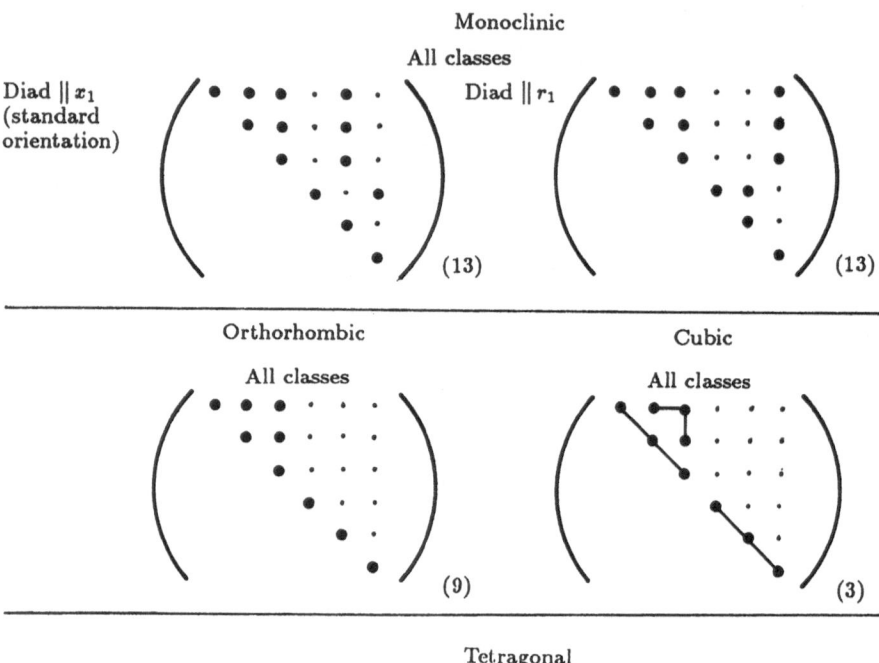

Monoclinic

All classes

Diad $\parallel x_1$ (standard orientation)   Diad $\parallel r_1$

(13)   (13)

Orthorhombic   Cubic

All classes   All classes

(9)   (3)

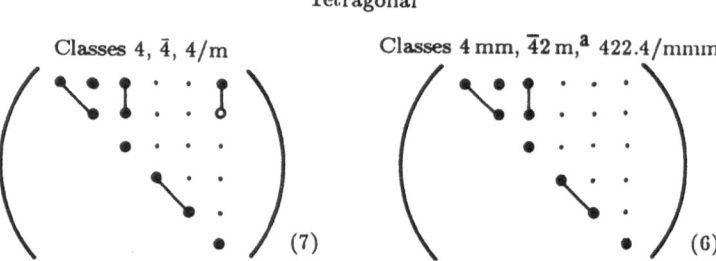

Tetragonal

Classes 4, $\bar{4}$, 4/m   Classes 4 mm, $\bar{4}2$ m,[a] 422.4/mmm

(7)   (6)

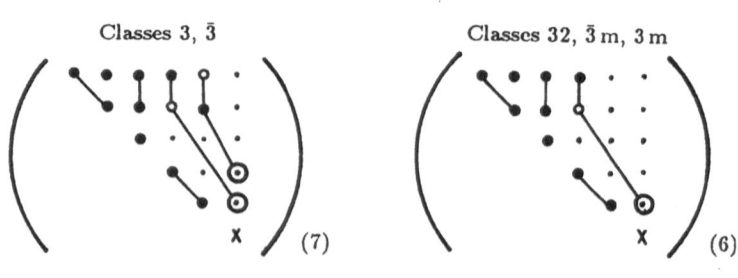

Trigonal

Classes 3, $\bar{3}$ Classes 32, $\bar{3}$ m, 3 m

X   X

(7)   (6)

Hexagonal

All classes

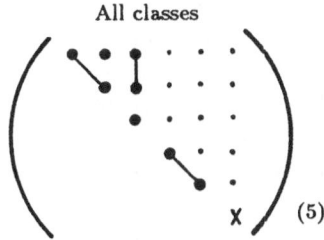

(5)

Isotropic

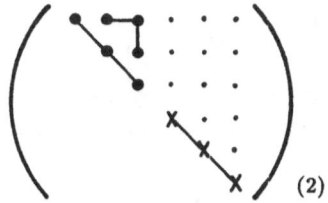

(2)

ᵃ The same matrix holds for both possible orientations of class $\bar{4}2$ m $(2\|\ x_1$ and $m \perp x_1)$ since the addition of a center of symmetry makes the two orientations indistinguishable

Table E.2. Form of the (e_{ij}) and (r_{ij}) piezoelectric and electro-optic coefficient matrices

	Key to Notation
•	zero modulus
●	nonzero modulus
●——●	equal moduli
●——○	moduli numerically equal, but opposite in sign
◉	a modulus equal to minus 2 times the heavy dot modulus to which it is joined

Centrosymmetrical classes

All moduli vanish

Noncentrosymmetrical classes

Triclinic

Class 1

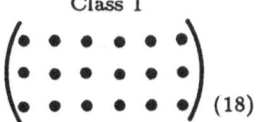

(18)

459

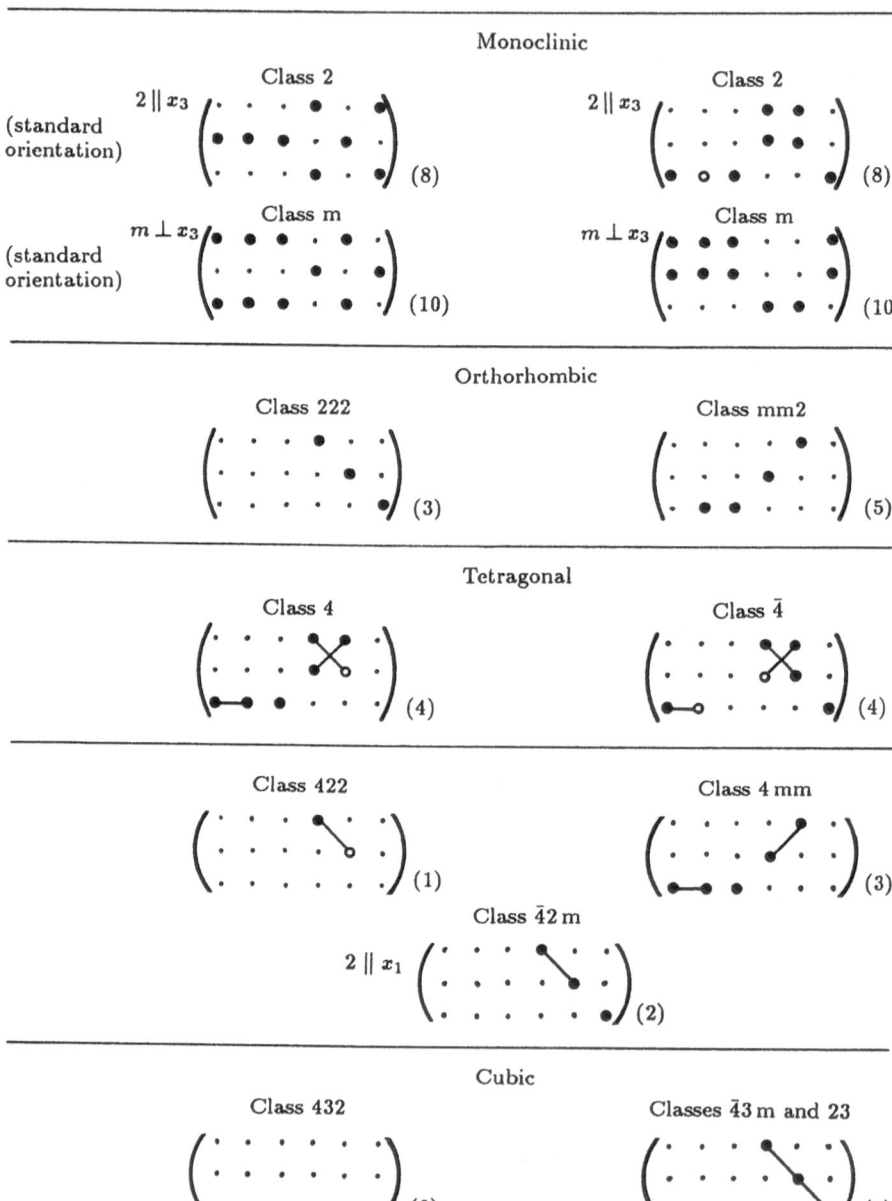

Monoclinic

Class 2 $2 \parallel x_3$ (8)   (standard orientation)

Class 2 $2 \parallel x_3$ (8)

Class m $m \perp x_3$ (10)   (standard orientation)

Class m $m \perp x_3$ (10)

Orthorhombic

Class 222 (3)

Class mm2 (5)

Tetragonal

Class 4 (4)

Class $\bar{4}$ (4)

Class 422 (1)

Class 4 mm (3)

Class $\bar{4}2\,m$ $2 \parallel x_1$ (2)

Cubic

Class 432 (0) — All moduli vanish

Classes $\bar{4}3\,m$ and 23 (1)

Table E.2. (Continued)

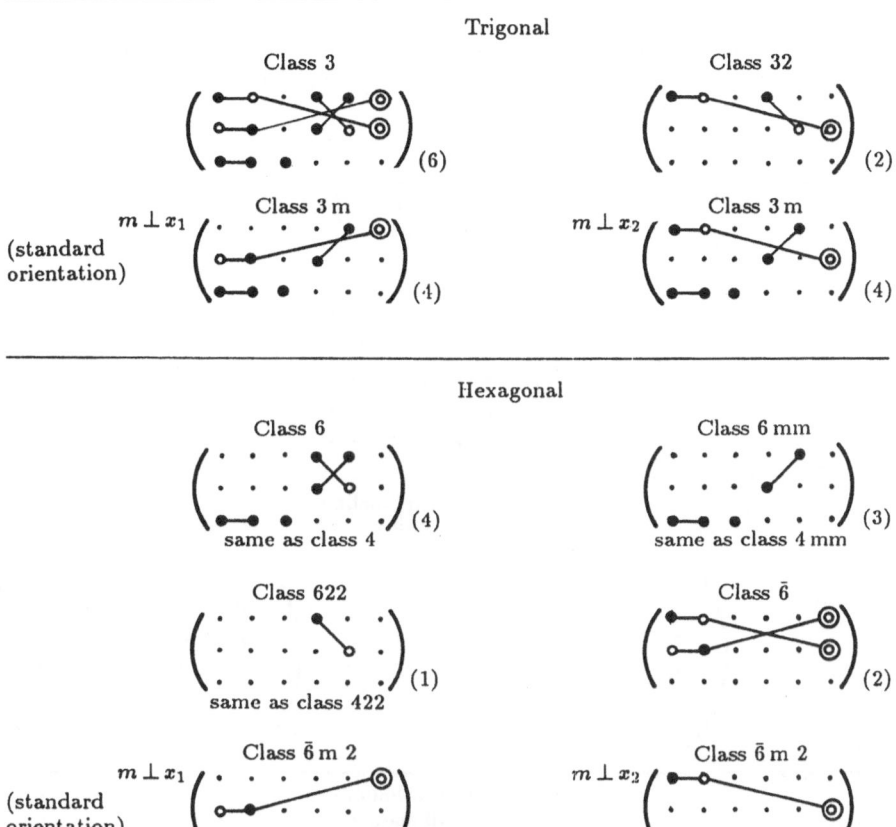

Trigonal

Class 3 (6)

Class 32 (2)

$m \perp x_1$ (standard orientation) — Class 3 m (4)

$m \perp x_2$ — Class 3 m (4)

Hexagonal

Class 6 (4) — same as class 4

Class 6 mm (3) — same as class 4 mm

Class 622 (1) — same as class 422

Class $\bar{6}$ (2)

$m \perp x_1$ (standard orientation) — Class $\bar{6}$ m 2 (1)

$m \perp x_2$ — Class $\bar{6}$ m 2 (1)

Table E.3. Forms of the p photoelastic matrices

Key to Notation
In both π and p matrices

- •   zero component
- ○   nonzero component
- ●——○   equal components
- ●——○   components numerically equal, but opposite in sign

In the π matrices

- ◉   a component equal to twice the heavy dot component to which it is joined
- ◎   a component equal to minus 2 times the heavy dot component to which it is joined
- × $(\pi_{11} - \pi_{12})$

461

Table E.3. (Continued)

In the p matrices

⊙   a component equal to the heavy dot component to which it is joined

⊚   a component equal to minus the heavy dot component to which it is joined

× $\frac{1}{2}(p_{11} - p_{12})$.

Triclinic
Both classes

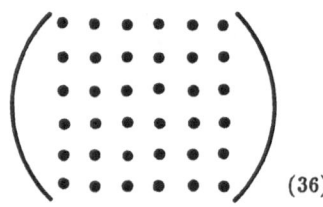

(36)

Monoclinic
All classes

Diad $\parallel x_2$
(standard
orientation)

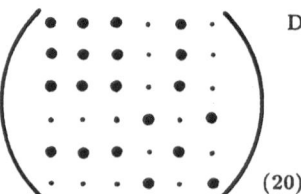

(20)

Diad $\parallel x_3$

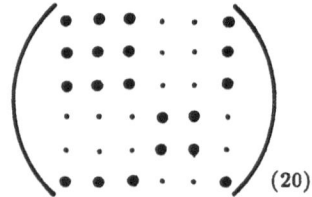

(20)

Orthorhombic
All classes

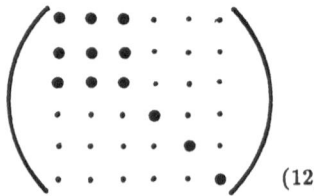

(12)

Tetragonal

Classes
4, $\bar{4}$, 4/m

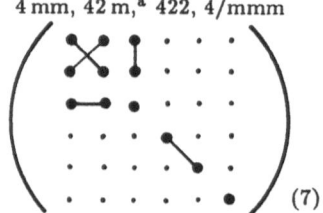

Classes
4 mm, $\bar{4}$2 m,[a] 422, 4/mmm

(10)

(7)

Table E.3. (Continued)

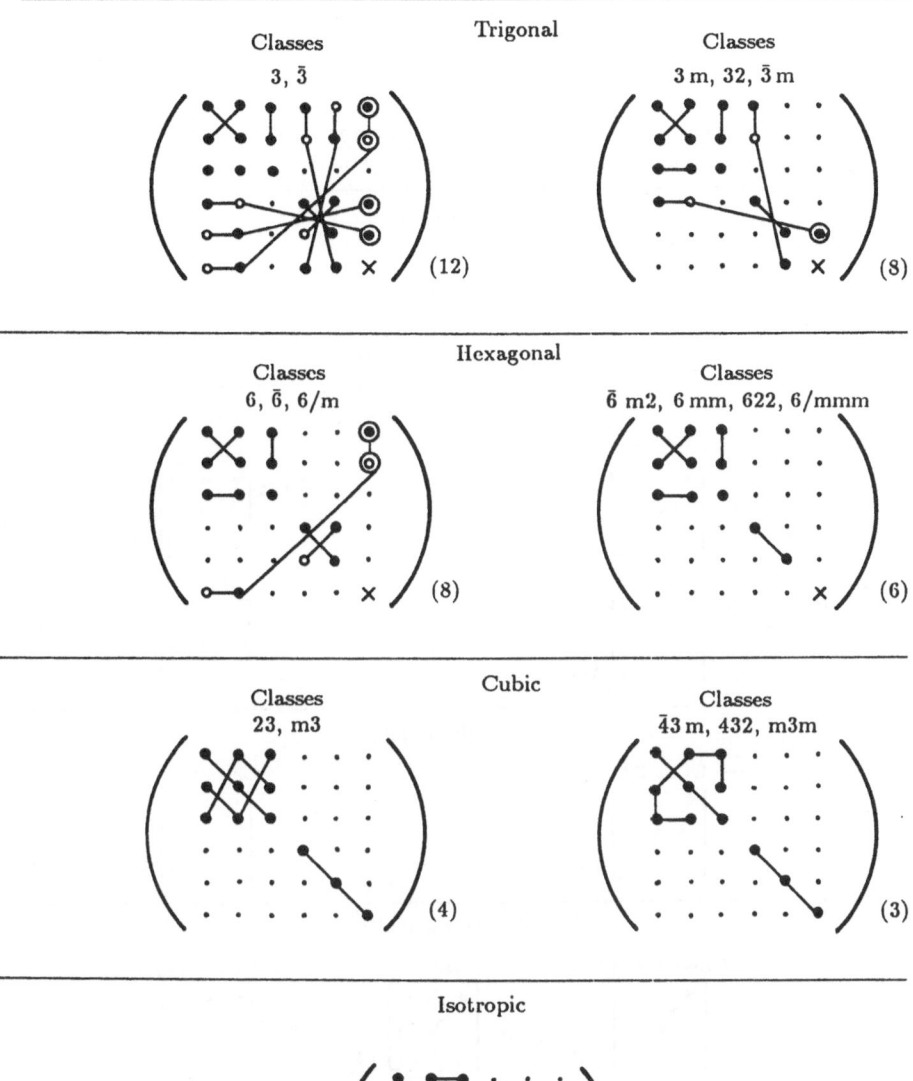

---

[a] The same matrix holds for both orientations of class $\bar{4}2m$ ($2 \parallel x_1$ and $m \perp x_1$), for the addition of a center of symmetry makes the two orientations indistinguishable.

Table E.4. Forms of the gyration tensor $[g_{ij}]$

Key to Notation

- · zero component
- • nonzero component
- •——• equal components
- •——○ components numerically equal, but opposite in sign.

All tensors are symmetrical about the leading diagonal.

Triclinic

Class 1

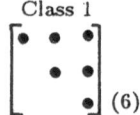

Monoclinic

Class 2

$2 \parallel x_2$
(standard
orientation)

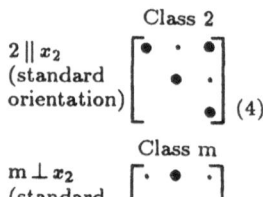

Class 2

$2 \parallel x_3$

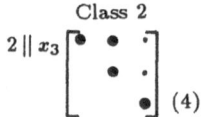

Class m

$m \perp x_2$
(standard
orientation)

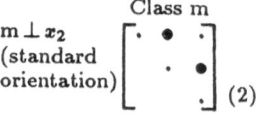

Class m

$m \perp x_3$

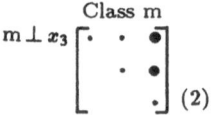

Orthorhombic

Class 222

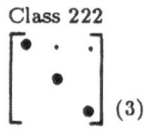

Class mm2

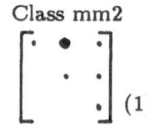

Tetragonal

Classes 4, 422

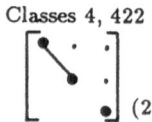

Class $\bar{4}$

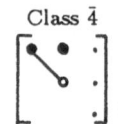

Class $\bar{4}2m$

$2 \parallel x_1$

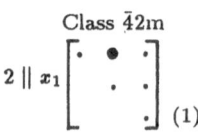

Trigonal and Hexagonal

Classes 3, 32, 6, 622

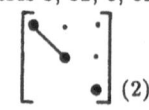

Table E.4. (Continued)

Cubic	Isotropic
Classes 432, 23	without center of symmetry

References

Chapter 1

1.1 P. Das, L.B. Milstein, D.R. Arsenault: "Adaptive Spread Spectrum Receiver Using SAW Technology", National Telecommunication Conference Record, 1977; IEEE Pub. No. 77CH1202-2 CSCB (1977) p. 35: 7-1-6

1.2 G.J. Saulnier, P. Das, L.B. Milstein: "A Digitally Implemented Adaptive LMS Suppression Filter for Direct Sequence Spread Spectrum Communications", in Proc. of Globecom '84; IEEE Pub. No. 84CH2064-4 (1984) p. 1544-1547

Chapter 2

2.1 M.V. Klein: *Optics* (Wiley, New York 1970)

2.2 P. Das: *Lasers and Optical Engineering* (Springer, New York 1991)

2.3 E. Snitzer: Cylindrical dielectric waveguide modes. J. Opt. Soc. Am. **51**, 491 (1961)

2.4 M. Gottlieb, C. Ireland, J.M. Ley: *Electro-Optic and Acousto-Optic Scanning and Deflection* (Marcel Dekker, New York 1983)

2.5 R.A. Toupin: J. Rational Mech. Anal. **5**, 849 (1956)

2.6 H.F.Tiersten, C.F. Tsai: On the interaction of the electromagnetic field with heat conducting deformable insulators. J. Math. Phys. **13**, 316 (1972)

2.7 M. Lax, D.F. Nelson: Phys. Rev. B **4**, 3694 (1971)

2.8 M.J. Fraiser: A survey of magneto-optic effects. IEEE Trans. MAG-4, 152 (1968)

2.9 J. Goodman: *Introduction to Fourier Optics* (McGraw-Hill, New York 1968)

2.10 E.N. Leith, J. Upatnieks: Reconstructed wavefronts and communication theory. J. Opt. Soc. Am. **52**, 1123 (1962)

2.11 D. Gabor: A new microscope principle. Nature **161**, 777 (1948)

2.12 R.J. Collier, C.B. Burckhardt, L.H. Lin: *Optical Holography* (Academic, New York 1971)

2.13 R.S. Elliot: *Electromagnetics* (McGraw-Hill, New York 1966)

2.14 A.H. Cherin: *An Introduction to Optical Fibers* (McGraw-Hill, New York 1983)

2.15 T. Okoshi: *Optical Fibers* (Academic, New York 1982)

2.16 G.R. Fowles: *Introduction to Modern Optics* (Holt, Rinehart and Winston, New York 1968)

2.17 A. Yariv: *Introduction to Optical Electronics*, 2nd ed. (Holt, Rinehart and Winston, New York 1976)

2.18 W.R. Klein, B.D. Cook: IEEE Trans. SU-14, 123 (1967)

2.19 R.W. Dixon: IEEE J. QE-3, 84 (1967)

Chapter 3

3.1 S. Darlington: US patent No. 2,678,997, May 18 (1954)

3.2 F.J. Harris: Proc. IEEE **66**, 51 (1978)

3.3 F. Itakura, S. Saito: In 7th Intl. Cong. Acoustics (Budapest) (1971) pp. 261–264

3.4 C.H. Sequin, M.F. Tompsett: *Charge Transfer Devices* (Academic, New York 1975)

3.5 D.A. Smith, C.M. Puckett, W.J. Butler: Active bandpass filtering with bucket brigade delay lines. IEEE J. SC-7, 421 (1972)

3.6 R.H. Tancrell: IEEE Trans. SU-21, 12 (1974)

3.7 M.H. White, I.A. Mack, G.M. Borsuk, D.R. Lampe, F.J. Kub: Charge coupled device (CCD) adaptive discrete analog signal processing. IEEE J. SC-14, 132 (1979)

3.8 E.L. O'Neil: *Introduction to Statistical Optics* (Addison-Wesley, Reading, MA 1963)

3.9 R.M.A. Azzam, N.M. Bashara: *Ellipsometry and Polarized Light* (North-Holland, Amsterdam 1977)

3.10   A.W. Rihaczek: *Principles of High Resolution Radar*, (McGraw-Hill, New York 1969)
3.11   J.R. Klauder: The design of radar signals having both high range resolution and high velocity resolution. Bell Syst. Tech. J. **39**, 809 (1960)
3.12   H.O. Bartlet, K.H. Brenner, A.W. Lohmann: The Wigner distribution function and its optical production. Opt. Commun. **2**, 32 (1981)

Chapter 4

4.1   W.S. Boyle, G.E. Smith: Bell Syst. Tech. J. **49**, 587 (1970)
4.2   G.F. Amelio, M.F. Tompsett, G.E. Smith: Bell Syst. Tech. J. **49**, 593 (1970)
4.3   F.L. Sangster, K. Teer: IEEE J. Solid-State Circuits, SC-4, 131 (1969)
4.4   W.E. Engeler, J.J. Tiemann, R.D. Baertsch: Appl. Phys. Lett. **17**, 469 (1970)
4.5   R.M. White, F.W. Voltmer: Appl. Phys. Lett. **7**, 314 (1965)
4.6   A.A. Oliner (ed.): *Acoustic Surface Waves*, Topics Appl. Phys., Vol. 24 (Springer, Berlin, Heidelberg 1978)
4.7   F.G. Marshall, C.O. Newton, E.G.S. Paige: IEEE Trans. SU-20, 124 (1973)
4.8   F.G. Marshall, C.O. Newton, E.G.S. Paige: IEEE Trans. SU-20, 134 (1973)
4.9   J. deKlerk: Phys. Today **25**, 32-39 (Nov. 1972)
4.10   R.D. Baertsch, W.E. Engeler, H.S. Goldberg, C.M. Puckette, IV, J.J. Tiemann: CCD 74 Int.Conf., Edinburgh, Proc. (1974) pp. 229–236
4.11   R.H. Walden, R. Krambeck, R.H. Strain, R.J. McKenna, N.L. Schryer, G.E. Smith: Bell Syst. Tech. J. **51**, 1635 (1972)
4.12   L.J.M. Esser: Electron. Lett. **8**, 620 (1972)
4.13   C.H. Sequin, M.F. Tompsett: *Charge Transfer Devices* (Academic, New York 1975)
4.14   L.B. Milstein, P. Das: IEEE Commun. Mag. **17**, 25-33 (1979)
4.15   W.R. Smith, H.M. Gerard, J.H. Collins, T.M. Reeder, H.J. Shaw: IEEE Trans. MTT-17, 865-873 (1969)
4.16   C.S. Hartmann, W.J. Jones, H. Vollers: IEEE Trans. SU-19, 378-384 (1972)
4.17   R.M. Hays, C.S. Hartmann: Proc. IEEE **64**, 652-671 (1976)
4.18   N.J. Berg, J.N. Lee: *Acousto-Optic Signal Processing* (Marcel Dekker, New York 1983)
4.19   A. Chatterjee, P.K. Das, L.B. Milstein: IEEE Trans. SU-32, 745-759 (1985)
4.20   A.S. Grove: *Physics and Technology of Semiconductor Devices* (Wiley, New York 1967)
4.21   G.S. Hobson: *Charge Transfer Devices* (Wiley, New York 1978)
4.22   R.D. Melon, J.K. Shott, J.T. Walker, J.D. Meindl: Proc. 1975 Naval Electronics Lab. Center Int. Conf. on the Applications of Charge Coupled Devices (1975) pp. 165-171
4.23   J.D.E. Beynon, D.R. Lamb: *Charge Coupled Devices and Their Applications* (McGraw-Hill, New York 1980)
4.24   S.N. Chakravarti, P. Das: Solid-State Electron. **23**, 747-753 (1980)
4.25   I. Deyhimy, J.S. Harris, R.C. Eden, S.J. Anderson, D.D. Edwall: IEDM Tech. Digest 619-621 (1979)
4.26   W.A. Reed, J.M. Owens, R.L. Carter: Circuits, Syst. Signal Process. **4**, 157-179 (1985)
4.27   J.P. Parekh, K.W. Chang, H.S. Tuan: Circuits, Syst. Signal Process. **4**, 9-39 (1985)
4.28   W.S. Ishak, E. Reese, E. Huiger: Circuits, Syst. Signal Process. **4**, 285-300 (1985)
4.29   J.P. Castera, P. Hartemann: Circuits, Syst. Signal Process. **4**, 181-200 (1985)
4.30   S.N. Suitzer, P.R. Emtage: Circuits, Syst. Signal Process. **4**, 227-252 (1985)
4.31   M.J. Hoskins, B.J. Hunsinger: 1982 Ultrasonics Symp. Proc., ed. by B.R. McAvoy (IEEE, New York 1982), pp. 456-460
4.32   H. Mathews: *Surface Wave Filters* (Wiley, New York 1977)
4.33   R.C. Williamson, H.I. Smith: IEEE Trans. SU-20, 113-123 (1973)
4.34   T.A. Martin: IEEE Trans. SU-20, 104-112 (1973)

Appendix

A.1   J.F. Nye: *Physical Properties of Crystals* (Oxford University Press, Oxford 1957)

Bibliography

Chapter 1

J.W. Goodman: Opt. News **10**, 25 (1984)

Chapter 2

B.A. Auld: *Acoustic Fields and Waves in Solids*, Vols. I and II (Wiley, New York 1973)

M.K. Barnoski: *Introduction to Integrated Optics* (Plenum, New York 1973)

M. Born, E. Wolf: *Principles of Optics*, 6th ed. (Pergamon, Oxford 1980)

R.E. Collin: *Foundations for Microwave Engineering* (McGraw-Hill, New York 1966)

H.J. Eichler, P. Günter, D.W. Pohl: *Laser-Induced Dynamic Gratings*, Springer Ser. Opt. Sci., Vol. 50 (Springer, Berlin, Heidelberg 1986)

S. Ezekiel, H.J. Arditty (eds.): *Fiber-Optic Rotation Sensors*, Springer Ser. Opt. Sci., Vol. 32 (Springer, Berlin, Heidelberg 1982)

R.G. Hunsperger: *Integrated Optics: Theory and Technology*, Springer Ser. Opt. Sci., Vol. 33 (Springer, Berlin, Heidelberg 1985)

K. Iizuka: *Engineering Optics*, 2nd ed., Springer Ser. Opt. Sci., Vol. 35 (Springer, Berlin, Heidelberg 1987)

J.D. Jackson: *Classical Electrodynamics*, 2nd ed. (Wiley, New York 1975)

F.A. Jenkins, H.E. White: *Fundamentals of Optics*, 4th ed. (McGraw-Hill, New York 1976)

D.K. Killinger, A. Mooradian (eds.): *Optical and Laser Remote Sensing*, Springer Ser. Opt. Sci., Vol. 39 (Springer, Berlin, Heidelberg 1983)

D.F. Nelson: *Electric, Optic and Acoustic Interactions in Dielectrics* (Wiley, New York 1979) Chap. 13

E.G. Neumann: *Single-Mode Fibers*, Springer Ser. Opt. Sci., Vol. 57 (Springer, Berlin, Heidelberg 1988)

H.-P. Nolting, R. Ulrich (eds.): *Integrated Optics*, Proc. 3rd European Conf., ECIO 85, Berlin 1985

E.L. O'Neill: *Introduction to Statistical Optics* (Addison-Wesley, Reading, MA 1963)

W. Schumann, J.-P. Zürcher, D. Cuche: *Holography and Deformation Analysis*, Springer Ser. Opt. Sci., Vol. 46 (Springer, Berlin, Heidelberg 1985)

K. Shimoda: *Introduction to Laser Physics*, 2nd ed., Springer Ser. Opt. Sci., Vol. 44 (Springer, Berlin, Heidelberg 1986)

T. Tamir (ed.): *Integrated Optics*, 2nd ed., Topics Appl. Phys., Vol. 7 (Springer, Berlin, Heidelberg 1979)

J.T. Verdeyen: *Laser Electronics* (Prentice Hall, Englewood Cliffs, NJ 1981)

Chapter 3

R.S. Berkowitz: *Modern Radar* (Wiley, New York 1965)

K.H. Brenner, A.W. Lohmann: Wigner distribution function display of complex ID signals. Opt. Commun. **42**, 310 (1982)

T.A.C.M. Claasen, W.F.G. Mecklenbrauker: The Wigner distribution – A tool for time-frequency signal analysis, Part I: Continuous-time signals. Philips J. Res. **35**, 217 (1980)

T.A.C.M. Claasen, W.F.G. Mecklenbrauker: The Wigner distribution – A tool for time-frequency signal analysis, Part II: Discrete-time signals. Philips J. Res. **35**, 276 (1980)

T.A.C.M. Claasen, W.F.G. Mecklenbrauker: The Wigner distribution – A tool for time-frequency signal analysis, Part III: Relations with other time-frequency signal transformations. Philips J. Res. **35**, 372 (1980)

W.L. Eversole, D.J. Meyer, R.J. Kansy: "A CCD Two-Dimensional Transform" in Proc. 1978 Int. Conf. on the Application of Charge Coupled Devices, p. 3B-31

B. Friedlander: Lattice filters for adaptive processing. Proc. IEEE **70**, 829 (1982)

J.D. Gaskill: *Linear Systems, Fourier Transforms and Optics* (Wiley, New York 1978)

A. Gelb (ed.): *Applied Optimal Estimation* (MIT, Cambridge, MA 1974)

J. Goodman: *Introduction to Fourier Optics* (McGraw-Hill, New York 1968)

L.J. Griffiths: "Adaptive Structures for Multiple-Input Noise Cancelling Applications", in Proc. IEEE Int. Conf. on Acoustics, Speech, and Signal Processing (1979) p. 925

A.W. Lohmann, B. Wirnitzer: Triple correlations, Proc. IEEE **72**, 889 (1984)

J. Mahoul: A class of all-zero lattice digital filters: Properties and applications. IEEE Trans. ASSP-26, 304 (1978)

A. Oppenheim, R.W. Schafer: *Digital Signal Processing* (Prentice Hall, Englewood Cliffs, NJ 1975)

A. Papoulis: *Systems and Transforms with Applications in Optics* (McGraw-Hill, New York 1968)

A. Papoulis: *Signal Analysis* (McGraw-Hill, New York 1977)

W.K. Pratt: *Digital Image Processing* (Wiley, New York 1978)

E.A. Robinson: A historical perspective of spectrum estimation. Proc. IEEE **70**, 885 (1982)

B. Saleh: *Photoelectron Statistics*, Springer Ser. Opt. Sci., Vol. 6 (Springer, Berlin, Heidelberg 1978)

Special Issue, *Spectral Estimation*. Proc. IEEE, Vol. 70 (Sept. 1982)

H.H. Szu, H.J. Caulfield: The mutual time-frequency content of two signals. Proc. IEEE **72**, 902 (1984)

B. Widrow, M. Hoff, Jr.: "Adaptive Switching Circuits", in IRE WESCON Conv. Rec. (1960) Pt. 4, pp. 96–104

B. Widrow, S.D. Stearns: *Adaptive Signal Processing* (Prentice Hall, Englewood Cliffs, NJ 1985)

Chapter 4

B.A. Auld: *Acoustic Fields and Waves in Solids,* Vols. I and II (Wiley, New York 1973)

D.F. Barbe (ed.): *Charge-Coupled Devices,* Topics Appl. Phys., Vol. 38 (Springer, Berlin, Heidelberg 1980)

M.J. Hoskins, H. Morkov, B.J. Hunsinger: Charge transport by surface acoustic waves in GaAs. Appl. Phys. Lett. **41**, 332 (1982)

M.J. Howes, D.V. Morgan: *Charge Coupled Devices and Systems* (Wiley, New York 1979)

G.S. Kino: *Acoustic Waves; Devices, Imaging and Analog Signal Processing* (Prentice Hall, Englewood Cliffs, NJ 1987)

T. Martin: "Low Sidelobe IMCON Pulse Compression", in Ultrasonics Symp. Proc., ed. by J. deKlerk, B. McAvoy (IEEE, New York 1976) pp. 411–414

R. Melen, D. Buss: *Charge Coupled Devices: Technology and Applications* (Wiley, New York 1977)

D.P. Morgan: *Surface Wave Devices for Signal Processing* (Elsevier, New York 1985)

T.E. Parker, G.K. Montress: Precision surface-acoustic-wave (SAW) oscillators. IEEE Trans. on Ultrasonics, Ferroelectrics and Frequency Control **35**, 342-364 (1988)

V.M. Ristic: *Principles of Acoustic Devices* (Wiley, New York 1983)

R.H. Tancrell: IEEE Trans. SU-21, 12 (1974)

W.C. Wang, P. Das: Ultrasonics Symp. Proc., ed. by J. deKlerk (IEEE, New York 1972) p. 316

R.S. Withers, S.A. Reible: Superconducting chirp transform spectrum analyser. IEEE Electron Dev. Lett. EDL-6, 261 (1985)

Appendices

M. Abramowitz, I.A. Stegun (eds.): *Handbook of Mathematical Functions* (National Bureau of Standards, Washington, DC 1964)

W.L. Brogan: *Modern Control Theory* (Quantum, New york 1974)

R.S. Elliot: *Electromagnetics* (McGraw-Hill, New York 1966)

A.K. Jain: Fast inversion of banded Toeplitz matrices by circular decomposition. IEEE Trans. ASSP-26, 121 (1978)

C. Lanczos: *Applied Analysis* (Prentice Hall, Englewood Cliffs, NJ 1964)

A. Papoulis: *Systems and Transforms with Applications in Optics* (McGraw-Hill, New York 1968)

Subject Index